Partial Differential Equations and Group Theory

Mathematics and Its Applications

Managing Editor:

M. HAZEWINKEL

Centre for Mathematics and Computer Science, Amsterdam, The Netherlands

Volume 293

Partial Differential Equations and Group Theory

New Perspectives for Applications

by

J.-F. Pommaret
Centre de Recherche en Mathématiques Appliquées (CERMA),
Ecole Nationale des Ponts et Chausées (ENPC),
Noisy-le-Grand, France

KLUWER ACADEMIC PUBLISHERS
DORDRECHT / BOSTON / LONDON

A C.I.P. Catalogue record for this book is available from the Library of Congress.

ISBN 978-90-481-4432-7

Published by Kluwer Academic Publishers,
P.O. Box 17, 3300 AA Dordrecht, The Netherlands.

Kluwer Academic Publishers incorporates
the publishing programmes of
D. Reidel, Martinus Nijhoff, Dr W. Junk and MTP Press.

Sold and distributed in the U.S.A. and Canada
by Kluwer Academic Publishers,
101 Philip Drive, Norwell, MA 02061, U.S.A.

In all other countries, sold and distributed
by Kluwer Academic Publishers Group,
P.O. Box 322, 3300 AH Dordrecht, The Netherlands.

Printed on acid-free paper

Contents

Foreword

Ordinary differential control theory (the classical theory) studies input/output relations defined by systems of ordinary differential equations (ODE). The various concepts that can be introduced (controllability, observability, invertibility, etc.) must be tested on formal objects (matrices, vector fields, etc.) by means of formal operations (multiplication, bracket, rank, etc.), but without appealing to the explicit integration (search for trajectories, etc.) of the given ODE. Many partial results have been recently unified by means of new formal methods coming from differential geometry and differential algebra. However, certain problems (invariance, equivalence, linearization, etc.) naturally lead to systems of partial differential equations (PDE).

More generally, *partial differential control theory* studies input/output relations defined by systems of PDE (mechanics, thermodynamics, hydrodynamics, plasma physics, robotics, etc.). One of the aims of this book is to extend the preceding concepts to this new situation, where, of course, functional analysis and/or a dynamical system approach cannot be used. A link will be exhibited between this domain of applied mathematics and the famous *'Bäcklund problem'*, existing in the study of solitary waves or *solitons*. In particular, we shall show how the methods of differential elimination presented here will allow us to determine compatibility conditions on input and/or output as a better understanding of the foundations of control theory. At the same time we shall unify differential geometry and differential algebra in a new framework, called *differential algebraic geometry*. A systematic approach, without mathematics, will be presented in the Introduction.

Independently of the preceding considerations, the *formal theory of systems of PDE* has been developed in the USA in the years 1960–1975 under impulses of D.C. Spencer. It intends to study the space of solutions of systems of PDE *without integrating the PDE explicitly*. However, though this pioneering work largely supersedes the classical approach of M. Janet and E. Cartan, it is still largely unknown by mathematicians and has never been applied by physicists.

The aim of the first part of this book is to provide a self-contained introduction to these methods in the case of linear and nonlinear systems of PDE, at the most basic graduate level. One must notice that these topics have not been taught elsewhere during the last ten years and, in any case, have never been applied. In particular, we shall show on specific examples coming from engineering physics how these new tools (diagram chasing, Spencer cohomology, etc.) exactly furnish the proper algorithms for applications. Meanwhile, these algorithms will be presented in such a way as to give the possibility to use *computer algebra* (MACSYMA, REDUCE, MATHEMATICA, MAPLE, etc.) for the study of problems involving PDE or purely algebraic equations.

On another side and, at least initially, in a purely algebraic framework, classical Galois theory has been elaborated in the second half of the 19th century, in order to reduce the search for roots of polynomials to the study of permutation groups. Then, at the beginning of the 20th century, a few French mathematicians, including E. Picard, E. Vessiot and J. Drach, tried to extend Galois theory to systems of linear ODE and even to systems of algebraic PDE, that is, polynomials in the derivatives of the unknowns. For this purpose they used the *theory of transformation groups* already introduced a few years earlier by the Norwegian mathematician S. Lie.

However, Spencer also introduced the formal theory of PDE in order to study the latter groups of transformations, today called *Lie pseudogroups* in contrast to *Lie groups* of transformations depending on a finite number of constant parameters. Hence, we may conceive that the already quoted techniques must also open a new way for computer algebra, linking the integration of PDE to the theory of Lie pseudogroups.

The aim of the second part of this book is to give a brief introduction to the Galois theory for systems of PDE, simply called '*differential Galois theory*'. We shall insist on the constructive aspect through many explicit examples. In particular, we shall show how to find the biggest Lie pseudogroup of invariance of a given system of PDE, and how to use it to perform *cascade integration* of that system by reducing integration of it to integration of a chain of simpler systems.

Finally, we shall sketch how to use the former group-theoretical methods, culminating in the construction of differential sequences (chains of differential operators such that the composition of two successive ones vanishes), in order to obtain new insights into mathematical physics. In particular, we shall prove that the foundations of continuum mechanics and thermodynamics is directly concerned with the Spencer operator and with group theory, along the ideas first put forward by the brothers E. and F. Cosserat in 1909. Similarly, we shall extend this point of view to electromagnetism and gravitation, along ideas sketched by H. Weyl in 1918. As a byproduct, the results thereby presented provide, in a simple manner and for the first time ever, the long awaited group-theoretical unification of finite element methods for the engineering sciences (elasticity, heat, electromagnetism).

We hope that this book will give a new impulse to the study of differential algebraic geometry and its applications to mathematical physics and engineering sciences.

The material of this book is taken from an intensive ERCIM (European Research Consortium for Informatics and Mathematics) course already given with success in France (Institut National de Recherche en Informatique et en Automatique, INRIA, Paris, 1990), Germany (Gesellschaft für Mathematik und Datenverarbeitung mbH, GMD, Bonn, 1992) and the Netherlands (Centrum voor Wiskunde en Informatica, CWI, Amsterdam, 1993). The applications to mathematical physics are taken from an intensive ICTP (International Centre for Theoretical Physics) course in Trieste, Italy, in 1991.

The development of the technical part closely follows the computational aspect, and most of the time local coordinates are used in order to help the reader to mature the intricate concepts. Moreover, all the results are illustrated by many explicit examples, placed at the end of each chapter.

It is a pleasure to express my deep gratitude to all the people involved in organizing these courses, in particular to J.P. Quadrat and A. Bensoussan from INRIA, M. Dresen and F. Schwarz from GMD, F. Snijders and M. Hazewinkel from CWI, J. Eells and A. Verjovski from ICTP. Finally I would like to thank S. Lazzarini for the many helpful comments while reading the manuscript.

Introduction

Once upon a time $\cdots$ a French professor of mathematics had two excellent students, far above the other ones and so good that, whenever one was not the first at an examination, the other was and vice versa. However, the minds of these two students were quite different: one was more attracted by the possibility to pass from explicit examples to high level abstract ideas, while the second hated to compute and was more concerned with logic through the use of computers. Of course, the professor was desperately wanting to know who was indeed the brightest student.

It happened that, on a dark night when he was not able to sleep, the Devil spoke to him and offered a way to select the best student on the basis of the following problem:

Devil's Problem. Let u, v, y be 3 functions of the cartesian coordinates x^1, x^2, x^3 of euclidean space, related by the following system of 2 PDE, where $\partial_{33}y = \partial^2 y/\partial x^3 \partial x^3$, $\ldots$:

$$\partial_{33}y - x^2 \partial_{11}y - u = 0, \qquad \partial_{22}y - v = 0.$$

1) If $u = v = 0$, the space of solutions of the resulting linear system of PDE for y is a vector space over the constants. What is its dimension?
2) Otherwise, what are the compatibility conditions that must be satisfied by u and v in order to ensure the existence of solutions for y?

Though this problem seemed easy to the professor at first sight, he called the two students and gave them a few days for providing the solutions to the two questions. After a week, the two answers were as follows.

First Work. For the first question, the system to be solved is

$$\partial_{33}y - x^2 \partial_{11}y = 0, \qquad \partial_{22}y = 0.$$

By sheer luck, the general solution of the second PDE can be easily obtained as

$$y = A(x^1, x^3)x^2 + B(x^1, x^3).$$

Substitution into the first PDE gives

$$(x^2)^2 \partial_{11}A + x^2(\partial_{11}B - \partial_{33}A) - \partial_{33}B = 0,$$

1

which leads to the following 3 PDE for A and B:

$$\partial_{11}A = 0, \qquad \partial_{11}B - \partial_{33}A = 0, \qquad \partial_{33}B = 0.$$

Again, by sheer luck, the first and third can be integrated, and one obtains

$$A = a(x^3)x^1 + b(x^3), \qquad B = \alpha(x^1)x^3 + \beta(x^1).$$

Taking into account the second PDE, one has

$$x^1 \partial_{33}a(x^3) + \partial_{33}b(x^3) - x^3 \partial_{11}\alpha(x^1) - \partial_{11}\beta(x^1) = 0.$$

Differentiating with respect to x^1 gives

$$\partial_{33}a(x^3) - x^3 \partial_{111}\alpha(x^1) - \partial_{111}\beta(x^1) = 0,$$

while differentiation with respect to x^3 gives

$$\partial_{333}a(x^3) = \partial_{111}\alpha(x^1) = p = \text{const},$$

since the first quantity is a function of x^3 only and the second quantity is a function of x^1 only. Similarly:

$$\partial_{111}\beta(x^1) = q = \text{const}.$$

Substitution into the previous PDE gives

$$\partial_{33}a(x^3) - px^3 - q = 0.$$

An easy integration now gives:

$$a(x^3) = \frac{p}{6}(x^3)^3 + \frac{q}{2}(x^3)^2 + rx^3 + s,$$

$$\alpha(x^1) = \frac{p}{6}(x^1)^3 + \frac{l}{2}(x^1)^2 + mx^1 + n,$$

$$\beta(x^1) = \frac{q}{6}(x^1)^3 + \frac{u}{2}(x^1)^2 + vx^1 + w.$$

Hence,

$$\partial_{33}b = x^3 \partial_{11}\alpha + \partial_{11}\beta - x^1 \partial_{33}a = lx^3 + u,$$

and finally

$$b(x^3) = \frac{l}{6}(x^3)^3 + \frac{u}{2}(x^3)^2 + fx^3 + g.$$

Since $y = a(x^3)x^1x^2 + b(x^3)x^2 + \alpha(x^1)x^3 + \beta(x^1)$, the general solution depends on the 12 constants $p, q, r, s, l, m, n, u, v, w, f, g$, and the space of solutions is therefore a 12-dimensional vector space over the constants, a fact not evident at all at first sight, and we have the additional property that all 6-order derivatives of y are null, because y is a polynomial of degree at most 5 in x^1, x^2, x^3. One can also conclude that all the 5-order derivatives can be known from lower-order ones.

For the second question it is easy to exhibit a 3rd order compatibility condition, as follows:

$$\partial_{233}y - x^2\partial_{112}y - \partial_{11}y = \partial_2 u,$$
$$\partial_{2233}y - x^2\partial_{1122}y - 2\partial_{112}y = \partial_{22}u,$$
$$\partial_{22233}y - x^2\partial_{11222}y - 3\partial_{1122}y = \partial_{222}u,$$

that is, *necessarily*

$$\partial_{233}v - x^2\partial_{112}v - 3\partial_{11}v - \partial_{222}u = 0.$$

It is much more difficult to exhibit another, higher-order, compatibility condition, not linearly dependent on the partial derivatives of the one already obtained $\cdots$, provided one is convinced of the existence of such a condition! One can proceed as follows. Introduce

$$\partial_{112}y = \frac{1}{2}(\partial_{33}v - x^2\partial_{11}v - \partial_{22}u) = w.$$

Then the previous compatibility condition becomes

$$\frac{1}{2}A \equiv \partial_2 w - \partial_{11}v = 0,$$

but we have successively

$$\partial_{11233}y = \partial_1(\partial_{1233}y - \partial_{111}y) + \partial_{1111}y$$
$$= \partial_{11}(\partial_{233}y - \partial_{11}y) + \partial_{1111}y$$
$$= \partial_{11}\left(\partial_2(\partial_{33}y - x^2\partial_{11}y) + x^2\partial_{112}y\right) + \partial_{1111}y,$$

which leads to

$$\partial_{1111}y = \partial_{33}w - \partial_{112}u - x^2\partial_{11}w$$

and to

$$\partial_{111133}y = \partial_{3333}w - \partial_{11233}u - x^2\partial_{1133}w$$
$$= \partial_{1111}u + x^2\partial_{111111}y$$
$$= \partial_{1111}u + x^2\partial_{1133}w - x^2\partial_{11112}u - (x^2)^2\partial_{1111}w.$$

Hence one obtains the following 6th order compatibility condition:

$$B \equiv \partial_{3333}w - 2x^2\partial_{1133}w + (x^2)^2\partial_{1111}w - \partial_{11233}u + x^2\partial_{11112}u - \partial_{1111}u$$
$$= 0.$$

The various derivatives of u appearing in B are:

$$\partial_{223333}u, \quad \partial_{112233}u, \quad \partial_{111122}u,$$
$$\partial_{11233}u, \quad \partial_{11112}u, \quad \partial_{1111}u,$$

and it follows that B cannot be factorized through derivatives of A, that is, B is *effectively a new 6th order compatibility condition for u and v.* Any higher-order computation becomes extremely tedious and, in any case, there does not seem to exist a general procedure for searching such compatibility conditions. One should also notice that even a very simple linear transformation of the independent variables x^1, x^2, x^3 could have hidden the possibility of integrating the PDE in the first question.

Finally, one should notice that the compatibility conditions A and B are not linearly differentially independent. After a straightforward calculation one obtains

$$\partial_{3333}A - 2x^2\partial_{1133}A + (x^2)^2\partial_{1111}A - 2\partial_2 B = 0,$$

which involves 7th order partial derivatives of u and v.

Second Work. After trying a few calculations myself, these became so tedious that I gave up and tried to use a computer according to the standard differentiation rules. My general idea was to start with the two given PDE, written as follows:

$$\partial_{33}y - x^2\partial_{11}y = u, \qquad \partial_{22}y = v,$$

that is, with a linear homogeneous partial differential operator as left hand side and with a second term. Then I used a symbolic package for successively differentiating them, in order to compute, term by term, the various derivatives of y at the origin $(0, 0, 0)$. After 10 hours of computation, I obtained the following scheme:

'PUZZLE'

order	8	7	6	5	4	3	2	1	0	system rank	total number	partial number	order
partial number	45	36	28	21	15	10	6	3	1				
total number	165	120	84	56	35	20	10	4	1				
symbol rank	41	32	24	17	11	6	2						
		75		43									
		58											
	103						2				2	2	0
		79				8					8	6	1
					20						20	12	2
				39							40	20	3
			66								70	30	4
		102									112	42	5
	147										168	56	6

On the horizontal lines are indicated, respectively, the order q of derivatives of y, the partial number of derivatives of strict order q, the total number of derivatives of y up to order q, the rank of the derived equations with respect to the top-order derivatives (the *symbol* of order q), and finally the rank of the system with respect to derivatives of orders 5 and 6 (43), 6 and 7 (58), 7 and 8 (75), 5 and 6 and 7 (79), 6 and 7 and 8 (103). On the vertical lines are indicated, respectively, the number of times one differentiates each equation, the number of equations obtained each time, and the total number of equations already produced. Along the diagonal one can read the rank of the various sets of equations with respect to the total number of derivatives that could appear at the corresponding order.

Presented as it is, I had the idea that the numbers obtained should constitute the parts of a 'puzzle', complicate enough to give the answer to the two equations $\cdots$ under the condition that we know how to organize this computational knowledge!

Everything went on easily up to differentiating the given PDE five times, using **MACSYMA** on **MULTICS**. Then the computations took a much longer time, and I had to use a SUN for 3 hours in order to obtain the *unexpected total rank* 147 *that will be shown to be crucial!* Although I do not know whether it is useful or not, I spent some extra money to differentiate 7 times, and obtained the total rank 202 in 9 hours! Any computations involving higher-order derivatives will be extremely expensive, and almost impossible to be done by hand, even in this simple example where the matrices involved are quite empty, and one should use powerful computers. However, the advantage of the method I describe is that it is intrinsic (I mean coordinate free) and can, with slight modifications, be used on any example.

Let me now explain why I was quite surprised when getting the total rank 147 while differentiating 6 times!

First of all, let me return to the first equation. The number of arbitrary derivatives at each order is 1 for order 0, 3 for order 1, then $6-2 = 10-6 = 15-11 = 21-17 = 28-24 = 36-32 = \cdots = 4$ at any order larger than or equal to 2. In fact, one has to know at the origin $y, \partial_1 y, \partial_2 y, \partial_3 y$, and, more generally, $\partial_{1\ldots1}y, \partial_{1\ldots12}y, \partial_{1\ldots13}y, \partial_{1\ldots123}y$. Hence one has to evaluate an infinite number of derivatives, namely 4 at each order ≥ 2, and the space of solutions is surely infinite dimensional. In any case it is evident that *a symbolic package is not able to 'integrate' ODE or PDE in general*, this possibility being restricted to a list of cases taken from handbooks or to cases that can be reduced to these by means of a few algorithms (decomposition of rational fractions, factorization, etc.).

It is much more delicate to answer the second question.

First of all, since $2-2 = 8-8 = 20-20 = 0$, there are no compatibility conditions of orders $0,1,2$ in u and v, but there is $40 - 39 = 1$ compatibility condition of order 3, namely:

$$A \equiv \partial_{233}v - x^2 \partial_{112}v - 3\partial_{222}u = 0.$$

At order 4 there are $70 - 66 = 4$ compatibility conditions, that is, the given one and its (independent) 3 first-order derivatives, hence no *essentially* new compatibility condition. A similar comment holds at order 5, because $112 - 102 = 10 = 1 + 3 + 6$.

It is extremely surprising that things change at order 6. Using Newton' method of the vanishing triangle for studying sequences of increasing integers, I found the following scheme:

```
              0       0       0
          1       1       1       1
      6       7       8       9      10
  6      12      19      27      36      46
2       8      20      39      66     102     148
```

In this scheme, each number is the sum of its two left neighbors, and therefore the final awaited rank is 148, though it is 147. Since the number of compatibility conditions produced from the ones already known should be $1 + 3 + 6 + 10 = 20$, the rank should be effectively $168 - 20 = 148$. The fact that it is 147 surely proves that there is one additional compatibility condition, say $B \equiv 0$, which could be found by computer with a slight extra work (like the preceding one, $A \equiv 0$). However, I have not been able to to stop the argument at higher order. For example, at order 7 in u and v one should have $(1+3+6+10+15)+(1+3) = 39$ compatibility conditions, and therefore a total rank of $240 - 39 = 201$ *if there are no more compatibility conditions and no compatibility condition among these compatibility conditions*! In the present situation one only finds $202-201 = 1$ differential identity between A and B. However, I have no reason to conclude this, because new compatibility conditions may appear after many more differentiations, as we saw for A and B. I hope that the future will see papers that answer this question in general by finding an algorithm for intrinsically solving such PDE problems. In a certain sense I have the feeling that this solution is a kind of 'puzzle' furnishing most of the information$\cdots$, for somebody knowing how to manage with the information through a convenient algorithm.

Finally, looking at the example once more, I discovered that the use of a power of x^2, say $(x^2)^2$, instead of x^2 as coefficient would considerably increase the difficulty of solving the problem.

It was only after reading the two contradictory works that the professor understood why the Devil had freely given the challenge. He went through a very sad period of his life, and did not teach anymore, being obsessed by solving such a problem in general. He decided to leave for the USA, where he had heard about certain recent new methods for studying systems of PDE. After a year at Princeton University, almost as a student, he finally was able to solve the problem himself, and to discover that both students were, at the same time, right $\cdots$ and wrong!

The story needs no comments and, contrary to customary usage, any analogy between the persons involved and real persons is $\cdots$ surely true.

The problem above was quoted for the first time by M. Janet in 1920 [85, 86], and will be treated by means of homological algebra techniques in Chapter III. In this Introduction we shall only present a short historical survey of the work done on PDE since the beginning of the 20th century, referring to [144] for more details.

In essence, the first ideas for applying algebraic type methods to systems of PDE are contained in the works of a few Frenchmen.

1) C. Riquier introduced, at the beginning of the 20th century, the concept of a '*cut*' determined by a system of PDE in the formal power series solutions in the analytic case [157, 158]. Roughly speaking, the given PDE and all their derivatives could be used (if possible, successively) to compute certain derivatives at a point, called '*principal*', as functions of the independent variables and other derivatives, called '*parametric*'. The trick is therefore to obtain a unique way of looking at the cut (two different ways of computing the same principal derivative must produce the same result) while '*packing*' the parametric derivatives through arbitrary formal power series of $0, 1, \ldots, n$ independent variables, respectively, in order to obtain knowledge about the '*degree of generality*' of the solution space, that is, the analytic functions of 0 (constants), $1, \ldots, n$ variables, respectively, that must be specified in order to have, at least formally (independently of convergence problems), a uniquely determined solution.

2) Conversely, if one has an infinite number of PDE, one could assume that eventually these come in fact from a finite subset of them, or a '*basis*', via differentiations.

Such a result, deriving from noetherian arguments, was first suggested and sketched by F. Tresse.

3) In 1920, M. Janet, having spent a few months with D. Hilbert and under the influence of J. Hadamard, for the first time ever understood the concept of *'involution'* for PDE, a concept that had already been proposed in a computational way by E. Cartan within the framework of exterior calculus. Note that Cartan does not refer to the work of Janet in his numerous letters to A. Einstein dating from 1930, mainly for personal reasons (see [145, 146] for details about this story). As a consequence, almost nobody paid attention to Janet's work, who turned to mechanics.

In fact, the ultimate achievement of Janet's work lies in the first construction of a finite-length differential sequence, namely the explicit determination of a finite chain of differential operators the composition of two successive ones being zero. Indeed, suppose we start with a linear partial differential operator $\mathcal{D}$ with respect to n independent variables. Following Janet, we assume that the corresponding system is 'passive' with respect to a complete set of leading monomials (in Chapter IV we will see a modern formulation). Then, if we add a second term, this second term must, in general, satisfy compatibility conditions expressed, as we saw, again by a linear partial differential operator $\mathcal{D}_1$, to which we can add a second term again, etc. The surprising result found by Janet, and adapted from Hilbert's syzygies, is that one stops at $\mathcal{D}_n$ or before, possibly at $\mathcal{D}$ itself, that is, $\mathcal{D}_n$ is bound to have no compatibility conditions. This situation is well known in the classical vector geometry of 3-dimensional space, where only three operators are known, namely *gradient*, *rotation* and *divergence*, with rot ∘ grad $\equiv 0$, div ∘ rot $\equiv 0$. The extension to n-dimensional space has been done by E. Cartan via the construction of the exterior derivative in the so-called *Poincaré sequence*. The reader will notice that here we have n operators, whereas in the general situation we have the $n + 1$ operators $\mathcal{D}, \mathcal{D}_1, \ldots, \mathcal{D}_n$. This property will be explained in an involved exercise, at the end of Chapter IV.

Although the purpose of this book is to present the neat mathematical tools without any reference to their historical roots, which have already been presented in [144, 145, 146], we shall nevertheless illustrate the procedure used by Janet on the Devil's example, and make some comments about it.

Writing, for simplicity, subindices instead of derivatives (a notation leading in fact to *jet coordinates*, as we shall see in Chapter II), we have

$$y_{33} - x^2 y_{11} = 0, \qquad y_{22} = 0.$$

Let us now totally order the derivatives by means of their order and the lexicographical ordering on the corresponding indices. We may then write the system anew, in a so-called *'solved form'*, such that each equation has a *leading term*, namely the highest term involved. We may associate in a unique way to each derivative a *monomial* in the variables x, such that $\partial_{23}y$ is written y_{23} and gives the monomial $x^2 x^3$. It follows that the given PDE and their derivatives may be used to determine a certain number of (solved) derivatives, called *principal*, by means of arbitrarily given ones, calledindxparametric derivatives *parametric*. The main problem, effecting crossed derivatives, is that two different expressions may be obtained for the same principal derivatives and, by subtraction, certain parametric derivatives may be found to be dependent. As a byproduct, *knowing a formal power series solution amounts to having a clear knowledge of the parametric and principal derivatives* (the '*cut*' in the sense of Riquier) obtained by taking into account the infinite number of PDE obtained by successive differentiations.

A first idea is to differentiate the PDE in such away that only distinct principal derivatives are involved. Such a rule has been established by Janet in order to select, for each equation in solved form, the *multiplicative variables* (variables with respect to which the given PDE may be differentiated) from the *nonmultiplicative variables* (variables with respect to which the given PDE may not be differentiated, in order to avoid finding the same leading derivative many times). The rule is as follows. The independent variable x^n is multiplicative for the monomials having maximal degree in x^n, and the independent variable x^i is said to be multiplicative for the monomial $x^\mu = (x^1)^{\mu_1} \ldots (x^n)^{\mu_n}$, with multi-index $\mu = (\mu_1, \ldots, \mu_n)$ (where n is the number of independent variables), if, among all existing monomials of the form $(x^1)^{\alpha_1} \ldots (x^i)^{\alpha_i}(x^{i+1})^{\mu_{i+1}} \ldots (x^n)^{\mu_n}$, the monomial is such that $\mu_i = \max_\alpha \alpha_i$. Multiplicative and nonmultiplicative variables can be written in a scheme, where nonmultiplicative variables are indicated by a dot $\bullet$.

In our case we can frame the leading terms and, not writing out the monomials for simplicity reasons, we obtain

$$
\boxed{\begin{array}{l} y_{33} \\ y_{22} \end{array}} \begin{array}{l} - x^2 y_{11} = 0 \\ = 0 \end{array} \boxed{\begin{array}{ccc} x^1 & x^2 & x^3 \\ x^1 & x^2 & \bullet \end{array}}
$$

A system of monomials is called '*complete*' is any multiple can be obtained by using only multiplicative variables. In our case the system of monomials is $\{(x^3)^2, (x^2)^2\}$, but $(x^2)^2 x^3$, although a multiple of $(x^2)^2$, cannot be obtained from neither $(x^3)^2$ nor $(x^2)^2$, because x^3 is not a multiplicative variable for $(x^2)^2$. Saturating the construction by adding each such monomial and proceeding inductively, one can prove that any set of monomials can be enlarged to a complete set, the completion being naturally extended to the PDE via the corresponding differentiations. In our case the completed system is

$$
\boxed{\begin{array}{l} y_{223} \\ y_{33} \\ y_{22} \end{array}} \begin{array}{l} = 0 \\ - x^2 y_{11} = 0 \\ = 0 \end{array} \boxed{\begin{array}{ccc} x^1 & x^2 & \bullet \\ x^1 & x^2 & \bullet \\ x^1 & x^2 & \bullet \end{array}}
$$

Hence, each multiple monomial can be obtained via *one and only one* differentiation process involving multiplicative variables only.

A system of PDE having such a complete set of monomials will be called '*passive*' if any other computation of a principal derivative is coherent with one obtained by using multiplicative variables only.

In our case, let us consider the monomial $(x^2)^2(x^3)^2$ as the principal derivative y_{2233}. On the one hand, using multiplicative variables we obtain

$$
y_{2233} - x^2 y_{1122} - 2y_{112} = 0.
$$

But y_{1122} is a principal derivative and, again using multiplicative variables, we obtain

$$y_{1122} = 0,$$

and therefore

$$y_{2233} - 2y_{112} = 0.$$

However, y_{2233} may be obtained by differentiating y_{223} with respect to x^3, and should give

$$y_{112} = 0,$$

a result not coherent with the previous one because y_{112} is a parametric derivative. Hence the system is not passive, and one should add the PDE $y_{112} = 0$ and start anew.

Therefore, Janet's algorithm proceeds as follows, in an inductive way:

1) Write out the solved form and determine the scheme corresponding to the leading terms.
2) Check if the system of leading monomials is complete, otherwise complete it.
3) Once the system of leading monomials (or derivatives) is complete, check the passivity by differentiating with respect to dots only.
4) If the system is passive, one has clearly made a cut between parametric and principal derivatives, the latter being known without any ambiguity from the parametric derivatives and the differentiated PDE. Otherwise, add the new PDE obtained; they involve only parametric derivatives. Start with step 1.

Janet then proved that this algorithm finishes in finitely many steps because of purely algebraic arguments going back to E. Noether and D. Hilbert.

We shall give all the steps for the preceding example, which is quite appealing for a few reasons, which are given as comments.

JANET ALGORITHM

first (given) system

$$y_{33} - x^2 y_{11} = 0 \qquad x^1 \; x^2 \; x^3$$
$$y_{22} \phantom{- x^2 y_{11}} = 0 \qquad x^1 \; x^2 \; \bullet$$

completed first system

$$y_{223} \phantom{- x^2 y_{11}} = 0 \qquad x^1 \; x^2 \; \bullet$$
$$y_{33} - x^2 y_{11} = 0 \qquad x^1 \; x^2 \; x^3$$
$$y_{22} \phantom{- x^2 y_{11}} = 0 \qquad x^1 \; x^2 \; \bullet$$

second system

$$y_{223} \phantom{- x^2 y_{11}} = 0 \qquad x^1 \; x^2 \; \bullet$$
$$y_{112} \phantom{- x^2 y_{11}} = 0 \qquad x^1 \; \bullet \; \bullet$$
$$y_{33} - x^2 y_{11} = 0 \qquad x^1 \; x^2 \; x^3$$
$$y_{22} \phantom{- x^2 y_{11}} = 0 \qquad x^1 \; x^2 \; \bullet$$

completed second system

$$y_{1123} \phantom{- x^2 y_{11}} = 0 \qquad x^1 \; \bullet \; \bullet$$
$$y_{223} \phantom{- x^2 y_{11}} = 0 \qquad x^1 \; x^2 \; \bullet$$
$$y_{112} \phantom{- x^2 y_{11}} = 0 \qquad x^1 \; \bullet \; \bullet$$
$$y_{33} - x^2 y_{11} = 0 \qquad x^1 \; x^2 \; x^3$$
$$y_{22} \phantom{- x^2 y_{11}} = 0 \qquad x^1 \; x^2 \; \bullet$$

third system

$$y_{1123} \phantom{- x^2 y_{11}} = 0 \qquad x^1 \; \bullet \; \bullet$$
$$y_{1111} \phantom{- x^2 y_{11}} = 0 \qquad x^1 \; \bullet \; \bullet$$
$$y_{223} \phantom{- x^2 y_{11}} = 0 \qquad x^1 \; x^2 \; \bullet$$
$$y_{112} \phantom{- x^2 y_{11}} = 0 \qquad x^1 \; \bullet \; \bullet$$
$$y_{33} - x^2 y_{11} = 0 \qquad x^1 \; x^2 \; x^3$$
$$y_{22} \phantom{- x^2 y_{11}} = 0 \qquad x^1 \; x^2 \; \bullet$$

completed third system

$$y_{11113} \phantom{- x^2 y_{11}} = 0 \qquad x^1 \; \bullet \; \bullet$$
$$y_{1123} \phantom{- x^2 y_{11}} = 0 \qquad x^1 \; \bullet \; \bullet$$
$$y_{1111} \phantom{- x^2 y_{11}} = 0 \qquad x^1 \; \bullet \; \bullet$$
$$y_{223} \phantom{- x^2 y_{11}} = 0 \qquad x^1 \; x^2 \; \bullet$$
$$y_{112} \phantom{- x^2 y_{11}} = 0 \qquad x^1 \; \bullet \; \bullet$$
$$y_{33} - x^2 y_{11} = 0 \qquad x^1 \; x^2 \; x^3$$
$$y_{22} \phantom{- x^2 y_{11}} = 0 \qquad x^1 \; x^2 \; \bullet$$

This last system is passive and the algorithm stops.

As a main criticism we can say

JANET's ALGORITHM IS NOT INTRINSIC.

Of course, such a comment will also apply to any other algorithm based on a similar ranking technique, that is, $\cdots$ to all the algorithms actually known and used for PDE! Meanwhile, this does not mean that this algorithm must be checked in local coordinates, but it does mean that this algorithm depends on *the* coordinate system chosen. To clarify this point, let us imagine two rooms equipped with the same computer, and modify the given system of PDE by means of a linear change of independent variables with constant coefficients. Let two students work independently on the two resulting systems, each one being in its own room with no communication between them. It is then clear that, even if certain results will (and must!) be found to be identical (e.g., the dimension of the space of solutions for a linear system of PDE), most of the steps will not correspond one to the other via the change of variables. As a byproduct, even if we know that the algorithms must stop, they need not follow the same steps and, most importantly, the way to go from the initial to the final system will not be understood. For example, if one deals with a Frobenius type system for interpreting a distribution of vector fields, the crossed derivatives will be effected by the machine $\cdots$ but the machine will never say that the procedure amounts to successively saturating brackets of vector fields.

The following example will prove that, surely, certain coordinate systems are better than others and that knowledge of the best ones is a *necessary* step towards an intrinsic formulation. Indeed, with the notations already used, let us consider the ordered system

$$\boxed{\begin{array}{l} y_{22} \\ y_{11} \end{array}} \begin{array}{l} - y_{12} = 0 \\ = 0 \end{array} \boxed{\begin{array}{ll} x^1 & x^2 \\ x^1 & \bullet \end{array}}$$

Its leading set of monomials, namely $\{(x^2)^2, (x^1)^2\}$ is not complete. However, if we interchange the coordinates x^1 and x^2, the new ordered system becomes

$$\boxed{\begin{array}{l} y_{22} \\ y_{12} \end{array}} \begin{array}{l} = 0 \\ - y_{11} = 0 \end{array} \boxed{\begin{array}{ll} x^1 & x^2 \\ x^1 & \bullet \end{array}}$$

and its set of leading monomials is now complete, although the system is not passive with respect to this set. Returning to the picture of the two different rooms, this simple example shows that the two procedures adopted for following Janet's algorithm will be different.

We also have to notice that, apart from the previous fact that *completeness is not an intrinsic property, passivity is not an intrinsic property* either, because it mixes up the order of the equations. Indeed, if *an* equation is differentiated with respect to *one* independent variable, in an intrinsic procedure any other equation or variable must play the same role and all other differentiated equations must be taken into account at the same time. Hence, the connection between completeness and passivity, though it may be useful, must be entirely revised.

However, Janet's purpose was also to study differential elimination theory by noting that, whenever one differentiates a nonlinear PDE, the highest-order derivatives will appear linearly. The idea thus was to introduce linear algebra techniques for dealing with orders higher than a given one, say q, while reducing to nondifferential business, using a finite number of indeterminates, the way of dealing with orders smaller than q. This approach gave rise to '*differential algebra*' in work of J.F. Ritt [89, 159, 160], who exclusively used Janet's work $\cdots$ without quoting him at all! (see [145] for details).

Of course, one should notice that both Janet and Ritt were in essence interested in the possibility of having constructive methods and finite algorithms that could at least be performed in a manual way. With the recent fast development of informatics and computer science, manual computations can be performed faster and faster by machines, a fact similarly giving rise to '*computer algebra*'!

Roughly speaking, differential algebra deals with the possibility of adding the word '*differential*' in front of all concepts in algebra, and computer algebra now deals with both pure algebra and differential algebra.

In classical algebra, a root of a polynomial equation $y^2 - 2 = 0$ is denoted by a symbol $\sqrt{2} = 1,414\ldots$. However, the infinite sequence of integers defining this number cannot be stored in a computer. Only two approaches can be adopted. In *numerical computation*, an approximation will be used by dealing with finitely many terms only. On the contrary, in *symbolic computation* one just stores a symbol $\sqrt{2}$ (any other symbol, like η, will be equally convenient) with the only property that, whenever its square occurs, this square must be replaced by 2. Of course, calculations are performed over the rational numbers, which can be stored exactly in the computer, and the polynomial must be *irreducible* over the rational numbers, that is, it cannot be factored, because otherwise one could start anew with one of the factors. However, the factorization $y^4 + 4 = (y^2 + 2y + 2)(y^2 - 2y + 2)$ is not evident, and a root of $y^2 + 1 = 0$ is denoted by i, since it has no numerical value (numbers involving i are called *complex numbers*). By the *fundamental theorem of algebra*, all the roots of polynomial equations can be found in a big reserve, the field $\mathbb{C}$ of complex numbers. More generally, algebraic sets are the solutions of systems of algebraic equations in many indeterminates, and a point is usually defined by a set of complex numbers, called coordinates. Again, an algebraic set is called *irreducible* or a *variety* if the ideal generated by the given defining polynomials is a prime ideal in the corresponding ring of polynomials. We recall that an ideal in a ring is *prime* if, whenever a product of two ring elements belongs to the ideal, at least one of the elements belongs to the ideal.

We shall now consider differential algebraic equations defined by *differential polynomials*, that is, polynomials in the derivatives of the unknowns with coefficients that are rational functions of a certain number of independent variables. If $n = 1$, we shall speak of *ordinary differential algebra*, for $n > 1$ of *partial differential algebra*.

Similarly, a root of the differential polynomial ODE $y_x - y = 0$ is denoted by the symbol $e^x = 1 + \frac{x}{1!} + \frac{x^2}{2!} + \ldots$, and y_x is used instead of the classical notation y' in order to exhibit the independent variable x and the unknown variable y separately. However, the infinite Taylor series describing the exponential of x cannot be stored in a computer, and one just stores the symbol e^x (any other symbol, like η, will be equally convenient) with the only property that, whenever its derivative occurs, this derivative must be replaced by the symbol itself. In particular, the computer cannot distinguish between e^x and $2e^x$, which are in fact two symbols having the same property.

The first difficulty arises when trying to extend the concept of irreducibility.

Indeed, let us consider the differential polynomial $P \equiv y_x^2 - 4y$ and the ideal $\mathfrak{a}$, the *differential ideal* generated by P and all its derivatives in the differential ring $\mathbb{Q}\{y\}$ of polynomials in y and its derivatives with rational coefficients. We obtain at once

$$d_x P \equiv 2y_x(y_{xx} - 2).$$

Hence, looking for the solutions of $P = 0 \Rightarrow d_x P = 0$, we must have either $y_x = 0$, and thus $y = 0$, or $y_{xx} - 2 = 0$, and thus $y = (x + a)^2$, with $a = $ const (no clear meaning, though any student can write out this result at once!). In any case, $y_x(y_{xx} - 2)$ is a product, and neither y_x nor $y_{xx} - 2$ belongs to $\mathfrak{a}$; therefore $\mathfrak{a}$ is not a prime differential ideal.

In fact, things are much more tricky. Indeed, according to a celebrated theorem of Hilbert [102, 164, 208], whenever a polynomial vanishes for all the roots of given polynomials, then a certain power of that polynomial must belong to the ideal generated by these given polynomials. Accordingly, an ideal is called *perfect* if, whenever a power of a polynomial belongs to the ideal, then the polynomial itself belongs to the ideal. Otherwise, one can prove (see Chapter VI) that all the polynomials having a power in an ideal $\mathfrak{a}$ generate an ideal $\sqrt{\mathfrak{a}}$, called the *root of the ideal* $\mathfrak{a}$, and one may add the word '*differential*' in front of these assertions. The importance of perfect ideals lies in the fact that they are always intersections of finitely many prime ideals, *even in the differential case*, a property which explains their practical importance.

In the preceding example, the differential ideal generated by P is surely not prime, and now we prove that it is not even perfect. Indeed, we have successively,

$$P \in \mathfrak{a} \Rightarrow y_x(y_{xx} - 2) \in \mathfrak{a}$$
$$\Rightarrow y_x y_{xxx} + y_{xx}(y_{xx} - 2) \in \mathfrak{a}$$
$$\Rightarrow (y_x)^2 y_{xxx} \in \mathfrak{a}$$
$$\Rightarrow yy_{xxx} \in \mathfrak{a}$$
$$\Rightarrow yy_{xxxx} + y_x y_{xxx} \in \mathfrak{a}$$
$$\Rightarrow y_x(y_{xxx})^2 \in \mathfrak{a}$$
$$\Rightarrow 2y_x\, y_{xxx}y_{xxxx} + y_{xx}(y_{xxx})^2 \in \mathfrak{a}$$
$$\Rightarrow 3y_{xx}(y_{xxx})^2 y_{xxxx} + (y_{xxx})^4 \in \mathfrak{a}$$
$$\Rightarrow (y_{xxx})^5 \in \mathfrak{a}$$
$$\Rightarrow y_{xxx} \in \sqrt{\mathfrak{a}}.$$

Hence $\mathfrak{a}$ is not perfect, but every solution of $P = 0$ must be such that $(y_{xxx})^5 = 0$, hence $y_{xxx} = 0$, a fact which is true for $y = 0$ and $y = (x + a)^2$. Hence $\sqrt{\mathfrak{a}}$ is the intersection of two prime differential ideals.

However, in general things are not so simple, and this will be the main difficulty in differential algebra when looking for a criterion or test that would allow us to

characterize differential ideals. To become convinced of this, the reader is encouraged
to consider the differential ideal generated by $P \equiv y_x^2 - 4y^3$. We have

$$d_x P \equiv 2y_x(y_{xx} - 6y^2),$$

and the first impression is to believe that this situation is quite similar to the preceding example. We will prove, though, that it is entirely different. Indeed, we have
successively

$$P \in \mathfrak{a} \Rightarrow y_x(y_{xx} - 6y^2) \in \mathfrak{a}$$
$$\Rightarrow y_{xx}(y_{xx} - 6y^2) + y_x(y_{xxx} - 12yy_x) \in \mathfrak{a}$$
$$\Rightarrow y_{xx}(y_{xx} - 6y^2)^2 \in \mathfrak{a}$$
$$\Rightarrow (y_{xx})^2(y_{xx} - 6y^2)^2 - 12y^2 y_{xx}(y_{xx} - 6y^2)^2 + 36y^4(y_{xx} - 6y^2)^2 \in \mathfrak{a}$$
$$\Rightarrow (y_{xx} - 6y^2)^4 \in \mathfrak{a}$$
$$\Rightarrow y_{xx} - 6y^2 \in \sqrt{\mathfrak{a}}.$$

So, every solution *must* satisfy $y_{xx} - 6y^2 = 0$. It follows that $\mathfrak{a}$ is neither prime nor
perfect, as in the preceding example, but $\sqrt{\mathfrak{a}}$ is prime. This proves that the differential
algebraic set of solutions of $P = 0$ is irreducible.

Now we will show the solution is even more tricky in the partial differential case.
Indeed, suppose we have two independent variables x^1 and x^2, and one unknown y,
and consider the differential ideal $\mathfrak{a}$ generated by the two differential polynomials

$$\begin{cases} P_1 \equiv y_{22} - \frac{1}{2}(y_{11})^2, \\ P_2 \equiv y_{12} - y_{11}. \end{cases}$$

We have

$$Q \equiv d_2 P_2 - d_1 P_1 + d_1 P_2 \equiv (y_{11} - 1)y_{111},$$

hence $\mathfrak{a}$ is not a prime ideal. However,

$$Q \in \mathfrak{a} \Rightarrow (y_{111})^2 + (y_{11} - 1)y_{1111} \in \mathfrak{a}$$
$$\Rightarrow (y_{111})^3 \in \mathfrak{a},$$

and so $\mathfrak{a}$ is not even perfect, but $\sqrt{\mathfrak{a}}$ is prime. It follows that the differential algebraic
set of solutions of $P_1 = 0$, $P_2 = 0$ is irreducible.

As a byproduct, we see that *passitivity interferes with irreducibility*, a fact discovered by Ritt. Hence, in Chapter VI our aim will be to prove that a modern treatment
of differential algebra cannot leave out the modern formal theory of PDE described
in Chapter III.

Subsequently, differential algebra was almost forgotten, until 1940, when, for group
theoretical reasons to be explained later, it was taken up again by E.R. Kolchin and
brought to a relatively modern state in the book [95], which can be considered as
the bible on this subject. However, the lack of the modern achievements of algebraic
geometry at the same time, with the use of tensor products of rings and fields, makes
this book rather old-fashioned, at least in comparison with a very compact elegant
exposition such as [24, 145]. Meanwhile, differential elimination has been studied by

Seidenberg [163] and Wu Wen-Tsun [202, 203]. We will say a few words about its main consequence, namely the existence of a tree of possibilities which will be studied and illustrated in Chapter VII.

Let us consider a nonlinear system of PDE for two sets of unknowns, also called a *differential correspondence*. We want to look for a system of PDE, called a *resolvent system*, that must be satisfied by one of the sets in such a way that the other set will exist. To this end we consider the given system as a linear system for the second set of unknowns, and manipulate it by means of the Janet–Ritt–Wu algorithm described earlier. In checking the passivity at various steps of the algorithm, we must use the implicit function theorem for finding the principal derivatives. Therefore, because of certain rank conditions, certain determinants appear, and, depending on the vanishing or nonvanishing of a determinant, two cases have to be treated separately. Proceeding inductively, we find a tree (graph) of possibilities, where one branch corresponds to the most generic situation and the final leaf of this branch therefore describes the most generic conditions to be satisfied by the first set of unknowns in order that the second set will exist.

A typical situation in Chapter VII will be the *inverse calculus of variations*. Indeed, if one has to extremize an action integral with given Lagrangian function on a curve, the curve must be a solution of the well-known second-order Euler–Lagrange equations. More precisely, on a curve $x \to (x, y(x), z(x))$ we may introduce the speed $(\dot{y} = dy/dx, \dot{z} = dz/dx)$ and acceleration $(\ddot{y} = d^2y/dx^2, \ddot{z} = d^2z/dx^2)$. Therefore the condition

$$\delta \int \Phi(x, y, z, \dot{y}, \dot{z}) \, dx = 0$$

implies, after integration by parts,

$$\frac{d}{dx}\left(\frac{\partial \Phi}{\partial \dot{y}}\right) - \frac{\partial \Phi}{\partial y} = 0, \qquad \frac{d}{dx}\left(\frac{\partial \Phi}{\partial \dot{z}}\right) - \frac{\partial \Phi}{\partial z} = 0,$$

where d/dx is the total derivative with respect to x. Hence, when the *Hessian condition*

$$\det \begin{pmatrix} \dfrac{\partial^2 \Phi}{\partial \dot{y} \partial \dot{y}} & \dfrac{\partial^2 \Phi}{\partial \dot{y} \partial \dot{z}} \\ \dfrac{\partial^2 \Phi}{\partial \dot{z} \partial \dot{y}} & \dfrac{\partial^2 \Phi}{\partial \dot{z} \partial \dot{z}} \end{pmatrix} \neq 0$$

holds, Cramer's rule in linear algebra implies that the Euler–Lagrange equations can be written as follows:

$$\ddot{y} = F(x, y, z, \dot{y}, \dot{z}), \qquad \ddot{z} = G(x, y, z, \dot{y}, \dot{z}).$$

Conversely, if such a second-order system is given, Φ must be a solution of the linear second-order system

$$\begin{cases} \dfrac{\partial^2 \Phi}{\partial x \partial \dot{y}} + \dot{y}\dfrac{\partial^2 \Phi}{\partial y \partial \dot{y}} + \dot{z}\dfrac{\partial^2 \Phi}{\partial z \partial \dot{y}} + F\dfrac{\partial^2 \Phi}{\partial \dot{y} \partial \dot{y}} + G\dfrac{\partial^2 \Phi}{\partial \dot{z} \partial \dot{y}} - \dfrac{\partial \Phi}{\partial y} = 0, \\[2ex] \dfrac{\partial^2 \Phi}{\partial x \partial \dot{z}} + \dot{y}\dfrac{\partial^2 \Phi}{\partial y \partial \dot{z}} + \dot{z}\dfrac{\partial^2 \Phi}{\partial z \partial \dot{z}} + F\dfrac{\partial^2 \Phi}{\partial \dot{y} \partial \dot{z}} + G\dfrac{\partial^2 \Phi}{\partial \dot{z} \partial \dot{z}} - \dfrac{\partial \Phi}{\partial z} = 0, \end{cases}$$

involving the 5 independent variables $(x, y, z, \dot{y}, \dot{z})$ and with coefficients depending also on (F, G).

Ordering the derivatives and applying Janet's algorithm reduces the question to one about ranks of matrices depending on F, G and their derivatives, and so to the tree almost completely exhibited in the nice reference [48, 49], where each leaf is illustrated by means of an explicit situation.

In nonlinear cases the problem is more difficult, because the elements of the determinants may also depend on the second set of unknowns and subsequent work must take care of this fact.

Rank problems and inequalities may therefore emerge even in the direct study of passitivity along Janet's algorithm. In Chapters III and VI we shall therefore sketch how to give an intrinsic version of the tree, and illustrate it.

Before returning to the story of PDE, we examine how Janet's algorithm could be useful in pure algebra.

One of the most important tools in nowadays constructive commutative algebra are *Gröbner bases*, or standard bases, called after the Tyrolean mathematician W. Gröbner (1899–1980). An algorithm for constructing explicit Gröbner bases has been found by B. Buchberger [30, 66, 169] and is now implemented in most of the symbolic packages presently available (e.g., MACSYMA, REDUCE, MATHEMATICA, and MAPLE). Roughly speaking, Gröbner bases are special kinds of generating sets for polynomial ideals. Indeed, with respect to the ordering of monomials considered earlier, with $1 < x^2 < \cdots < x^n$, we may consider the set of highest monomials, called *initials*, in a given set of polynomials $P = \{P_1, \ldots, P_r\}$. We denote this set by IN P, and the set of all possible multiplies of them by (IN P). Of course, this latter set is contained in the set IN $\mathfrak{a}$ of initials of all polynomials contained in the ideal $\mathfrak{a}$ generated by P. We may therefore *define* a Gröbner basis for a polynomial ideal $\mathfrak{a}$ as a set $P = \{P_1, \ldots, P_r\}$ of polynomials in $\mathfrak{a}$ that generate $\mathfrak{a}$ in such a way that IN $\mathfrak{a}$ = (IN P).

For example, $\{x^1 x^2 - 1, x^2 - x^1\}$ is *not* a Gröbner basis for the polynomial ideal $\mathfrak{a} \subset \mathbb{Q}[x^1, x^2]$ generated by these two polynomials with respect to the total ordering with $x^1 < x^2$. Indeed, $(x^1 x^2 - 1) - x^1(x^2 - x^1) = (x^1)^2 - 1$ belongs to $\mathfrak{a}$ and $(x^1)^2$ is *not* a multiple of $x^1 x^2$ or x^2. However, $\{x^2 - x^1, (x^1)^2 - 1\}$ also generate $\mathfrak{a}$, because $x^1 x^2 - 1 = x^1(x^2 - x^1) + ((x^1)^2 - 1)$, and is a Gröbner basis of $\mathfrak{a}$, since one can easily verify that any polynomial in $\mathfrak{a}$ has as leading term a multiple of x^2 or $(x^1)^2$.

We shall prove that *there is no conceptual difference at all between the Buchberger algorithm and the Janet algorithm!*

To this end we associate to any monomial a derivative of a given unknown y, exactly as has been done in the reverse way by Janet. For example, $x^1 x^2 \leftrightarrow \partial_{12} y = y_{12}$. By linearity, any polynomial in $k[x^1, \ldots, x^n]$ is transformed into a linear differential polynomial in $k\{y\}$, where the field k is now regarded as a trivial differential field. An elementary way is to say that the given polynomial equations $P_1 = 0, \ldots, P_n = 0$ are transformed into a linear system of PDE in one unknown with constant coefficients $\cdots$ that is the simplest situation one can encounter! It follows from the construction that leading derivatives correspond to leading monomials, or initials. Let us apply Janet's algorithm to this system of PDE. With respect to a given ordering it will provide *one* (completeness) *and only one* (passivity) way to compute all the principal derivatives. Translating this fact back into terms of pure polynomials, we precisely find the definition of a Gröbner basis!

We shall illustrate this on the preceding example:

$$\text{correspondence} \begin{cases} x^1 x^2 - 1 \leftrightarrow y_{12} - y, \\ x^2 - x^1 \leftrightarrow y_2 - y_1. \end{cases}$$

Applying Janet's algorithm, with ordering $x^1 < x^2$, we obtain

$$\boxed{\begin{array}{l} y_{12} \\ y_2 \end{array}} \begin{array}{l} -y = 0 \\ -y_1 = 0 \end{array} \qquad \boxed{\begin{array}{cc} x^1 & x^2 \\ \bullet & x^2 \end{array}}$$

The corresponding system of monomials is complete, nevertheless the system is *not* passive, because

$$\partial_1(\partial_2 y - \partial_1 y) - (\partial_{12} y - y) = -\partial_{11} y + y.$$

Hence we have to consider the new system

$$\boxed{\begin{array}{l} y_{12} \\ y_{11} \\ y_2 \end{array}} \begin{array}{l} -y = 0 \\ -y = 0 \\ -y_1 = 0 \end{array} \qquad \boxed{\begin{array}{cc} x^1 & x^2 \\ x^1 & \bullet \\ \bullet & x^2 \end{array}}$$

which is passive with respect to its leading monomials. It now remains to remove a redundant PDE (like the top one) by looking at the principal derivatives through this computation starting from the use of the bottom (lower rank in the ordering) equations. We are therefore left with the following complete system:

$$\boxed{\begin{array}{l} y_{11} \\ y_2 \end{array}} \begin{array}{l} -y = 0 \\ -y_1 = 0 \end{array} \qquad \boxed{\begin{array}{cc} x^1 & \bullet \\ x^1 & x^2 \end{array}}$$

This system is passive, because

$$\partial_2(\partial_{11} y - y) - \partial_{11}(\partial_2 y - \partial_1 y) = \partial_{111} y - \partial_2 y$$
$$= \partial_1(\partial_{11} y - y) - (\partial_2 y - \partial_1 y).$$

As a byproduct, a new way to look at the Janet–Ritt algorithm, which we will soon sketch, can be transferred onto the Buchberger algorithm, and this is the reason why in this book we shall not speak about Gröbner bases anymore.

Returning to our historical survey, it was therefore a challenge to modify the previous computational approach (Janet, Ritt, Thomas, Kolchin, Wu) in such a way as to deal with intrinsic (coordinate-free) methods and results. Such an improvement has been made during 1965–1975 by D.C. Spencer and coworkers in the USA [73, 172], quite independently of the previous evolution. However, the Americans did not care too much about the constructive aspect (they did not even know Janet's work!) and did not made any link with differential algebra and its applications. This link was laid for the first time in our book [145], with promising applications to most domains of mathematical physics; it predates the main novel aspect of the present book.

Although Spencer's tools relay on difficult mathematics, we shall try to sketch the underlying motivations by comparison with Janet's method presented above.

Roughly speaking, Janet's algorithm uses two major, successive, ingredients:

- *completeness* as a purely algebraic tool;
- *passivity* as a purely differential tool.

Here, by 'algebraic' we mean that we do not have to differentiate the equations, while 'differential' means that differentiation of the equations is essential. The ingredients are independent.

Spencer's main idea has been to find intrinsic counterparts of these two ingredients, and, at the same time, a way to combine them.

A first step has been to consider the derivatives of the unknowns as new indeterminates, and thus to transfer *'partial differential'* equations into *'simple'* equations. This was indeed the basic motivation for *jet theory*.

A second step has been to consider only top-order derivatives, and to notice that, even for nonlinear systems of order q, the derived equations contain the derivatives of strict order $q + 1$ in a linear way, a result leading to pure linear algebra with the δ-sequence and its cohomology. In particular, 2-*acyclicity* of the top-order part, or the *symbol*, was shown to be a cornerstone superseding the *involution* of Hilbert, Cartan and Janet, while appealing to intrinsic computations only. This fact was missed by Janet, although he had most of the material concerning involution in hand. However, Janet was stopped by the necessity of using a special coordinate system for verifying involution. Such a very special coordinate system is nowadays called a δ-*regular coordinate system*, and it is indeed a magic fact that, as we will show in Chapter III, certain concepts have a purely intrinsic meaning and definition, although they can be tested in special coordinate systems only. For this reason the methods *seem* similar, while they are, in fact, very different. One should notice that acyclicity is lacking in differential algebra; one additional reason to say that the 'difficulty in differential algebra is not algebra but PDE'.

We now turn to *formal integrability*, an intrinsic concept which supersedes that of passivity. First we define a *prolongation procedure*, in the following intrinsic way. If we start with a given system of PDE of order q, we can substitute indeterminates instead of derivatives and use the resulting equations to define a manifold, $\mathcal{R}_q$, whenever certain rank conditions are satisfied. For the moment we shall not worry too much about these conditions. Now, starting from $\mathcal{R}_q$ we can substitute derivatives instead of indeterminates, then differentiate r times all the PDE obtained, with respect to all independent variables, and finally substitute indeterminates to (eventually) define a manifold, denoted by $\rho_r(\mathcal{R}_q)$ or $\mathcal{R}_{q+r}$ if there is no possible confusion with a family of systems, which is called the r-*prolongation* of $\mathcal{R}_q$. Of course, $\mathcal{R}_{q+r}$ may involve more equations of order q than $\mathcal{R}_q$ itself, so that the projection $\mathcal{R}_q^{(r)}$ of $\mathcal{R}_{q+r}$ on $\mathcal{R}_q$ may be such that we have a strict inclusion $\mathcal{R}_q^{(r)} \subset \mathcal{R}_q$. If this happens we will be unable to compute the principal derivatives or to determine the Taylor expansion of the unknowns at a point. For example, in Janet's example $\mathcal{R}_2$ is determined by the two PDE $y_{33} - x^2 y_{11} = 0$, $y_{22} = 0$, but $\mathcal{R}_4$ involves $y_{2233} - x^2 y_{1122} - 2y_{112} = 0$, $y_{2233=0}$, $y_{1122} = 0$. Therefore, among the equations defining $\mathcal{R}_4$ one finds $y_{112} = 0$, which is not found among the equations defining $\mathcal{R}_3$. Hence $\mathcal{R}_2^{(1)} = \mathcal{R}_2$, but there is a strict inclusion $\mathcal{R}_3^{(1)} \subset \mathcal{R}_3$.

As a byproduct, systems of PDE can be roughly classified into two categories:

- *'good' systems*: allowing us to successively compute principal derivatives from parametric derivatives, without 'feedback' information;
- *'bad' systems*: not allowing us to successively compute principal derivatives from parametric derivatives because of disturbing feedback information.

Spencer called the first class of systems *formally integrable systems*. They are such that $\mathcal{R}_{q+r}^{(1)} = \mathcal{R}_{q+r}, \forall r \geq 0$. In general, 'bad' systems can be reduced to good systems having the same solutions. This intrinsic concept supersedes passivity, and nothing can be known about the space of solutions of a 'bad' system of PDE before transforming the system into a formally integrable system having the same solutions.

We shall see that a way to proceed is to inductively use a criterion first given by Spencer [172] and extended by H. Goldschmidt to nonlinear systems [29, 73]. This criterion can be used as a constructive test, and it says that whenever $\mathcal{R}_q^{(1)} = \mathcal{R}_q$ and a purely algebraic property of the symbol M_q of $\mathcal{R}_q$ is satisfied, namely 2-acyclicity, then $\mathcal{R}_q$ is formally integrable. However, in practice it is still not known whether 2-acyclicity can be verified in a constructive way, and one has to use involutiveness of M_q instead.

We give a short correspondence scheme between the old and the new approaches:

old approach	new approach
monomials	symbol
completeness	2-acyclicity
passivity	formal integrability

Returning to Janet's algorithm, the crucial, but highly nontrivial, theorem that has to be used for obtaining an intrinsic version amounts to saying that $\rho_r(\mathcal{R}_q^{(1)}) = \mathcal{R}_{q+r}^{(1)}$ whenever M_q is 2-acyclic. It follows from this theorem that the way to obtain a formally integrable system from a given system lies in a succession of prolongations ('*up*') and projections ('*down*'), which, in general, terminates after finitely many steps, in such a way that, starting with a system $\mathcal{R}_q$, we obtain a final system $\mathcal{R}_{q+r}^{(s)}$ for certain integers r, s. In the Janet example, we shall see that $r = 3$, $s = 2$.

Another extremely important point in applications to differential algebra, and thus to engineering sciences, is the strange link between rank conditions and acyclicity. Indeed, we shall see that the symbol M_{q+r} of $\mathcal{R}_{q+r}$ depends only on the symbol M_q of $\mathcal{R}_q$ and is, in general, a family of vector spaces defined above each point of $\mathcal{R}_q$. It may happen that the dimensions of these vector spaces vary from point to point. Therefore it is very important to have a way of stabilizing this dimension, in particular, to make it constant over $\mathcal{R}_q$. The fundamental result towards the solution of this problem is a theorem saying that M_{q+r} is a vector bundle over $\mathcal{R}_q$ (that is, all vector spaces have the same dimension) whenever M_{q+1} is a vector bundle over $\mathcal{R}_q$ and M_q is 2-acyclic.

Finally, in view of Hilbert's theorem for polynomials, only perfect ideals are of importance in applications, and it is necessary to have a test for verifying whether a differential ideal is perfect or not. Surprisingly too, the answer will be given by translating the preceding formal integrability criterion into the language of algebraic geometry. We shall find that '*manifolds*' correspond to '*perfect ideals*', while vector bundles can be defined almost similarly. At that time the *translation obtained provides a test for a differential ideal to be perfect or prime*, and also gives a new intrinsic procedure for studying differential ideals.

We shall combine the above results with the so-called *symbol prolongation theorem*, which says that a prolongation M_{q+r} of an arbitrary symbol M_q becomes involutive for r large enough in order to give an intrinsic formulation of the 'trees' already found in the differential-algebraic framework. Indeed, let us start with a certain number of given differential polynomials of order q, where we may assume that they generate a perfect (nondifferential) ideal at order q (otherwise we have to take the root of this ideal using nondifferential techniques, a problem outside the scope of this book). Then we apply the test by looking at the symbol M_{q+1} of order $q + 1$. There are two possibilities: it is a vector bundle or it is not. Since we may assume that q is large enough for 2-acyclicity of involutivity, we may try to apply the criterion for an ideal to be perfect on the condition that certain determinants, say $D_1, \ldots, D_a$, do not simultaneously vanish to ensure that M_{q+1} has a well-defined (hence minimal) generic dimension. If so, the differential ideal is perfect, and nothing more can be said about it. If not, this means that there are only two possibilities: the dimension of M_{q+1} is the generic dimension or it is greater than this, if $D_1 = 0, \ldots, D_a = 0$. This is the first branching process producing the tree in an intrinsic way. It is clearly reflected in the simple examples of ordinary differential algebra already examined by us. In general, however, we shall see that there is no reason why the tree should be finite, since the 'up and down' procedure need not terminate.

We now turn to group theory.

The word '*group*' appeared for the first time in a mémoire of E. Galois presented to the French Academy of Sciences in 1830. However, at that time group theory was only concerned with finite groups, more precisely with permutation groups, through the pioneering work of A. Cauchy. After the importance of Galois theory had been disclosed by J. Liouville in 1845, the concept of a group slowly passed from algebra to geometry with the work of F. Klein, culminating in his '*Erlangen Program*' of 1872.

Klein's idea was to relate a geometry to its underlying group via the behavior of certain invariants of the group. For example, euclidean geometry is based on the invariance of the square of a line element, and any deviation from this invariance can be interpreted as a *deformation*, a reason for linking elasticity theory with group theory in its early stage.

In fact, the groups considered, which are nowadays called *Lie groups of transformations*, consist of families of transformations depending on a certain number of constant *parameters*. They were exhaustively considered by the Norwegian mathematician S. Lie in 1880 [109]. He provided three main general theorems, to be recalled in Chapter V.

Only ten years later, Lie discovered that these groups of transformations are just particular instances of a more important class of groups, then called infinite groups but nowadays called *Lie pseudogroups*. We shall try to describe these groups in an intuitive way by using a comparison with robotics which will subsequently prove to be crucial in applications to the domain of robotics. Indeed, let us consider a human body as an elaborated robot. Looking at the movement of an arm, we notice that the main point is not to describe the rotations that can be effected by each part of the arm (elbow, hand, etc.), but to discover that the various parts of the arm can move according to rotation only, rotation of one part with respect to the next part. Passing to the mathematical framework, we would like to consider a Lie group of transformations as a certain typical family, independently of possibly changing values of the parameters. For example, the transformations $y = ax + b$, where $a, b = \text{const}$, can be regarded as an affine transformation of the real line. However, we also notice that they are all the (invertible) solutions of the simple ODE $y'' = 0$, which has a meaning independently of the values of the parameters. Although this example may seem trivial because, conversely, we are able to immediately and explicitly integrate this ODE, the case of projective transformations is rather less trivial. Indeed, let us consider the transformation

$$y = \frac{ax + b}{cx + d}.$$

We obtain in succession:

$$y' = \frac{ad - bc}{(cx + d)^2}, \quad y'' = -2c\frac{ad - bc}{(cx + d)^3}, \quad y''' = 6c^2\frac{ad - bc}{(cx + d)^4},$$

and therefore we discover that the projective transformations are solutions of the Schwarzian ODE

$$\frac{y'''}{y'} - \frac{3}{2}\left(\frac{y''}{y'}\right)^2 = 0.$$

Conversely, it is not quite evident that all the (invertible) solutions of this ODE are projective transformations. One can use an integration by parts, setting $z = y''/y'$ to find $z' - \frac{1}{2}z^2 = 0$, which leads to

$$\frac{1}{z} = -\frac{1}{2}(x + \alpha) \quad (\alpha = \text{const}) \Rightarrow z = \frac{-2}{x + \alpha},$$

and therefore to

$$y' = \frac{-\beta}{(x + \alpha)^2} \quad (\beta = \text{const}),$$

that is, to

$$y = \frac{\beta}{x + \alpha} + \gamma \quad (\gamma = \text{const}).$$

This decomposes a projective transformation into the product of a translation, an inversion and a dilation.

As a byproduct we discover that certain groups of transformations consist of the (invertible) solutions of a system of ODE or PDE. The above procedure is quite general as it applies to any Lie group of transformations. One only needs to differentiate sufficiently many times in order to eliminate the parameters among the independent variables, the dependent variables and their derivatives. More generally, a Lie pseudo-group of transformations is a group of transformations that are solutions of a system of PDE. As we have seen, such a system of *defining equations* may be nonlinear and of high order.

For example, holomorphic transformations of the plane, $Z = f(z)$ with $z = x + iy$ and $Z = X + iY$, are solutions of the well-known Cauchy–Riemann system

$$\frac{\partial Y}{\partial y} - \frac{\partial X}{\partial x} = 0, \qquad \frac{\partial Y}{\partial x} + \frac{\partial X}{\partial y} = 0,$$

and are quite important in fluid mechanics for the study of flow around an obstacle. Similarly, the movement of an incompressible fluid is modeled by volume-preserving transformations that are solutions of the Jacobian equation

$$\frac{\partial(y^1, \ldots, y^n)}{\partial(x^1, \ldots, x^n)} = 1$$

in n-dimensional space. In both cases, we notice that we do not have a manner for expressing the general solution by integrating the systems using arbitrary constants or functions of 1 to n variables. As a byproduct, new techniques must be used, even if we treat Lie groups of transformations in this new framework, because we are no longer able to use the concept of parameters $\cdots$, a concept leading to the deformation of *Lie groups* and its implications in modern physics.

Let us dwell somewhat on this comment, as it will be crucial for applications to physics. In Lie groups or pseudogroups of transformations, the transformations cannot be separated from the ambient space. Of course, this is particularly true for a Lie pseudogroup, because the defining equations are given in terms of dependent and independent variables belonging just to the same space. In the case of a Lie group of transformations we endow the parameters, say a, with a Lie group structure G; letting X be the space, this amounts to giving an *action* of the group on the space or, equivalently, the graph of this action:

$$X \times G \to X \times X$$
$$(x, a) \to (x, ax),$$

where $y = ax$ can also be written as $y = f(x; a)$ if we want to indicate the functional dependence. Hence, it is quite a surprise to note that certain modern applications of group theory to physics, namely 'gauge theory', introduces a space X, usually our 4-dimensional space-time, and a Lie group G *which does not act on* X! This will lead us to revise the foundations of electromagnetism, in Chapter VIII. To do this, and also to critisize gauge theory in its classical formulation of C.N. Yang and R.L. Mills [1, 5, 27, 51, 204, 205, 206], we shall need a very important concept, namely the *gauging of a group*. Roughly speaking , and independently of any action, if we have a manifold X and a Lie group G as another manifold, we may consider maps $X \to G$, i.e. from X to G. Another formulation is to consider the group parameters a no longer as constants, but as functions $a(x)$ on X. Let us show that this idea is related to the foundation of elasticity $\cdots$ if only we have an action! Indeed, let us take a deformable body in an initial position with coordinates x, and translate it to a final position $y = x + a$ with $a = $ const. It is clear that the body will not have been deformed. On the other hand, let us deform the body, so that each point of it passes from an initial position with coordinates x to a final position with coordinates y, say in euclidean space (for simplicity). We can introduce a vector field $u(x)$, originating at each initial point and ending at each final point. It is called the *displacement field*, and we can write $y = x + u(x)$. We then discover that elasticity theory has its foundation in the above concept of gauging. Hence we see that if such a concept can be found from another source, a new interpretation has been provided. In Chapter VIII we shall see how this remark will allow us to give, for the first time ever, a group-theoretical unification of elasticity, heat and electromagnetism, by gauging a group with 15 parameters. At the same time we shall see that the procedure of gauging has to do with an action, just as in elasticity, and that the fact that the action disappears in gauge theory is a pure miracle. It will follow that the (pure) Lagrangians of mathematical physics *must* be functions on 1-forms and *not* on 2-forms, contrary to what is nowadays believed in classical gauge theory. There are even experimental reasons for this, which, although the name of J.C. Maxwell is associated with them, are not very well known among gauge people. Indeed, at the beginning of the 19th century, Brewster discovered a field-matter coupling now called '*photoelasticity*. The phenomenological law was established by Maxwell himself and, independently and simultaneously, by the physicist Neumann in the middle of the 19th century. The experiment can be easily reproduced today, it costs but a few dollars, and, although it was quite fashionable for engineers at the beginning of the 20th century, it is now mainly practical routine in civil engineering schools and a way to attract students because of the beautiful colors it provides. The experiment goes as follows. Let us separate the two glasses of a pair of polaroid sunglasses, place them parallel, and turn them in such a way that only vertical electric vibrations can pass through the first glass, called the *polarizer*, and only horizontal electric vibrations can pass through the second glass, called the *analyzer*. If a light ray passes through this construct, an observer will see nothing. The same observation is made when a piece of 5 mm thick transparent plastic material ('altuglass') is put in between the polarizer and the analyzer, parallel to both. Things change drastically if the plastic is subjected to forces and reacts according to its strain-stress constitutive relations. Then an observer sees beautiful interferences, and this can be used to infer nondestructive information about the elasticity inside the material. The explanation brought about by Maxwell is that, given that the stress and strain tensors are simultaneously symmetric and diagonalizable because of the constitutive relations, one may decompose the incident vertical electric vibration coming from the polarizer into the two proper directions in the plane parallel to the plastic sheet. The velocity of light being different because of a coupling phenomenon to be discussed in detail in Chapter VIII, an interference pattern emerges through the analyzer, according to the *Maxwell–Neumann formula*

$$\sigma_1 - \sigma_2 = \frac{k\lambda}{eC},$$

where σ_1, σ_2 are the principal stress values, k is a relative integer, λ is the wavelength in the case of monochromatic light, e is the thickness of the plastic plate, and C is a constant, which is in fact the coupling constant of the phenomenon (and is therefore called the *photoelastic constant*) and depends only on the material used.

It follows from this experiment, and this can be confirmed by a thermodynamical description, that strain and electromagnetic field appear *exactly* at the same conceptual level. Now strain only deals with the first-order derivatives of the displacement vector. Also, energy of deformation and kinetic energy are on the same conceptual level. Suppose that a rigid body moves according to the law of motion

$$x = A(t)x_0 + B(t),$$

where x_0 denotes the initial position of a point of the body at time t_0, x denotes the final position of this point at time t, $B(t)$ is the vector describing a time-dependent translation, and $A(t)$ is an orthogonal matrix describing a time-dependent rotation. We may also suppose that the body has a built-in frame attached to it in such a way that it coincides at time t_0 with the absolute frame used for fixing the position. The projection of the speed onto that frame is therefore given by the vector

$$A^{-1}(t)\dot{x} = A^{-1}(t)\dot{A}(t)x_0 + A^{-1}(t)\dot{B}(t).$$

In the simplest case of a top, i.e. a rigid body with a fixed point, $B(t) = 0$ and the kinetic energy becomes a (quadratic) function of the component $A^{-1}(t)\dot{A}(t)$ of the 1-form $A^{-1}(t)\dot{A}(t)\,dt$. In general, 'dt' is omitted, since no confusion may arise. Finally, differentiating the orthogonality relation

$$A(t)^t A(t) = I,$$

where 't' means transposition, we obtain the skew-symmetry condition

$$A^{-1}(t)\dot{A}(t) + {}^t(A^{-1}(t)\dot{A}(t)) = 0.$$

It follows that $A^{-1}(t)\dot{A}(t)\,dt$ is a 1-form with values in the skew-symmetric matrices whenever $A(t)$ is a gauging of the rotation group. This situation is generalized in classical gauge theory by a gauging of the Lie group G, with 1-forms having values in the Lie algebra $\mathcal{G}$ of G.

Therefore we discover at once that the geometrical and variational framework of classical gauge theory is in complete contradiction with photoelasticity if the electromagnetic field, which is clearly a 2-form even at the engineering level, is interpreted as a 2-form with vales in $\cdots$ a 1-dimensional Lie algebra, namely that of the 1-dimensional unitary Lie group. However, it is not possible to obtain a 2-form from a 1-form with values in the Lie algebra of a Lie group *whenever the group does not act on the base manifold*. On the other hand, if the group does act on the base manifold, classical gauge theory cannot be used because $\cdots$ the action disappears. The solution of this dilemma can only be achieved if, starting with the same objects, namely a manifold and a Lie group which now has an action, we can work out *another mathematical framework*, in which strain, speed and electromagnetic field are interpreted as 1-forms. As a byproduct, Lagrangian functions should be functions on 1-forms and not a word should be left from classical gauge theory. Since this 15-parameter group that we shall exhibit is the conformal group of space-time, namely the group of transformations preserving the Minkowski metric $ds^2 = dx^2 + dy^2 + dz^2 - c^2\,dt^2$ up to a functional multiplier, we are faced with the following alternative:

Electromagnetism has to do with $U(1)$ or with the conformal group.

The two interpretations are so different, both mathematically and conceptually, that only the future can tell which one is correct. Nevertheless, the reader should not forget that using the conformal group to describe electromagnetism was precisely an idea of H. Weyl in 1918 [138, 196, 197]. In fact, he failed, and this dead-end gave rise to classical gauge theory, not because his idea was bad, but mainly because the mathematics that has to be used for properly describing groups of nonlinear transformations was developed only in 1975, in work of Spencer.

As we shall see, it is rather astonishing to note that, at about the same time, although independently, the brothers E. and F. Cosserat used Spencer's later machinery on the '*simple*' group of '*rigid*' transformations [41, 42], while H. Weyl only dreamt about such a new machinery for the '*complicate*' group of '*conformal*' transformations [196]. It seems that there is not merely a group-theoretical link between these two approaches, i.e. *there must be a common group-theoretical reason for using another mathematical framework*. As a byproduct, the positive answer to this question explains, in a way coherent with our historical survey, why *most of nowadays mathematical physics is based on a confusion between two differential sequences*: the Janet sequence (1920) and the Spencer sequence (1975).

The collected work of Lie, despite its publication by his student F. Engel at the end of the 19th century, did not attract a large audience. At the beginning of the 20th century, only two Frenchmen built upon it: E. Cartan (1869–1951) in 1905 [32] and E. Vessiot (1865–1952) in 1903 [191]. Both had been students at the École Normale Supérieure in Paris. This common origin, combined with the fact that Vessiot was head of the École Normale Supérieure for two periods of ten years while Cartan was a University teacher in Besançon, explains why Cartan never spoke about Vessiot's work, and vice versa. Also, Vessiot tried, his life long and in vain, to extend Galois' work on polynomials and permutation groups to PDE and pseudogroups. For this reason he was convinced that Cartan's approach could never been used for this purpose, while Cartan never paid attention to the Galois theory for PDE.

Lie had shown that any Lie group can be studied by means of 1-forms on the group that are (left or right) invariant under the action of the group, and therefore it was a challenge to extend to Lie pseudogroups the *structure equations*, also called *Maurer–Cartan equations*, that they satisfy. To this end Cartan invented *exterior calculus*, which gave a nice and compact way to write down the Maurer–Cartan equations and the defining equations for pseudogroups [171]. However, the generalization he proposed for Lie pseudogroups was much too sophisticated, involving exterior calculus on large spaces containing the derivatives of transformations as new variables (these are now called *jet coordinates*, as we have seen). Also, the extension to the new framework of the finite set of *structure constants* describing the bracket of Lie algebras was not so convincing. Nevertheless, he used these methods with success in order to solve certain classification problems.

As noticed by contemporaries, like U. Amaldi in 1909 (see the comments in [8, 145,

146]), the way followed by Vessiot at that time looked so completely different, that no single link seemed to exist between the two approaches. In fact, Vessiot closely followed the work of Lie, and therefore dealt with *differential invariants*, that is, functions of the derivatives of a transformation $y = f(x)$ that are invariant under any transformation $\bar{y} = g(y)$ belonging to the pseudogroup. For example, while it is easy to discover that y''/y' is invariant under $\bar{y} = ay + b$, it is no so easy to discover that $(y'''/y') - \frac{3}{2}(y''/y')^2$ is invariant under $\bar{y} = (ay + b)/(cy + d)$. Equating a differential invariant to its value whenever one sets $y = x$ in it, a PDE is obtained which admits as solution all transformations of the Lie pseudogroup. Also, the derivative of a differential invariant is again a differential invariant, and it can be proved that there always exists a fundamental set of differential invariants generating the others. As a byproduct the defining equations of a Lie pseudogroup can be written in *Lie form*:

$$\Phi^\tau(y_q) = \omega^\tau(x),$$

where y_q denotes y and its derivatives (in old language) or jet coordinates (in new language) up to order q. A first property, discovered by Vessiot, is that the differential invariants transform into differential invariants under any transformation $\bar{x} = \varphi(x)$ extended to derivatives. For example,

$$y_x = y_{\bar{x}}\partial_x\varphi(x), \qquad y_{xx} = y_{\bar{x}\bar{x}}(\partial_x\varphi(x))^2 + y_{\bar{x}}\partial_{xx}\varphi(x)$$

leads to

$$\frac{y_{xx}}{y_x} = \frac{y_{\bar{x}\bar{x}}}{y_{\bar{x}}}\partial_x\varphi(x) + \frac{\partial_{xx}\varphi(x)}{\partial_x\varphi(x)}$$

and to the following transformation law $(x, u) \rightarrow (\bar{x}, \bar{u})$:

$$\bar{x} = \varphi(x), \qquad u = \bar{u}\partial_x\varphi(x) + \frac{\partial_{xx}\varphi(x)}{\partial_x\varphi(x)}.$$

This is precisely the way of defining a bundle by patching coordinates; we will call such bundles *natural bundles* [126]. The simplest natural bundles known to engineers are *tensor bundles*, which involve first-order derivatives of $\varphi(x)$ only and have linear transformation rules $u \rightarrow \bar{u}$. A section $u = \omega(x)$ of such a bundle is usually called a (field of) *geometric object*(s). Clearly, this is a generalization of vector or tensor fields.

However, Vessiot's main contribution has been the clarification of a result poorly stated by Lie in [109, 191] without any proof. Namely, any geometric object must satisfy differential *integrability conditions* of the form

$$I(j_1(\omega)) = c(\omega)$$

which are invariant under any change of coordinates, and where $c(\omega)$ is a function of ω depending on a certain number of constants related by algebraic conditions:

$$J(c) = 0,$$

which are called *generalized Jacobi identities* because they will be shown to generalize the *Jacobi relations*

$$c^\mu_{\rho\sigma}c^\nu_{\mu\tau} + c^\mu_{\tau\rho}c^\nu_{\mu\sigma} + c^\mu_{\sigma\tau}c^\nu_{\mu\rho} = 0$$

for the structure constants of a Lie algebra. The latter differential conditions are necessary and sufficient for formal integrability of the system of defining equations of the Lie pseudogroup under consideration.

A simple example will clarify this point. Let a geometric object consist of a pair $\omega = (\alpha, \beta)$ with α a 1-form and β a 2-form. Since any diffeomorphism of the base manifold can be lifted to the forms (we are dealing with well-known tensor bundles), consider the Lie pseudogroup of transformations $y = f(x)$ preserving α and β and let $\dim X = 2$. A simple computation which is left as an exercise to the reader shows that these transformations also preserve the 2-form $d\alpha$, and therefore the quotient $d\alpha/\beta$, which is a scalar function. It should give rise to a zero-order equation, unless it reduces to a constant c, leading to the differential condition

$$d\alpha = c\beta$$

in order to have formal integrability. In effect, $\alpha = x^2 \, dx^1$, $\beta = dx^1 \wedge dx^2$ gives $c = -1$, and determines the Lie pseudogroup

$$y^1 = f(x^1), \qquad y^2 = \frac{x^2}{\frac{\partial f(x^1)}{\partial x^1}},$$

while $\alpha = dx^1$, $\beta = dx^1 \wedge dx^2$ gives $c = 0$ and determines the, completely different, Lie pseudogroup

$$y^1 = x^1 + a, \qquad y^2 = x^2 + h(x^1).$$

In our first book [144] we have brought Vessiot's results up-to-date. However, while Vessiot's results were never recognized and acknowledged by his contemporaries, similarly our results seem not known and have not been acknowledged by some experts, even in recent publications [29, 75, 173]. The purpose of this book is not only to rectify this state of affairs, but also to prove that these results can be useful in applications.

Let us briefly describe the kind of 'revolution' brought about in mathematical physics by these results.

For this we shall treat in parallel, along the lines outlined above, two well-known examples.

Let us start with a Lie group G acting simply transitively on a manifold X, that is, we assume that the graph of the action $X \times G \to X \times X$ is an isomorphism, in particular, $\dim X = \dim G = n$, and X may be equal to G. Consider a basis $\omega^\tau = \omega_i^\tau(x) \, dx^i$ of the differential 1-forms on X with values in the Lie algebra $\mathcal{G}$ of G such that $\det \omega = \det(\omega_i^\tau(x)) \neq 0$. By construction, the previous Lie group of transformations, viewed as a Lie pseudogroup of transformations, is defined by stating that the transformations preserve the forms ω^τ, for $\tau = 1, \ldots, n$. The simplest example of this is given by the translations $y^i = x^i + a^i$ and the forms $dx^1, \ldots, dx^n$. The n^2 defining equations of the Lie pseudogroup therefore take the Lie form

$$\omega_k^\tau(y)y_i^k = \omega_i^\tau(x),$$

where, as usual, we understand that a transformation is a solution whenever the following *identity* holds:

$$\omega_k^\tau(f(x))\partial_i f^k(x) = \omega_i^\tau(x) \qquad \forall x \in X.$$

Introducing the inverse matrix α of ω, we find

$$y_i^k = \alpha_\tau^k(y)\omega_i^\tau(x).$$

Using the successively crossed derivatives, implicit differentiation and substitution of the derivatives, we obtain a zero-order PDE:

$$\alpha_\rho^k(y)\alpha_\sigma^l(y)\left(\frac{\partial\omega_l^\tau(y)}{\partial y^k} - \frac{\partial\omega_k^\tau(y)}{\partial y^l}\right) = \alpha_\rho^i(x)\alpha_\sigma^j(x)\left(\frac{\partial\omega_j^\tau(x)}{\partial x^i} - \frac{\partial\omega_i^\tau(x)}{\partial x^j}\right).$$

As a byproduct, the system of defining equations is formally integrable if and only if

$$\alpha_\rho^i(x)\alpha_\sigma^j(x)(\partial_i\omega_j^\tau(x) - \partial_j\omega_i^\tau(x)) = c_{\rho\sigma}^\tau,$$

where the $n^2(n-1)/2$ righthand terms are constants. These equations, which are in fact exactly the Maurer–Cartan equations if $X = G$ and which we thus give this name in case no confusion is possible, are usually written as follows:

$$\partial_i\omega_j^\tau(x) - \partial_j\omega_i^\tau(x) = c_{\rho\sigma}^\tau\omega_i^\rho(x)\omega_j^\sigma(x).$$

Again closing this system by taking the exterior derivative of the lefthand terms and substituting, we recover the Jacobi conditions for the structure constants.

Let us now consider a nondegenerate *metric* ω, that is, a 2-tensor field $\omega_{ij}(x) = \omega_{ji}(x)$ such that $\det\omega = \det(\omega_{ij}(x)) \neq 0$, and let us consider the pseudogroup of isometries of ω defined by the following system of $n(n+1)/2$ PDE of the form

$$\omega_{kl}(y)y_i^k y_j^l = \omega_{ij}(x),$$

where we have to stipulate that every (local) isometry $y = f(x)$ satisfies the identities

$$\omega_{kl}(f(x))\partial_i f^k(x)\partial_j f^l(x) = \omega_{ij}(x) \qquad \forall x \in X.$$

We can define the *Christoffel symbols* γ by the following formula, where (ω^{ij}) denotes the matrix inverse to (ω_{ij}):

$$\gamma_{ij}^k(x) = \frac{1}{2}\omega^{kr}(x)(\partial_i\omega_{rj}(x) + \partial_j\omega_{ir}(x) - \partial_r\omega_{ij}(x)).$$

The $n^2(n+1)/2$ different prolonged equations can therefore be written as follows:

$$y_{ij}^k + \gamma_{rs}^k(y)y_i^r y_j^s = \gamma_{ij}^r(x)y_r^k,$$

and we have to multiply by the inverse of the matrix (y_i^r) to obtain the corresponding Lie form. Note that the Jacobian determinant $\det(\partial_i f^k(x))$, and therefore $\det(y_i^r)$, must be different from 0 because we are dealing with invertible transformations. A similar situation has been encountered with the defining equation of affine or projective 1-dimensional transformations, where we assumed $y' \neq 0$. To find the crossed derivatives in this case is slightly more complicated. Indeed, differentiating each of the $n^2(n+1)/2$ PDE with respect to each independent variable would give $n^3(n+1)/2$ PDE, in which we would eliminate the $n^2(n+1)(n+2)/6$ third-order derivatives (or jets if we prolonged once). Hence we would obtain $n^3(n+1)/2 - n^2(n+1)(n+2)/6 = n^2(n^2-1)/3$ first-order PDE, expressing the invariance of the so-called *curvature tensor*:

$$\rho_{lij}^k(x) = \partial_i\gamma_{lj}^k(x) - \partial_j\gamma_{li}^k(x) + \gamma_{ri}^k(x)\gamma_{lj}^r(x) - \gamma_{rj}^k(x)\gamma_{li}^r(x),$$

which has effectively $n^2(n^2-1)/3$ components only when $\gamma_{ij}^k(x) = \gamma_{ji}^k(x)$. Now, if γ is interpreted in terms of ω, ρ becomes a quasilinear function of the second derivatives of ω; it is called the *Riemann tensor* and has only $n^2(n^2-1)/12$ components. This fact can be explained by counting indices and using intricate combinatorial arguments as in most textbooks, but we shall sketch another approach, to be followed in Chapter IV. Namely, if we start with the $n(n+1)/2$ PDE defining an isometry and differentiate twice with respect to the $n(n+1)/2$ different pairs of independent variables, we indeed obtain only $n^2(n+1)^2/4 - n^2(n+1)(n+2)/6 = n^2(n^2-1)/12$ new first-order PDE. In this way the condition of formal integrability, that is, the condition under which these new $n^2(n^2-1)/12$ PDE are algebraic consequences of the given $n(n+1)/2$ PDE, has been found by Eisenhart in [58], using tricky, specific, manipulations in differential algebra. Such a condition amounts to the so-called *constant curvature condition*:

$$\rho_{l,ij}^k(x) = c(\delta_j^k \omega_{li}(x) - \delta_i^k \omega_{lj}(x)),$$

where c is an *arbitrary* constant and δ_i^k is the Kronecker symbol, equal to 0 if $k \neq i$ and to 1 if $k = i$. Finally we notice that, if in the formula

$$\frac{n^2(n+1)^2}{4} - \frac{n^2(n+1)(n+2)}{6} = \frac{n^2(n^2-1)}{12}$$

we substitute skewsymmetric instead of purely symmetric terms, we again find, surprisingly,

$$\frac{n^2(n-1)^2}{4} - \frac{n^2(n-1)(n-2)}{6} = \frac{n^2(n^2-1)}{12}$$

Hence, there are only two possibilities: this is accidental coincidence or it is not. If it is not accidental coincidence, as will be proved in Chapter III, the *proper concept of curvature must be revised from the very beginning!*

Now we will give another reason for modifying the existing concept of curvature. We have convinced the reader that the Maurer–Cartan equations with the structure constants $c_{\rho\sigma}^\tau$ and the constant curvature condition with the single constant c both originate from the same problem of formal integrability, but applied to two different Lie pseudogroups or, equivalently, to two different fields of geometric objects over X or *structures* over X. We copy these conditions, with corresponding terms right below each other:

$$\partial_i \gamma_{lj}^k - \partial_j \gamma_{li}^k + \boxed{\gamma_{ri}^k \gamma_{lj}^r - \gamma_{rj}^k \gamma_{li}^r} = c(\delta_j^k \omega_{li} - \delta_i^k \omega_{lj}),$$

$$\partial_i \omega_j^\tau - \partial_j \omega_i^\tau = c_{\rho\sigma}^\tau \boxed{\omega_i^\rho \omega_j^\sigma}.$$

As a first consequence, the fact that the $c_{\rho\sigma}^\tau$ are structure constants of a Lie algebra is purely accidental, since a single c cannot be related to any Lie algebra. However, *there is no conceptual difference at all between c and the $c_{\rho\sigma}^\tau$*.

Let us introduce the Lie group $GL(n)$ of invertible square $(n \times n)$-matrices, with as Lie algebra the set $M(n)$ of all square $(n \times n)$-matrices equipped with the natural bracket $[A, B] = AB - BA$, $\forall A, B \in M(n)$.

If one considers a 1-form on X with values in the Lie algebra $\mathcal{G}$ of a Lie group G, say $A = (A_i^\tau(x)\, dx^i)$, then, since G *does not act on* X, any change of coordinates in

X, or indeed any transformation of X, interferes with only the index i, and *not* with the index τ.

As a byproduct, if, following Cartan and successors, the Christoffel symbols are regarded as a connection, that is, a 1-form on X with values in $M(n)$, then we cannot explain why, under a change of variables, these symbols behave like a second-order geometric object (this is an exercise), where *all* the indices are involved. However, it is standard to exhibit curvature as a Maurer–Cartan equation for 1-frames with values in $M(n)$ It follows that the framed quadratic term $\gamma\gamma - \gamma\gamma$ is identified with the quadratic term $\omega\omega\omega$, which is *nonsensical* in light of our previous computations and the first example of a geometric object (α, β), where the integrability condition $d\alpha = c\beta$ has a righthand term which is linear in the object, as in the constant curvature condition, $\cdots$ and not quadratic, as in the Maurer–Cartan equations.

Although this result is striking enough and needs no comment, the reader will discover that it is still ignored in the recent edition of the classical textbook [173].

Of course, if the concept of curvature needs revision, any domain of mathematical physics using it in a crucial way may need revision too. In particular, this is true for gauge theory (electromagnetism) and general relativity (gravitation), but also for continuum mechanics, in which curvature is nothing else but the set of compatibility conditions for the strain tensor.

Two major questions then arise:

How is it possible that such a confusion took place?
How is it possible that such a confusion still takes place?

As earlier said, people have paid attention not to the work of Vessiot but to that of Cartan, and Cartan's idea was $\cdots$ to extend the Maurer–Cartan equations to Lie pseudogroups. Missing the Vessiot structure equations was therefore Cartan's first misunderstanding. In fact, he became aware of problems with his exterior calculus approach for two reasons.

The first reason is that he knew that for classical structures and geometric objects, the structure equations he proposed (which looked like the Maurer–Cartan equations) were too complicated indeed. Therefore in 1922 [33, 34] he turned to the theory of 'generalized spaces' by 'decoupling' the group G and the manifold X. Indeed, noting that the Maurer–Cartan equations still hold true for 1-forms $A = (A_i^\tau(x)\, dx^i)$ on X with values in the Lie algebra $\mathcal{G}$ of G, one may introduce a second term and write

$$\partial_i A_j^\tau(x) - \partial_j A_i^\tau(x) - c_{\rho\sigma}^\tau A_i^\rho(x) A_j^\sigma(x) = F_{ij}^\tau(x)$$

in order to define 2-forms $F = (F_{ij}^\tau(x)\, dx^i \wedge dx^j,\ i \leq j)$ on X with values in the same Lie algebra and called $\cdots$ *curvature*! A reader with knowledge of mathematical physics will recognize the so-called Yang–Mills *potentials* A and Yang–Mills *fields* F in the gauge-theoretical framework, with now $i = 1, \ldots, \dim X$ and $\tau = 1, \ldots, \dim \mathcal{G}$. The confusion with curvature appears gradually, along with the construction of certain differential sequences. At this moment we have a sufficient amount of material for constructing the following sequence, called '*gauge sequence*' in the sequel:

$$X \times G \longrightarrow T^* \otimes \mathcal{G} \xrightarrow{\text{MC}} \wedge^2 T^* \otimes \mathcal{G}$$

$$a \longrightarrow a^{-1} da = A$$

$$A \longrightarrow dA - \tfrac{1}{2}[A, A] = F$$

By definition, $X \times G$ means the set of (local) maps a from X to G, also called gaugings, $T^* \otimes \mathcal{G}$ is the set of 1-forms on X with values in $\mathcal{G}$, and $\wedge^2 T^* \otimes \mathcal{G}$ is the set of 2-forms on X with values in $\mathcal{G}$. The first operator can be easily described by the following picture:

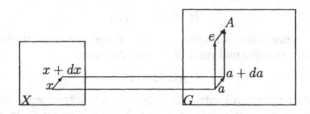

Take a point x and a neighbor $x + dx$ on X, then apply the map a to these points to obtain points a and $a + da$ on G, and pull back these points by the left action of $a(x)$ on G to obtain the identity $e \in G$ and a neighboring point A close to e. This makes it possible to define $a^{-1} da = A$ by a limiting process in the tangent space of G at e, which is precisely $\mathcal{G}$. The second operator, called MC, has already been described in local coordinates, and our notation is only a symbolic standard equivalent of it.

The second reason is that Janet discovered the *Riemann embedding problem* before Cartan did (and Cartan never forgave him, as Janet told the author!). This problem is the question whether any riemannian space (a space equipped with a metric structure) can be embedded in a euclidean space of higher dimension. The system to be solved is:

$$\omega_{kl}(y) y_i^k y_j^l = \overline{\omega}_{ij}(x),$$

with now a different range for the indices (i, j) and (k, l), where ω is the euclidean and $\overline{\omega}$ the given metric. Janet discovered a solution $y = f(x)$, at least locally, whenever the dimension of the euclidean space is at least $n(n + 1)/2$, and such a limiting condition produces no differential condition on $\overline{\omega}$, even if the resulting system is not formally integrable.

We shall now prove that Vessiot's results fit perfectly with a nonlinear version of the beginning part of the Janet sequence which we have described already. Indeed, we start with *any* transformation $y = f(x)$ and denote the derivatives up to order q by $y_q = j_q(f)(x)$. E.g., in the previous examples we have only used

$$j_2(f) \colon x \to (f^k(x), \partial_i f^k(x), \partial_{ij} f^k(x)).$$

Substitution into the basis of differential invariants used for obtaining the Lie form, we should get

$$\Phi(j_q(f)(x)) = \overline{\omega}(x),$$

where we have suppressed the index τ for simplicity. Of course, by construction, $\overline{\omega} = \omega$ if and only if $x \to y = f(x)$ is a transformation of our given Lie pseudogroup. It follows from the definition of the geometric objects ω and $\overline{\omega}$ that we have

$$j_q(f)^{-1}(\omega) = \overline{\omega},$$

since we can lift any (local) transformation to sections and transform them using the patching rules for the natural bundle. This is well known for tensor fields (and the reader may verify this on all the examples already considered). Suppose now that the structure constants appearing in the Vessiot structure equations

$$I(j_1(\omega)) = c(\omega)$$

are fixed by ω. Since these structure equations are invariant under any change of coordinates, or any transformation indeed, we also have

$$I(j_1(\overline{\omega})) = c(\overline{\omega})$$

with the same structure constants. As a byproduct, introducing the set $\text{aut}(X)$ of all local diffeomorphisms of X, any $f \in \text{aut}(X)$ will give an $\overline{\omega}$, in general different from ω but nevertheless satisfying the *same* structure equations. Among the various $f \in \text{aut}(X)$ only those belonging to our Lie pseudogroup Γ will give exactly the *same* ω, which of course satisfy the *same* structure equations. This is the beginning of a nonlinear differential sequence, called a *nonlinear Janet sequence* because the Vessiot structure equations play *at the same time* the role of *integrability conditions* (see the examples already given) and of *compatibility conditions* for the second righthand term $\overline{\omega}$ of the new Lie form introduced for *any* $f \in \text{aut}(X)$. Without entering into the details of Chapter V, the nonlinear Janet sequence can be written as follows:

$$0 \to \Gamma \to \text{aut}(X) \underset{\omega \circ \pi}{\overset{\Phi \circ j_1}{\rightrightarrows}} \mathcal{F} \underset{c}{\overset{I \circ j_1}{\rightrightarrows}} \mathcal{F}_1$$

$$f \diagdown_{\omega}\nearrow \quad \Phi(j_q(f)) = j_q(f)^{-1}(\omega)$$

$$\overline{\omega} \diagdown_{c(\overline{\omega})}\nearrow \quad I(j_1(\overline{\omega}))$$

where $\mathcal{F}$ and $\mathcal{F}_1$ are natural bundles, ω and $\overline{\omega}$ are geometric objects, or *sections*, of $\mathcal{F}$, and $c(\omega)$ and $c(\overline{\omega})$ are sections of $\mathcal{F}_1$; the double arrows are separately described. If one passes to the infinitesimal limit $t \to 0$ for a parameter t with $y = x + t\xi(x) + \ldots$, one can linearize the defining equations close to the identity transformation, just like physicists do when passing from volume preserving transformations to divergence-free vector fields. In that case $\overline{\omega} = \omega_t = \omega + t\Omega + \ldots$ whenever $f_t = \exp(t\xi)$ is a diffeomorphism reducing to the identity for $t = 0$. We shall prove that for two vector field solutions ξ and η of the linearized system, their bracket, defined by

$$([\xi, \eta])^k = \xi^r \partial_r \eta^k - \eta^r \partial_r \xi^k,$$

is again a vector field solution. Therefore, the set Θ of solutions is a Lie algebra of vector fields, namely (with a slight abuse of language) $\Theta \subset T$ is the Lie algebra of $\Gamma \subset \text{aut}(X)$. Therefore, in the limit, the nonlinear Janet sequence becomes the following linear Janet sequence:

$$0 \longrightarrow \Theta \longrightarrow T \xrightarrow{\mathcal{D}} F_0 \xrightarrow{\mathcal{D}_1} F_1,$$

which can be extended to the right, unlike the nonlinear sequence, and where $\mathcal{D}$ and $\mathcal{D}_1$ are the linearizations of the defining equations and of the compatibility/integrability conditions, respectively.

This is particularly clear for the case of isometries, by introducing the standard Lie derivative for tensor fields. Indeed, in that case $F_0 = S_2 T^*$ is a symmetric tensor bundle and $\mathcal{D}\xi = \mathcal{L}(\xi)\omega$. The linear PDE defining Θ are called *Killing equations* and are defined as follows in terms of the Lie derivative:

$$(\mathcal{L}(\xi)\omega)_{ij} \equiv \omega_{rj}(x)\partial_i\xi^r + \omega_{ir}(x)\partial_j\xi^r + \xi^r\partial_r\omega_{ij}(x) = 0.$$

For any vector field ξ, the righthand term should be Ω_{ij}, which is nonzero in general. For ω the euclidean metric of 3-space, Ω is twice the so-called strain tensor and ξ is called the *displacement field* while $\mathcal{D}_1$ represents the compatibility conditions for small strain. Among them we find:

$$\partial_{11}\Omega_{22} + \partial_{22}\Omega_{11} - 2\partial_{12}\Omega_{12} = 0.$$

For $n = 2$ the number of such compatibility conditions is $n^2(n^2 - 1)/12 = 1$ and the one listed is the only one. The reader may similarly treat all other examples already presented.

Hence, for the moment we have two nonlinear differential sequences, the *nonlinear Janet sequence*, which fits perfectly with Vessiot's results, and the *gauge sequence*, which fits perfectly with Cartan's results. It is clear that the Maurer–Cartan condition and the constant curvature condition, presented in Vessiot's framework, only fit with the nonlinear Janet sequence, ... though they have been used by Cartan for arriving at the idea of the gauge sequence. He knew the first condition from his work on Lie groups, and the second from his work on general relativity.

It remains to break this vicious circle. Surprisingly again, the solution will come from continuous mechanics, as will be clear from a careful examination of the results obtained by E. and F. Cosserat in [41].

Indeed, let us base continuum mechanics, and in particular elasticity, on the concept of invariance of deformation under a rigid motion $\bar{y} = Ay + B$ when the body undergoes an arbitrary displacement $x \to y = f(x)$. We immediately obtain the starting construction of the nonlinear Janet sequence, where $n = 1, 2, 3$ in dependence on the kind of elasticity under consideration, ω is the standard euclidean metric, and $\bar{\omega}$ is the strain tensor. In this case $c = 0$, and we recover the classical compatibility conditions of the strain tensor. For 'small' displacements and the so-called 'linear' elasticity with small strain tensor, we find all the results that can be found in textbooks, with the conceptual improvement brought about by the (linear/nonlinear) Janet sequence. However, when *dualizing* these concepts, problems do immediately appear. Indeed, the *general physical concepts must include stress* σ^{ij} *as well as couple-stress* μ^{rij} because the surface forces and couples cannot be separated. Writing out the tensor equilibrium, we find

$$\begin{cases} \partial_i \sigma^{ij} = f^j, \\ \partial_r \mu^{rij} + \sigma^{ij} - \sigma^{ji} = m^{ij}. \end{cases}$$

Again, these equations are in most textbooks. In them, (f^j) is the volume density of force and $(m^{ij} = -m^{ji})$ is the volume density of couples. One should not forget, and any student knows this, that at first sight there is no reason for the stress tensor (density) σ^{ij} to be symmetric. This is a consequence of only a strong *constitutive condition*, stating that $\mu^{rij} = 0$, $m^{ij} = 0$ and *therefore* $\sigma^{ij} = \sigma^{ji}$. The discovery, in 1950, of *liquid crystals* gives a good example of media in which the stress can be nonsymmetric because of electromagnetic forces.

We can use finite elements in elasticity because elastic problems can be formulated as variational problems according to the two principles of thermodynamics. In particular, if an elastic experiment is done slowly in such a way that the temperature remains constant, the work done by the exterior forces and couples is the total differential of the free energy, which only depends locally on the $n(n+1)/2$ independent components of the strain tensor ϵ_{ij} for $i \leq j$. If $\varphi(\epsilon_{ij})$ is the local density of free energy, then we have:

$$\sigma^{ij} = \frac{\partial \varphi}{\partial \epsilon_{ij}} \qquad \text{for } i \leq j,$$

which is obtained from the variation of the free energy and an integration by parts. This result we be studied in detail in Chapter VIII, and can be found in all textbooks on mathematical elasticity theory [52, 69, 106]. It follows that the use of the strain tensor only does not allow us to recover *all* concepts and equations already exhibited. In particular, σ^{ij} is only defined for $i \leq j$ and the constitutive condition $\sigma^{ij} = \sigma^{ji}$ does not follow from the variational formulation, as is often believed. Hence, a theory of elasticity depending on the strain tensor alone is coherent with only a symmetric stress tensor. Equivalently, if the stress tensor is not symmetric, a theory of elasticity cannot be founded on the strain tensor alone. The discovery of the Cosserat brothers has been that this dilemma can be overcome by new group-theoretical arguments. Roughly speaking, starting with the *same* group of rigid motions, there can be *another* differential-geometric framework, with more fields than just strain, such that a dualization can incorporate both stress (possibly nonsymmetric) *and* couple stress. Of course, such an approach cannot make use of the invariance point of view and thus cannot rely on the Janet sequence. Therefore we are faced with the following question:

> Starting with the same Lie (pseudo)group of transformations,
> can we find another differential sequence?

Of course, for Lie groups of transformations (and only for them!) we have found the gauge sequence. However, this sequence has two main drawbacks: first, it works for Lie groups only, and secondly, the group action is not taken into account. The second drawback is in contradiction with the intuitive approach of mechanics based on rigid body kinematics or dynamics, as translations and rotations *act* on the ambient 3-space.

For the above very particular situation, such a sequence has been discovered in the first place by E. and F. Cosserat.

Using the formal theory of systems of PDE elaborated by him earlier, Spencer succeeded in answering the question in 1970 [101], but one could say that only in 1975 the mathematical framework had been supplied with full details [76]. Unfortunately, the spaces and sequences involved cannot be described by using classical tools, and cannot be sketched in this Introduction. We will only say that, in the case of a Lie group of transformations, the resulting *nonlinear Spencer sequence* (in which the group action is used in a crucial way) is isomorphic to the corresponding gauge sequence (in which, strange enough, the group action disappears). Even more surprisingly, E. and F. Cosserat were aware of both sequences, since they knew the gauge sequence for the dynamics of rigid bodies and exhibited the above isomorphism in [38], where it is called the *'fundamental formula of kinematics'*.

The solution to the vicious circle is now obvious, since the second operator in the nonlinear Spencer sequence incorporates both *curvature* and *torsion* in the sense of Cartan, $\cdots$ but with additional terms that could not be discovered by a classical approach. One should also note that the exterior derivative must be replaced by the *Spencer operator*. It follows that the (non)linear Spencer sequence and the (non)linear Janet sequence are, in general, *completely different*, that is, the spaces and operators involved in them seem *completely different* in all circumstances, $\cdots$ except in the two leading to the Maurer–Cartan equations and the constant curvature condition, described earlier.

To recapitulate, a deep confusion still prevails even today between the (non)linear Janet sequence and the (non)linear Spencer sequence, $\cdots$ since neither of them is known in practice. Moreover, for Lie groups of transformations there is another confusion: between the nonlinear Spencer sequence and the nonlinear gauge sequence; it leads to an incorrect model of electromagnetism.

We must say that it is quite extraordinary that the brothers Cosserat were able to escape from these delicate computations without almost any mistake. Conversely, if in 1970 Spencer would have asked a student to compute the nonlinear Spencer sequence for the simplest case of motion of a rigid body, then the student would have had to repeat exactly *all* the formulas established by E. and F. Cosserat in 1909!

We will prove that the tool Cartan dreamt about is precisely the nonlinear Spencer sequence, with derivatives as in the Vessiot structure equations, with respect to independent variables *only* and *not* with respect to all jet variables together. Consequently, we can sketch the historical background leading to the quoted confusion as follows:

$$
\text{Lie} \left\langle
\begin{array}{ccc}
\text{Cartan} & \longrightarrow & \text{Spencer} \\
\uparrow ? & \text{Cosserat} & \uparrow ? \\
\text{Vessiot} & \longrightarrow & \text{Janet}
\end{array}
\right.
$$

In this diagram, the '?' mean that the links are of a cohomological nature and cannot be discovered by local coordinate computations only, as we shall see on examples.

Extending the construction of the Spencer sequence to the 15-parameter Lie group of conformal transformations of the Minkowski metric will provide at once the required group-theoretical unification for the finite elements of elasticity, heat and electromagnetism.

We sketch the main ideas using the gauge sequence.

Any engineering textbook dealing with finite elements and variational methods for elasticity, heat and electromagnetism *necessarily* includes the following variations:

$$\delta\vec{\xi}, \quad \delta\vec{\omega}, \quad \delta T, \quad \delta A,$$

for, respectively, translations, rotations, temperature, and electromagnetic 4-potential. They dualize the stress equation, the couple-stress equation, the heat equation, and the Maxwell equations for the electromagnetic inductions $(\vec{H}, \vec{D})$. Working with space-time, as must be done for electromagnetism, in accordance with special relativity at a college level, and counting the number of variations involved, we find:

$$4 + 6 + 1 + 4 = 15.$$

Therefore, gauging the translations provides classical elasticity, as we have said, gauging the rotations too provides Cosserat elasticity, gauging the dilatations too provides thermodynamics, while gauging the additional 4 nonlinear transformations too provides electrodynamics along the ideas of Weyl. Surprisingly, the corresponding equations will only depend on the Spencer operator, and thus on the tower of group inclusions with increasing dimensions:

translation group	$\subset$	Poincaré group	$\subset$	Weyl group	$\subset$	conformal group
4	<	10	<	11	<	15

As a striking consequence, let us sketch how *this gauging procedure implies the existence of an absolute zero for the absolute temperature*. Indeed, dilatation of *both* space *and* time is not accessible to intuition, but nevertheless only depends on the structure of the multiplicative group with the following topology:

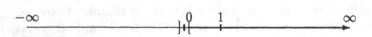

Gauging the connected component of the identity will produce a scalar field with zero as limiting value from below (an open set on the real line), no upper bound and a reference value (identity). This is not only just the topological structure of absolute temperature T, but it is also unchanged under the transformation $T \to 1/T$, which is known to be quite useful in statistical mechanics!

We end this historical survey by saying that the above unification *could not* have been achieved by means of classical techniques, even though the gauge sequence has been sufficient for an intuitive sketch of the results. Indeed, as already mentioned, by definition, in the gauge sequence the Lie group involved (here, the conformal group) does not act on the base manifold (here, space-time). Therefore the electromagnetic field, as a 2-form, can only be recovered from the nonlinear Spencer sequence, which thus *cannot be avoided*. Hence, it is only *this* sequence that allows us to combine, in a unified framework, the concepts and equations to be found in engineering sciences.

We now turn briefly to the contents of Chapter VI, the Galois theory for systems of PDE, or simply 'differential Galois theory'.

One might say that, for a few Frenchmen, the sudden fame of the Galois theory for polynomials in the second half of the 19th century offered, at the end of that century, the idea of enlarging it from pure algebra to differential algebra.

The first tentatives of E. Picard [140, 141] and E. Vessiot [192], now known under the name '*Picard–Vessiot theory*', brought about a procedure for dealing with linear ODE and linear transformations with 'constant' coefficients, instead of with roots of polynomials and permutation groups. Before discussing this, let us examine the first difficulty encountered in such a group-theoretical setting.

For this we start with a differential field K, say the rational functions in x, and consider the linear ODE

$$y_x - \omega y = 0, \qquad \omega \in K.$$

Suppose we want to look for the biggest *group of invariance* of this ODE, namely the set of transformations $\overline{y} = g(y)$ preserving this ODE. They must therefore be such that

$$\overline{y}_x - \omega \overline{y} = 0, \qquad \omega \in K.$$

Using the chain rule for differentiation, we find

$$y\frac{\partial \overline{y}}{\partial y} - \overline{y} = 0,$$

which defines the Lie group of transformations

$$\overline{y} = ay, \qquad a = \text{const}.$$

In this particular case we notice that the Lie group of transformations is obtained as a Lie pseudogroup and that we are lucky to be able to integrate the defining equation, which is defined over $\mathbb{Q}$ *for every choice of* K. Hence, in this case '*constant*' means '*does not depend on y*', with no reference to K at all. The problem of what the kind of constants to be usable is, is of course completely open, as becomes clear when using computer algebra, where 'a' is written for 'const'. Therefore we immediately discover the advantage of the Lie pseudogroup point of view: it allows us to consider all transformations simultaneously.

Let us now explain the main confusion of Picard in his first attempt to generalize classical Galois theory. We present the results along the line followed by Vessiot in the fascinating reference [192] which won him the 'Grand Prix de l'Académie des Sciences' For simplicity we shall deal with a polynomial of degree two and a second-order linear ODE, and the extension to the general situation is left to the reader as an exercise.

Let K be a field. Consider the polynomial equation

$$P \equiv (y)^2 - \omega^1 y + \omega^2 = 0, \qquad \omega^1, \omega^2 \in K,$$

where the indices of the coefficients are placed superior for reasons that will become clear later on. Looking for the two roots η^1, η^2 of P, we may assume, without loss of generality, that they are different, that is, we may assume that the *discriminant function* $\Delta \equiv y^1 - y^2$ has value $\delta = \eta^1 - \eta^2 \neq 0$. We recall the relation $\delta^2 = (\omega^1)^2 - 4\omega^2 \in K$. Now we write two copies of the same equation, but for different indeterminates y^1, y^2. We obtain

$$\begin{cases} P_1 \equiv (y^1)^2 - \omega^1 y^1 - \omega^2 = 0, \\ P_2 \equiv (y^2)^2 - \omega^1 y^2 - \omega^2 = 0. \end{cases}$$

If a group G acts on a space X in such a way that the graph $X \times G \to X \times X$ of the action is an isomorphism, we say that X is a *principal homogeneous space* for G. Similarly, following Vessiot, an *automorphic system* is a system of (partial differential) equations such that its space of solutions is a principal homogeneous space for a certain (pseudo)group of transformations. Notice that the preceding system is a linear system for ω^1, ω^2 that can be solved when $\delta \neq 0$, to give

$$\begin{cases} \Phi^1 \equiv y^1 + y^2 = \omega^1, \\ \Phi^2 \equiv y^1 y^2 = \omega^2. \end{cases}$$

It is easily verified that such a system is an automorphic system for the group S_2 of permutations:

$$\begin{cases} \overline{y}^1 = y^1, \\ \overline{y}^2 = y^2, \end{cases} \quad \text{and} \quad \begin{cases} \overline{y}^1 = y^2, \\ \overline{y}^2 = y^2, \end{cases}$$

and that it is written in Lie form with a fundamental set (Φ^1, Φ^2) of invariants of S_2 on the lefthand side. It can be considered as the linear group of transformations generated by the matrices

$$\begin{pmatrix} 1 & 0 \\ 0 & 1 \end{pmatrix} \quad \text{and} \quad \begin{pmatrix} 0 & 1 \\ 1 & 0 \end{pmatrix}$$

or even as the Lie pseudogroup defined by the following two zero-order equations in Lie form

$$\overline{y}^1 + \overline{y}^2 = y^1 + y^2, \qquad \overline{y}^1 \overline{y}^2 = y^1 y^2.$$

Therefore, using Vessiot's trick, *we have reduced the search for the two roots of P to the search for a single solution of an automorphic system.*

We may wonder why, in this case, the Galois group of P is a subgroup of the group S_2 of permutations of two objects, as it is in general. To understand this crucial fact, let us consider the ideal $\mathfrak{a}$ generated in $K[y^1, y^2]$ by $y^1 + y^2 - \omega^1$ and $y^1 y^2 - \omega^2$. Note that this ideal, although perfect whenever $\delta \neq 0$ (an exercise), is not, in general, prime. Indeed, if $\delta^2 \in K$ has a root $\delta \in K$, we obtain in succession:

$$\begin{aligned} \Delta^2 - \delta^2 &= (y^1 - y^2)^2 - \delta^2 \\ &= (y^1 + y^2)^2 - 4y^1 y^2 - \delta^2 \\ &= ((y^1 + y^2 - \omega^1) + \omega^1)^2 - 4((y^1 y^2 - \omega^2) + \omega^2) - \delta^2 \\ &= (y^1 + y^2 - \omega^1)^2 + 2\omega^1(y^1 + y^2 - \omega^1) - 4(y^1 y^2 - \omega^2) \\ &= (y^1 - y^2 - \delta)(y^1 - y^2 + \delta) \in \mathfrak{a}, \end{aligned}$$

a result proving that $\mathfrak{a}$ need not be prime. In that case the new system

$$\begin{cases} y^1 + y^2 = \omega^1, \\ y^1 y^2 = \omega^2, \qquad \omega^1, \omega^2, \delta \in K, \\ y^1 - y^2 = \delta, \end{cases}$$

is equivalent to

$$y^1 = \frac{1}{2}(\omega^1 + \delta), \qquad y^2 = \frac{1}{2}(\omega^1 - \delta),$$

and is therefore still an automorphic system for the alternating subgroup $A_2 \subset S_2$ consisting of the even permutations, which in this case reduces to the identity. We may give two interpretations. On the one hand, one can say that sometimes additional equations can be added to an automorphic system in such a way that the new system is compatible with the old one, and one may wonder whether the new system is automorphic for a smaller group (because it contains more equations, the space of solutions is smaller, hence so is the group of invariance). On the other hand, one can say that the ideal is not prime but that it is an intersection of prime ideals, since it is perfect, and one is left with the same question.

The counterexample

$$y^4 + 4 = (y^2 + 2y + 2)(y^2 - 2y + 2)$$

throws some light on the preceding discussion, by exhibiting a principal homogeneous space for the group of 4th roots of unity, which splits into two (irreducible) components corresponding to the two prime ideals generated by $y^2 + 2y + 2$ and $y^2 - 2y + 2$ which, nevertheless, are *not* determining principal homogeneous spaces for subgroups.

In modern language, the first interpretation amounts to saying that $\mathfrak{a}$ need not be maximal, while the second interpretation amounts to saying that $\mathfrak{a}$ is not prime. By pure coincidence, *in the purely algebraic case only*, a prime ideal is also maximal and a maximal ideal is always prime (an exercise), so the two interpretations are equivalent.

A main result in the Galois theory for a polynomial P of degree m is that the prefect ideal $\mathfrak{a}$ is an intersection $\mathfrak{p}_1 \cap \ldots \mathfrak{p}_r$ of finitely many prime ideals, *each defining a principal homogeneous space for various isomorphic subgroups of* S_m. Any one of these, say Γ, can be called the Galois group of P, and such a determination can be done by an efficient computer algorithm. If $\mathfrak{p} \subset K[y^1, \ldots, y^m]$ is the corresponding prime ideal, we can introduce the algebraic extension $L = Q(K[y^1, \ldots, y^m]/\mathfrak{p})$ of K. Regarding L as a finite-dimensional vector space over K and denoting by $|\Gamma|$ the number of elements in Γ, we have

$$\dim_K L = |L/K| = |\Gamma| < \infty.$$

Finally, calling any field in between L and K an *intermediate field*, the *fundamental theorem of Galois theory* establishes a bijective, order-reversing correspondence between intermediate fields and subgroups of Γ. This is used to break the initial problem of finding the roots of P into various 'simpler' problems relative to subgroups of its Galois group.

Suppose now that we have an ordinary differential field K and that we are looking for two linearly independent solutions of the second-order linear ODE

$$P \equiv y_{xx} - \omega^1 y_x + \omega^2 y = 0, \qquad \omega^1, \omega^2 \in K.$$

We similarly write two copies of the same ODE, and obtain the system of ODE

$$\begin{cases} P_1 \equiv y_{xx}^1 - \omega^1 y_x^1 + \omega^2 y^1 = 0, \\ P_2 \equiv y_{xx}^2 - \omega^1 y_x^2 + \omega^2 y^2 = 0. \end{cases}$$

Assuming that the Wronskian determinant

$$W = \begin{vmatrix} y^1 & y_x^1 \\ y^2 & y_x^2 \end{vmatrix}$$

has value $w \neq 0$ for two solutions η^1, η^2 amounts to saying that η^1 and η^2 are linearly independent functions over the 'constants' (same comment as before!). Hence we may solve the linear system in ω^1, ω^2 obtained, to get

$$\begin{cases} \Phi^1 \equiv \dfrac{\begin{vmatrix} y^1 & y_{xx}^1 \\ y^2 & y_{xx}^2 \end{vmatrix}}{\begin{vmatrix} y^1 & y_x^1 \\ y^2 & y_x^2 \end{vmatrix}} = \omega^1, \\[6mm] \Phi^2 \equiv \dfrac{\begin{vmatrix} y_x^1 & y_{xx}^1 \\ y_x^2 & y_{xx}^2 \end{vmatrix}}{\begin{vmatrix} y^1 & y_x^1 \\ y^2 & y_x^2 \end{vmatrix}} = \omega^2. \end{cases}$$

The biggest group of invariance may be similarly computed, by extending the transformations $\overline{y} = g(y)$ of the plane to the various derivatives, and the defining equations are easily seen to be

$$\overline{y}^l = \frac{\partial \overline{y}^l}{\partial y^k} y^k, \qquad \frac{\partial^2 \overline{y}^r}{\partial y^k \partial y^l} = 0.$$

Again, by sheer coincidence, this system can be integrated, and we obtain

$$\overline{y}^l = a_k^l y^k, \qquad a_k^l = \text{const},$$

with the same comment as for the first-order case considered earlier. Since we have a Lie group of transformations with 4 arbitrary parameters and we have 4 indeterminates y^1, y^2, y_x^1, y_x^2, the space of solutions of the preceding system is a principal homogeneous space for GL(2) with the standard action. As Vessiot noticed, *the problem of finding two linearly independent solutions of the given ODE is again reduced to the problem of finding a single solution of an automorphic system.* Such a system is already in Lie form, and Φ^1, Φ^2 is a fundamental set of differential invariants.

But now the differential ideal generated by P_1, P_2 in $K\{y^1, y^2\}$ is a prime differential ideal, $\mathfrak{p}$, because P_1 and P_2 are linear in the differential indeterminates. In fact, we have

$$L = Q(K\{y^1, y^2\}/\mathfrak{p}) \simeq K(y^1, y^2, y_x^1, y_x^2)$$

as an equivalent proof of that property. Therefore we *already* have an irreducible principal homogeneous space for GL(2). Nevertheless, in this situation, even though a maximal differential ideal must be prime, conversely, the prime differential ideal $\mathfrak{p}$ need not be maximal; $\cdots$ however, the concept of a prime differential ideal in a differential extension appeared only in 1930 in work of Ritt! For this reason, when Picard, followed by Vessiot, defined his concept of so-called irreducibility by looking for compatible ODE that could be added to P_1, P_2 in such a way that the group of invariance is made smaller, he clearly made a *conceptual mistake*! Nevertheless even though maximal ideals cannot be determined in a constructive way and maximality of an ideal cannot be constructively tested (though it is a nice mathematical problem to find out whether a prime ideal is maximal or not!), let us consider two maximal ideals, $\mathfrak{m}_1$ and $\mathfrak{m}_2$, both containing $\mathfrak{p}$.

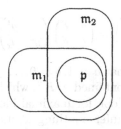

A first question to ask is whether they again define irreducible principal homogeneous spaces, and a second question to ask is whether their groups of invariance are isomorphic. Surprisingly, the answer to both questions is '*yes*' $\cdots$ under the condition that the field C of constants of K is algebraically closed (typically, $K = \mathbb{C}(x)$, thus $C = \mathbb{C}$). For a very simple counterexample with $K = \mathbb{Q}$ or $\mathbb{R}$, let $\mathfrak{p}$ be generated by $P \equiv y_x$, where $m = 1$, and let $\mathfrak{m}_1$ be generated by $y - 1$ and $\mathfrak{m}_2$ by $y^2 + 1$. Then $\mathfrak{m}_1$ gives rise to an automorphic system for the identity transformation, while $\mathfrak{m}_2$ gives rise to an automorphic system for the group $\bar{y} = \pm y$. Hence the Picard–Vessiot theory can never contain classical Galois theory as a subtheory, although a permutation group can be replaced by a group of matrices with entries 0 and 1. Also, notice that if $\mathfrak{m} \subset K\{y\}$ is a maximal differential ideal, then the field D of constants of the differential extension $L = Q(K\{y\}/\mathfrak{m})$ is an algebraic extension of C. Indeed, any constant $c \in L$ satisfies at least the first-order ODE $d_x c = 0$. If c would be transcendental over K, that is, c does not satisfy any polynomial equation over K, then we could take a representative of c in $K\{y\}$ and add it to $\mathfrak{m}$ (this amounts to setting $c = 0$), a fact contradicting the maximality property of $\mathfrak{m}$. If c satisfies a polynomial equation over K, we may assume of course that $c \neq 0$ and divide the polynomial by powers of c before differentiating. It follows that every coefficient of this polynomial belongs to

$C \subset K$, and c is algebraic over C. Of course, if now C is algebraically closed (take $C = \mathbb{C}$), then $C = D$. Hence, the condition that K and L must have the same field of constants, which is presented by Kolchin and others as a purely technical condition, is in fact a crucial assumption in the theory. However, this point is still missed by most authors, since only a *very careful* reading of [95] will reveal it (one needs to read backwards: VI.5, Proposition 13, p. 412; IV.5, Corollary 2 to Proposition 2, p. 152; III.10, Propositions 6 and 7, p. 142; II.1, Theorem 1, p. 86; and related results!).

It also follows that the determination of the so-called differential Galois group of the linear ODE $P = 0$ cannot be done by means of a computer. But there is an even worse fact. Consider the simplest possible second-order ODE, say $y_{xx} = 0$. The corresponding automorphic system is defined by $y_{xx}^1 = 0$, $y_{xx}^2 = 0$, and is therefore irreducible. Here we are faced with the following two possibilities, which are equally bad. On the one hand the corresponding differential extension L/K with $L = K(y^1, y^2, y_x^1, y_x^2)$ is *not* a Picard–Vessiot extension, because $C \neq D$. As a byproduct, Kolchin's theory cannot be applied and there need not be a bijective order-reversing correspondence between the intermediate differential fields and the linear subgroups of GL(2). Indeed, consider $K \subset K' = K(y_x^1) \subset L$ and extend GL(2) to jet derivatives as follows:

$$\begin{pmatrix} \overline{y}^1 \\ \overline{y}^2 \end{pmatrix} = \begin{pmatrix} a & b \\ c & d \end{pmatrix} \begin{pmatrix} y^1 \\ y^2 \end{pmatrix} \Rightarrow \begin{pmatrix} \overline{y}^1 & \overline{y}_x^1 \\ \overline{y}^2 & \overline{y}_x^2 \end{pmatrix} = \begin{pmatrix} a & b \\ c & d \end{pmatrix} \begin{pmatrix} y^1 & y_x^1 \\ y^2 & y_x^2 \end{pmatrix}.$$

The subgroup of GL(2) preserving y_x^1 is such that $a = 1, b = 0, \ldots$, but then it also preserves $K'' = K(y^1)$, although K' is strictly contained in K'', which is preserved by the same subgroup. On the other hand, if we look for maximal ideals in $K\{y^1, y^2\}$ containing y_{xx}^1 and y_{xx}^2, in general we cannot infer anything about the differential Galois group in a constructive way.

In Chapter VI we will not only give an alternative approach avoiding these difficulties, but we shall also extend it to the most general situation involving PDE and Lie pseudogroups, in a manner sketched by J. Drach [50] and Vessiot [192, Chapter III], but missed in [95] and therefore in [96]. The main idea is to provide a new definition of principal homogeneous space (PHS) for Lie pseudogroups, including a criterion for testing such a property in a constructive way. For example, we shall prove that the system $y^2 y_x^1 = \omega \in K$ is an automorphic system for the Lie pseudogroup of invariance $\overline{y}^1 = g(y^1)$, $\overline{y}^2 = y^2/(\partial g/\partial y^1)$ in a formal way. *In this case, though, if $y^1 = f^1(x)$ and $\overline{y}^1 = \overline{f}^1(x)$, in general one may find $\overline{y} = g(y)$ such that $\overline{f}^1 = g \circ f^1$.* As a byproduct, the contents of Chapter VI can be summarized in the short comment

DIFFERENTIAL GALOIS THEORY = PDE + PHS.

We finally turn to the contents of Chapter VII, namely the application of the formal theory of systems of PDE and group theory to *control theory*, which seems to be the most promising area for future trends. In this Introduction, we shall only sketch the main problems and ideas, using very simple examples in order to motivate the technical exposition in Chapter VII.

As already said in the Foreword, *ordinary differential (classical) control theory* studies input/output relations defined by systems of ODE. We give one simple example of the kind of problems to be encountered in such a framework. With one input u and two outputs y^1, y^2, we consider the control system

$$\begin{cases} y_x^1 = y^2 + u, \\ y_x^2 = \alpha y^1 + u, \end{cases} \qquad \alpha = \text{const}.$$

In the plane (y^1, y^2), for each given $u = u(x)$ we may consider the vector field

$$(y^2 + u)\frac{\partial}{\partial y^1} + (\alpha y^1 + u)\frac{\partial}{\partial y^2}$$

and look for the trajectories parametrized by x, such that the corresponding vector at any point is tangent to the trajectory passing through this point. As a first basic concept, the control system is said to be *controllable* if there is an input making it possible to pass from any given *initial* point to any given *final* point in a finite time. This is in fact a formal problem that cannot be subjected to numerical analysis, since we do not look for trajectories passing 'very close' to the final point but for those 'arriving exactly' at this point. Hence, in the preceding system, the question cannot be solved by an ordinary computer and we may have to turn to computer algebra, even though the definition above seems to be of a functional nature. Indeed, by chance, in textbooks [84, 110] one can find a criterion saying that a linear control system with r inputs and m outputs and of the form $y = Ay + Bu$, with A an $(m \times m)$-matrix and B an $(m \times r)$-matrix, is controllable if and only if

$$\text{rk}(B, AB, \dots, A^{m-1}B) = m.$$

Of course, this formal criterion, involving only the determination of the rank of a matrix, can be tested by means of computer algebra. A similar criterion can be found for nonlinear control systems of the form $y_x = a(y) + b(y)u$, where now a and b are vector fields [78].

The transition from ODE to PDE seems evident. Indeed, *partial differential control theory* studies input/output relations defined by systems of PDE. Thus, in compact notation, such a control system of order q can be written as

$$\Phi^\tau(x, u_q, y_q) = 0,$$

with an arbitrary number n of independent variables x, an arbitrary number r of inputs u and an arbitrary number m of outputs y. Of course, in this framework there are no longer vector fields, dynamical systems or trajectories. It follows that either the concepts introduced in classical control theory, such as controllability, are specific to that theory and cannot be generalized, or there are general concepts for PDE which give the classical concepts when applied to ODE. Of course, if the second option holds, this will immediately show that the foundations of classical control theory must be revised. This is precisely the purpose of Chapter VII for *all* the concepts of classical control theory, where in this Introduction we have restricted to controllability.

This first problem cannot be separated from a second one. Indeed, there may be ways to observe a control system through different choices of the outputs. In the quoted example we may choose $y = y^1$ for observing the system, and we then obtain

$$y \to y^1 = y, \qquad y^2 = y_x - u,$$

but now y must be a solution of the second-order ODE

$$y_{xx} - \alpha y - u - u_x = 0.$$

Of course, integrating this ODE amounts to integrating the preceding system of two first-order ODE, and vice versa. Hence we are led to the following question:

Does there exist a *formal definition* of controllability which:
- does not depend on the way of observing the system;
- can be extended from ODE to PDE;
- is coherent with the classical definition.

In a more precise form, applying the criterion to the linear system, we find

$$A = \begin{pmatrix} 0 & 1 \\ \alpha & 0 \end{pmatrix}, \ B = \begin{pmatrix} 1 \\ 1 \end{pmatrix} \Rightarrow \det(B, AB) = \det \begin{pmatrix} 1 & 1 \\ 1 & \alpha \end{pmatrix} = \alpha - 1.$$

So, the control system is controllable if and only if $\alpha \neq 1$. Indeed, for $\alpha = 1$ we set $z = y^2 - y^1$ and obtain, by subtraction, $\dot{z} + z = 0$. Therefore, if the initial point is on the diagonal $y^1 = y^2$, on which $z = 0$, then z must remain zero and the final point cannot be reached if it does not lie on the diagonal. The reader immediately discovers that, at first sight, it does not seem easy to find out why $\alpha = 1$ has a special meaning for this second-order ODE!

We now sketch the solution of this problem and explain its differential algebraic origin. To this end, we shall refer to problems in fluid mechanics, where liquids are moving and one wants to obtain knowledge about temperature, pressure and speed of the medium, given conservation of mass, the Navier–Stokes equations and the heat equation. Such problems will be considered in great detail in Chapter VII, and we refer to the excellent reference [36] on hydromagnetic and hydrodynamic stability. Next to a theoretical approach one can use an experimental approach using various instruments. In fact, a measuring instrument is an apparatus able to use a physical device (expansion of liquid under heating for a thermometer, thermocouple, manometer, $\cdots$) in order to associate a scalar number, or measure, to a given physical quantity. E.g., a thermometer measures temperature T, a manometer measures pressure p, an optical device can measure the various components of speed $\vec{v}$, $\cdots$. Eventually, taking into account more complicated phenomena, one can obtain information on each component of the *vortex vector* $\vec{\omega} = \frac{1}{2}\vec{\nabla} \wedge \vec{v} = \frac{1}{2} \operatorname{rot} \vec{v}$, of the gradient of temperature, or of the rate of deformation $V_{ij} = \frac{1}{2}(\partial_i v_j + \partial_j v_i)$, which is responsible for viscosity, $\cdots$. In any case, the reader should be convinced, provided she/he has some familiarity with engineering physics, that the scalar nature of a measure is unavoidable. In accordance with this experimental approach, we may thus define an *observable* as any scalar function of temperature, pressure, speed, electromagnetic field, $\cdots$ and their derivatives. Once an observable has been defined, *there are only two possibilities*:

- the observable is *free*, that is, it does not satisfy any ODE or PDE by itself;
- the observable is *constrained* by at least one ODE or PDE for itself.

As we shall see through examples, a first consequence of 'decoupling' in this manner the behavior of one physical quantity (say, temperature) from others (say, pressure and speed, for example) is that the constraining ODE or PDE are in general of quite high order and quite nonlinear. In practice, the least example is on the edge of human possibility, and in future research computer algebra has to be used in a systematic way for such a purpose. Of course, the preceding problem is just a problem of differential elimination, but before 'doing' the elimination we would like to know *a priori* if some observable is free or constrained, or eventually if some observables are free and the others are constrained.

Let us now turn to control theory. 'Tradition' and 'intuition' seem to combine in the following basic picture, which can be found in any textbook on control theory:

$$\xrightarrow{u} \boxed{\text{black box}} \xrightarrow{y}$$

The 'black box' may be a complicate electronical circuit filled with resistances, capacities and coils! In that case the input may be a difference of potential between two slots, and the output may be an intensity of courant somewhere. Nevertheless, in the light of our generalization, various unknowns are linked by the control system, and one can separate them into two blocks, called input and output respectively. Hence one draws arrows from input to output, while reverse arrows may be useful as well. One can only say that if the input is given , the remaining system determines the output (*direct* way) and that, conversely, if the output is given, the remaining system determines the input (*inverse* way). In particular, for our initial control system, subtracting the two ODE we find

$$y_x^2 - y_x^1 + y^2 - \alpha y^1 = 0.$$

Therefore, in general, the outputs cannot be arbitrarily given. Similarly, in Chapter VII we shall encounter many examples, such as the second set of Maxwell equations, in which the inputs cannot be arbitrarily given. This behavior is particularly clear when one *matches* control systems one after the other, the outputs of one system becoming the inputs of the next. Recapitulating the above said, we may conceive of a control system as a *differential correspondence*, linking input and output.

In this framework an observable becomes a function of inputs, outputs and their derivatives. Of course, if one can find a constrained observable, one cannot 'control' it, since the inputs no longer appear in the ODE or PDE it must satisfy. Hence, for a control system to be controllable it is necessary that all observables be free. As we shall see in Chapter VII, sufficiency of this is highly nonevident! Hence, a control system is said to be '*controllable*' if all observables are free. Of course, such a definition depends on the input/output description only, works for both ODE and PDE, as well as for control systems explicitly depending on the independent variables (in general, space-time) or not depending on them.

The detailed computation will be given in Chapter VII, but here we can already check a consequence of this on a very simple example. Indeed, when $\alpha = 1$ the observable $z = y^2 - y^1$ is constrained y the OE $z_x + z = 0$, but the observable $y^1 = y$

is free, as it satisfies $y_{xx} - \alpha y - u - u_x = 0$ only, which involves, next to y, also u. Similarly, the observable $z = y_x - y - u$ satisfies $z_x + z = 0$ whenever $\alpha = 1$, and this new definition of controllability has nothing to do with the classical one, which is based on a property of the trajectories. Nevertheless, we shall prove that in the classical cases for which criteria are known, our formal definition can be tested $\cdots$ by the same criteria! Hence it is coherent with classical control theory.

Finally we prove that this definition could have been found absolutely independent of control theory (another reason for suppressing the arrows!), by using differential algebra. Indeed, consider a partial differential control system defined by a system of algebraic PDE in the inputs and outputs. This is not such a great loss of generality, because most of the engineering examples are of this kind. We may assume that the given differential polynomials have coefficients in a certain differential field K, also called a *ground differential field*, which usually is the field of rational functions in the independent variables. Consider the larger differential field N, also called the *input/output differential field*, consisting of all rational functions over K of the inputs, outputs and their derivatives, restricted by the condition that the inputs and outputs *together* are solutions of the control system. Of course, such a situation only makes sense $\cdots$ if it can be tested and we recover the criterion for being a prime differential ideal presented earlier.

Now, in any textbook on algebra, whenever one deals with two fields K and N satisfying $K \subset N$, one may look for the *algebraic closure K'* of K in N. It consists of the elements of N satisfying a polynomial equation over (with coefficients in) K. By adding the word '*differential*' in front of the purely algebraic concepts appearing in the previous sentences, we arrive at the following. Whenever one has two differential fields K and N satisfying $K \subset N$, one may look for the *differential algebraic closure K'* of K in N. It consists of the elements of N satisfying at least one algebraic ODE or PDE over K. In Chapter VI we shall prove that K' is indeed a differential field. Hence the control system is (totally) controllable if $K' = K$, (totally) uncontrollable if $K' = N$, and, otherwise, in general, partially controllable only. In our example, K' is generated by $z = y^2 - y^1$ whenever $\alpha = 1$, and we have thus split up the 'large' differential extension N/K into two 'smaller' ones, namely N/K' and K'/K. Of course, K'/K is totally uncontrollable. We shall prove that N/K' is totally controllable. Indeed, in Chapter VI we shall prove that an element of N not in K' and satisfying an algebraic ODE or PDE over K' must satisfy an algebraic ODE or PDE over K also, because each element of K' is differentially algebraic over K. Hence this element should be in K', which is therefore its proper algebraic closure: this is precisely the concept of *minimum realization* in classical control theory! Of course we should be fair in admitting that differential algebra can only be a 'guide' towards new concepts, because the single input/single output second-order control system $uy_{xx} - u_x = 0$ has a *nonalgebraic* observable $z = y_x - \log u$ satisfying nevertheless the ODE $z_x = 0$. On the other hand, the control system $uy_{xx} - u_x = 1$ is totally controllable. Though this may seem surprising, in Chapter VII we shall use this example to see that the study of controllability is directly related to the study of $\cdots$ formal integrability, and therefore we have been able to 'loop the loop'.

Before concluding this introductory survey on control theory, we would like to situate our work with regard to others on similar subjects [112].

In 1970, J.L. Lions tried to extend to concept of controllability to PDE in so-called *distributed parameters systems*. The motivation came from a direct application to the vibration of long antennas folding out of satellites but the main idea can be understood on the simple example of a vibrating drum. Indeed, the displacement of the skin of a vibrating drum is a solution of a second-order PDE. Hence we may say that this system is controllable if there exists an input action on the boundary of the drum making it possible to pass from any *initial* state of vibration into any *final* state of vibration (say, rest!) in a finite time. The answer can be given after very delicate functional analysis of the problem. Nevertheless, though this problem is a quite useful one, we do not believe that the spirit of the concept of controllability is kept. Indeed, in the case of a linear ordinary control system, the existing formal criterion can be tested by means of a computer algebra package and does not depend at all on the initial data or on integration of the control system. On the other hand, in the present situation controllability is not at all a 'built in' property of the control system, but typically depends on its integration. So we do not agree with the fact that the corresponding approach is the only one to be followed when extending control theory from ODE to PDE, contrary to what people working in the field claim. In particular, one should notice that *all* publications on distributed parameters systems, and there are many, contain the same type of equations and results. Nevertheless, in view of the usefulness of this approach, we believe that future research will exhibit a link between this functional approach and our formal point of view.

We now end this, necessarily lengthy, Introduction, and turn to the tools needed to overcome all the problems quoted in the preceding pages.

CHAPTER I

Homological Algebra

The purpose of this Chapter is to present in a self-contained way a few of the results of homological algebra that can only be found in a few specialized textbooks [130, 131]. We warn the reader that all these results are essential in the remaining of the book, $\cdots$ although at first sight they do not look like to be related with the problems quoted in the Introduction. In fact, we shall touch at once upon the novelty of this book, namely the use of unexpected tools for studying engineering problems.

All vector spaces (or modules) $A, B, \ldots$ considered in this Chapter will be over a given field k. Also, for simplicity, we shall use a few abbreviations:

$$\begin{array}{lll} \text{rk} = \text{rank}, & \text{nb} = \text{number}, & \text{dim} = \text{dimension}, \\ \text{ker} = \text{kernel}, & \text{im} = \text{image}, & \text{coker} = \text{cokernel}. \end{array}$$

All maps $\Phi, \Psi, \ldots$ will be taken linear maps (or homomorphisms).

We start by presenting an interpretation of the famous *Cramer's rule* in a slightly different language.

Definition 1. A *sequence of vector spaces* is a chain of vector spaces and maps between them such that the composition of two maps is zero (the zero map!).

Let $\Phi \colon A \to B$ be a map. We put

$$\ker \Phi = \{a \in A : \Phi(a) = 0\},$$
$$\text{im}(\Phi) = \{b \in B : \exists a \in A, \ \Phi(a) = b\},$$
$$\text{coker} \, \Phi = B \setminus \text{im} \, \Phi.$$

Accordingly, we give

Definition 2. A sequence

$$A_{r-1} \xrightarrow{\Phi_r} A_r \xrightarrow{\Phi_{r+1}} A_{r+1}$$

is *exact at* A_r if $\text{im} \, \Phi_r = \ker \Phi_{r+1}$. It is called *exact* if it is exact at every place.

In a coherent way, when we want to specify the properties of a map $\Phi \colon A \to B$, we say that Φ is *injective* if and only if the sequence

$$0 \longrightarrow A \xrightarrow{\Phi} B$$

49

is exact, and Φ is *surjective* if and only if the sequence

$$A \xrightarrow{\Phi} B \longrightarrow 0$$

is exact.

Definition 3. A sequence of the form

$$0 \longrightarrow A \xrightarrow{\Phi} B \xrightarrow{\Psi} C \longrightarrow 0$$

is said to be a *short exact sequence*.

More generally, combining the preceding definitions, we may introduce the following exact sequence, called the *ker-coker sequence*:

$$0 \longrightarrow \ker \Phi \longrightarrow A \xrightarrow{\Phi} B \longrightarrow \operatorname{coker} \Phi \longrightarrow 0,$$

where $\ker \Phi \to A$ is the natural injection and $B \to \operatorname{coker} \Phi$ is the natural projection appearing in the two short exact sequences into which the above sequence splits:

$$0 \longrightarrow \ker \Phi \longrightarrow A \xrightarrow{\Phi} \operatorname{im} \Phi \longrightarrow 0,$$
$$0 \longrightarrow \operatorname{im} \Phi \longrightarrow B \longrightarrow \operatorname{coker} \Phi \longrightarrow 0.$$

Defining $\operatorname{rk} \Phi = \dim \operatorname{im} \Phi$ and counting the dimensions, we immediately see that

$$\dim \ker \Phi = \dim A - \dim \operatorname{im} \Phi,$$
$$\dim \operatorname{coker} \Phi = \dim B - \dim \operatorname{im} \Phi.$$

By subtracting we arrive at the quite useful formula

$$\dim \ker \Phi - \dim A + \dim B - \dim \operatorname{coker} \Phi = 0.$$

It says that the 'alternate' sum of the dimensions in a ker-coker sequence is zero.

We shall establish a relation between this way of notation and the usual college approach, while indicating the link with symbolic manipulations.

Indeed, in a basis for A, any element $a \in A$ can be written $a = (a^1, \ldots, a^r)$, where $r = \dim A$. Similarly, any element $b \in B$ can be written $b = (b^1, \ldots, b^s)$, where $s = \dim B$. The linear map Φ can be written $\Phi_i^k a^i = b^k$ for $i = 1, \ldots, r; \ k = 1, \ldots, s$. Looking for $\ker \Phi$ amounts to solving the linear system $\Phi_i^k a^i = 0$, while one can use the symbolic command rk to find $\operatorname{rk} \Phi = \dim \operatorname{im} \Phi$. Finally, looking for $\operatorname{coker} \Phi$ amounts to looking for the maximum number of linearly independent *compatibility conditions* of the form $\lambda_k^\tau b^k = 0$ whenever $b^k = \Phi_i^k a^i$. Hence, the single knowledge of $\operatorname{rk} \Phi$, given by a computer, *at the same time* provides the numbers

$$\dim \ker \Phi = \dim A - \operatorname{rk} \Phi, \qquad \dim \operatorname{coker} \Phi = \dim B - \operatorname{rk} \Phi.$$

This is just a modern setting of the so-called Cramer's rule in college textbooks.

More generally, the following proposition holds, which is very useful in computations.

Proposition 4. *In a long exact sequence*

$$0 \longrightarrow A_0 \xrightarrow{\Phi_1} A_1 \longrightarrow \ldots \xrightarrow{\Phi_n} A_n \longrightarrow 0$$

the following formula holds:

$$\sum_{i=0}^{n} (-1)^i \dim A_i = 0.$$

PROOF. We split the exact sequence into the following nested chain of short exact sequences:

$$0 \longrightarrow \ker \Phi_{i+1} \longrightarrow A_i \longrightarrow \operatorname{im} \Phi_{i+1} \longrightarrow 0.$$

Using exactness, we have $\ker \Phi_{i+1} = \operatorname{im} \Phi_i$, and therefore $\dim A_i = \dim \operatorname{im} \Phi_i + \dim \operatorname{im} \Phi_{i+1}$, leading at once to the formula. $\square$

The reason for introducing this framework will become clear if we now regard A and B as function spaces and Φ as a differential operator. Roughly speaking, the idea of a *differential sequence* is to try to copy in this new situation what we did in the old situation; this was just the idea of Spencer. More precisely, in the next Chapters we shall try to reduce problems involving PDE to problems in pure linear algebra. In particular, looking for $\ker \Phi$ will amount to looking for the 'space of solutions' of Φ, while looking for $\operatorname{coker} \Phi$ will amount to looking for compatibility conditions that must be satisfied by a second member, as we have seen in the Introduction.

Now we join sequences in two dimensions.

Definition 5. A commutative *square*

$$
\begin{array}{ccc}
A & \xrightarrow{\alpha} & A' \\
\Phi \downarrow & & \uparrow \Phi' \\
B & \xrightarrow{\beta} & B'
\end{array}
$$

is a set of maps such that $\Phi' \circ \alpha = \beta \circ \Phi$. A *commutative diagram* is then a juxtaposition of commutative squares.

The main useful result is provided by the following theorem, which is usually called the *snake theorem*.

Theorem 6. *In the following commutative diagram:*

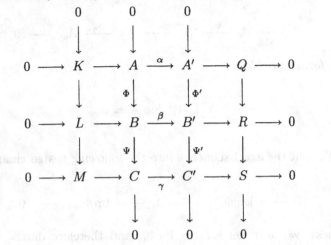

where the two central columns are exact and K, L, M (Q, R, S) *are the respective kernels (cokernels) of* α, β, γ, *there exists a long exact sequence*

$$0 \longrightarrow K \longrightarrow L \longrightarrow M \longrightarrow Q \longrightarrow R \longrightarrow S \longrightarrow 0$$

with a connecting map $K \to Q$.

PROOF. We give full details of the proof in order to avoid repeating these later in similar situations.

Initially, we start with a single commutative square, and obtain the induced dotted arrows:

$$0 \longrightarrow K \longrightarrow A \overset{\alpha}{\longrightarrow} A' \longrightarrow Q \longrightarrow 0$$
$$0 \longrightarrow L \longrightarrow B \underset{\beta}{\longrightarrow} B' \longrightarrow R \longrightarrow 0$$

Indeed, on the left any $a \in \ker \alpha = K \subset A$ is such that $\alpha(a) = 0$. Then $(\Phi' {\circ} \alpha)(a) = \Phi'(\alpha(a)) = (\beta {\circ} \Phi)(a) = \beta(\Phi(a)) = 0$ and therefore $\Phi(a) \in \ker \beta = L \subset B$. On the right any $q \in Q$ can be lifted to an $a' \in A'$, to which we can apply Φ' and project to $r \in R$. This map is well defined because if $a'_1, a'_2 \in A'$ project to the same $q \in Q$, by the exactness of the top sequence (or the fact that $Q = \operatorname{coker} \alpha$), there is $a \in A$ such that $a'_2 - a'_1 = \alpha(a)$. Therefore, setting $b'_1 = \Phi'(a'_1)$, $b'_2 = \Phi'(a'_2)$ we find $b'_2 - b'_1 = \Phi'(\alpha(a)) = \beta(\Phi(a))$, proving that b'_1 and b'_2 have the same projection $r \in R$.

A similar comment may be applied to the second commutative square, with $\Psi' {\circ} \beta = \gamma {\circ} \Psi$.

The injectivity of $K \to L$ follows at once from the injectivity of $\Phi \colon A \to B$, while the surjectivity of $R \to S$ follows at once from the surjectivity of $\Psi' \colon B' \to C'$. Similarly, the fact that $K \to L \to M$ is a sequence follows immediately from the fact that $A \to B \to C$ is a sequence, while the fact that $Q \to R \to S$ is a sequence follows immediately from the fact that $A' \to B' \to C'$ is a sequence. Hence it remains only to construct the connecting map $M \to Q$, and to prove the exactness of the long sequence.

Any $m \in M$ can be considered as a $c \in C$ such that $\gamma(c) = 0$. Since Ψ is surjective, one can find $b \in B$ such that $\Psi(b) = c$ and thus

$$0 = \gamma(c) = \gamma(\Psi(b)) = (\gamma \circ \Psi)(b) = (\Psi' \circ \beta)(b) = \Psi'(\beta(b)).$$

According to the exactness of the central columns, there is $a' \in A'$ such that $b' = \Phi'(a')$. Hence the projection $q \in Q$ of $a' \in A'$ goes to the projection of b' onto R, which is zero because $b' \in \operatorname{im} \Phi'$. It follows that $q \in \ker(Q \to R)$, and it remains to prove that $q \in Q$ is well defined.

Indeed, let $b_1, b_2 \in B$ be such that $\Phi(b_1) = \Phi(b_2) = c$. Then $\Phi(b_2 - b_1) = 0$ and, by the exactness of the central columns, there is $a \in A$ such that $b_2 - b_1 = \Phi(a)$. Introduce $b_1' = \beta(b_1)$ and $b_2' = \beta(b_2)$. We have

$$b_2' - b_1' = \beta(b_2 - b_1) = \beta(\Phi(a)) = (\beta \circ \Phi)(a) = (\Phi' \circ \alpha)(a) = \Phi'(\alpha(a)).$$

Hence, if $b_1' = \Phi'(a_1')$ and $b_2' = \Phi'(a_2')$, then $b_2' - b_1' = \Phi'(a_2' - a_1') = \Phi'(\alpha(a))$, and thus $\Phi'(a_2' - a_1' - \alpha(a)) = 0$. By the injectivity of Φ' it follows that $a_2' - a_1' = \alpha(a)$. Therefore a_1' and a_2' have the same projection $q \in Q$, a result completing the construction of the connecting map $M \to Q$.

A short way to describe the preceding chase is indicated in the following diagram:

$$
\begin{array}{ccccc}
a & \cdots & a' & \longrightarrow & q \\
\vdots & & \downarrow & & \\
b & \longrightarrow & b' & & \\
\downarrow & & \downarrow & & \\
m & \longrightarrow & c & \cdots & 0
\end{array}
$$

Finally we will prove the exactness of the sequence. For this, any $l \in \ker(L \to M)$ can be identified with a $b \in B$ such that $b \in \ker \Psi$ and $\beta(b) = 0$. Therefore, by the exactness of the central column, $b \in \operatorname{im} \Phi$ and there is $a \in A$ such that $b = \Phi(a)$. Now

$$\beta(\Phi(a)) = (\beta \circ \Phi)(a) = (\Phi' \circ \alpha)(a) = \Phi'(\alpha(a)) = 0,$$

and thus $\alpha(a) = 0$ because Φ' is injective. Therefore there is a unique $k \in K$ that can be identified with $a \in A$ and having image $l \in L$ under the map $K \to L$ induced by $\Phi \colon A \to B$. I.e., $\operatorname{im}(K \to L) = \ker(L \to M)$.

Let $m \in \ker(M \to Q)$, that is, let $m \in M$ be such that $q \in Q$ is $q = 0$ in the connecting map. It follows that $a' = \alpha(a)$ and

$$b' = \Phi'(a') = \Phi'(\alpha(a)) = (\Phi' \circ \alpha)(a) = (\beta \circ \Phi)(a) = \beta(\Phi(a)).$$

But by construction, $b' = \beta(b)$ and thus $\beta(b - \Phi(a)) = 0$, that is, $b = l + \Phi(a)$ for some $l \in L$. Therefore

$$c = \Psi(b) = \Psi(l) + \Psi(\Phi(a)) = \Psi(l) + (\Psi \circ \Phi)(a) = \Psi(l) = m,$$

because the vertical column is a sequence and thus $m \in \operatorname{im}(L \to M)$, that is, $3 \operatorname{im}(L \to M) = \ker(M \to Q)$.

Let now $q \in \ker(Q \to R)$. One can lift $q \in Q$ to $a' \in A'$ such that $\Phi'(a') = \beta(b)$ for some $b \in B$. It follows that

$$\Psi'(\Phi'(a')) = (\Psi' \circ \Phi')(a') = \Psi'(\beta(b)) = (\Psi' \circ \beta)(b) = (\gamma \circ \Psi)(b) = \gamma(\Psi(b)) = 0,$$

which implies $m = \Psi(b) \in M$. But this is just the way to construct the connecting map by going backwards. Thus, our q is the image of this m under the connecting map, that is, $\operatorname{im}(M \to Q) = \ker(Q \to R)$.

Finally, let $r \in \ker(R \to S)$. We can lift $r \in R$ to $b' \in B'$ with $\Psi'(b') = \gamma(c)$ for some $c \in C$. Then there is $b \in B$ with $c = \Psi(b)$, and we obtain

$$\Psi'(b') = \gamma(c) = \gamma(\Psi(b)) = (\gamma \circ \Psi)(b) = (\Psi' \circ \beta)(b) = \Psi'(\beta(b)).$$

It follows that $\Psi'(b' - \beta(b)) = 0$, and therefore, by the exactness of the central columns, there is $a' \in A'$ with $b' = \Phi'(a') + \beta(b)$. Hence the image in R of the projection of a' into Q under the induced map $Q \to R$ is the projection of $\Phi'(a') = b' - \beta(b)$ in R, that is, it is the same as the projection r of b' into R. Thus, $\operatorname{im}(Q \to R) = \ker(R \to S)$.

For future applications the reader must keep in mind that the connecting map relates the *bottom left* to the *upper right* in the last diagram, a possibility not evident at first sight. $\square$

Let us now consider a nonexact sequence:

$$A \xrightarrow{\ \Phi\ } B \xrightarrow{\ \Psi\ } C.$$

Since it is a sequence, we have $\Psi \circ \Phi = 0$, and hence $\operatorname{im} \Phi \subseteq \ker \Psi$. Therefore, to compare $\operatorname{im} \Phi$ and $\ker \Psi$ we can introduce the quotient vector space $\ker \Psi / \operatorname{im} \Phi$. This leads to a *cohomology* theory, which we will describe using more specific notations.

Definition 7. For a long sequence

$$\cdots \xrightarrow{\ \Phi_r\ } A_r \xrightarrow{\ \Phi_{r+1}\ } \cdots$$

we introduce the *coboundary space* $B_r = \operatorname{im} \Phi_r$ at A_r, the *cocycle space* $Z_r = \ker \Phi_{r+1}$ at A_r and the *cohomology space* $H_r = Z_r / B_r$ at A_r.

It follows from Definitions 2 and 7 that a long sequence is exact if and only if all the cohomology spaces are null.

The effective computation can be rather delicate, and the next theorem will be of some help in later computations.

Theorem 8. *In the following commutative diagram:*

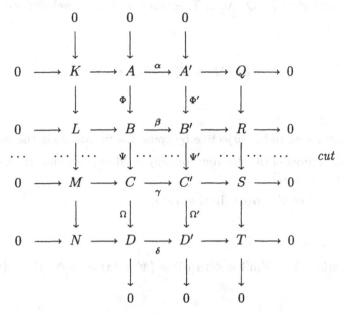

where the two central vertical columns are exact and K, L, M, N (Q, R, S, T) are the respective kernels (cokernels) of $\alpha, \beta, \gamma, \delta$, there exists an isomorphism

$$\ker(M \to N)/\operatorname{im}(L \to M) = H(M) \simeq \ker(Q \to R).$$

PROOF. We shall give two, independent proofs.

1) The first proof is direct and follows the same lines as the one of the construction of the connecting map in Theorem 6. For this reason we do not repeat all the details, but give the construction diagram:

$$
\begin{array}{ccccc}
a & \cdots & a' & \longrightarrow & q \\
& & \downarrow & & \\
l & \cdots & b & \to & b' \\
\vdots & & \downarrow & & \vdots \\
& & m & \longrightarrow & c & \cdots & 0 \\
& & & & \vdots & & \\
& & & & 0 & &
\end{array}
$$

Indeed, $m \in \ker(M \to N)$ can be represented by $c \in C$ with $\gamma(c) = 0$, $\Omega(c) = 0$. The second condition shows that, by the exactness of the central vertical column, there is $b \in B$ such that $c = \Psi(b)$ and we define $b' = \beta(b)$, with

$$\Psi'(b') = \Psi'(\beta(b)) = (\Psi' \circ \beta)(b) = (\gamma \circ \Psi)(b) = \gamma(\Psi(b)) = \gamma(c) = 0.$$

Using again the exactness of the central columns, there is $a' \in A'$ such that $b' = \Phi'(a')$, and we project a' to $q \in Q$. As in Theorem 6 we have a well-defined map

$$\ker(M \to N) \longrightarrow \ker(Q \to R),$$

which is easily seen to be surjective by going backwards along the preceding chase.

To obtain a proof of the theorem, we only need to prove that the kernel of this map is exactly $\mathrm{im}(L \to M)$.

Indeed, $q = 0 \Leftrightarrow a' = \alpha(a)$. In that case,

$$\beta(b) = b' = \Phi'(a') = \Phi'(\alpha(a)) = (\Phi' \circ \alpha)(a) = (\beta \circ \Phi)(a) = \beta(\Phi(a)),$$

and therefore $\beta(b - \Phi(a)) = 0$. Thus, there is $l \in L$ such that $b = l + \Phi(a)$. Since $c = \Psi(b)$ we obtain $m = \Psi(l)$. By saying also that the map $\Psi \colon L \to M$ is induced by $\Psi \colon B \to C$, we obtain a proof of the theorem.

2) The second proof uses two successive applications of Theorem 6. After cutting the main diagram as indicated by the dotted line, we obtain the following two diagrams:

$$
\begin{array}{ccccccccc}
 & & 0 & & 0 & & 0 & & \\
 & & \downarrow & & \downarrow & & \downarrow & & \\
0 \longrightarrow & K & \longrightarrow & A & \overset{\alpha}{\longrightarrow} & A' & \longrightarrow & Q \longrightarrow 0 \\
 & \downarrow & & \Phi\downarrow & & \downarrow\Phi' & & \downarrow & \\
0 \longrightarrow & L & \longrightarrow & B & \overset{\beta}{\longrightarrow} & B' & \longrightarrow & R \longrightarrow 0 \\
 & \downarrow & & \Psi\downarrow & & \downarrow\Psi' & & & \\
0 \longrightarrow & Z(M) & \longrightarrow & \mathrm{im}\,\Psi & \longrightarrow & \mathrm{im}\,\Psi' & & & \\
 & & & \downarrow & & \downarrow & & & \\
 & & & 0 & & 0 & & &
\end{array}
$$

and

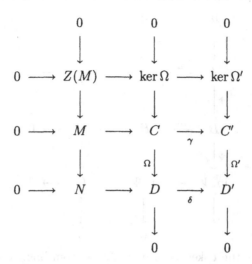

where now the bottom row of the first diagram is identical with the top row of the second diagram, because the exactness of the two central columns of the main diagram in the theorem implies $\operatorname{im} \Psi = \ker \Omega$ and $\operatorname{im} \Psi' = \ker \Omega'$.

Applying the snake Theorem 6 to the first diagram, we obtain the long exact sequence

$$0 \longrightarrow K \xrightarrow{\Phi} L \xrightarrow{\Psi} Z(M) \longrightarrow Q \longrightarrow R,$$

and thus the short exact sequence

$$0 \longrightarrow B(M) \longrightarrow Z(M) \longrightarrow \ker(Q \to R) \longrightarrow 0.$$

This gives the result. $\square$

The results above constitute all we need in the sequel. Nevertheless, to aid the reader in mastering these new techniques, we propose an exercise. We advise the reader to fill in the details of the exercise and only give a few hints.

Exercise 9. In the diagram of Theorem 8, prove the following isomorphism:

$$\ker(R \to S)/\operatorname{im}(Q \to R) = H(R) \simeq \operatorname{coker}(M \to N).$$

PROOF. A direct proof can be given, as in the previous Theorem, and here we only sketch the intrinsic part. Using the same cut as before, but looking at the righthand side of the diagrams, we obtain the following commutative and exact diagram:

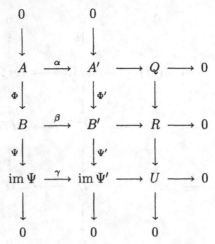

where U is defined as the cokernel of γ in the bottom line. Using this diagram we therefore obtain the exact sequence

$$Q \to R \to U \to 0,$$

and using the main diagram of Theorem 8 we obtain the exact sequence

$$R \longrightarrow S \longrightarrow T \longrightarrow 0.$$

Using the equalities $\operatorname{im} \Psi = \ker \Omega$, $\operatorname{im} \Psi' = \ker \Omega'$ already introduced in the proof of Theorem 8, we obtain the following commutative and exact diagram:

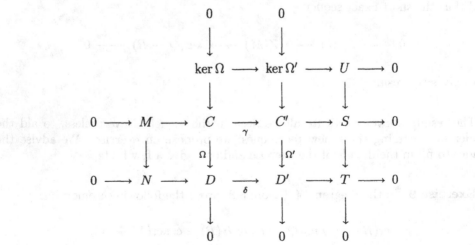

Using Theorem 6 we obtain the long exact sequence

$$M \xrightarrow{\ \Omega\ } N \longrightarrow U \longrightarrow S \longrightarrow T \longrightarrow 0,$$

and therefore the long exact sequence

$$0 \longrightarrow \operatorname{coker}(M \to N) \longrightarrow U \longrightarrow S \longrightarrow T \longrightarrow 0.$$

Combining the three preceding exact sequences obtained, we arrive at the following commutative and exact diagram:

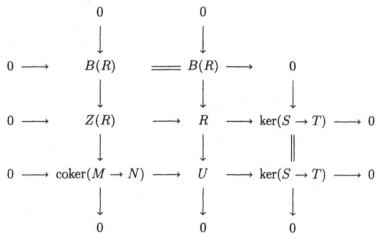

This proves the desired isomorphism. □

We end this Chapter with a simple yet useful result, which is left to the reader as an exercise.

Exercise 10. Consider the following commutative diagram:

$$
\begin{array}{ccccccc}
& 0 & & 0 & & 0 & & 0 \\
& \downarrow & & \downarrow & & \downarrow & & \downarrow \\
0 \longrightarrow & K & \longrightarrow & A & \xrightarrow{\alpha} & A' & \xrightarrow{\alpha'} & A'' \\
& \downarrow & & \Phi\downarrow & & \downarrow\Phi' & & \downarrow\Phi'' \\
0 \longrightarrow & L & \longrightarrow & B & \xrightarrow{\beta} & B' & \xrightarrow{\beta'} & B'' \\
& \downarrow & & \Psi\downarrow & & \downarrow\Psi' & & \downarrow\Psi'' \\
0 \longrightarrow & M & \longrightarrow & C & \xrightarrow{\gamma} & C' & \xrightarrow{\gamma'} & C''' \\
& \downarrow & & \downarrow & & \downarrow & & \downarrow \\
& 0 & & 0 & & 0 & & 0
\end{array}
$$

where the columns are assumed to be exact. Then the horizontal cohomologies at A', B', C' are related by the following exact sequence:

$$0 \longrightarrow H(A') \longrightarrow H(B') \longrightarrow H(C').$$

In particular, if the top row and the bottom row are exact, then so is the middle row.

HINT. Cut the diagram into two blocks by a vertical dotted line passing between A and A', B and B', and C and C'. Construct the resulting right image sequence in the left diagram and the left kernel sequence in the right diagram. Finally connect them by inclusion and chase.

CHAPTER II

Jet Theory

In this Chapter we only give results that are absolutely necessary for understanding the remaining Chapters. Most of the time we will use local coordinates, in order to be able to study explicit situations. Nevertheless, the intrinsic exposition will point to concepts that are independent of a local description.

All manifolds and maps are assumed to be differentiable.

A. Fibered Operations

Let X be a manifold of dimension n with local coordinates $(x^1, \ldots, x^n)$ on an open set $U \subset X$. For simplicity reasons we shall not present details about the covering, and we denote a change of coordinates by $\bar{x} = \varphi(x)$ because changes of coordinates will be used only for knowing how certain objects transform. Similarly, we shall use the same notation for a point $x \in X$ and its local coordinates $(x^i)_{i=1,\ldots,n}$. The context will always clearly indicate whether we are dealing with a single point represented in two coordinate systems, or with a map between two points in the same coordinate system.

Let $\mathcal{E}$ be another manifold, of dimension $n + m$, and let $\pi: \mathcal{E} \to X$ be a surjective map.

Definition 1. We say that $\mathcal{E}$ is a *fibered manifold* over X with projection π if we can cover $\mathcal{E}$ by coordinate charts (U, Φ) that can be projected onto coordinate charts (V, φ) of X using the following commutative diagram:

$$
\begin{array}{ccc}
\mathcal{E} \supset U & \xrightarrow{\Phi} & \mathbb{R}^n \times \mathbb{R}^m \\
\downarrow{\scriptstyle \pi} & \downarrow & \downarrow \\
X \supset V & \xrightarrow[\varphi]{} & \mathbb{R}^n
\end{array}
$$

In practice we shall adopt local coordinates (x^i, y^k) with $i = 1, \ldots, n$; $k = 1, \ldots, m$ on $\mathcal{E}$ and this definition then amounts to saying that changes of coordinates on $\mathcal{E}$ can be chosen to be of the form

$$\bar{x} = \varphi(x), \qquad \bar{y} = \psi(x, y).$$

This type of change will cover most of the situations encountered in mechanics and physics (vectors, tensors, densities, etc.) as well as the new situations we shall meet in Chapter V, involving geometric objects as a generalization of tensors.

61

Example 2. If Y is another manifold, of dimension m, then $\pi\colon X \times Y \to X$ equipped with the projection into the first factor is called a *trivial fibered manifold*. Any map $f\colon X \to Y$ will be identified with its graph $f\colon X \to X \times Y$, $x \to (x, y = f(x))$.

Definition 3. X is called the *base manifold* and $\mathcal{E}$ the *total manifold*. For any point $x \in X$, the submanifold $\mathcal{E}_x = \pi^{-1}(x) \subset \mathcal{E}$ is called the *fiber* over x, and we let $\mathcal{E} \mid U = \pi^{-1}(U)$ be the *restriction* of $\mathcal{E}$ to $U \subset X$.

Definition 4. A *local section* of $\mathcal{E}$ over an open set $U \subset X$ is a map $f\colon U \to \mathcal{E}$ such that $(\pi \circ f)(x) = x$, $\forall x \in U$. We let $U = \operatorname{dom} f$ be the *domain* of f, and a section f is called a *global section* if $X = \operatorname{dom} f$.

Example 5. The graph of a map as in Example 2 is the simplest example of a section.

Now we give some examples of nontrivial fibered manifolds. We will frequently use these examples in the sequel.

Example 6. A fibered manifold is called a *vector bundle* if we can choose the coordinate transformations to be of the form

$$\overline{x} = \varphi(x), \qquad \overline{y} = A(x)y \quad \text{with } \det A \neq 0,$$

where A is an $m \times m$ square invertible matrix.

To distinguish vector bundles, we denote vector bundles over X by capital letters $E, F, \ldots$, and $\dim E$ will denote the dimension of a typical fiber as a vector space over $\mathbb{R}$. In that case, the tensor product $E \otimes_X F$ of two vector bundles E and F over X is a vector bundle over X, with coordinate transformations given by the matrix $A \otimes B$ if A and B give the changes of coordinates in E and F, respectively. When the underlying base space X is clear, we will simply write $E \otimes F$. We have $\dim E \otimes F = \dim E \times \dim F$.

Among the well-known typical examples, the *tangent bundle* $T = T(X)$ has changes of coordinates $(x, u) \to (\overline{x}, \overline{u})$ of the form

$$\overline{x} = \varphi(x), \qquad \overline{u}^j = \partial_i \varphi^j(x) u^i.$$

Similarly, the *cotangent bundle* $T^* = T^*(X)$ has changes of coordinates $(x, u) \to (\overline{x}, \overline{u})$ of the form

$$\overline{x} = \varphi(x), \qquad u_i = \partial_i \varphi^j(x) \overline{u}_j,$$

where we have put the indices in upper(lower) position according to the *contravariant* (*covariant*) property of this simple type of tensor. Such a rule can be extended to *tensor bundles*, just by copying these definitions. A section of a tensor bundle is more generally known as a *tensor field*. Also, $\otimes, S, \wedge$ denote the tensor, the symmetric, and the exterior product, respectively. In the case of exterior products, a section ω of $\wedge^r T^*$ will be written as $\omega = \omega_I(x)\,dx^I$, where $I = \{i_1 < \cdots < i_r\}$ is a multi-index and $dx^I = dx^{i_1} \wedge \cdots \wedge dx^{i_r}$ indicates the decomposition with respect to a basis of the exterior algebra. For example, when $n = 3$ and $r = 2$, a 2-form can be written as follows:

$$\omega = \omega_{12}(x)\,dx^1 \wedge dx^2 + \omega_{13}(x)\,dx^1 \wedge dx^3 + \omega_{23}(x)\,dx^2 \wedge dx^3.$$

Equivalently: $\omega = (\omega_{12}(x), \omega_{13}(x), \omega_{23}(x))$. Of course, an r-form can always be written as

$$\omega = \frac{1}{r!}\omega_{i_1\ldots i_r}(x)\, dx^{i_1} \wedge \cdots \wedge dx^{i_r}$$

if we do not want to separate the independent components. For example, defining $\omega_{12} = -\omega_{21}$ leads to

$$\frac{1}{2}(\omega_{12}(x)\, dx^1 \wedge dx^2 + \omega_{21}(x)\, dx^2 \wedge dx^1)$$

$$= \frac{1}{2}(\omega_{12}(x)\, dx^1 \wedge dx^2 - \omega_{12}(x)\, dx^2 \wedge dx^1)$$

$$= \frac{1}{2}\left(\omega_{12}(x)\, dx^1 \wedge dx^2 + \omega_{12}(x)\, dx^1 \wedge dx^2\right) = \omega_{12}(x)\, dx^1 \wedge dx^2.$$

We insist on this *convention of summation*, and we will use it often in the sequel.

Remark 7. In mathematical physics there is often some ambiguity between tensor bundles and their sections. However, in practice it is quite useful to have a unique way of writing both. Similarly, we will denote a fibered manifold $\mathcal{E}$ and its set (one should say, sheaf) of sections by the same letter $\mathcal{E}$. In general, the context will always make clear what kind of interpretation must be used. E.g., if we have to differentiate, the sections point of view must be adopted, of course. Hence, similarly to writing $\omega \in \wedge^2 T^*$ for an r-form as a field of forms, we shall sometimes write $f \in \mathcal{E}$ for a section, keeping in mind Examples 2 and 5 so that we can write $f\colon (x^i) \to (x^i, y^k = f^k(x))$ or simply $f\colon x \to (x, y = f(x))$ when we need an explicit local description. At first reading this may seem a notation which is difficult to adapt to, but when experience grows it will soon become evident that this is a worthwhile notation, as it saves a lot of place in formula writing.

Example 8. A fibered manifold is called an *affine bundle* if the typical fiber is an affine space and if we can choose the coordinate transformations to be of the form

$$\bar{x} = \varphi(x), \qquad \bar{y} = A(x)y + B(x) \quad \text{with } \det A(x) \neq 0,$$

where A is an $m \times m$ square invertible matrix and B is a vector.

It follows from Example 6 that to any affine bundle $\mathcal{E}$ over X we can associate a vector bundle E over X, with coordinate changes

$$\bar{x} = \varphi(x), \qquad \bar{v} = A(x)v \quad \text{with } \det A \neq 0.$$

Therefore we make the following definition.

Definition 9. We shall say that the affine bundle $\mathcal{E}$ is *modeled* on the vector bundle E over X. We denote this situation by a *dotted arrow*, as follows:

$$E \cdots \to \mathcal{E} \xrightarrow{\pi} X.$$

In practice, if (x, y_1) and (x, y_2) are two points of $\mathcal{E}$ over the same point $x \in X$, then the *difference* $(x, y_2) - (x, y_1) \overset{\text{def}}{=} (x, y_2 - y_1)$ has a meaning, being a point of E over $x \in X$.

The three following basic definitions are often taken together:

Definition 10. If $\pi\colon \mathcal{E} \to X$ and $\pi'\colon \mathcal{E}' \to X$ are two fibered manifolds over X, we let $\mathcal{E} \times_X \mathcal{E}'$ be the *fibered product* of $\mathcal{E}$ and $\mathcal{E}'$ over X. It is the fibered manifold over X consisting of all pairs of points in $\mathcal{E}$ and $\mathcal{E}'$ having the same projection into X, according to the following commutative diagram:

$$
\begin{array}{ccc}
(x,y,y')\quad \mathcal{E} \times_X \mathcal{E}' & \longrightarrow & \mathcal{E}' \quad (x,y') \\
\downarrow \qquad\quad \downarrow & & \downarrow{\scriptstyle\pi'} \quad \downarrow \\
(x,y)\qquad \mathcal{E} & \xrightarrow{\ \pi\ } & X \quad (x)
\end{array}
$$

Definition 11. If $\varphi\colon X \to X'$ is a map and $\pi'\colon \mathcal{E}' \to X'$ is a fibered manifold over X', we let $\varphi^{-1}(\mathcal{E}')$ be the *reciprocal image* or *pullback* of $\mathcal{E}'$ under φ over X, i.e. the set of all pairs of points in X and $\mathcal{E}'$ having the same image in X under φ and π', according to the following commutative diagram:

$$
\begin{array}{ccc}
\{(x,x',y')|\ \varphi(x) = x'\} \quad \varphi^{-1}(\mathcal{E}') & \longrightarrow & \mathcal{E}' \quad (x',y') \\
\downarrow & \downarrow & \downarrow{\scriptstyle\pi'} \quad \downarrow \\
(x) & X \xrightarrow{\ \varphi\ } & X' \quad (x')
\end{array}
$$

Remark 12. Although $\mathcal{E} \times_X \mathcal{E}' = \pi^{-1}(\mathcal{E}') = \pi'^{-1}(\mathcal{E})$, in general the context will indicate which interpretation is the most appropriate.

Definition 13. If $\pi\colon \mathcal{E} \to X$ and $\pi'\colon \mathcal{E}' \to X'$ are two fibered manifolds, a *fibered morphism* $\Phi\colon \mathcal{E} \to \mathcal{E}'$ over $\varphi\colon X \to X'$ is a pair of maps (Φ, φ) making the following diagram commutative:

$$
\begin{array}{ccc}
(x,y)\quad \mathcal{E} & \xrightarrow{\ \Phi\ } & \mathcal{E}' \quad (x' = \varphi(x), y' = \Phi(x,y)) \\
\downarrow{\scriptstyle\pi} & \downarrow{\scriptstyle\pi'} & \downarrow \\
(x)\quad X & \xrightarrow{\ \varphi\ } & X' \qquad (x' = \varphi(x))
\end{array}
$$

If $X = X'$ and $\varphi = \mathrm{id}_X$, we simply say that Φ is a morphism, and denote it by the upper row.

A morphism of vector bundles will be defined as above, adding the condition that it preserves the vector space structure of the fibers, i.e. in local coordinates we must have $y' = A(x)y$. Similarly, a morphism of affine bundles will be defined by adding the condition that it preserves the affine space structure of the fibers, i.e. in local coordinates we must have $y' = A(x)y + B(x)$.

The most useful example of a vector bundle morphism is given in the following example.

Example 14. For a map $f\colon X \to Y$ we define the *tangent map* $T(f)\colon T(X) \to T(Y)$ over f by the following formula, in local coordinates:

$$(x^i, u^i) \to (y^k = f^k(x), v^k = \partial_i f^k(x)u^i).$$

Definition 15. We say that a fibered manifold $\mathcal{R} \to X$ is a *fibered submanifold* of $\pi \colon \mathcal{E} \to X$ if $\mathcal{R}$ is a submanifold of $\mathcal{E}$ and the inclusion is a morphism.

The following situation sometimes gives rise to fibered submanifolds.

Definition 16. If $\Phi \colon \mathcal{E} \to \mathcal{E}'$ is a morphism and f' is a section of $\mathcal{E}'$ over X, then we define the *kernel* of Φ with respect to f' as the set of points of $\mathcal{E}$ having the same image in $\mathcal{E}'$ under Φ and under $f' \circ \pi$, i.e.

$$\ker_{f'} \Phi = \{(x, y) \in \mathcal{E} \mid \Phi(x, y) = f'(x)\}.$$

In practice it is convenient to describe this definition as follows:

$$0 \longrightarrow \ker_{f'} \Phi \longrightarrow \mathcal{E} \overset{\Phi}{\underset{f' \circ \pi}{\rightrightarrows}} \mathcal{E}'$$

$$\pi \downarrow \qquad \pi' \Big\Vert f'$$

$$X = X$$

using a double arrow. We shall examine the conditions under which such a set projects onto X. Indeed, consider the m' equations $\Phi(x, y) = f'(x)$ in the m unknowns y. If we could solve these equations for a certain number of y equal to the rank of the jacobian matrix $\partial\Phi(x, y)/\partial y$, using the implicit function theorem, the remaining equalities must become identities after substitution, for otherwise we would obtain, among the equations defining $\ker_{f'} \Phi$, a number of equations involving only x, and this would contradict the surjectivity of the composite map $\ker_{f'} \Phi \to \mathcal{E} \overset{\pi}{\to} X$. In fact, chasing in the commutative 'triangle' of the preceding diagram, we discover that the above-said is nothing else than the practical (functional) way to verify that $f'(X) \subset \operatorname{im} \Phi$, although this is not completely evident at first sight. Hence, if this condition is satisfied, then the fibered submanifold property depends only on the constant rank of the jacobian matrix considered above.

When $\mathcal{E}'$ is a vector bundle, the kernel will always be taken with respect to the zero section.

The next definition will be used in a crucial way when dealing with nonlinear systems of PDE in a manner similar to linear systems. It will also be quite important for giving a modern interpretation of variational calculus. Using Example 14 and the above remark we state

Definition 17. We define the *vertical bundle* $V(\mathcal{E})$ as a vector bundle over $\mathcal{E}$ by means of the following short exact sequence:

$$0 \longrightarrow V(\mathcal{E}) \longrightarrow T(\mathcal{E}) \overset{T(\pi)}{\longrightarrow} T(X) \times_X \mathcal{E} \longrightarrow 0.$$

We now explain the details of the construction using local coordinates, since this technique will be of constant use in the whole book.

First we have the following commutative diagram:

$$(x, y; u, v) \quad T(\mathcal{E}) \xrightarrow{T(\pi)} T(X) \quad (x; u)$$

$$\downarrow \qquad \downarrow \qquad\qquad \downarrow \qquad \downarrow$$

$$(x, y) \qquad \mathcal{E} \xrightarrow[\pi]{} X \qquad (x)$$

Then we pullback $T(X)$ as a vector bundle over X to $T(X) \times_X \mathcal{E} = \pi^{-1}(T(X))$ as a vector bundle over $\mathcal{E}$.

$$(x, y; u) \quad T(X) \times_X \mathcal{E} \longrightarrow T(X) \quad (x; u)$$

$$\downarrow \qquad\qquad \downarrow \qquad\qquad \downarrow \qquad \downarrow$$

$$(x, y) \qquad \mathcal{E} \xrightarrow[\pi]{} X \qquad (x)$$

We may therefore describe the sequence as follows:

$$0 \longrightarrow (x, y; 0, v) \longrightarrow (x, y; u, v) \longrightarrow (x, y; u) \longrightarrow 0,$$

because it becomes a short exact sequence of vector spaces over each point $(x, y) \in \mathcal{E}$, and we recover the notation of Chapter I. For later use we give the coordinate transformations of $V(\mathcal{E}) \subset T(\mathcal{E})$:

$$T(\mathcal{E}) \quad \begin{cases} \overline{v}^l = \dfrac{\partial \psi^l(x, y)}{\partial x^i} u^i + \dfrac{\partial \psi^l(x, y)}{\partial y^k} v^k, \\[2mm] \overline{u}^j = \dfrac{\partial \varphi^j(x)}{\partial x^i} u^i, \\[2mm] \overline{y} = \psi(x, y), \\[1mm] \overline{x} = \varphi(x), \end{cases}$$

$$V(\mathcal{E}) \quad \begin{cases} \overline{v}^l = \dfrac{\partial \psi^l(x, y)}{\partial y^k} v^k, \\[2mm] \overline{y} = \psi(x, y), \\[1mm] \overline{x} = \varphi(x). \end{cases}$$

For simplicity, in the sequel we shall not write out explicitly the reciprocal images when the context is clear. Therefore we shall write the latter short exact sequence as follows:

$$0 \longrightarrow V(\mathcal{E}) \longrightarrow T(\mathcal{E}) \xrightarrow{T(\pi)} T(X) \longrightarrow 0.$$

For an affine bundle we obtain

$$V(\mathcal{E}) \quad \begin{cases} \overline{v} = A(x)v, \\ \overline{y} = A(x)y + B(x), \\ \overline{x} = \varphi(x)., \end{cases}$$

and we obtain the following commutative diagram:

$$
\begin{array}{ccccc}
(x,y;v) & V(\mathcal{E}) & \longrightarrow & E & (x;v) \\
& \downarrow & \downarrow & \downarrow\ \downarrow & \\
(x,y) & \mathcal{E} & \xrightarrow{\ \pi\ } & X & (x)
\end{array}
$$

where E is the model of $\mathcal{E}$ and $V(\mathcal{E}) = E \times_X \mathcal{E}$. For reasons that will become clear in the sequel, in the case of affine bundles we shall sometimes write E for $V(\mathcal{E})$, or identify them.

Similarly, whenever $\Phi \colon \mathcal{E} \to \mathcal{E}'$ is a morphism, we can extend it to $T(\Phi) \colon T(\mathcal{E}) \to T(\mathcal{E}')$ and restrict it to $V(\Phi) \colon V(\mathcal{E}) \to V(\mathcal{E}')$ according to the following formulas, in local coordinates:

$$
\begin{aligned}
\Phi & \qquad y' = \Phi(x,y), \\
T(\Phi) & \qquad v' = \frac{\partial \Phi(x,y)}{\partial x}u + \frac{\partial \Phi(x,y)}{\partial y}v, \\
V(\Phi) & \qquad v' = \frac{\partial \Phi(x,y)}{\partial y}v,
\end{aligned}
$$

where we do not write out indices, for simplicity reasons.

A first use of the *vertical machinery* is given in the following definition:

Definition 18. We shall say that $\Phi \colon \mathcal{E} \to \mathcal{E}'$ is a *monomorphism* (*epimorphism*) of fibered manifolds if $V(\Phi) \colon V(\mathcal{E}) \to V(\mathcal{E}')$ is a monomorphism (epimorphism) of vector bundles over $\mathcal{E}$, i.e. the restriction to each fiber over a point of $\mathcal{E}$ is a monomorphism (epimorphism), namely an injective (surjective) map in the sense of Chapter I.

Of course, in the Definition we have followed our rule not to indicate the reciprocal images, and more precisely we should have written $V(\Phi) \colon V(\mathcal{E}) \to \Phi^{-1}(V(\mathcal{E}'))$.

A second use of the vertical machinery will give a surprise, in that it explains the most basic definitions of variational calculus. Recall that, e.g. in analytical mechanics, one has to vary the position $q(t)$, where t is time, by introducing $\delta q(t)$ as a variation. Let us return to the definition of vertical bundle, and look at a section of $V(\mathcal{E})$ over a section f of $\mathcal{E}$, as in the following picture:

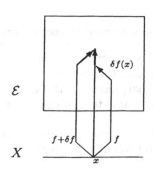

We see that $\delta f(x)$, at each point $x \in X$, is a vertical vector. Hence $(f, \delta f)$ is a section of $V(\mathcal{E})$ over a section f of $\mathcal{E}$. It follows that the results of variational calculus do not need the notion of 'small' variation (as mechanicians often believe!), for two reasons. The first is that 'small' has no meaning independently of a unit of measurement, and the second is that according to the comment above it is nonsense from a mathematical point of view.

Having in mind Definition 16 of kernel with respect to a section, we shall try to define for fibered manifolds the concepts of sequence and exactness that were so useful in Chapter I.

Definition 19. We say that a chain of fibered manifolds

$$\mathcal{E} \xrightarrow{\Phi} \mathcal{E}' \xrightarrow{\Psi} \mathcal{E}''$$

is a *sequence* with respect to a section f'' of $\mathcal{E}''$ if $\operatorname{im} \Phi \subset \ker_{f''} \Psi$.

In local coordinates we therefore have

$$\Psi(x, \Phi(x, y)) \equiv f''(x) \qquad \forall (x, y) \in \mathcal{E}.$$

Differentiating with respect to y, we find

$$\frac{\partial \Psi}{\partial y'}(x, \Phi(x, y)) \cdot \frac{\partial \Phi}{\partial y}(x, y) \equiv 0 \qquad \forall (x, y) \in \mathcal{E}.$$

It follows that, in effect, we have a sequence

$$V(\mathcal{E}) \xrightarrow{V(\Phi)} V(\mathcal{E}') \xrightarrow{V(\Psi)} V(\mathcal{E}'')$$

of vector bundles pulled back over $\mathcal{E}$ by reciprocal image.

As a byproduct, the next definition is natural and coherent with the preceding one.

Definition 20. The double arrow sequence

$$\mathcal{E} \xrightarrow{\Phi} \mathcal{E}' \underset{f'' \circ \pi'}{\overset{\Psi}{\rightrightarrows}} \mathcal{E}''$$

is said to be *exact* if

(1) $\operatorname{im} \Phi = \ker_{f''} \Psi$;
(2) the corresponding vertical sequence of vector bundles is exact.

As already said, whenever $\mathcal{E}''$ is a vector bundle, the kernel is automatically taken with respect to the zero section, and this convention is in agreement with the definitions of monomorphism and epimorphism. In the simplest case of affine bundles, the following proposition provides a simple test of exactness.

Proposition 21. *A sequence of affine bundles is exact if and only if the corresponding sequence of model vector bundles is exact.*

PROOF. We set $\Phi(x,y) = A(x)y + B(x)$, $\Psi(x,y') = a(x)y' + b(x)$, and obtain

$$\Psi(x, \Phi(x,y)) = a(x)A(x)y + a(x)B(x) + b(x) \equiv f''(x).$$

Therefore, since we are dealing with a sequence, we must have

$$a(x)A(x) \equiv 0, \qquad a(x)B(x) + b(x) \equiv f''(x) \qquad \forall x \in X,$$

and it follows that the corresponding chain of model vector bundles is a sequence. Accordingly, we have the following commutative diagram:

$$
\begin{array}{ccccc}
E & \xrightarrow{V(\Phi)} & E' & \xrightarrow{V(\Psi)} & E'' \\
\vdots & & \vdots & \nearrow & \vdots \\
\mathcal{E} & \xrightarrow{\Phi} & \mathcal{E}' & \xrightarrow{\Psi} & \mathcal{E}'' \\
\pi \downarrow & & \pi' \downarrow & & \pi'' \Updownarrow f'' \\
X & = & X & = & X
\end{array}
$$

Let the top row be exact and pick a point $(x, y') \in \mathcal{E}'$ such that $a(x)y' + b(x) = f''(x)$. For any point $(x, y) \in \mathcal{E}$ we have $a(x)\Phi(x,y) + b(x) = f''(x)$. Hence $a(x)(y' - \Phi(x,y)) = 0$, with $(x, y' - \Phi(x,y)) \in E'$. Since the top row is exact, we can find $(x, v) \in E$ such that $y' - \Phi(x,y) = y' - A(x)y - B(x) = A(x)v$, and thus $y' = A(x)(y+v) + B(x) = \Phi(x, y+v)$ is obtained by modifying $(x, y) \in \mathcal{E}$ to $(x, y+v) \in \mathcal{E}$. Therefore $(x, y') \in \operatorname{im} \Phi$, and one way is proved.

The converse way is quite similar and left to the reader as an exercise (compare Exercise I.10). $\square$

Corollary 22. *In the situation of the previous Proposition, there is an exact sequence*

$$\mathcal{E} \xrightarrow{\Phi} \mathcal{E}' \longrightarrow E''.$$

PROOF. As we have seen in the proof of this Proposition, the existence of a sequence implies the existence of a distinguished section f'' of the affine bundle $\mathcal{E}''$. Hence, to any point $(x, y'') \in \mathcal{E}''$ we can associate the point $(x, y'' - f''(x)) \in E''$. In a more direct way, to any point $(x, y') \in \mathcal{E}'$ we can associate an arbitrary point $(x, y) \in \mathcal{E}$, consider the point $(x, y' - \Phi(x,y)) \in E'$ and map it to E'' by $V(\Psi)$. The image is well defined, because two points (x, y_1) and (x, y_2) of $\mathcal{E}$ are such that $(x, y_2 - y_1) \in E$, and therefore $(x, y' - \Phi(x, y_2)) - (x, y' - \Phi(x, y_1)) = (x, A(x)(y_1 - y_2)) \in \ker V(\Psi)$. Hence we have avoided the use of a section of $\mathcal{E}''$. $\square$

We shall now apply all these results to a typical, and particularly useful, situation, namely *jet theory*.

B. Jet Bundles

We shall start with a short calculation in order to explain the main motivation of this section and the novelty brought about by jet theory which could *never* have been discovered via any 'classical' approach.

Let $\pi\colon \mathcal{E} \to X$ be a fibered manifold as in the previous Section, and consider a local section f of $\mathcal{E}$ over $U \subset X$. We remind the reader how to transform the first and second 'derivatives' of f under a change of coordinates $(x,y) \to (\overline{x} = \varphi(x), \overline{y} = \psi(x,y))$. Indeed, in the initial coordinate system the local expression of the section is $y = f(x)$, while in the final coordinate system this local expression of the (same) section is $\overline{y} = \overline{f}(\overline{x})$, and they are related by

$$\overline{f}(\varphi(x)) \equiv \psi(x, f(x)) \qquad \forall x \in U.$$

Differentiating with respect to x^i gives:

$$\frac{\partial \overline{f}^u}{\partial \overline{x}^k}(\varphi(x)) \frac{\partial \varphi^k}{\partial x^i}(x) \equiv \frac{\partial \psi^u}{\partial x^i}(x, f(x)) + \frac{\partial \psi^u}{\partial y^k}(x, f(x)) \frac{\partial f^k}{\partial x^i}(x),$$

while differentiating again with respect to x^j gives

$$\begin{aligned}
\frac{\partial^2 \overline{f}^u}{\partial \overline{x}^r \partial \overline{x}^s}(\varphi(x)) &\frac{\partial \varphi^r}{\partial x^i}(x) \frac{\partial \varphi^s}{\partial x^j}(x) + \frac{\partial \overline{f}^u}{\partial \overline{x}^r}(\varphi(x)) \frac{\partial^2 \varphi^r}{\partial x^i \partial x^j}(x) \\
&\equiv \frac{\partial^2 \psi^u}{\partial x^i \partial x^j}(x, f(x)) + \frac{\partial^2 \psi^u}{\partial x^i \partial y^l}(x, f(x)) \frac{\partial f^l}{\partial x^j}(x) + \frac{\partial^2 \psi^u}{\partial x^j \partial y^k}(x, f(x)) \frac{\partial f^k}{\partial x^i}(x) \\
&\quad + \frac{\partial^2 \psi^u}{\partial y^k \partial y^l}(x, f(x)) \frac{\partial f^k}{\partial x^i}(x) \frac{\partial f^l}{\partial x^j}(x) + \frac{\partial \psi^u}{\partial y^k}(x, f(x)) \frac{\partial^2 f^k}{\partial x^i \partial x^j}(x),
\end{aligned}$$

for $i, j, r, s = 1, \ldots, n$ and $k, l, u = 1, \ldots, m$.

Note that differentiating more often leads to absolutely tedious, more and more complex, formulas.

Roughly speaking, in the first place jet theory has been created for dealing with this type of calculation for arbitrary order of differentiation, and to provide results that cannot be checked in local coordinates because the method of simply writing the expressions out on paper is an almost impossibility.

However, there is another purpose of jet theory, which is not widely known and is equally important for applications. We shall clarify this point using the above calculations.

Indeed, patching together changes of coordinates as before and introducing new co-ordinates, called *jet coordinates*, having the same indices as the corresponding derivatives, while transforming in the same way, we may successively exhibit fibered manifolds over $\mathcal{E}$ and X, denoted by $J_1(\mathcal{E})$ and $J_2(\mathcal{E})$ and called the *jet bundle* of $\mathcal{E}$ of order 1 and of order 2. We give the details of the coordinate transformations.

$$
J_2(\mathcal{E}) \left\{ J_1(\mathcal{E}) \left\{ \mathcal{E} \left\{ X \left\{
\begin{aligned}
&\overline{y}^u_{rs}\partial_i\varphi^r\partial_j\varphi^s + \overline{y}^u_r\partial_{ij}\varphi^r = \partial_{ij}\psi^u + \frac{\partial^2\psi^u}{\partial x^i\partial y^l}y^l_j \\
&\qquad\qquad + \frac{\partial^2\psi^u}{\partial x^j\partial y^k}y^k_i + \frac{\partial^2\psi^u}{\partial y^k\partial y^l}y^k_i y^l_j + \frac{\partial\psi^u}{\partial y^k}y^k_{ij}, \\
&\overline{y}^u_r\partial_i\varphi^r(x) = \partial_i\psi^u(x,y) + \frac{\partial\psi^u(x,y)}{\partial y^k}y^k_i, \\
&\overline{y} = \psi(x,y), \\
&\overline{x} = \varphi(x),
\end{aligned}
\right.\right.\right.\right.
$$

and the various projections

$$
\begin{array}{lll}
J_2(\mathcal{E}) & (x^i, y^k, y^k_i, y^k_{ij}) & (x, y_2) \\[2pt]
\Big\downarrow{\scriptstyle\pi^2_1} & & \\[2pt]
J_1(\mathcal{E}) & (x^i, y^k, y^k_i) & (x, y_1) \\[2pt]
\Big\downarrow{\scriptstyle\pi^1_0} & & \\[2pt]
\mathcal{E} & (x^i, y^k) & (x, y) \\[2pt]
\Big\downarrow{\scriptstyle\pi} & & \\[2pt]
X & (x^i) & (x)
\end{array}
$$

where the condensed notation y_q denotes all the jet coordinates up to order q, with $y_0 = y$. Of course, among the sections of $J_q(\mathcal{E})$ we have the 'derivatives' up to order q of a section f of $\mathcal{E}$, and we will denote these by the symbol j_q. For example:

$$
j_2(f)\colon (x) \to (x, f^k(x), \partial_i f^k(x), \partial_{ij} f^k(x)).
$$

However, the novelty is that some of the sections f_q of $J_q(\mathcal{E})$ will not be of the type $j_q(f)$ for some section f of $\mathcal{E}$. E.g.,

$$
f_2\colon (x) \to (x, f^k(x), f^k_i(x), f^k_{ij}(x))
$$

whenever $\partial_i f^k(x) - f^k_i(x) \neq 0$, $\partial_i f^k_j(x) - f^k_{ij}(x) \neq 0$, $\partial_r f^k_{ij}(x) - f^k_{ijr}(x) \neq 0, \ldots$, etc. We shall see later on that the various 'differences' allowing us to distinguish the jet of a bundle section from a section of the jet bundle will give rise to the Spencer operator. The crucial point is that these new sections, not coming from derivatives of a section, cannot be dealt with by any classical tool, and will play a key role in applications to physics.

For the sceptical reader, let us explain now how the apparently simple calculations make possible a completely new understanding of gravitation. Indeed, briefly recapitulating our previous comment, we could roughly say that sections of jet bundles have 'something to do' with jets of bundle sections $\cdots$ but that they are 'different'; this difference being described by certain $\cdots$ differences!

At this moment, let us ask a college student to describe the movement of a heavy mass (say, a ball) in the constant gravitational field $\vec{g}$ on the surface of the Earth. He/She will successively write:

$$\frac{dx^k}{dt} = v^k, \qquad \frac{dv^k}{dt} = g^k, \qquad \frac{\partial g^k}{\partial x^i} = 0.$$

The first equation is the kinematic definition of the speed, the second is Newton's law equating acceleration and gravitation while showing that the trajectory does not depend on the mass, and the third describes a locally constant gravitational field which is a good approximation on the surface of the Earth.

We rewrite these equations as follows:

$$\frac{dx^k}{dt} - v^k = 0, \qquad \frac{dv^k}{dt} - g^k = 0, \qquad \frac{dg^k}{dx^i} - 0 = 0.$$

We recognize the Spencer operator already described. Of course, this approach will shake a lot of intuitive belief. In particular, the 'speed' appears no longer as a tangent vector but as a first jet, and the gravitation no longer as a vector but as a second-order jet. Hence we may choose space-time with $m = n = 4$ and look for a situation in which only 4 second-order jets do exist while third-order jets should vanish identically. The solution will be given $\cdots$ in the last Chapter.

Nevertheless, we stress the importance of the difference $\partial_i f^k_j - f^k_{ij} \neq 0$ by giving a simple but useful interpretation of the difference $(dv^k/dt) - g^k \neq 0$. Indeed, a reader having some familiarity with aviation should know that an *accelerometer does not measure the acceleration or the gravitation independently, but only their difference* (an accelerometer is an apparatus consisting of a mass attached to a spring). In particular, a movement is said to be '*free*' if and only if this difference vanishes. A well-known example is the '*free fall*' of a lift, considered by A. Einstein in general relativity. Of course, it follows that, *if the origin of gravitation lies in second-order jets, no classical argument can give an interpretation.*

In any case, we understand that it is important to be able to describe the properties of jet bundles of arbitrary order, and we shall introduce a specific notation for this purpose.

Let $\mu = (\mu_1, \ldots, \mu_n)$ be a multi-index.

Definition 1. We let $|\mu| = \mu_1 + \cdots + \mu_n$ be the length of μ, and set $\mu + 1_i = (\mu_1, \ldots, \mu_{i-1}, \mu_i + 1, \mu_{i+1}, \ldots, \mu_n)$.

We may use multi-indices to write out derivatives, as follows:

$$\partial_\mu f(x) = \frac{\partial^{|\mu|} f(x)}{(\partial x^1)^{\mu_1} \ldots (\partial x^n)^{\mu_n}} \qquad \text{with } \partial_0 f(x) = f(x).$$

Here, for $n = 2$ and $\mu = (1,1)$ we have $|\mu| = 2$ and the following notations are identified:

$$\frac{\partial^2 f}{\partial x^1 \partial x^2} = \partial_{12} f = \partial_{(1,1)} f.$$

There is no rule for preferring one notation above others, and certain formulas are quite simple when written with multi-indices whereas in others the standard notation is simpler.

Generalizing the results obtained earlier, we may introduce the fibered manifold $J_q(\mathcal{E})$ over X with local coordinates (x^i, y^k_μ) for $0 \leq |\mu| \leq q$, or simply (x, y_q), transforming like the derivatives up to order q under fibered changes of coordinates of $\mathcal{E}$, and we may identify $J_0(\mathcal{E})$ and $\mathcal{E}$.

Definition 2. $J_q(\mathcal{E})$ is called the *q-jet bundle* of $\mathcal{E}$.

Recall that the number of derivatives of strict order q of a function of n variables is $(q + n - 1)!/q!\,(n - 1)!$, and that the total number of derivatives up to order q of such a function is $(q + n)!/q!\,n!$. It follows that we have

$$\dim J_q(\mathcal{E}) = n + m\frac{(q + n)!}{q!\,n!}.$$

We may introduce the canonical projection

$$\pi^{q+r}_q \colon J_{q+r}(\mathcal{E}) \to J_q(\mathcal{E}) \colon (x, y_{q+r}) \to (x, y_q) \qquad \forall q, r \geq 0.$$

We denote a section of $J_q(\mathcal{E})$ by $f_q \colon (x) \to (x, f^k_\mu(x))$, and associate to any section f of $\mathcal{E}$ the section $j_q(f) \colon (x) \to (x, \partial_\mu f^k(x))$ of $J_q(\mathcal{E})$.

Jet theory is based on the following three fundamental results only. Since we do not know of any other proof of them, we closely follow [144, 162].

Proposition 3. $J_q(\mathcal{E})$ *is an affine bundle over* $J_{q-1}(\mathcal{E})$, *modeled on* $S_qT^* \otimes V(\mathcal{E})$.

PROOF. For $q = 1$ we have

$$\overline{y}^l_j = \frac{\partial x^i}{\partial \overline{x}^j} \frac{\partial \overline{y}^l}{\partial y^k} y^k_i + (x, y).$$

Taking into account this result, for $q = 2$ we have

$$\overline{y}^l_{j_1 j_2} = \frac{\partial x^{i_1}}{\partial \overline{x}^{j_1}} \frac{\partial x^{i_2}}{\partial \overline{x}^{j_2}} \frac{\partial \overline{y}^l}{\partial y^k} y^k_{i_1 i_2} + (x, y_1).$$

By induction, for any q we have

$$\overline{y}^l_{j_1 \ldots j_q} = \frac{\partial x^{i_1}}{\partial \overline{x}^{j_1}} \cdots \frac{\partial x^{i_q}}{\partial \overline{x}^{j_q}} \frac{\partial \overline{y}^l}{\partial y^k} y^k_{i_1 \ldots i_q} + (x, y_{q-1}).$$

Now S_qT^* can be pulled back over $J_{q-1}(\mathcal{E})$ under π^{-1}, and $V(\mathcal{E})$ can be pulled back over $J_{q-1}(\mathcal{E})$ under $(\pi^{q-1}_0)^{-1}$. However, as usual, pullbacks are not indicated for $\otimes$. $\square$

The resulting sequence

$$S_qT^* \otimes V(\mathcal{E}) \cdots \to J_q(\mathcal{E}) \xrightarrow{\pi^q_{q-1}} J_{q-1}(\mathcal{E})$$

$$(x, y_q) \longrightarrow (x, y_{q-1})$$

will be constantly used in the two following Chapters.

Lemma 4. *There is a canonical inclusion*

$$J_{q+1}(\mathcal{E}) \subset J_1(J_q(\mathcal{E})).$$

PROOF. We first describe this in local coordinates, using equations that make it possible to completely 'symmetrize' the indices involved:

$$\begin{cases} y^k_{\mu+1_i,j} - y^k_{\mu+1_j,i} = 0, & |\mu| = q-1, \\ y^k_{\mu,j} - y^k_{\mu+1_j} = 0, & 0 \le |\mu| \le q-1. \end{cases}$$

The first equations are *not* linearly independent (an exercise) and it is not at all easy to select a good basis. The number of equations of the first type is $m\frac{(n+q-2)!}{(q-1)!\,(n-1)!}\frac{n(n-1)}{2}$, while the number of (independent) equations of the second type is $mn\frac{(n+q-1)!}{(q-1)!\,n!}$. However, the total number of independent relations should be $\dim J_1(J_q(\mathcal{E})) - \dim J_{q+1}(\mathcal{E})$, i.e.

$$m\frac{(n+1)(n+q)!}{q!\,n!} - m\frac{(n+q+1)!}{(q+1)!\,n!} = m\frac{q(n+q)!}{(q+1)!\,(n-1)!}.$$

It follows that the number of independent equations of the first type is

$$m\frac{q(n+q)!}{(q+1)!\,(n-1)!} - m\frac{(n+q-1)!}{(q-1)!\,(n-1)!} = m\frac{q(q+n-1)!}{(q+1)!\,(n-2)!}.$$

We verify that this number is equal to

$$\begin{aligned}
&\dim J_1(J_q(\mathcal{E})) - \dim J_{q+1}(\mathcal{E}) - n\dim J_{q-1}(\mathcal{E}) \\
&= [n\dim J_q(\mathcal{E}) + \dim J_q(\mathcal{E})] - [\dim(S_{q+1}T^* \otimes V(\mathcal{E})) + \dim J_q(\mathcal{E})] - n\dim J_{q-1}(\mathcal{E}) \\
&= n\dim J_q(\mathcal{E}) - n\dim J_{q-1}(\mathcal{E}) - \dim(S_{q+1}T^* \otimes V(\mathcal{E})) \\
&= n\dim(S_q T^* \otimes V(\mathcal{E})) - \dim(S_{q+1}T^* \otimes V(\mathcal{E})) \\
&= nm\dim S_q T^* - m\dim S_{q+1}T^* \\
&= nm\frac{(q+n-1)!}{q!\,(n-1)!} - m\frac{(n+q)!}{(q+1)!(n-1)!} \\
&= m\frac{q(q+n-1)!}{(q+1)!\,(n-2)!},
\end{aligned}$$

a number which is strictly less than $mn\frac{(q+n-2)!}{2(q-1)!\,(n-2)!}$. $\square$

We thus discover that, although this lemma is often presented as simple, it has quite intrinsic combinatorial arguments behind it. After the next proposition we shall explain the intrinsic aspect of the preceding computations, since the underlying algebra will only be revealed in the next Chapter.

Proposition 5. *There is a canonical isomorphism* $V(J_q(\mathcal{E})) \simeq J_q(V(\mathcal{E}))$ *of vector bundles over* $J_q(\mathcal{E})$.

PROOF. For $q = 0$ we have $V(J_0(\mathcal{E})) = V(\mathcal{E}) = J_0(V(\mathcal{E}))$. To aid the reader, we give the computation for $q = 1$. From

$$\overline{y}_r^u \partial_i \varphi^r = \partial_i \psi^u(x, y) + \frac{\partial \psi^u}{\partial y^k}(x, y) y_i^k$$

we get

$$\overline{v}_r^u \partial_i \varphi^r = \frac{\partial^2 \psi^u}{\partial x^i \partial y^l}(x, y) v^l + \frac{\partial^2 \psi^u}{\partial y^k \partial y^l}(x, y) y_i^k v^l + \frac{\partial \psi^u}{\partial y^k}(x, y) v_i^k.$$

Now, from

$$\overline{v}^u = \frac{\partial \psi^u}{\partial y^k}(x, y) v^k$$

we get

$$v_{,r}^u \partial_i \varphi^r = \frac{\partial^2 \psi^u}{\partial x^i \partial y^k}(x, y) v^k + \frac{\partial^2 \psi^u}{\partial y^k \partial y^l}(x, y) y_i^l v^k + \frac{\partial \psi^u}{\partial y^k}(x, y) v_{,i}^k$$

and we can easily verify that the two formulas are identical, by interchanging a few indices and setting, in particular, $v_{,i}^k = v_i^k$.

By induction we finally prove that this result remains true for any order. Indeed, suppose that $V(J_{q-1}(\mathcal{E})) \simeq J_{q-1}(V(\mathcal{E}))$ is obtained by setting only $v_{,\mu}^k = v_\mu^k$ for $0 \leq |\mu| \leq q - 1$. Using Lemma 4, we have

$$V(J_1(J_{q-1}(\mathcal{E}))) \simeq J_1(V(J_{q-1}(\mathcal{E}))) \simeq J_1(J_{q-1}(V(\mathcal{E}))).$$

The vertical inclusion formula becomes

$$\begin{cases} v_{\mu+1_j,i}^k - v_{\mu+1_i,j}^k = 0, & |\mu| = q - 2, \\ v_{\mu,j}^k - v_{\mu+1_j}^k = 0, & 0 \leq |\mu| \leq q - 2, \end{cases}$$

with the same identification as before, leading therefore to a unique symbol 'v' for the restriction to $J_q(\mathcal{E})$. $\square$

As an interesting exercise, the reader may also try to prove the inclusion directly from the formulas [144, 146].

From Proposition 3 we immediately obtain the following commutative and exact diagram of vector bundles and affine bundles over $J_q(\mathcal{E})$:

$$0 \longrightarrow S_{q+1}T^* \otimes E \longrightarrow T^* \otimes J_q(E) \longrightarrow T^* \otimes J_q(E)/S_{q+1}T^* \otimes E \longrightarrow 0$$

$$0 \longrightarrow J_{q+1}(\mathcal{E}) \longrightarrow J_1(J_q(\mathcal{E}))$$

$$J_q(\mathcal{E}) \quad\quad === \quad\quad J_q(\mathcal{E})$$

Setting $C_1(E) = T^* \otimes J_q(E)/S_{q+1}T^* \otimes E$ and taking into account Corollary A.22, we obtain the following short exact sequence:

$$0 \longrightarrow J_{q+1}(\mathcal{E}) \longrightarrow J_1(J_q(\mathcal{E})) \longrightarrow C_1(E) \longrightarrow 0.$$

Again using Proposition 3, but now for vector bundles over $J_q(\mathcal{E})$, we obtain the following commutative and exact diagram, in which we have set $E = V(\mathcal{E})$ for simplicity:

$$
\begin{array}{ccccccc}
& 0 & & 0 & & 0 & \\
& \downarrow & & \downarrow & & \downarrow & \\
0 \longrightarrow & S_{q+1}T^* \otimes E & \longrightarrow & T^* \otimes S_q T^* \otimes E & \longrightarrow & T^* \otimes S_q T^* \otimes E \big/ S_{q+1}T^* \otimes E & \longrightarrow 0 \\
& \| & & \downarrow & & \downarrow & \\
0 \longrightarrow & S_{q+1}T^* \otimes E & \longrightarrow & T^* \otimes J_q(E) & \longrightarrow & C_1(E) & \longrightarrow 0 \\
& \downarrow & & \downarrow & & \downarrow & \\
& 0 & \longrightarrow & T^* \otimes J_{q-1}(E) & = & T^* \otimes J_{q-1}(E) & \longrightarrow 0 \\
& & & \downarrow & & \downarrow & \\
& & & 0 & & 0 &
\end{array}
$$

The exactness of the right column gives the intrinsic explanation of the combinatorial calculation in the proof of Lemma 4. However, only in the next Chapter can we explain the algebraic meaning of the top exact sequence, using Spencer δ-cohomology.

The third important result is again of a combinatorial nature.

Proposition 6. *For all* $r, s \geq 0$ *we have*

$$
J_{r+s}(\mathcal{E}) = J_r(J_s(\mathcal{E})) \cap J_{r+1}(J_{s-1}(\mathcal{E})) \subset \underbrace{J_1(\dots(J_1(\mathcal{E}))\dots)}_{r+s \ times}.
$$

PROOF. We decompose a set of $r + s$ integers into two subsets, using the following decomposition:

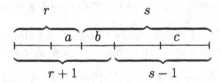

Permuting two integers belonging to the first r integers, or to the last s integers, does not change the corresponding component because of symmetry. Hence, the only problem in proving full symmetry lies in permuting two integers such as a and c. Let b be the integer in place $r + 1$. Permutation of a and c can be done as follows, in a manner respecting the full symmetry:

$$
\begin{aligned}
abc &\rightarrow bac && \text{in } J_{r+1}, \\
bac &\rightarrow bca && \text{in } J_s, \\
bca &\rightarrow cba && \text{in } J_{r+1}.
\end{aligned}
$$

Hence we have symmetry under permutation of indices. $\quad\square$

It follows from this Proposition that there is a canonical inclusion $J_{q+r}(\mathcal{E}) \subset J_r(J_q(\mathcal{E}))$, for all $q, r \geq 0$. However, a combinatorics similar to that of Lemma 4 cannot be given in this general situation.

We now give a first description of the (non-linear) Spencer operator.

To this end we consider the diagram following the proof of Proposition 5. Let f_{q+1} be a section of $J_{q+1}(\mathcal{E})$ projecting onto the section f_q of $J_q(\mathcal{E})$. By Lemma 4 it follows that both $j_1(f_q)$ and f_{q+1} can be regarded as sections of $J_1(J_q(\mathcal{E}))$ over the same section f_q of $J_q(\mathcal{E})$. By Proposition 3, $J_1(J_q(\mathcal{E}))$ is an affine bundle over $J_q(\mathcal{E})$, modeled on $T^* \otimes V(J_q(\mathcal{E})) = T^* \otimes J_q(V(\mathcal{E})) = T^* \otimes J_q(E)$ by Proposition 5. Hence, $j_1(f_q) - f_{q+1}$ is a section of $T^* \otimes J_q(E)$ and we may define an operator

$$D \colon J_{q+1}(\mathcal{E}) \to T^* \otimes J_q(E) \colon f_{q+1} \to j_1(f_q) - f_{q+1}.$$

In local coordinates we have

$$(Df_{q+1})^k_{\mu,i} = \partial_i f^k_\mu - f^k_{\mu+1_i}.$$

Definition 7. D is called the *Spencer operator*.

The kernel of D consists of the sections such that

$$f_{q+1} = j_1(f_q) = j_2(f_{q-1}) = \cdots = j_{q+1}(f).$$

Thus, we have

Proposition 8. *There is an exact sequence*

$$0 \longrightarrow \mathcal{E} \xrightarrow{\ j_{q+1}\ } J_{q+1}(\mathcal{E}) \xrightarrow{\ D\ } T^* \otimes J_q(E).$$

The following example will immediately show that D has two bad properties:

(1) D decreases the order of jets from $q + 1$ to q, although we would prefer 'stabilization' of the order of jets;

(2) D does not describe all the compatibility conditions for j_{q+1}, although we would prefer to have an operator exhibiting all such compatibility conditions.

Example 9. Consider the case $q = 0$:

$$j_1 \quad \begin{cases} f^k(x) = A^k(x), \\ \partial_i f^k(x) = A^k_i(x), \end{cases}$$

where on purpose we have written the righthand terms in a form avoiding any knowledge of jet theory (we could have written $(f^k(x), f^k_i(x))$ instead of $(A^k(x), A^k_i(x))$). The reader can readily see that the full set of compatibility conditions is:

$$\partial_i A^k(x) - A^k_i(x) = 0, \qquad \partial_i A^k_j(x) - \partial_j A^k_i(x) = 0,$$

and D reproduces the first set only. Again, D is not 'formally integrable' in the sense of the introduction, because the second set of first-order compatibility conditions follows from the first set by using crossed derivatives. Counting the *total number* of compatibility conditions, we find

$$(m \times n) + \left(m \times \frac{n(n-1)}{2} \right) = \frac{mn(n+1)}{2} = \dim C_1(E),$$

setting $C_1(E) = T^* \otimes J_1(E) / S_2 T^* \otimes E$ for $q = 1$ in the definition following the proof of Proposition 5.

The next Proposition explains and generalizes this result.

Proposition 10. *The sequence in Proposition 8 projects onto an exact sequence*

$$0 \longrightarrow \mathcal{E} \xrightarrow{j_q} J_q(\mathcal{E}) \xrightarrow{D_1} C_1(E),$$

where D_1 stabilizes the order of jets and exhibits all the compatibility conditions of j_q.

PROOF. We can lift a section f_q of $J_q(\mathcal{E})$ to a section f_{q+1} of $J_{q+1}(\mathcal{E})$ and apply D to get a section of $T^* \otimes J_q(E)$. We can project this section to a section of $C_1(E) = T^* \otimes J_q(E) / S_{q+1} T^* \otimes E$. Now, if f'_{q+1} is another lift of f_q, we notice that Df_{q+1} and Df'_{q+1} differ by a section of $S_{q+1} T^* \otimes E$, which is in fact $f'_{q+1} - f_{q+1}$ over f_q by Proposition 3. Hence Df_{q+1} and Df'_{q+1} have the same image in $C_1(E)$. The decomposition of $C_1(E)$ given in the diagram above Proposition 6 generalizes the decomposition of the total set of compatibility conditions into two subsets as in Example 9. $\square$

We conclude this Chapter with a technical proposition, exhibiting a *functorial* property of 'J_q'. To understand this property we have to define $J_q(\Phi)$ for a morphism Φ of fibered manifolds. We give this definition in two steps.

First, let Φ be a function defined on $J_q(\mathcal{E})$. If f_q is a section of $J_q(\mathcal{E})$, we may consider $\Phi \circ f_q(x)$, or $\Phi(x, f_q(x))$ whenever $y_q = f_q(x)$ in local coordinates. In this case we may define, *in local coordinates*, an operator d_i by the formula

$$d_i \Phi \circ j_{q+1}(f)(x) = \partial_i(\Phi \circ j_q(f)(x)) \qquad (\forall x \in X)$$

for any section f of $\mathcal{E}$ on an open set U containing x.

Definition 11. $d_i \Phi$ is called the *formal derivative* of Φ with respect to x^i.

In local coordinates again, we have

$$d_i = \frac{\partial}{\partial x^i} + y^k_{\mu+1_i} \frac{\partial}{\partial y^k_\mu} \qquad 0 \leq |\mu| \leq q,$$

and, using the definition, the reader can easily verify the following properties:

(1) $d_i(\Phi_1 + \Phi_2) = d_i \Phi_1 + d_i \Phi_2$;
(2) $d_i(\Phi_1 \cdot \Phi_2) = \Phi_2 \cdot d_i \Phi_1 + \Phi_1 \cdot d_i \Phi_2$;
(3) $d_i \circ d_j = d_j \circ d_i$.

As for ordinary derivatives, we can use the third property to define formal derivatives of higher order, and set $d_\nu = (d_1)^{\nu_1} \ldots (d_n)^{\nu_n}$ for any multi-index $\nu = (\nu_1, \ldots, \nu_n)$.

The second step is to give an intrinsic version of the preceding results. For this we define, for any morphism $\Phi \colon \mathcal{E} \to \mathcal{E}'$, the morphism $J_q(\Phi) \colon J_q(\mathcal{E}) \to J_q(\mathcal{E}')$ as: 'collect all the formal derivatives of Φ up to order q'.

In local coordinates, if $y' = \Phi(x, y)$ describes Φ, then $J_q(\Phi)$ is described by the formula

$$y_\nu^l = d_\nu \Phi(x, y_q), \qquad 1 \leq |\nu| \leq q.$$

The following example will allow us to revisit the definition of jet bundle.

Example 12. We compute $J_2(\Phi)$:

$$d_i \Phi = \partial_i \Phi + y_i^k \frac{\partial \Phi}{\partial y^k},$$

$$d_{ij} \Phi = \partial_{ij} \Phi + \frac{\partial^2 \Phi}{\partial x^j \partial y^k} y_i^k + \frac{\partial^2 \Phi}{\partial x^i \partial y^l} y_j^l + \frac{\partial \Phi}{\partial y^k} y_{ij}^k.$$

Here we recognize the righthand terms of the formulas that have been used to define $J_2(\mathcal{E})$ by patching transformations of jet coordinates.

The next proposition is the main step towards a key theorem in the following Chapter.

Proposition 13. *If*

$$\mathcal{E} \xrightarrow{\Phi} \mathcal{E}' \underset{f'' \circ \pi'}{\overset{\Psi}{\rightrightarrows}} \mathcal{E}''$$

is an exact sequence of fibered manifolds with respect to a section f'' of $\mathcal{E}''$, then

$$J_q(\mathcal{E}) \xrightarrow{J_q(\Phi)} J_q(\mathcal{E}') \underset{j_q(f'') \circ \pi'}{\overset{J_q(\Psi)}{\rightrightarrows}} J_q(\mathcal{E}'')$$

is an exact sequence of jet bundles with respect to the section $j_q(f'')$ of $J_q(\mathcal{E}'')$.

PROOF. By assumption we have $\Psi \circ \Phi \circ f \equiv f''$. Using local coordinates with $y = f(x)$ and $y' = \Phi \circ f(x)$, we can formally differentiate this identity q times, and obtain, using the chain rule, the identity $J_q(\Psi) \circ J_q(\Phi) \circ f_q \equiv j_q(f'')$ for any section f_q of $J_q(\mathcal{E})$, and not only for sections of the type $j_q(f)$. The proof then proceeds with an inductive chase in the following commutative diagram of affine bundles, by combining

Proposition A.21 and Exercise I.10:

$$S_q T^* \otimes E \xrightarrow{V(\Phi)} S_q T^* \otimes E' \xrightarrow{V(\Psi)} S_q T^* \otimes E''$$

$$\begin{array}{ccccc}
\vdots & & \vdots & & \vdots \\
J_q(\mathcal{E}) & \xrightarrow{J_q(\Phi)} & J_q(\mathcal{E}') & \overset{J_q(\Psi)}{\underset{j_q(f'')\circ\pi'}{\rightrightarrows}} & J_q(\mathcal{E}'') \\
\downarrow{\scriptstyle \pi_{q-1}^q} & & \downarrow{\scriptstyle \pi_{q-1}^q} & & \downarrow{\scriptstyle \pi_{q-1}^q} \\
J_{q-1}(\mathcal{E}) & \xrightarrow{J_{q-1}(\Phi)} & J_{q-1}(\mathcal{E}') & \overset{J_{q-1}(\Psi)}{\underset{j_{q-1}(f'')\circ\pi'}{\rightrightarrows}} & J_{q-1}(\mathcal{E}'')
\end{array}$$

Rather than giving all details of this chase, we indicate the squares that must be successively used:

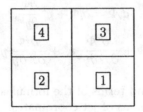

We point out that the exactness of the top row of the last diagram follows at once from the remark ending Example 12, which proves that it is the tensor by $S_q T^*$ of the sequence

$$E \xrightarrow{V(\Phi)} E' \xrightarrow{V(\Psi)} E'',$$

which is exact by assumption (see Definition A.20). $\square$

CHAPTER III

Nonlinear Systems

The purpose of this Chapter is to show that the preceding new techniques allow us to treat nonlinear systems of arbitrary order in exactly the same manner as we have treated linear systems, by using the vertical machinery. The main, central, result in this Chapter is a difficult theorem with no classical counterpart and having no classical proof in principle. The development of this Chapter is along the lines that have to be followed in any explicit example. The Janet example and many others will be treated in full detail at the end of this Chapter. For this reason we do not give examples in the theoretical part.

A. Prolongations

Our first objective is to transform a differential problem into a nondifferential problem by using the results of Chapter II, and to subsequently treat the latter problem by means of the linear algebra techniques presented in Chapter I.

The basic idea is very simple, once one is acquainted with jet theory. Indeed, any system of PDE involves a certain number of equations depending on a certain number of derivatives of unknown functions. Forgetting for the moment that we are looking for solutions, let us substitute jet coordinates for derivatives with the same indices. The PDE are now transformed into pure *equations* relating the jet coordinates to each other. Of course, because of the jacobian rank condition (the implicit function theorem), the resulting equations need not define a manifold that can be identified with a submanifold of a jet bundle of a certain order. Finally, instead of returning to derivatives, differentiating a certain number of times and substituting jet coordinates again, we will give a direct procedure that should appear naturally in this nonlinear framework. Of course, we would like this procedure to allow us to understand in an intrinsic way the 'old' approach presented in the Introduction.

Let $\mathcal{E}$ be a fibered manifold.

Definition 1. A *nonlinear system* of order q on $\mathcal{E}$ is a fibered submanifold $\mathcal{R}_q \subset J_q(\mathcal{E})$. Similarly, an *affine system* of order q on an affine bundle $\mathcal{E}$ is an affine subbundle $\mathcal{R}_q \subset J_q(\mathcal{E})$.

Let now E be a vector bundle over X. In this case we shall use upright capital letters as notation.

Definition 2. A *linear system* of order q on E is a vector subbundle $R_q \subset J_q(E)$.

Certain sections of $\mathcal{E}$ will have a special meaning.

Definition 3. A local (global) *solution* of $\mathcal{R}_q$ is a local (global) section of $\mathcal{E}$ such that $j_q(f)$ is a local (global) section of $\mathcal{R}_q$.

Definition 4. The *r-prolongation* of $\mathcal{R}_q$ is the *subset* (be careful!):

$$\rho_r(\mathcal{R}_q) = J_r(\mathcal{R}_q) \cap J_{q+r} \subset J_r(J_q(\mathcal{E})).$$

In this Definition we have used the functorial property of J_q saying that

$$\mathcal{R}_q \subset J_q(\mathcal{E}) \Rightarrow J_r(\mathcal{R}_q) \subset J_r(J_q(\mathcal{E})),$$

and the canonical inclusion $J_{q+r}(\mathcal{E}) \subset J_r(J_q(\mathcal{E}))$.

In the sequel, whenever there is no confusion possible with another system or family of systems, we will simply write $\mathcal{R}_{q+r}$ for $\rho_r(\mathcal{R}_q)$.

Of course, using that d_i commutes with d_j for all $i,j = 1,\ldots,n$, and noting that the first prolongation amounts to taking the formal derivatives of the equations, we obtain the following lemma.

Lemma 5. *If $\mathcal{R}_q \subset J_q(\mathcal{E})$ is a system of order q and $\rho_r(\mathcal{R}_q) \subset J_{q+r}(\mathcal{E})$ is a system of order $q+r$, then $\rho_{r+s}(\mathcal{R}_q) = \rho_s(\rho_r(\mathcal{R}_q)) = \mathcal{R}_{q+r+s}$, regarded as a subset, in general, of $J_{q+r+s}(\mathcal{E})$.*

Note that differentiation or elimination of the derivatives involved are the only natural intrinsic operations that can be performed on a system, that is:

one can only prolongate or project.

Definition 6. We define the following subsets:

$$\mathcal{R}_{q+r}^{(s)} = \pi_{q+r}^{q+r+s}(\mathcal{R}_{q+r+s}) \subseteq \mathcal{R}_{q+r}.$$

Indeed, prolongation can only bring *new* low-order equations and thus leads to a subset, being defined by *more* equations.

Definition 7. A system $\mathcal{R}_q \subset J_q(\mathcal{E})$ is said to be *formally compatible* (*transitive*) if $\pi: \mathcal{R}_{q+r} \to X$ is surjective for all $r \geq 0$ (respectively, $\pi_0^{q+r}: \mathcal{R}_{q+r} \to \mathcal{E}$ is surjective for all $r \geq 0$).

Of course, if prolongations or projections are not fibered manifolds (be careful, we have to verify two properties!), then one cannot use the implicit function theorem to separate principal jet coordinates from parametric jet coordinates as in the classical approach. Therefore the following definition is essential in practice, or, at least, it is essential to restrict the kind of equations considered.

Definition 8. A system $\mathcal{R}_q \subset J_q(\mathcal{E})$ is said to be *regular* if $\mathcal{R}_{q+r}^{(s)}$ is a fibered manifold, for all $r, s \geq 0$.

Later we shall see that, in applications, we need yet another property, for otherwise certain inductive constructions will not be effective.

We see that lack of surjectivity of the various projections prevents us (as in the Janet example) from knowing exactly how to distinguish principal from parametric jet coordinates, even if the system is regular. So, we see that the distinction between 'good' and 'bad' systems made in the Introduction is made clearer by the following definition

Definition 9. A system $\mathcal{R}_q \subset J_q(\mathcal{E})$ is said to be *formally integrable* if $\mathcal{R}_{q+r}$ is a fibered manifold for all $r \geq 0$ and if the maps $\pi_{q+r}^{q+r+s}: \mathcal{R}_{q+r+s} \to \mathcal{R}_{q+r}$ are epimorphisms for all $r, s \geq 0$.

Of course, this definition is not very useful in practice because it amounts to checking an infinite number of conditions. Hence, for applications it is crucial to develop an effective test. Surprisingly, such a test will involve a lot of linear algebra. Before describing it, we shall solve a problem which is useful in later applications.

Consider a system $\mathcal{R}_q \subset J_q(\mathcal{E})$ and a system $\mathcal{R}_{q+1} \subset J_{q+1}(\mathcal{E})$. Our problem is to recognize the inclusion $\mathcal{R}_{q+1} \subset \rho_1(\mathcal{R}_q)$.

Proposition 10. $\mathcal{R}_{q+1} \subset \rho_1(\mathcal{R}_q)$ *if and only if* $\pi_q^{q+1}(\mathcal{R}_{q+1}) \subset \mathcal{R}_q$ *and* $D\mathcal{R}_{q+1} \subset T^* \otimes R_q$.

PROOF. This is the first application of the Spencer operator. We have to prove that it restricts to $D: \mathcal{R}_{q+1} \to T^* \otimes R_q$, where, as usual, $R_q = V(\mathcal{R}_q)$. We shall use local coordinates for sections:

$$\mathcal{R}_q \qquad \Phi^\tau(x, f_q(x)) = 0, \qquad \forall x \in X,$$

$$\mathcal{R}_{q+1} \qquad \partial_i \Phi^\tau(x, f_q(x)) + f_{\mu+1_i}^k(x)\frac{\partial \Phi^\tau}{\partial y_\mu^k}(x, f_q(x)) = 0, \qquad \forall x \in X.$$

We differentiate the equations of $\mathcal{R}_q$ with respect to x^i:

$$\partial_i \Phi^\tau(x, f_q(x)) + \partial_i f_\mu^k(x)\frac{\partial \Phi^\tau}{\partial y_\mu^k}(x, f_q(x)) = 0, \qquad \forall x \in X.$$

By subtraction we obtain

$$(\partial_i f_\mu^k(x) - f_{\mu+1_i}^k(x))\frac{\partial \Phi^\tau}{\partial y_\mu^k}(x, f_q(x)) = 0, \qquad \forall x \in X.$$

By definition we have

$$R_q \qquad \frac{\partial \Phi^\tau}{\partial y_\mu^k}(x, y_q)v_\mu^k = 0, \qquad \forall(x, y_q) \in \mathcal{R}_q,$$

and we may pull this back over X using f_q^{-1}, according to the following diagram of reciprocal images:

$$
\begin{array}{ccc}
f_q^{-1}(V(\mathcal{R}_q)) & \longrightarrow & V(\mathcal{R}_q) \\
\downarrow & & \downarrow \\
X & \underset{\pi}{\overset{f_q}{\rightleftarrows}} & \mathcal{R}_q
\end{array}
$$

Hence we only have to identify the resulting equations. By the functorial property of the vertical symbol 'V':

$$\mathcal{R}_q \subset J_q(\mathcal{E}) \Rightarrow V(\mathcal{R}_q) \subset V(J_q(\mathcal{E})) \simeq J_q(V(\mathcal{E})).$$

The 'only if' part follows by reversing the arguments. Finally, the equations defining $\mathcal{R}_q$ surely are part of the equations defining $\mathcal{R}_{q+1}$. $\square$

As we have seen in the Introduction, systems of PDE are not always defined by sets of PDE, but sometimes as kernels.

Definition 11. A *nonlinear differential operator* of order q from $\mathcal{E}$ to $\mathcal{E}'$ is a map $\mathcal{D} = \Phi \circ j_q : \mathcal{E} \to \mathcal{E}'$ between the sets of (local) sections of $\mathcal{E}$ and $\mathcal{E}'$.

Hence, to define $\mathcal{D}$ we need a morphism $\Phi : J_q(\mathcal{E}) \to \mathcal{E}'$, and without loss of generality we may assume the Φ is an epimorphism. Having a section f' of $\mathcal{E}'$, we may define $\mathcal{R}_q$ by the kernel of Φ with respect to f', in the following double arrow sequence

$$0 \to \mathcal{R}_q \to J_q(\mathcal{E}) \underset{f' \circ \pi}{\overset{\Phi}{\rightrightarrows}} \mathcal{E}'.$$

Taking into account the functorial property of 'J_q', expressed by Proposition II.B.13, we can state:

Definition 12. The r-prolongation of Φ,

$$\rho_r(\Phi) : J_{q+r}(\mathcal{E}) \to J_r(\mathcal{E}'),$$

is defined by the composition

$$J_{q+r}(\mathcal{E}) \longrightarrow J_r(J_q(\mathcal{E})) \xrightarrow{J_r(\Phi)} J_r(\mathcal{E}').$$

As before, we may define $\rho_r(\mathcal{R}_q) = \mathcal{R}_{q+r}$ as the kernel of $\rho_r(\Phi)$ with respect to $j_r(f')$ in the following double arrow sequence:

$$0 \longrightarrow \mathcal{R}_{q+r} \longrightarrow J_{q+r}(\mathcal{E}) \underset{j_r(f') \circ \pi}{\overset{\rho_r(\Phi)}{\rightrightarrows}} J_r(\mathcal{E}').$$

Such double arrow sequences will be extremely convenient for studying PDE defining Lie pseudogroups, provided these are in Lie form.

We shall try to copy this kernel-like definition for a general system of order q. To this end, passing to the vertical level, we have an inclusion $R_q \subset J_q(E)$, which can be pulled back over $\mathcal{R}_q$ using the reciprocal image. Hence we may define a vector bundle F_0 over $\mathcal{R}_q$ by the short exact sequence

$$0 \longrightarrow R_q \longrightarrow J_q(E) \longrightarrow F_0 \longrightarrow 0.$$

For a linear system $R_q \subset J_q(E)$ over a vector bundle E, F_0 is a vector bundle over X; however, nothing has to be changed in the way of writing out the previous sequence.

Combining the inclusion $S_qT^* \otimes E \subset J_q(E)$ of vector bundles, pulled back over $\mathcal{R}_q$ using reciprocal images, with the epimorphism $J_q(E) \to F_0$ of vector bundles over $\mathcal{R}_q$, we may define a *family of vector spaces* M_q over $\mathcal{R}_q$ by the following short exact sequence

$$0 \longrightarrow M_q \longrightarrow S_qT^* \otimes E \longrightarrow F_0.$$

We combine these two sequences in a commutative diagram over $\mathcal{R}_q$:

$$
\begin{array}{ccccccccc}
 & & 0 & & 0 & & & & \\
 & & \downarrow & & \downarrow & & & & \\
0 & \longrightarrow & M_q & \longrightarrow & S_qT^* \otimes E & \longrightarrow & F_0 & & \\
 & & \downarrow & & \downarrow & & \| & & \\
0 & \longrightarrow & R_q & \longrightarrow & J_q(E) & \longrightarrow & F_0 & \longrightarrow & 0
\end{array}
$$

The reader may give some thoughts to the facts that, in general, the morphism $S_qT^* \otimes E \to F_0$ is *not* an epimorphism, and the dimension of M_q may vary from point to point over $\mathcal{R}_q$.

Definition 13. The symbol of $\mathcal{R}_q$ is the family $M_q = R_q \cap S_qT^* \otimes E$ of vector spaces over $\mathcal{R}_q$.

In local coordinates we have

$$M_q \qquad \frac{\partial \Phi^\tau}{\partial y_\mu^k}(x, y_q)v_\mu^k = 0, \qquad |\mu| = q, \quad (x, y_q) \in \mathcal{R}_q.$$

Now we prove that the symbol M_{q+r} of $\mathcal{R}_{q+r}$ depends only on M_q by a direct prolongation procedure. Indeed, in local coordinates we have

$$\mathcal{R}_{q+r} \qquad d_\nu \Phi^\tau = 0, \qquad 0 \le |\nu| \le r.$$

Noting that we have successively

$$d_i \Phi^\tau = \frac{\partial \Phi^\tau}{\partial y_\mu^k} y_{\mu+1_i}^k + (x, y_q) \qquad \text{with } |\mu| = q,$$

$$d_{ij} \Phi^\tau = \frac{\partial \Phi^\tau}{\partial y_\mu^k} y_{\mu+1_i+1_j}^k + (x, y_{q+1}) \qquad \text{with } |\mu| = q.$$

We obtain at once:

$$M_{q+r} \qquad \frac{\partial \Phi^\tau}{\partial y_\mu^k}(x, y_q)v_{\mu+\nu}^k = 0, \qquad |\mu| = q, \quad |\nu| = r, \quad (x, y_q) \in \mathcal{R}_q.$$

We arrive at the same formula by defining the r-prolongation $\rho_r(M_q)$ of M_q to be the kernel of the complete morphism

$$S_{q+r}T^* \otimes E \longrightarrow S_r T^* \otimes S_q T^* \otimes E \longrightarrow S_r T^* \otimes F_0,$$

where the right morphism is induced by the morphism $S_q T^* \otimes E \to F_0$ already defined. Hence, if no confusion is possible, we may identify $\rho_r(M_q)$ and M_{q+r} while keeping in mind that M_{q+r} depends on M_q only.

We are now ready for learning the new linear algebra techniques that are necessary for proceeding with the study of formal integrability.

B. Spencer Cohomology

We advice the reader to go ahead quite carefully in this Section, since it is delicate and intricate at both the theoretical and the combinatorial level. It is only after having solved the many examples proposed at the end of this Chapter that the reader will be familiar with the new tools. Since it does not seem possible to present the material in a different way, we closely follow [144], but put more emphasis on computational usefulness. Therefore we may say that the novelty of this book does not lie in the theoretical part, but rather in the possibility of learning the material by way of the explicit examples put together at the end of this Chapter.

Definition 1. The *Spencer δ-map*

$$\wedge^s T^* \otimes S_{q+1} T^* \otimes E \xrightarrow{\ \delta\ } \wedge^{s+1} T^* \otimes S_q T^* \otimes E$$

is defined by the composition

$$\wedge^s T^* \otimes S_{q+1} T^* \otimes E \longrightarrow \wedge^s T^* \otimes T^* \otimes S_q T^* \otimes E \longrightarrow \wedge^{s+1} T^* \otimes S_q T^* \otimes E,$$

where the epimorphism on the right is induced by the exterior product $\wedge^s T^* \otimes T^* \to \wedge^{s+1} T^*$.

In local coordinates we have the, essential, formula

$$(\delta\omega)^k_\mu = dx^i \wedge \omega^k_{\mu+1_i}.$$

It is obtained by choosing local coordinates $\omega^k_\mu = v^k_{\mu,I}\, dx^I$ with $dx^I = dx^{i_1} \wedge \cdots \wedge dx^{i_r}$ and $i_1 < \cdots < i_r$ for r-forms with values in $S_q T^* \otimes E$.

The usefulness of the δ-map is expressed in the following lemma.

Lemma 2. $\delta \circ \delta = 0$,

PROOF. $(\delta \circ \delta \omega)^k_\mu = dx^i \wedge (\delta\omega)^k_{\mu+1_i} = dx^i \wedge dx^j \wedge \omega^k_{\mu+1_i+1_j} = 0.$ $\quad\square$

The resulting sequences are called δ-*sequences*.
Leaving out E for simplicity, we give a few examples:

$$0 \longrightarrow S_2 T^* \xrightarrow{\ \delta\ } T^* \otimes T^* \xrightarrow{\ \delta\ } \wedge^2 T^* \longrightarrow 0$$

$$v_{i,j} \longrightarrow v_{i,j} - v_{j,i}$$

$$0 \longrightarrow S_3 T^* \xrightarrow{\ \delta\ } T^* \otimes S_2 T^* \xrightarrow{\ \delta\ } \wedge^2 T^* \otimes T^* \xrightarrow{\ \delta\ } \wedge^3 T^* \longrightarrow 0.$$

We leave it as an exercise to the reader to prove that these sequences are exact, and that therefore the alternating sum of the dimensions is zero, by Proposition I.4. More generally, we have

Proposition 3. *δ-sequences are exact.*

PROOF. The proof proceeds by induction with respect to n, starting with the case $n = 1$, which is trivial. For simplicity we do not write E, and split $\omega_\mu = v_{\mu,I} \, dx^I = \omega' + \omega''$ into two parts, where ω' does not contain dx^n. We also define the maps δ_i by the formula $(\delta_i \omega)_\mu = \omega_{\mu+1_i}$, so that we have $\delta = dx^i \wedge \delta_i$. Let $\omega \in \bigwedge^r T^* \otimes S_q T^*$ be such that $\delta \omega = 0$. From the case $n = 1$ we find $\tau \in \bigwedge^{r-1} T^* \otimes S_q T^*$ such that $\omega'' = dx^n \wedge \delta_n \tau$. Consider $\overline{\omega} = \omega - \delta\tau$. It does not contain dx^n. Now $\delta\overline{\omega} = \delta\omega - \delta^2\tau = 0$, and we can find σ such that $\overline{\omega} = \delta\sigma$ by the induction hypothesis for $n - 1$ variables. Finally, $\omega = \overline{\omega} + \delta\tau = \delta(\sigma + \tau)$, as required. $\square$

We now restrict δ-sequences to symbols.

Proposition 4. *The δ-map restricts to*

$$\bigwedge^s T^* \otimes M_{q+1} \, 2 \xrightarrow{\;\delta\;} \bigwedge^{s+1} T^* \otimes M_q \,.$$

PROOF. We set

$$\begin{array}{lll} M_q & A^{\tau\mu}_k v^k_\mu = 0, & |\mu| = q, \\ M_{q+1} & A^{\tau\mu}_k v^k_{\mu+1_i} = 0, & |\mu| = q. \end{array}$$

We do not indicate the point over which the symbol is considered. From Definition 1,

$$A^{\tau\mu}_k (\delta\omega)^k_\mu = dx^i \wedge (A^{\tau\mu}_k \omega^k_{\mu+1_i}) = 0,$$

and the restriction follows. $\square$

Although full δ-sequences are exact, there is no reason why their restrictions should be exact.

Definition 5. We let $H^s_{q+r} = H^s_{q+r}(M_q)$ be the cohomology of the δ-sequence at $\bigwedge^s T^* \otimes M_{q+r}$. It depends on M_q only. M_q is said to be *s-acyclic* if $H^1_{q+r} = \cdots = H^s_{q+r} = 0$ for all $r \geq 0$, *involutive* if it is n-acyclic, and of *finite type* if $M_{q+r} = 0$ for a certain $r \geq 0$.

Noticing that prolongations of the given equations are always quasilinear in the top-order jet coordinates, we see that a finite-type symbol makes it possible to obtain all the jet coordinates of a certain order from the jet coordinates of lower order. The following result will be useful in applications.

Proposition 6. *If M_q is involutive and of finite type, then $M_q = 0$.*

PROOF. We know that the corresponding δ-sequences are exact. Hence, if $M_{q+r} = 0$ for a certain $r \geq 0$, then

$$\cdots \xrightarrow{\;\delta\;} \bigwedge^{n-1} T^* \otimes M_{q+r} \xrightarrow{\;\delta\;} \bigwedge^n T^* \otimes M_{q+r-1} \longrightarrow 0$$

is such an exact δ-sequence. Hence, for $r \geq 1$ we have the exact sequence

$$0 \longrightarrow \wedge^n T^* \otimes M_{q+r-1} \longrightarrow 0,$$

and thus $M_{q+r-1} = 0$ too. By induction we arrive at $M_q = 0$. $\square$

Although a symbol need not be $2, 3, \ldots, n$-acyclic, it must be 0- and 1-acyclic, as stated in the following proposition.

Proposition 7. *Every symbol is 0- and 1-acyclic.*

PROOF. By the formula in the proof of Proposition 4 we have the following commutative diagram with exact rows:

$$
\begin{array}{ccccccc}
& & 0 & & 0 & & 0 \\
& & \downarrow & & \downarrow & & \downarrow \\
0 \longrightarrow & & M_{q+r+1} & \longrightarrow & S_{q+r+1}T^* \otimes E & \longrightarrow & S_{r+1}T^* \otimes F_0 \\
& & \downarrow{\scriptstyle\delta} & & \downarrow{\scriptstyle\delta} & & \downarrow{\scriptstyle\delta} \\
0 \longrightarrow & & T^* \otimes M_{q+r} & \longrightarrow & T^* \otimes S_{q+r}T^* \otimes E & \longrightarrow & T^* \otimes S_r T^* \otimes F_0 \\
& & \downarrow{\scriptstyle\delta} & & \downarrow{\scriptstyle\delta} & & \\
0 \longrightarrow & & \wedge^2 T^* \otimes S_{q+r-1}T^* \otimes E & = & \wedge^2 T^* \otimes S_{q+r-1}T^* \otimes E & &
\end{array}
$$

The proposition therefore follows from the snake Theorem (Chapter I). $\square$

Since there is no test known for checking s-acyclicity, $s = 2, 3, \ldots, n-1$, in general, we now turn to a careful technical study of involutivity. We shall both *improve* and *simplify* the exposition as compared to previous books [144, 146], in order to enable the reader to use computer algebra. For this, *in a given system of local coordinates*, we define the subspaces

$$(S_q T^*)^i = \{v \in S_q T^* \mid \delta_1 v = 0, \ldots, \delta_i v = 0\},$$

with the conventions $(S_q T^*)^0 = S_q T^*$ and $(S_q T^*)^n = 0$. More generally, we can say that a jet coordinate v_μ^k belongs to *class 1* if $\mu_1 \neq 0$ and to *class i* if $\mu_1 = \cdots = \mu_{i-1} = 0$, $\mu_i \neq 0$. Hence we see that $(S_q T^*)^i$ *is obtained by equating to zero all the* v_μ^k *of class* $1, \ldots, i$. We similarly define $(M_q)^i = M_q \cap (S_q T^*)^i \otimes E = M_q \cap (S_q T^* \otimes E)^i$. We now order the v_μ^k lexicographically. We can solve the linear system defining M_q over a point of $\mathcal{R}_q$ with respect to the maximum number of components of class n, and replace these in the other equations so that we no longer deal with components of class n but only with those of classes $1, \ldots, n-1$. We can solve the remaining equations with respect to the maximum number of components of class $n-1$, replace these in the other equations, etc., until we have equations containing components of class 1 only, which we then solve with respect to the maximum number of them. Classifying the resulting solved equations according to their leading terms, we obtain in this way the following scheme:

β_q^n equations of class n | $1 \dots\dots\dots n$
$\dots\ \dots\ \dots$
β_q^i equations of class i | $1 \dots i\bullet \dots \bullet$
$\dots\ \dots\ \dots$
β_q^1 equations of class 1 | $1\bullet \dots\dots \bullet$

We have the following exact sequences:

$$0 \longrightarrow (M_q)^i \longrightarrow (M_q)^{i-1} \xrightarrow{\delta_i} (S_{q-1}T^* \otimes E)^{i-1}.$$

We shall now proceed backwards. Indeed, if we want to determine $(M_q)^1$, we have to put equal to zero all components of class 1 in the equations defining M_q, and this will take out the β_q^1 equations of class 1, i.e.:

$$\dim(M_q)^1 = \dim(S_q T^* \otimes E)^1 - (\beta_q^2 + \cdots + \beta_q^n).$$

More generally,

$$\dim(M_q)^i = \dim(S_q T^* \otimes E)^i - (\beta_q^{i+1} + \cdots + \beta_q^n).$$

Of course, the number of components of *strict* class i is equal to $\dim(S_{q-1}T^* \otimes E)^{i-1} = m(q + n - i - 1)!/(q-1)!\,(n-i)!$, and we therefore obtain the inequalities

$$0 \leq \beta_q^i \leq m\frac{(q+n-i-1)!}{(q-1)!\,(n-i)!}.$$

Similarly, we can define the numbers

$$\alpha_q^i = \dim(M_q)^{i-1} - \dim(M_q)^i,$$

and we have the relations

$$0 \leq \alpha_q^i \leq m\frac{(q+n-i-1)!}{(q-1)!\,(n-i)!},$$

$$\alpha_q^i + \beta_q^i = m\frac{(q+n-i-1)!}{(q-1)!\,(n-i)!},$$

which can be deduced from the kernel/cokernel sequence for δ_i in the preceding sequence, using results of Chapter I. Also, by adding the equalities defining α_q^i we obtain

$$\dim M_q = \alpha_q^1 + \cdots + \alpha_q^n.$$

Finally, by Proposition 4 we have the following exact sequences:

$$0 \longrightarrow (M_{q+1})^i \longrightarrow (M_{q+1})^{i-1} \xrightarrow{\delta_i} (M_q)^{i-1},$$

and we deduce the inequalities

$$\dim(M_{q+1})^{i-1} - \dim(M_{q+1})^i \leq \dim(M_q)^{i-1}.$$

Adding these inequalities we arrive at the following key inequality

$$\dim M_{q+1} \le \alpha_q^1 + 2\alpha_q^2 + \cdots + n\alpha_q^n.$$

We now give another interpretation of this inequality, which is quite important for practical calculations. Indeed, looking at the preceding scheme, we give the following definition

Definition 8. Independent variables whose index is effectively written down are said to be *multiplicative variables*, while independent variables indexed by dots are said to be *nonmultiplicative variables*.

Hence, $x^1, \ldots, x^n$ are multiplicative variables for all equations of class n, while $x^1, \ldots, x^i$ are multiplicative variables for all equations of class i, and only x^1 is a multiplicative variable for all equations of class 1.

Now, an easy inspection of the equations of the various classes proves that the prolongations with respect to multiplicative variables have different leading terms of order $q + 1$, and are thus linearly independent equations among all the equations defining M_{q+1}. Indeed, if μ is of class i, then $\mu + 1_r$ with $r \le i$ is of class $r \le i$. Hence it can only be the prolongation of a component of class r with a multiplicative variable of index $\ge r$. This is a contradiction, unless $r = i$, in which case there is nothing to prove.

It follows that there is an inequality

$$\dim M_{q+1} \le m \frac{(q+n)!}{(q+1)!\,(n-1)!} - (\beta_q^1 + 2\beta_q^2 + \cdots + n\beta_q^n).$$

The right term is equal to

$$m \frac{(q+n)!}{(q+1)!\,(n-1)!} - m \sum_{i=1}^{n} i \frac{(q+n-i-1)!}{(q-1)!\,(n-i)!} + \alpha_q^1 + 2\alpha_q^2 + \cdots + n\alpha_q^n,$$

i.e. to $\alpha_q^1 + 2\alpha_q^2 + \cdots + n\alpha_q^n$, and we recover the previous inequality.

Collecting the results, we have

Proposition 9. *There is a coordinate system, called a δ-regular coordinate system, in which the following properties are equivalent.*

1) M_q *is involutive.*
2) *The following sequences are exact:*

$$0 \longrightarrow (M_{q+1})^i \longrightarrow (M_{q+1})^{i-1} \xrightarrow{\;\delta_i\;} (M_q)^{i-1} \longrightarrow 0.$$

3) $\dim M_{q+1} = \alpha_q^1 + 2\alpha_q^2 + \cdots + n\alpha_q^n.$
4) *Prolongation with respect to dots does not bring new equations.*

PROOF. The fact that 1) implies 2) is rather technical and not useful in applications. Hence we shall only prove the other equivalences. They can all be regarded as (equivalent) tests for checking involutivity of M_q.

The equivalence of 3) and 4) has been proved above. The equivalence between 2) and 3) follows from the fact that we have summed inequalities, and the resulting inequality becomes an equality if and only if each of the summand inequalities is an equality or, equivalently, if and only if the sequences are exact. □

If M_q is involutive, then so is M_{q+r} (by definition). The reader may try to prove the following corollary.

Corollary 10 (Hilbert polynomial).

$$\dim M_{q+r} = \alpha_q^1 + \cdots + \frac{(r+i-1)!}{r!\,(i-1)!}\alpha_q^i + \cdots + \frac{(r+n-1)!}{r!\,(n-1)!}\alpha_q^n.$$

Guided by this formula, we make the following definition:

Definition 11. The numbers $\alpha_q^1, \ldots, \alpha_q^n$ participating in the previous Proposition are called the *characters* of M_q. They are intrinsically defined.

We shall now study the effect of a linear change of coordinates on the characters. The many examples presented at the end of this Chapter illustrate that, in effective calculations, this is a delicate, yet important, point.

Since the characters depend on the coordinate system, to get as close as possible to $\dim M_{q+1}$ we may choose the coordinate system in such a way that $\alpha_q^n, \alpha_q^{n-1} + \alpha_q^n, \ldots, \alpha_q^2 + \cdots + \alpha_q^n, \alpha_q^1 + \cdots + \alpha_q^n$ be minimal. Since the last number equals $\dim M_q$, this amounts to choosing $\alpha_q^1, \alpha_q^1 + \alpha_q^2, \ldots, \alpha_q^1 + \cdots + \alpha_q^n$ maximal. Since $\alpha_q^1 + \cdots + \alpha_q^n = \dim M_q - \dim(M_q)^1$, this amounts to choosing $\dim(M_q)^1$ minimal, i.e. the defining system under consideration should have maximal rank.

Performing an invertible change of coordinates $\overline{x}^j = a_i^j x^i$ amounts to changing the coordinates and successive annihilation of the components of classes $1, \ldots, n$, i.e. to taking 'a' as indeterminates and to consideration of the systems

$$(M_q)^1 \quad \begin{cases} A_k^{\tau\mu} v_\mu^k = 0, & |\mu| = q, \\ v_{ri_2\ldots i_q} a_1^r = 0, \\ \cdots\cdots \\ v_{ri_2\ldots i_q} a_i^r = 0. \end{cases}$$

The various maximum rank conditions will impose algebraic inequalities on (a_i^j) that must be satisfied by 'good' changes leading to δ-regular coordinates. In actual practice this is the hard step, and one may eventually use a random choice of matrix entries, provided the matrix is invertible.

We finally notice that each additional set of equations of the same type cannot increase the total rank more than the preceding set, and we obtain

$$\dim M_q - \dim(M_q)^1 \geq \dim(M_q)^1 - \dim(M_q)^2 \geq \cdots \geq \dim(M_q)^{n-1} \geq 0,$$

i.e.

Lemma 12. $\alpha_q^1 \geq \alpha_q^2 \geq \cdots \geq \alpha_q^n \geq 0.$

After these considerations, it remains to study the dependence of the characters on the choice of the point of $\mathcal{R}_q$ over which we study the symbol M_q and its prolongations. In particular, it should be nice to know conditions under which the characters do not depend on the point or, equivalently, the $(M_q)^i$ are vector bundles over $\mathcal{R}_q$. By Corollary 10, in the case of involution the characters can be determined from $\dim M_q, \ldots, \dim M_{q+n-1}$, and, in any case, if the characters do not depend on the point of $\mathcal{R}_q$, then M_{q+r} is a vector bundle over $\mathcal{R}_q$ for $r \geq 0$. The following important proposition solves this problem [72, 172].

Proposition 13. *If* M_q *is 2-acyclic and* M_{q+1} *is a vector bundle over* $\mathcal{R}_q$, *then* M_{q+r} *is a vector bundle over* $\mathcal{R}_q$, $\forall r \geq 1$.

PROOF. We define a vector bundle F_1 over $\mathcal{R}_q$ by the following exact kernel/cokernel sequence:

$$0 \longrightarrow M_{q+1} \longrightarrow S_{q+1}T^* \otimes E \longrightarrow T^* \otimes F_0 \longrightarrow F_1 \longrightarrow 0.$$

We now have the following commutative diagram, with exact columns:

$$
\begin{array}{ccccccccc}
 & 0 & & 0 & & 0 & & 0 & \\
 & \downarrow & & \downarrow & & \downarrow & & \downarrow & \\
0 \longrightarrow & M_{q+r+1} & \longrightarrow & S_{q+r+1}T^* \otimes E & \longrightarrow & S_{r+1}T^* \otimes F_0 & \longrightarrow & S_r T^* \otimes F_1 & \longrightarrow 0 \\
 & \Big\downarrow{\scriptstyle\delta} & & \Big\downarrow{\scriptstyle\delta} & & \Big\downarrow{\scriptstyle\delta} & & \Big\downarrow{\scriptstyle\delta} & \\
0 \longrightarrow & T^* \otimes M_{q+r} & \longrightarrow & T^* \otimes S_{q+r}T^* \otimes E & \longrightarrow & T^* \otimes S_r T^* \otimes F_0 & \longrightarrow & T^* \otimes S_{r-1}T^* \otimes F_1 & \longrightarrow 0 \\
 & \Big\downarrow{\scriptstyle\delta} & & \Big\downarrow{\scriptstyle\delta} & & \Big\downarrow{\scriptstyle\delta} & & & \\
0 \longrightarrow & \Lambda^2 T^* \otimes M_{q+r-1} & \longrightarrow & \Lambda^2 T^* \otimes S_{q+r-1}T^* \otimes E & \longrightarrow & \Lambda^2 T^* \otimes S_{r-1}T^* \otimes F_0 & & & \\
 & \Big\downarrow{\scriptstyle\delta} & & \Big\downarrow{\scriptstyle\delta} & & & & & \\
0 \longrightarrow & \Lambda^3 T^* \otimes S_{q+r-2}T^* \otimes E & \Longrightarrow & \Lambda^3 T^* \otimes S_{q+r-2}T^* \otimes E & & & & & \\
 & \uparrow & & \uparrow & & & & & \\
 & 0 & & 0 & & & & &
\end{array}
$$

By induction we may assume that $M_{q+1}, \ldots, M_{q+r}$ are vector bundles over $\mathcal{R}_q$. Since $M_{q+r+1} = \ker(S_{q+r+1}T^* \otimes E \to S_{r+1}T^* \otimes F_0)$, it follows that $\dim M_{q+r+1}$ is an upper semicontinuous function on $\mathcal{R}_q$, i.e. it has minimal 'generic' value. The same is true for $\ker(S_{r+1}T^* \otimes F_0 \to S_r T^* \otimes F_1)$. However, by the exactness of the top row of the diagram (stemming from the exactness of the vertical left column under the assumption of 2-acyclicity), the sum of the dimensions of the two kernels is constant, and equal to $\dim(S_{q+r+1}T^* \otimes E) = m(q+r+n)!/(q+r+1)!\,(n-1)!$. Hence the two kernels are vector bundles over $\mathcal{R}_q$; in particular, M_{q+r+1} is a vector bundle over $\mathcal{R}_q$. $\square$

Remark 14. If M_q is also involutive, then $M_{q+1}, \ldots, M_{q+n}$ are vector bundles over $\mathcal{R}_q$ and it follows that the characters are uniquely defined over $\mathcal{R}_q$, and M_q is a vector bundle over $\mathcal{R}_q$ because $\dim M_q = \alpha_q^1 + \cdots + \alpha_q^n$. Alternatively, we can count the dimensions, and notice that their alternate sum equals zero, in the following exact sequence:

$$0 \longrightarrow M_{q+n} \xrightarrow{\ \delta\ } T^* \otimes M_{q+n-1} \xrightarrow{\ \delta\ } \cdots \xrightarrow{\ \delta\ } \wedge^{n-1}T^* \otimes M_{q+1} \xrightarrow{\ \delta\ } \wedge^n T^* \otimes M_q \longrightarrow 0.$$

Since $\dim \wedge^n T^* \otimes M_q = \dim M_q$, we obtain a manner of expressing $\dim M_q$ as a function of $\dim M_{q+1}, \ldots, \dim M_{q+n}$.

These results are extremely important in practice since they give a finite test for deciding about infinitely many rank conditions.

It remains to find out what to do if the preceding tests show that the symbol is not involutive. The answer is given by the following theorem.

Theorem 15. *For r large enough, M_{q+r} becomes involutive.*

We do not give a proof of this, rather technical, Theorem, and refer the reader to [144]. In any case, the proof is not at all constructive. This Theorem can be complemented by a corollary, giving an upper bound for the number of prolongations needed, as a function of the triplet (m, n, q). However, this bound is not effective at all because usually it is far too high to be implemented.

Corollary 16. *A bound q' of involutivity can be determined by the following inductive formulas:*

$$\begin{cases} q'(0, m, 1) = 0, \\ q'(n, m, 1) = m\binom{q'(n-1,m,1)+n}{n-1} + q'(n-1, m, 1) + 1, \\ q'(n, m, q) = q'\left(n, m\binom{q+n-1}{n}, 1\right). \end{cases}$$

C. Formal Integrability

In the effective study of formal integrability, in the sense of Definition A.9, we have to consider two points at the same time:

- rank conditions for obtaining fibered manifolds;
- surjectivity of the restricted prolongations.

The purpose of this Section is to show how to cope with these conditions, and to find out what to do when they do not hold. The hard step in this is a surprising theorem combing the results of the two previous sections.

To motivate its formulation, let us consider the following commutative diagram of projections, in which we assume that $\mathcal{R}_q$ is regular:

$$
\begin{array}{ccc}
\mathcal{R}_{q+1} & \xleftarrow{\pi_{q+1}^{q+r+1}} & \mathcal{R}_{q+r+1} \\
{\scriptstyle \pi_q^{q+1}}\downarrow & & \downarrow{\scriptstyle \pi_{q+r}^{q+r+1}} \\
\mathcal{R}_q^{(1)} & \xleftarrow{\pi_q^{q+r}} & \mathcal{R}_{q+r}^{(1)} \\
\cap & & \cap \\
\mathcal{R}_q & \xleftarrow{\pi_q^{q+r}} & \mathcal{R}_{q+r}
\end{array}
$$

Going backwards along the arrows, we find, of course, $\rho_r(\mathcal{R}_q) = \mathcal{R}_{q+r}$, $\rho_r(\mathcal{R}_{q+1}) = \mathcal{R}_{q+r+1}$, in accordance with Lemma A.5. We also have:

$$
\begin{aligned}
\mathcal{R}_{q+r}^{(1)} = \pi_{q+r}^{q+r+1}(\mathcal{R}_{q+r+1}) &= \pi_{q+r}^{q+r+1}(J_r(\mathcal{R}_{q+1}) \cap J_{q+r+1}(\mathcal{E})) \\
&\subseteq \pi_{q+r}^{q+r+1}(J_r(\mathcal{R}_{q+1})) \cap J_{q+r}(\mathcal{E}) \\
&= J_r(\mathcal{R}_q^{(1)}) \cap J_{q+r}(\mathcal{E}) \\
&= \rho_r(\mathcal{R}_q^{(1)}),
\end{aligned}
$$

because the projection of an intersection is contained in the projections of the intersection terms, and we have the following diagram:

$$
\begin{array}{ccc}
J_{q+r+1}(\mathcal{E}) & \longrightarrow & J_r(J_{q+1}(\mathcal{E})) \\
{\scriptstyle \pi_{q+r}^{q+r+1}}\downarrow & & \downarrow{\scriptstyle J_r(\pi_q^{q+1})} \\
J_{q+r}(\mathcal{E}) & \longrightarrow & J_r(J_q(\mathcal{E}))
\end{array}
$$

One can also simply say that the equations defining $\mathcal{R}_{q+1}$ include the equations defining $\mathcal{R}_q^{(1)}$, which are of order q. Hence, the equations defining $\mathcal{R}_{q+r+1}$ include the r-prolongation of the equations defining $\mathcal{R}_q^{(1)}$, which are of order $q+r$. It follows that the set of equations defining $\mathcal{R}_{q+r}^{(1)}$ includes the set of equations defining $\rho_r(\mathcal{R}_q^{(1)})$, and therefore $\mathcal{R}_{q+r}^{(1)} \subseteq \rho_r(\mathcal{R}_q^{(1)})$.

The only problem left to investigate is that of equality. The answer is given by the following theorem, which cannot be proved by classical techniques. It is even difficult to verify it on concrete examples (see Section D).

Theorem 1. Let $\mathcal{R}_q \subset J_q(\mathcal{E})$ be a system of order q on $\mathcal{E}$ such that $\mathcal{R}_{q+1}$ is a fibered manifold. If M_{q+1} is a vector bundle over $\mathcal{R}_q$ and M_q is 2-acyclic, then $\rho_r(\mathcal{R}_q^{(1)}) = \mathcal{R}_{q+r}^{(1)}$, $\forall r \geq 0$.

PROOF. The proof is delicate, since it uses a quite unusual tridimensional chase. For this reason we shall give full details and we advise the reader to follow the proof on any one of the examples given in the next Section.

First of all, using an implicit function theorem, after inspection of the jacobian matrix of the equations defining $\mathcal{R}_{q+1}$ we readily deduce that $\mathcal{R}_q^{(1)}$ is a fibered manifold with $\dim M_{q+1} = \dim \mathcal{R}_{q+1} - \dim \mathcal{R}_q^{(1)}$. The result of the Theorem is coherent with Definition A.4 concerning prolongations.

Since M_{q+1} is a vector bundle over $\mathcal{R}_q$, we may define the vector bundle F_1 over $\mathcal{R}_q$ by the following exact sequence:

$$0 \longrightarrow M_{q+1} \longrightarrow S_{q+1}T^* \otimes E \longrightarrow T^* \otimes F_0 \longrightarrow F_1 \longrightarrow 0,$$

and by Proposition B.13 we have the following exact sequence of vector bundles over $\mathcal{R}_q$:

$$0 \longrightarrow M_{q+r+1} \longrightarrow S_{q+r+1}T^* \otimes E \longrightarrow S_{r+1}T^* \otimes F_0 \longrightarrow S_r T^* \otimes F_1,$$

because M_q is 2-acyclic. In this sequence the morphisms are the algebraic r-prolongations of the morphisms previously defined in the exact sequence for $r = 0$.

The idea of the proof is to study separately the point sets $\mathcal{R}_{q+r}^{(1)}$ and $\rho_r(\mathcal{R}_q^{(1)})$, and to prove that they coincide. The main tool will be a tridimensional chase in the diagram of affine bundles:

and its specialization for $r = 0$:

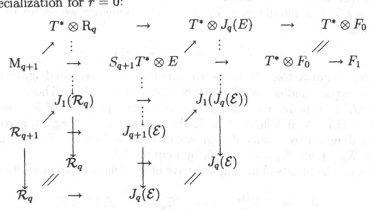

Chasing in the second diagram we obtain the exact sequence

$$\mathcal{R}_{q+1} \xrightarrow{\pi_q^{q+1}} \mathcal{R}_q \underset{0}{\overset{\kappa}{\longrightarrow}} F_1,$$

where the kernel of the double arrow is taken with respect to the zero section of the vector bundle F_1 over $\mathcal{R}_q$ and the map κ is considered as a section of F_1 over $\mathcal{R}_q$.

To this end we lift any point $a \in \mathcal{R}_q$ to a point $b \in J_1(\mathcal{R}_q)$ as well as to a point $c \in J_{q+1}(\mathcal{E})$. The images d of b and e of c in $J_1(J_q(\mathcal{E}))$ have the same projection, a, in $J_q(\mathcal{E})$. Since $J_1(J_q(\mathcal{E}))$ is an affine bundle over $J_q(\mathcal{E})$ modeled on $T^* \otimes J_q(E)$, the difference $d - e \in T^* \otimes J_q(E)$, and we may project it to F_1, using the composition

$$T^* \otimes J_q(E) \longrightarrow T^* \otimes F_0 \longrightarrow F_1.$$

If we choose other lifts $b' \in J_q(\mathcal{R}_q)$ and $c' \in J_{q+1}(\mathcal{E})$, then the image d' of b' is such that $d' - d \in \mathrm{im}(T^* \otimes \mathrm{R}_q \to T^* \otimes J_q(E))$. Similarly, the image e' of e is such that $e' - e \in \mathrm{im}(S_{q+1}T^* \otimes E \to T^* \otimes J_q(E))$. It follows that $d' - e' = (d - e) + (d' - d) - (e' - e)$ has the same projection in F_1, a result proving that we have a well-defined map $\kappa \colon \mathcal{R}_q \to F_1$. This map is called the *curvature* of $\mathcal{R}_q$. If $a \in \ker \kappa$, we can find $\alpha \in T^* \otimes \mathrm{R}_q$ and $\beta \in S_{q+1}T^* \otimes E$ such that $d - e = \mathrm{im}\, \alpha - \mathrm{im}\, \beta$, because the upper sequences are exact by construction. It follows that $d - \mathrm{im}\, \alpha = e - \mathrm{im}\, \beta$, and hence $\mathrm{im}(b - \alpha) = \mathrm{im}(c - \beta)$, i.e. a change of lifts gives a point $b - \alpha = c - \beta \in \mathcal{R}_{q+1}$ projecting to a. We have thus obtained the following exact sequence:

$$0 \longrightarrow \mathcal{R}_q^{(1)} \longrightarrow \mathcal{R}_q \underset{0}{\overset{\kappa}{\longrightarrow}} F_1$$

of fibered manifolds and vector bundles.

We can now repeat the construction of the curvature within the first diagram, taking into consideration the fact that here $\mathcal{R}_{q+r}$ and $\mathcal{R}_{q+r+1}$ are, in general, merely point sets in $J_{q+r}(\mathcal{E})$ and $J_{q+r+1}(\mathcal{E})$, respectively. Nevertheless, we can similarly obtain the exact sequence

$$\mathcal{R}_{q+r+1} \xrightarrow{\pi_{q+r}^{q+r+1}} \mathcal{R}_{q+r} \underset{0}{\overset{\kappa_r}{\longrightarrow}} S_r T^* \otimes F_1,$$

where the kernel of the double arrow is now taken with respect to the zero section of $S_r T^* \otimes F_1$ as a vector bundle over $\mathcal{R}_q$. From this sequence we derive the following exact sequence:

$$0 \longrightarrow \mathcal{R}_{q+r}^{(1)} \longrightarrow \mathcal{R}_{q+r} \overset{\kappa_r}{\underset{0}{\longrightarrow}} S_r T^* \otimes F_1.$$

The key step is to notice that all spaces participating in the second diagram are fibered manifolds or vector bundles over $\mathcal{R}_q$, by the assumptions of the Theorem. Hence we may apply' 'J_r' to this diagram and use the functorial property of it already proved in Proposition II.B.13. It follows that the first commutative diagram is contained in the resulting diagram by means of various canonical inclusions. We also notice that $\mathcal{R}_{q+r+1} \subset J_r(\mathcal{R}_{q+1})$ and $\mathcal{R}_{q+r} \subset J_r(\mathcal{R}_q)$, by Lemma A.5.

Combining all the results obtained, we arrive at the following commutative diagram:

$$
\begin{array}{ccccc}
0 \longrightarrow & \mathcal{R}_{q+r}^{(1)} & \longrightarrow & \mathcal{R}_{q+r} & \overset{\kappa_r}{\underset{0}{\longrightarrow}} & S_r T^* \otimes F_1 \\
& \downarrow & & \downarrow & & \downarrow \\
0 \longrightarrow & J_r(\mathcal{R}_q^{(1)}) & \longrightarrow & J_r(\mathcal{R}_q) & \overset{J_r(\kappa)}{\underset{0}{\longrightarrow}} & J_r(F_1)
\end{array}
$$

However, we have

$$
\begin{aligned}
\rho_r(\mathcal{R}_q^{(1)}) &= J_r(\mathcal{R}_q^{(1)}) \cap J_{q+r}(\mathcal{E}) \\
&= J_r(\mathcal{R}_q^{(1)}) \cap J_r(\mathcal{R}_q) \cap J_{q+r}(\mathcal{E}) \\
&= J_r(\mathcal{R}_q^{(1)}) \cap \mathcal{R}_{q+r},
\end{aligned}
$$

and a chase in the preceding diagram, using the injectivity of the right vertical arrow, proves that this intersection is equal to $\mathcal{R}_{q+r}^{(1)}$ as a set. $\square$

Definition 2. κ_r is called the *r-curvature* of $\mathcal{R}_q$, and $\kappa = \kappa_0$ is simply called the *curvature* of $\mathcal{R}_q$.

To aid the reader, we shall give local coordinates for κ, and point out that it is very difficult to work out local coordinates for κ_r and/or to give an interpretation of these. To this end we need the following lemma.

Lemma 3. $\forall f \in C^\infty(U)$ *with* $U \subset \mathbb{R}^n$ *and* $f(0) = 0$:

$$f(x) = x^i \int_0^1 \partial_i f(tx)\, dt.$$

PROOF. $f(x) = f(x) - f(0) = [f(tx)]_0^1 = \int_0^1 \frac{d}{dt} f(tx)\, dt$ and $\frac{d}{dt} f(tx) = x^i \partial_i f(tx)$. $\square$

The idea for constructing κ is to notice that the first prolongation $d_i \Phi^\tau = 0$ of the equations $\Phi^\tau = 0$ defining $\mathcal{R}_q$ become quasilinear in the jets of strict order $q + 1$. Since F_1 is defined by a ker/coker exact sequence, we recover Cramer's rule of Chapter I. Hence we have to look for $\dim F_1$ linearly independent combinations that make it possible to eliminate the jet coordinates of strict order $q + 1$, say of the form $A_\tau^{\alpha i}(x, y_q)\, d_i \Phi^\tau$. Since F_1 is a vector bundle over $\mathcal{R}_q$, these expressions can be considered as functions on $\mathcal{R}_q$. If they do not vanish on $\mathcal{R}_q$, they will define $\mathcal{R}_q^{(1)}$. Otherwise, by the preceding Lemma and the assumption that $\mathcal{R}_q$ is a fibered manifold, we can find $\dim F_1$ identities of the form

$$A_\tau^{ai}(x, y_q)\, d_i \Phi^\tau + B_\tau^a(x, y_q)\Phi^\tau \equiv 0, \qquad \forall (x, y_q) \in J_q(\mathcal{E}),$$

where $i = 1, \ldots, n;\ \tau = 1, \ldots, \dim F_0;\ a = 1, \ldots, \dim F_1$.

This is a practical procedure for checking the surjectivity of the map $\pi_q^{q+1} \colon \mathcal{R}_{q+1} \to \mathcal{R}_q$.

The first application of the previous theorem is the following corollary, which also goes by the name of *criterion of formal integrability* [73, 144].

Corollary 4. *Let $\mathcal{R}_q \subset J_q(\mathcal{E})$ be a system of order q on $\mathcal{E}$ such that $\mathcal{R}_{q+1} \subset J_{q+1}(\mathcal{E})$ is a fibered manifold. If $\pi_q^{q+1} \colon \mathcal{R}_{q+1} \to \mathcal{R}_q$ is an epimorphism and M_q is 2-acyclic, then $\mathcal{R}_q$ is formally integrable.*

PROOF. First we have

$$\dim M_{q+1} = \dim \mathcal{R}_{q+1} - \dim \mathcal{R}_q.$$

It follows that M_{q+1} is a vector bundle over $\mathcal{R}_q$. Hence we may apply the Theorem, with $\mathcal{R}_q^{(1)} = \mathcal{R}_q$, and obtain $\mathcal{R}_{q+r}^{(1)} = \mathcal{R}_{q+r}$. Finally, by Proposition B.13, M_{q+r+1} is also a vector bundle over $\mathcal{R}_q$, for any $r \geq 0$, and we have:

$$\dim M_{q+r+1} = \dim \mathcal{R}_{q+r+1} - \dim \mathcal{R}_{q+r}.$$

It follows by induction that for all $r \geq 0$, $\mathcal{R}_{q+r}$ is a fibered manifold, and $\mathcal{R}_q$ is formally integrable. $\square$

Remark 5. It is quite strange that the 2-acyclicity of M_q plays a role in the study of the two types of conditions contained in the definition of formal integrability.

Since it is, in general, not possible to test 2-acyclicity, but it is possible to test involution, we give the following definition.

Definition 6. A system is said to be *involutive* if it is formally integrable with an involutive symbol.

For applications to the next corollary and Chapter V, we state the following definition.

Definition 7. A system $\mathcal{R}_q \subset J_q(\mathcal{E})$ is said to be *strongly regular* if it is regular and the symbol $M_{q+r}^{(s)}$ of $\mathcal{R}_{q+r}^{(s)}$ is induced from a vector bundle over X, for all $r, s \geq 0$.

Finally, from Janet's example mentioned in the Introduction (and we revisit it in the next section) we know that, as long as a system is not formally integrable, we cannot have any information about formal power series solutions. Therefore the following corollary will be important for applications.

Corollary 8. *If $\mathcal{R}_q \subset J_q(\mathcal{E})$ is a strongly regular system, we can effectively find two integers, $r, s \geq 0$, such that the system $\mathcal{R}_{q+r}^{(s)}$ is formally integrable (involutive), with the same solutions as $\mathcal{R}_q$.*

PROOF. Since the proof is rather technical, we advise the reader to first take a look at a few examples in the next section. Afterwards, the proof will seem quite natural. Of course, this corollary is an intrinsic version of Janet's algorithm from 1920, studied in the Introduction.

First of all, we state a property of symbols.

Lemma 9. $\delta(M_{q+r+1}^{(s)}) \subset T^* \otimes M_{q+r}^{(s)}$.

PROOF. An argument similar as in the proof of Theorem 1 shows that

$$
\begin{aligned}
\mathcal{R}_{q+r+1}^{(s)} &= \pi_{q+r+1}^{q+r+s+1}(\mathcal{R}_{q+r+s+1}) \\
&= \pi_{q+r+1}^{q+r+s+1}(J_1(\mathcal{R}_{q+r+s}) \cap J_{q+r+s+1}(\mathcal{E})) \\
&\subseteq \pi_{q+r+1}^{q+r+s+1}(J_1(\mathcal{R}_{q+r+s})) \cap J_{q+r+1}(\mathcal{E}) \\
&= J_1(\mathcal{R}_{q+r}^{(s)}) \cap J_{q+r+1}(\mathcal{E}) \\
&= \rho_1(\mathcal{R}_{q+r}^{(s)}).
\end{aligned}
$$

Since the restriction to the symbol of the linearized Spencer operator is $-\delta$, the lemma follows from Proposition A.10. Of course, we also have to notice the inclusions $M_{q+r}^{(s+1)} \subset M_{q+r}^{(s)}, \forall r, s \geq 0$. $\square$

To prove the corollary we proceed inductively.

Starting with $\mathcal{R}_q$ we may assume, by Theorem B.15, that M_q is involutive, or at least 2-acyclic. If $\mathcal{R}_q^{(1)} = \mathcal{R}_q$, Corollary 4 implies, since M_{q+1} is a vector bundle over $\mathcal{R}_q$ by assumption, that the algorithm stops with $r = s = 0$. Otherwise we start anew, with $\mathcal{R}_q^{(1)}$ instead of $\mathcal{R}_q$ and $M_q^{(1)}$ instead of M_q.

If M_q is not involutive, or at least 2-acyclic, we again use Theorem B.15 to find r such that $\rho_r(\mathcal{R}_q^{(1)})$ has a symbol that is involutive, or at least 2-acyclic. By the assumptions concerning M_q, Theorem 1 implies that $\rho_r(\mathcal{R}_q^{(1)}) = \mathcal{R}_{q+r}^{(1)}$. Then $\rho_{r+1}(\mathcal{R}_q^{(1)}) = \mathcal{R}_{q+r+1}^{(1)}$, and we find

$$
\begin{aligned}
\pi_{q+r}^{q+r+1}(\mathcal{R}_{q+r+1}^{(1)}) &= \pi_{q+r}^{q+r+1} \circ \pi_{q+r+1}^{q+r+2}(\mathcal{R}_{q+r+2}) \\
&= \pi_{q+r}^{q+r+2}(\mathcal{R}_{q+r+2}) \\
&= \mathcal{R}_{q+r}^{(2)}.
\end{aligned}
$$

So, if $\mathcal{R}_{q+r}^{(2)} = \mathcal{R}_{q+r}^{(1)}$, Corollary 4 implies that the algorithm stops with $s = 1$.

Suppose we have inductively constructed a system $\mathcal{R}_{q_s}^{(s)}$ with a symbol that is involutive, or at least 2-acyclic, and such that $\rho_r(\mathcal{R}_{q_s}^{(s)}) = \mathcal{R}_{q_s+r}^{(s)}$. It remains to study the next step.

If $\pi_{q_s}^{q_s+1}(\rho_1(\mathcal{R}_{q_s}^{(s)})) = \pi_{q_s}^{q_s+1}(\mathcal{R}_{q_s+1}^{(s)}) = \mathcal{R}_{q_s}^{(s+1)}$ coincides with $\mathcal{R}_{q_s}^{(s)}$, then the algorithm stops at step s. Otherwise we start anew, with $\mathcal{R}_{q_s+1}^{(s+1)}$, which has a symbol that is involutive, or at least 2-acyclic, and we successively have:

$$\rho_r(\mathcal{R}_{q_s+1}^{(s+1)}) = \rho_r((\mathcal{R}_{q_s+1}^{(s)})^{(1)})$$

$$= (\rho_r((\mathcal{R}_{q_s+1}^{(s)}))^{(1)}$$

$$= (\mathcal{R}_{q_s+1+r}^{(s)})^{(1)} = \mathcal{R}_{q_s+1+r}^{(s+1)},$$

because of Theorem 1 and the fact that $M_{q_s+1}^{(s)}$ is the involutive prolongation of $M_{q_s}^{(s)}$.

Finally, Lemma 9 and Noetherian arguments outside the scope of this introductory book (see, e.g., [144, p. 82]) show that there is $s_0 \geq 1$, sufficiently large, such that $M_{q+r}^{(s)} = M_{q+r}^{(s_0)}$ for all $s \geq s_0, r \geq 0$. In fact, the lemma implies that $M_{q+r+1}^{(s)} \subseteq \rho_1(M_{q+r}^{(s)})$, and the inclusion becomes an equality for r sufficiently large and s fixed. Also, as already noticed, $M_{q+r}^{(s+1)} \subseteq M_{q+r}^{(s)}$ for all $r, s \geq 0$, and the inclusion becomes an equality for s sufficiently large, say from s_0 onwards.

Hence, for $s \geq s_0$ we may assume that $M_{q_{s_0}}^{(s_0)}$ is involutive, and we only have to consider the chain $\mathcal{R}_{q_{s_0}}^{(s_0)} \supset \mathcal{R}_{q_{s_0}}^{(s_0+1)} \supset \ldots$, which necessarily stops because the dimensions are finite. By Corollary 4 the algorithm stops. $\square$

Once he had brought a system to passive form, Janet was able to find out what the parametric derivatives are; in particular, their number at each order. The intrinsic version of this result comes from the use of formal integrability. Indeed, introducing the projection $\mathcal{R}_{q-1} = \pi_{q-1}^q(\mathcal{R}_q) \subset J_{q-1}(\mathcal{E})$ while taking into account Remark B.14, we immediately obtain from Corollary B.10 the formula:

$$\dim \mathcal{R}_{q+r} = \dim \mathcal{R}_{q-1} + \sum_{s=0}^{r} \dim M_{q+s}$$

$$= \dim \mathcal{R}_{q-1} + \sum_{s=0}^{r} \sum_{i=1}^{n} \frac{(s+i-1)!}{s!\,(i-1)!} \alpha_q^i$$

$$= \dim \mathcal{R}_{q-1} + \sum_{i=1}^{n} \frac{(r+i)!}{r!\,i!} \alpha_q^i.$$

hence the parametric data can be divided into α_q^n arbitrary series of n variables, $\ldots$, α_q^i arbitrary series of i variables, $\ldots$, α_q^1 arbitrary series of 1 variable, and a few constants. More precisely, we can prove [103, 144]:

Theorem 10 (Janet, Cartan, Kähler). *If $\mathcal{R}_q$ is an analytic system, then there is precisely one analytic solution $y^k = f^k(x)$ such that:*

1) $j_{q-1}(f)(x_0) \in \pi_{q-1}^q(\mathcal{R}_q) \subset J_{q-1}(\mathcal{E})$ *is a given point.*
2) *For $i = 1, \ldots, n$ the α_q^i parametric derivatives of order q and class i are equal to α_q^i given analytic functions of $x^1, \ldots, x^i$, when $x^{i+1} = x_0^{i+1}, \ldots, x^n = x_0^n$.*

Remark 11. To be able to apply this theorem it is essential to note that the quoted data must be given in a δ-regular coordinate system. We shall see that *space-time coordinates are not always δ-regular with respect to time*, even though physicists often believe that one obtains good data by fixing the time to its initial value (Cauchy data). In fact, a strange phenomenon takes place for such data, in that there are m arbitrary functions of $n - 1$ (space) variables. This situation corresponds to $\alpha_q^n = 0$ (minimum value) and $\alpha_q^{n-1} = mq$ (maximum value), which implies in a coherent way $\alpha_{q+r}^n = 0$, $\alpha_{q+r}^{n-1} = mq$, for all $r \geq 0$. The problem is then to find $\alpha_q^1 \geq \cdots \geq \alpha_q^{n-2} \geq mq > 0$. However, in an evolution equation of this type, the principal derivatives should be among the derivatives of $y^1, \ldots, y^m$ with respect to x^n. Hence $\beta_q^1 = \cdots = \beta_q^{n-1} = 0$, and it follows that the α_q^i all take the maximum values $m(q+n-i-1)!/(q-1)!\,(n-i)!$ for $i = 1, \ldots, n - 2$. It remains to understand why, in certain cases, we need only m functions of $n - 1$ variables on the whole. E.g., for the well-known *Korteweg-de Vries equation* $y_{xxx} + 12yy_x + y_t = 0$, with $n = 2$, $m = 1$, $q = 3$, x =space, t =time, we have to prescribe either $y(0,t), y_x(0,t), y_{xx}(0,t)$, i.e. 3 arbitrary functions of t, or $y(x,0)$ only, i.e. 1 arbitrary function of x. It seems that the only way to escape from the above dilemma is to reduce the system to a first-order transitive system via the inclusion $\mathcal{R}_{q+1} \subset J_1(\mathcal{R}_q)$ and to start afresh [144].

Remark 12. In the nonlinear case, if the assumption of strong regularity is dropped and only that of regularity is retained, the symbols may need to be defined over higher and higher jet bundles and the Noetherian argument may fail to work. However, we do not know of any explicit example of this type.

Finally, if the regularity assumption is also dropped, it is impossible to find any practical algorithm. Nevertheless, we shall see that in the differential algebraic case it is sometimes possible to say more, and sometimes a branched process can be started, depending on rank conditions as described in the Introduction.

D. Examples

The purpose of this Section is to illustrate the preceding results by means of explicit computations. Both linear and nonlinear examples will be treated. Since the constructions of the next Chapter will only apply to involutive linear systems, many of the linear systems considered in this Section will be treated again in Section C of Chapter IV.

Example 1 (Janet's example revisited). In Janet's example we have $n = 3$, $m = 1$, $q = 2$. Since it deals with a linear system, we may regard it as a vector subbundle $R_2 \subset J_2(E)$, with $X = \mathbb{R}^3$ and $E = X \times \mathbb{R}$. Our objective is to recover the results in the Introduction by means of a purely intrinsic (coordinate-free) approach. In particular, under a change of coordinates (a permutation will do!) the various steps of the algorithm will correspond and the various dimensions of the bundles remain the same. This is the first time that such an example is treated in this intrinsic way [155].

Before starting the exposition we give some notation, to be used in the case of linear systems.

First of all, a linear system of order q on a vector bundle E can be regarded as a vector subbundle $R_q \subset J_q(E)$, and we may define the vector bundle $F_0 = J_q(E)/R_q$ with canonical projection $\Phi \colon J_q(E) \to F_0$; it can also be given as the kernel of an epimorphism $\Phi \colon J_q(E) \to F_0$. In any case, we start from the short exact sequence

$$0 \longrightarrow R_q \longrightarrow J_q(E) \overset{\Phi}{\longrightarrow} F_0 \longrightarrow 0.$$

Introducing $J_{q+r}^q(E) = \ker \pi_q^{q+r}$, we have $J_q^{q-1}(E) = S_q T^* \otimes E$, and we obtain the following exact commutative diagram, denoted by $\mathrm{diag}(q, r, s)$:

$$
\begin{array}{ccccccccc}
& & 0 & & 0 & & 0 & & \\
& & \downarrow & & \downarrow & & \downarrow & & \\
0 & \longrightarrow & R_{q+r+s}^{q+r} & \longrightarrow & J_{q+r+s}^{q+r}(E) & \longrightarrow & J_{r+s}^r(F_0) & & \\
& & \downarrow & & \downarrow & & \downarrow & & \\
0 & \longrightarrow & R_{q+r+s} & \longrightarrow & J_{q+r+s}(E) & \overset{\rho_{r+s}(\Phi)}{\longrightarrow} & J_{r+s}(F_0) & \longrightarrow & Q_{r+s} & \longrightarrow & 0 \\
& & \downarrow & & {\scriptstyle \pi_{q+r}^{q+r+s}}\downarrow & & \downarrow & & \downarrow & & \\
0 & \longrightarrow & R_{q+r} & \longrightarrow & J_{q+r}(E) & \overset{\rho_r(\Phi)}{\longrightarrow} & J_r(F_0) & \longrightarrow & Q_r & \longrightarrow & 0 \\
& & \downarrow & & \downarrow & & \downarrow & & \\
& & 0 & & 0 & & 0 & &
\end{array}
$$

By definition, $M_{q+r} = R_{q+r}^{q+r-1}$, and we have the following short exact sequences, denoted by $\mathrm{seq}(q, r, s)$:

$$
0 \longrightarrow R_{q+r+s}^{q+r} \longrightarrow R_{q+r+s} \overset{\pi_{q+r}^{q+r+s}}{\longrightarrow} R_{q+r}^{(s)} \longrightarrow 0.
$$

Returning to the diagram in the Introduction obtained by the second student (the one using a computer), we notice that the numbers in the horizontal lines are, respectively,

$$
q, \quad \dim S_q T^* \otimes E, \quad \dim J_q(E), \quad \mathrm{rk}\, \sigma_r(\Phi),
$$

where $\sigma_r(\Phi): S_{q+r} T^* \otimes E \to S_r T^* \otimes F_0$ is the restriction of $\rho_r(\Phi): J_{q+r}(E) \to J_r(F_0)$. Of course, $\dim M_{q+r} = \dim S_{q+r} T^* \otimes E - \mathrm{rk}\, \sigma_r(\Phi)$. The numbers in the vertical columns are, respectively,

$$
r, \quad \dim S_r T^* \otimes F_0, \quad \dim J_r(F_0),
$$

while the diagonal numbers are $\mathrm{rk}\, \rho_r(\Phi)$, and we have

$$
\dim R_{q+r} = \dim J_{q+r}(E) - \mathrm{rk}\, \rho_r(\Phi).
$$

We are now ready to inductively apply Theorem C.1 and Corollary C.4 in order to find $r = 3$, $s = 2$ in Corollary C.8. Moreover, we shall prove that the resulting involutive system $R_5^{(2)}$ is of finite type, a result which is not at all evident at first sight, and that

$\dim R_5^{(2)} = 12$. We start with the commutative diagram $(2, -1, 1)$:

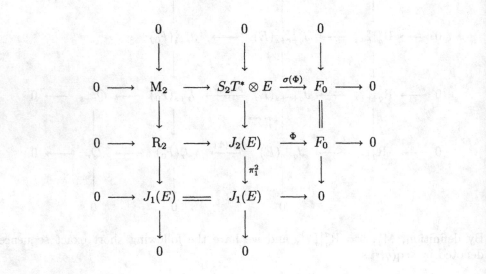

The first prolongation of this diagram is the following exact commutative diagram diag(2, 0, 1):

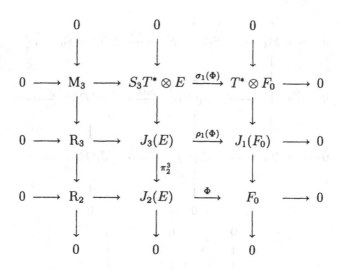

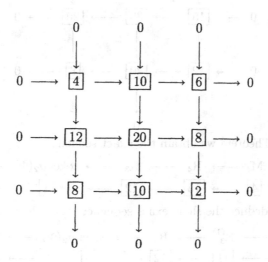

The surjectivity of $\pi_2^3: R_3 \to R_2$ follows from seq(2, 0, 1). We immediately arrive at the snake Theorem.

Things are different with $\mathrm{diag}(2,1,1)$, stemming from the fact that $\pi_3^4 \colon R_4 \to R_3$ is *not* surjective:

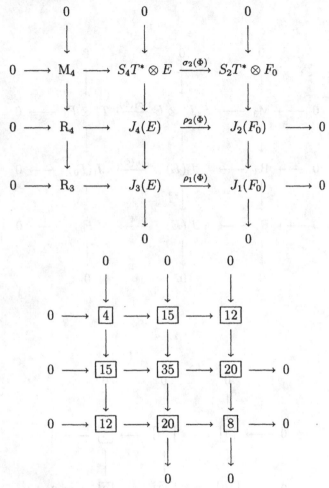

Using the snake Theorem we obtain the exact sequence

$$0 \longrightarrow M_4 \longrightarrow R_4 \longrightarrow R_3 \longrightarrow \mathrm{coker}\,\sigma_2(\Phi) \longrightarrow 0,$$
$$0 \longrightarrow \boxed{4} \longrightarrow \boxed{15} \longrightarrow \boxed{12} \longrightarrow \boxed{1} \longrightarrow 0,$$

from which we can deduce the short exact sequence

$$0 \longrightarrow R_3^{(1)} \longrightarrow R_3 \longrightarrow \mathrm{coker}\,\sigma_2(\Phi) \longrightarrow 0,$$
$$0 \longrightarrow \boxed{11} \longrightarrow \boxed{12} \longrightarrow \boxed{1} \longrightarrow 0.$$

Although this sequence looks like the curvature map participating in the proof of Theorem C.1, we see that it is different because $\sigma_1(\Phi)$ is surjective and therefore $F_1 = 0$.

The link with the Spencer cohomology of M_2 must be given via Corollary C.4. Indeed, since $\pi_2^3\colon R_3 \to R_2$ is surjective but $\pi_3^4\colon R_4 \to R_3$ is not, surely M_2 is *not* 2-acyclic. We shall effectively prove that $\operatorname{coker}\sigma_2(\Phi) \approx H_2^2(M_2)$, and obtain $\dim H_2^2(M_2) = 1$ by two different methods. To this end we apply Spencer cohomology to the top row of the previous diagram:

$$
\begin{array}{ccccccc}
& & 0 & & 0 & & 0 \\
& & \downarrow & & \downarrow & & \downarrow \\
0 \longrightarrow & & M_4 & \longrightarrow & S_4T^* \otimes E & \xrightarrow{\sigma_2(\Phi)} & S_2T^* \otimes F_0 & \longrightarrow \operatorname{coker}\sigma_2(\Phi) \to 0 \\
& & \delta\downarrow & & \delta\downarrow & & \delta\downarrow \\
0 \longrightarrow & & T^* \otimes M_3 & \longrightarrow & T^* \otimes S_3T^* \otimes E & \xrightarrow{\sigma_1(\Phi)} & T^* \otimes T^* \otimes F_0 & \longrightarrow \quad 0 \\
& & \delta\downarrow & & \delta\downarrow & & \delta\downarrow \\
0 \longrightarrow & & \wedge^2 T^* \otimes M_2 & \longrightarrow & \wedge^2 T^* \otimes S_2 T^* \otimes E & \xrightarrow{\sigma(\Phi)} & \wedge^2 T^* \otimes F_0 & \longrightarrow \quad 0 \\
& & \delta\downarrow & & \delta\downarrow & & \downarrow \\
0 \longrightarrow & & \wedge^3 T^* \otimes T^* \otimes E & = & \wedge^3 T^* \otimes T^* \otimes E & \longrightarrow & 0 \\
& & \downarrow & & \downarrow & & \\
& & 0 & & 0 & &
\end{array}
$$

$$
\begin{array}{ccccccc}
0 & & 0 & & 0 & & \\
\downarrow & & \downarrow & & \downarrow & & \\
0 \longrightarrow \boxed{4} & \longrightarrow & \boxed{15} & \longrightarrow & \boxed{12} & \longrightarrow \boxed{1} & \longrightarrow 0 \\
\downarrow & & \downarrow & & \downarrow & & \\
0 \longrightarrow \boxed{12} & \longrightarrow & \boxed{30} & \longrightarrow & \boxed{18} & \longrightarrow 0 & \\
\downarrow & & \downarrow & & \downarrow & & \\
0 \longrightarrow \boxed{12} & \longrightarrow & \boxed{18} & \longrightarrow & \boxed{6} & \longrightarrow 0 & \\
\downarrow & & \downarrow & & \downarrow & & \\
0 \longrightarrow \boxed{3} & = & \boxed{3} & \longrightarrow & 0 & & \\
\downarrow & & \downarrow & & & & \\
0 & & 0 & & & &
\end{array}
$$

Keeping in mind that $n = 3$, the second column on the left is exact, and therefore the last δ is surjective. We leave it to the reader to chase in the diagram and prove that the last δ in the first column is also surjective. The desired result follows now from Exercise I.9 and the relation

$$\dim H_2^2(M_2) = \dim Z_2^2 - \dim B_2^2 = (12 - 3) - (12 - 4) = 9 - 8 = 1.$$

The question is now to find out if we can start anew with $R_3^{(1)}$ in Theorem C.1. For this, we have to study the symbol M_2 and its various prolongations, until we hit upon an involutive symbol.

We start with M_2:

$$M_2 \quad \begin{cases} v_{33} - x^2 v_{11} = 0 \\ v_{22} \quad\quad\;\; = 0 \end{cases} \quad \begin{array}{|ccc|} \hline 1 & 2 & 3 \\ 1 & 2 & \bullet \\ \hline \end{array}$$

We use Proposition B.9 3) to test involutivity. Prolongation with respect to the dot should give

$$v_{223} = 0.$$

However, no prolongation of the first equation with respect to the multiplicative variables could provide such an equation, and hence M_2 is *not* involutive.

Let us study M_3:

$$M_3 \quad \begin{cases} v_{333} - x^2 v_{113} = 0 \\ v_{233} - x^2 v_{112} = 0 \\ v_{223} \quad\quad\;\; = 0 \\ v_{222} \quad\quad\;\; = 0 \\ v_{133} - x^2 v_{111} = 0 \\ v_{122} \quad\quad\;\; = 0 \end{cases} \quad \begin{array}{|ccc|} \hline 1 & 2 & 3 \\ 1 & 2 & \bullet \\ 1 & 2 & \bullet \\ 1 & 2 & \bullet \\ 1 & \bullet & \bullet \\ 1 & \bullet & \bullet \\ \hline \end{array}$$

We have proved that M_2 is *not* involutive because

$$\dim M_3 = 10 - 6 = 4 \neq 10 - 5 = 5.$$

We may repeat such a computation proving that M_3 is involutive with

$$\dim M_4 = 15 - (3 + 6 + 2) = 4,$$

which can be done by computer or from the explicit knowledge of M_4:

$$M_4 \quad \begin{cases} v_{3333} - x^2 v_{1113} = 0 \\ v_{2333} - x^2 v_{1123} = 0 \\ v_{2233} \quad\quad\;\; = 0 \\ v_{2223} \quad\quad\;\; = 0 \\ v_{2222} \quad\quad\;\; = 0 \\ v_{1333} - x^2 v_{1113} = 0 \\ v_{1233} - x^2 v_{1112} = 0 \\ v_{1223} \quad\quad\;\; = 0 \\ v_{1222} \quad\quad\;\; = 0 \\ v_{1133} - x^2 v_{1111} = 0 \\ v_{1122} \quad\quad\;\; = 0 \end{cases} \quad \begin{array}{|ccc|} \hline 1 & 2 & 3 \\ 1 & 2 & \bullet \\ 1 & 2 & \bullet \\ 1 & 2 & \bullet \\ 1 & 2 & \bullet \\ 1 & \bullet & \bullet \\ 1 & \bullet & \bullet \\ 1 & \bullet & \bullet \\ 1 & \bullet & \bullet \\ 1 & \bullet & \bullet \\ 1 & \bullet & \bullet \\ \hline \end{array}$$

Of course, M_4 is involutive because M_3 is involutive, and we have

$$\dim M_5 = 21 - (3 + 8 + 6) = 4.$$

At this point we notice that (x^1, x^2, x^3) is δ regular, but (x^2, x^1, x^3) is not δ-regular, as can be immediately verified:

$$M_3 \begin{cases} v_{333} - x^1 v_{223} = 0 \\ v_{233} - x^1 v_{222} = 0 \\ v_{133} - x^1 v_{122} = 0 \\ v_{113} \qquad\quad = 0 \\ v_{112} \qquad\quad = 0 \\ v_{111} \qquad\quad = 0 \end{cases} \quad \begin{array}{ccc} 1 & 2 & 3 \\ 1 & 2 & \bullet \\ 1 & \bullet & \bullet \\ 1 & \bullet & \bullet \\ 1 & \bullet & \bullet \\ 1 & \bullet & \bullet \end{array}$$

Of course we have

$$\dim M_4 = 4 < 15 - (3 + 2 + 4) = 6.$$

By Theorem C.1, we can start anew with $R_3^{(1)}$, which is thus such that $\rho_r(R_3^{(1)}) = R_{3+r}^{(1)}$, $\forall r \geq 0$.

In particular, if we want to apply the criterion of formal integrability to $R_3^{(1)}$, we have to study $M_3^{(1)}$:

$$M_3^{(1)} \begin{cases} v_{333} - x^2 v_{113} = 0 \\ v_{233} \qquad\quad = 0 \\ v_{223} \qquad\quad = 0 \\ v_{222} \qquad\quad = 0 \\ v_{133} - x^2 v_{111} = 0 \\ v_{122} \qquad\quad = 0 \\ v_{112} \qquad\quad = 0 \end{cases} \quad \begin{array}{ccc} 1 & 2 & 3 \\ 1 & 2 & \bullet \\ 1 & 2 & \bullet \\ 1 & 2 & \bullet \\ 1 & \bullet & \bullet \\ 1 & \bullet & \bullet \\ 1 & \bullet & \bullet \end{array}$$

The explanation for adding the last equation is as follows: Prolongating $y_{33} - x^2 y_{11} = 0$ twice with respect to x^2 gives $y_{2233} - x^2 y_{1122} - 2 y_{112} = 0$. Prolongating $y_{22} = 0$ twice with respect to x^3 and x^2 in succession gives $y_{2233} = 0$ and $y_{1122} = 0$. Therefore the additional equation $y_{112} = 0$ appears among the equations defining $R_3^{(1)}$, and the corresponding equation $v_{112} = 0$ appears among the equations defining $M_3^{(1)}$. However, $M_3^{(1)}$ is not involutive, since prolongation of $v_{112} = 0$ with respect to x^3 gives $v_{1123} = 0$, which cannot be obtained by prolongation with respect to the multiplicative variables.

Hence we should look at $\rho_1(M_3^{(1)}) = M_4^{(1)}$, which is the symbol of $\rho_1(R_3^{(1)}) = R_4^{(1)}$.

$$
M_4^{(1)} \begin{cases}
v_{3333} & = 0 \\
v_{2333} & = 0 \\
v_{2233} & = 0 \\
v_{2223} & = 0 \\
v_{2222} & = 0 \\
v_{1333} - x^2 v_{1113} & = 0 \\
v_{1233} & = 0 \\
v_{1223} & = 0 \\
v_{1222} & = 0 \\
v_{1133} & = 0 \\
v_{1123} & = 0 \\
v_{1122} & = 0 \\
v_{1112} & = 0
\end{cases}
\qquad
\begin{array}{|ccc|}
\hline
1 & 2 & 3 \\
1 & 2 & \bullet \\
1 & 2 & \bullet \\
1 & 2 & \bullet \\
1 & 2 & \bullet \\
1 & \bullet & \bullet \\
1 & \bullet & \bullet \\
1 & \bullet & \bullet \\
1 & \bullet & \bullet \\
1 & \bullet & \bullet \\
1 & \bullet & \bullet \\
1 & \bullet & \bullet \\
1 & \bullet & \bullet \\
\hline
\end{array}
$$

Using the computer we obtain $\dim R_6^4 = 21 + 28 - 43 = 6$. We have the short exact sequences

$$
0 \longrightarrow M_5^{(1)} \longrightarrow R_5^{(1)} \xrightarrow{\pi_4^5} R_4^{(2)} \longrightarrow 0,
$$

$$
0 \longrightarrow M_6 \longrightarrow R_6 \xrightarrow{\pi_5^6} R_5^{(1)} \longrightarrow 0,
$$

$$
0 \longrightarrow R_6^4 \longrightarrow R_6 \xrightarrow{\bar\pi_4^6} R_4^{(2)} \longrightarrow 0,
$$

and hence

$$
\begin{aligned}
\dim M_5^{(1)} &= \dim R_5^{(1)} - \dim R_4^{(2)} \\
&= (\dim R_6 - \dim M_6) - (\dim R_6 - \dim R_6^4) \\
&= \dim R_6^4 - \dim M_6 = 6 - 4 = 2,
\end{aligned}
$$

and $\dim M_5^{(1)} = 21 - (3 + 8 + 8) = 2$ by prolongation with respect to the multiplicative variables only. It follows that $M_4^{(1)}$ is involutive, and we may try to apply the criterion to $R_4^{(1)}$. Since $\rho_1(R_4^{(1)}) = R_5^{(1)}$, by Theorem C.1 we have $\pi_4^5(R_5^{(1)}) = R_4^{(2)}$, and we have already verified that $\dim R_4^{(2)} = 12$. However, there is also the short exact sequence

$$
0 \longrightarrow M_5 \longrightarrow R_5 \xrightarrow{\pi_4^5} R_4^{(1)} \longrightarrow 0,
$$

and thus $\dim R_4^{(1)} = \dim R_5 - \dim M_5 = (56 - 39) - 4 = 13$. It is then easy to verify that $R_4^{(2)}$ is obtained from $R_4^{(1)}$ by adding the single equation $y_{1111} = 0$.

We may start anew with $R_4^{(2)}$. Its symbol $M_4^{(2)} \subset M_4^{(1)}$ is:

$$M_4^{(2)} \begin{cases} v_{3333} & = 0 \\ v_{2333} & = 0 \\ v_{2233} & = 0 \\ v_{2223} & = 0 \\ v_{2222} & = 0 \\ v_{1333} - x^2 v_{1113} = 0 \\ v_{1233} & = 0 \\ v_{1223} & = 0 \\ v_{1222} & = 0 \\ v_{1133} & = 0 \\ v_{1123} & = 0 \\ v_{1122} & = 0 \\ v_{1112} & = 0 \\ v_{1111} & = 0 \end{cases} \quad \begin{array}{ccc} 1 & 2 & 3 \\ 1 & 2 & \bullet \\ 1 & 2 & \bullet \\ 1 & 2 & \bullet \\ 1 & 2 & \bullet \\ 1 & \bullet & \bullet \\ 1 & \bullet & \bullet \\ 1 & \bullet & \bullet \\ 1 & \bullet & \bullet \\ 1 & \bullet & \bullet \\ 1 & \bullet & \bullet \\ 1 & \bullet & \bullet \\ 1 & \bullet & \bullet \\ 1 & \bullet & \bullet \end{array}$$

$M_4^{(2)}$ is not involutive, because prolongation of $v_{1111} = 0$ with respect to the non-multiplicative variable x^3 gives $v_{11113} = 0$, which cannot be obtained by prolongation with respect to the multiplicative variables. Also, $M_4^{(2)}$ is the symbol of $R_4^{(2)}$, and thus $\rho_1(M_4^{(2)})$ is the symbol of $\rho_1(R_4^{(2)}) = \rho_1((R_4^{(1)})^{(1)}) = (\rho_1(R_4^{(1)}))^{(1)} = (R_5^{(1)})^{(1)} = R_5^{(2)}$, i.e. $\rho_1(M_4^{(2)}) = M_5^{(2)}$. We now have the short exact sequences

$$0 \longrightarrow M_5^{(2)} \longrightarrow R_5^{(2)} \xrightarrow{\pi_4^5} R_4^{(3)} \longrightarrow 0,$$

$$0 \longrightarrow R_7^5 \longrightarrow R_7 \xrightarrow{\pi_5^7} R_5^{(2)} \longrightarrow 0,$$

$$0 \longrightarrow R_7^4 \longrightarrow R_7 \xrightarrow{\pi_4^7} R_4^{(3)} \longrightarrow 0,$$

and hence

$$\begin{aligned} \dim M_5^{(2)} &= \dim R_5^{(2)} - \dim R_4^{(3)} \\ &= (\dim R_7 - \dim R_7^5) - (\dim R_7 - \dim R_7^4) \\ &= \dim R_7^4 - \dim R_7^5 \\ &= (21 + 28 + 36 - 79) - (28 + 36 - 58) = 6 - 6 = 0, \end{aligned}$$

and again we can check that $M_4^{(2)}$ is not involutive, because

$$\dim M_5^{(2)} < 21 - (3 + 8 + 9) = 1.$$

Hence we can start anew with $R_5^{(2)}$, because $M_5^{(2)} = 0$ is trivially involutive (being of finite type). We successively find:

$$\rho_1(R_5^{(2)}) = \rho_1((R_5^{(1)})^{(1)}) = (\rho_1(R_5^{(1)}))^{(1)} = (R_6^{(1)})^{(1)} = R_6^{(2)},$$

and $\pi_5^6(R_6^{(2)}) = \pi_5^8(R_8) = R_5^{(3)}$. Hence, using $\mathrm{diag}(2,3,3)$ we obtain the short exact sequence

$$0 \longrightarrow R_8^5 \longrightarrow R_8 \xrightarrow{\pi_5^8} R_5^{(3)} \longrightarrow 0,$$

and we have

$$\begin{aligned}
\dim R_5^{(3)} &= \dim R_8 - \dim R_8^5 \\
&= (165 - 147) - (28 + 36 + 45 - 103) \\
&= 18 - 6 = 12.
\end{aligned}$$

However, also

$$\begin{aligned}
\dim R_5^{(2)} &= \dim R_7 - \dim R_7^5 \\
&= (120 - 102) - (28 + 36 - 58) \\
&= 18 - 6 = 12.
\end{aligned}$$

Hence, $R_5^{(3)} \subseteq R_5^{(2)}$, and $\dim R_5^{(3)} = \dim R_5^{(2)} = 12$, implying that $R_5^{(3)} = R_5^{(2)}$. Corollary C.4 finally implies that $R_5^{(2)}$ is involutive and has zero symbol.

Now we can understand the mistake of the second student: he believed that the various parametric/principal derivatives depend only on the symbols M_{2+r}, $r \geq 0$.

In Example 6 we shall understand why the preceding calculations will easily give the compatibility conditions, and why these conditions are of order 6 over F_0.

Example 2. $n = 2$, $m = 1$, $q = 2$.
We want to study the system

$$R_2 \quad \begin{cases} y_{11} + y_{22} = 0, \\ x^1 y_1 + x^2 y_2 = 0. \end{cases}$$

First of all we verify whether it is a fibered manifold. We look at its symbol

$$M_2 \quad \left\{ v_{22} + v_{11} = 0 \;\; \boxed{1 \;\; 2} \right.$$

Since it is defined by an equation with constant coefficients, it is a vector bundle over $X = \mathbb{R}^2$ with $\dim M_2 = 3 - 1 = 2$. Hence R_2 is a vector bundle if and only if its projection in $J_1(E)$ is a vector bundle. Since it is defined by the equation $x^1 y_1 + x^2 y_2 = 0$, we see that the maximum rank condition is satisfied if and only if we leave out the origin $(0,0)$ of $\mathbb{R}^2$. Thus, from now on, we work over $\mathbb{R} \setminus \{(0,0)\}$, and study the regularity of R_2 over this open set. We have $\dim R_2 = \dim J_2(E) - \dim F_0 = 6 - 2 = 4$, but $\sigma(\Phi) = \sigma_0(\Phi) \colon S_2 T^* \otimes E \to F_0$ is not an epimorphism, because R_2 is defined by first- *and* second-order equations. Note that M_3 is a vector bundle (constant coefficients) and that M_2 is trivially involutive.

Hence we may study $R_2^{(1)}$:

$$R_2^{(1)} \begin{cases} x^1 y_{11} + x^2 y_{12} + y_1 = 0, \\ x^1 y_{12} + x^2 y_{22} + y_2 = 0, \\ y_{11} + y_{22} = 0, \\ x^1 y_1 + x^2 y_2 = 0. \end{cases}$$

First of all we have to find out whether it is a vector bundle. For this we look at the symbol:,

$$M_2^{(1)} \begin{cases} x^1 v_{11} + x^2 v_{12} = 0, \\ x^1 v_{12} + x^2 v_{22} = 0, \\ v_{11} + v_{22} = 0. \end{cases}$$

The determinant of the jacobian matrix with respect to v_{11}, v_{12}, v_{22} is:

$$\det \begin{pmatrix} x^1 & x^2 & 0 \\ 0 & x^1 & x^2 \\ 1 & 0 & 1 \end{pmatrix} = (x^1)^2 + (x^2)^2.$$

Hence $M_2^{(1)}$ is a vector bundle over $\mathbb{R}^2 \setminus \{(0,0)\}$, with $\dim M_2^{(1)} = 3 - 3 = 0$, i.e. $M_2^{(1)} = 0$ is trivially involutive (being of finite type). Under the same condition, $R_2^{(1)}$ is a vector bundle, since its first-order defining equation is the same as that of R_2, and $\dim R_2^{(1)} = 6 - 4 = 2$.

We are now in a good position to apply the criterion of formal integrability, and by Theorem C.1 we have to introduce $\rho_1(R_2^{(1)}) = R_3^{(1)}$. Finally, $\pi_2^3(R_3^4) = \pi_2^4(R_4) = R_2^{(2)}$, and by computer we may verify that $\dim R_2^{(2)} = 2$, i.e. $R_2^{(1)}$ is involutive. We can also follow a direct way, by extending the scheme of multiplicative/nonmultiplicative variables to the full set of equations defining the system, the top-order part being in solved form with corresponding scheme as in Section B while the lower part of the scheme contains dots only. Hence we only need check that prolongation of the lower-order equations does not lead to any new equation. In our case, setting $r^2 = (x^1)^2 + (x^2)^2$, we find:

$$R_2^{(1)} \begin{cases} y_{22} - \dfrac{x^2}{(r)^2} y_1 + \dfrac{x^2}{(r)^2} y_2 = 0 & \boxed{1 \quad 2} \\ y_{12} + \dfrac{x^2}{(r)^2} y_1 + \dfrac{x^1}{(r)^2} y_2 = 0 & \boxed{1 \quad \bullet} \\ y_{11} + \dfrac{x^1}{(r)^2} y_1 - \dfrac{x^2}{(r)^2} y_2 = 0 & \boxed{1 \quad \bullet} \\ x^1 y_1 + x^2 y_2 = 0 & \boxed{\bullet \quad \bullet} \end{cases}$$

and we can easily check this property.

Such a full scheme is extremely useful for the applications in the next Chapter.

It remains to find out how to construct formal power series solutions over the origin. Indeed, using the equations defining R_2 we obtain

$$y_{11}(0) + y_{22}(0) = 0.$$

Using the equations defining $R_2^{(1)}$ we obtain

$$y_1(0) = 0, \qquad y_2(0) = 0.$$

Prolongation of the top-order equations again leads to

$$y_{11}(0) = 0, \qquad y_{12}(0) = 0, \qquad y_{22}(0) = 0,$$

etc. hence, the only analytic solution defined at the origin is $y = \text{const}$. However, this solution is too specific to have a general meaning.

Example 3. $n = 3$, $m = 1$, $q = 3$.
We consider the following constant-coefficient system:

$$R_3 \quad \begin{cases} y_{112} = 0, \\ y_{33} = 0, \\ y_{22} = 0. \end{cases}$$

Its symbol M_3 is involutive, since it is defined by a single equation. However, the computation of the characters must be done after a change of coordinates, since the given coordinates are not δ-regular. Hence we may look for $R_3^{(1)}$, which is easily seen to be obtained by adding the prolongations of the second-order equations:

$$R_3^{(1)} \quad \begin{cases} y_{333} = 0 \\ y_{233} = 0 \\ y_{223} = 0 \\ y_{222} = 0 \\ y_{133} = 0 \\ y_{122} = 0 \\ y_{112} = 0 \\ y_{33} = 0 \\ y_{22} = 0 \end{cases}$$

1	2	3
1	2	•
1	2	•
1	2	•
1	•	•
1	•	•
1	•	•
•	•	•
•	•	•

The scheme on the right is determined as in the previous exercise. This system is formally integrable, although its symbol $M_3^{(1)}$ is neither involutive nor 2-acyclic. Indeed, the prolongations of the second-order equations already are part of the third-order equations, and the constant-coefficient situation shows that we have only to prolongate the third-order equations as often as we please. We leave it to the reader to prove that $M_3^{(1)}$ is not involutive, but that $M_4^{(1)}$ is involutive, with characters $\alpha_4^1 = 2, \alpha_4^2 = 0, \alpha_4^3 = 0$, giving $\dim M_{4+r}^{(1)} = 2$, for all $r \geq 0$.

We prove that $M_3^{(1)}$ is not 2-acyclic. In fact, we only have to prove that the first corresponding Spencer δ-sequence is not exact:

$$0 \longrightarrow M_5^{(1)} \xrightarrow{\delta} T^* \otimes M_4^{(1)} \xrightarrow{\delta} \wedge^2 T^* \otimes M_3^{(1)} \xrightarrow{\delta} \wedge^3 T^* \otimes S_2 T^* \otimes E.$$

We give the various dimensions:

$$0 \longrightarrow \boxed{2} \longrightarrow \boxed{6} \longrightarrow \boxed{9} \longrightarrow \boxed{6}.$$

Since the beginning part of the sequence is exact by Proposition B.7, we only have to prove that the kernel of the last δ has dimension strictly greater than $6 - 2 = 4$ or, equivalently, that its image has dimension strictly less than $9 - 4 = 5$. This kernel is determined by the following 6 equations:

$$\begin{cases} v_{331,23} + v_{332,31} + v_{333,12} = 0, \\ v_{231,23} + v_{232,31} + v_{233,12} = 0, \\ v_{221,23} + v_{222,31} + v_{223,12} = 0, \\ v_{131,23} + v_{132,31} + v_{133,12} = 0, \\ v_{121,23} + v_{122,31} + v_{123,12} = 0, \\ v_{111,23} + v_{112,31} + v_{113,12} = 0. \end{cases}$$

In the definition of $M_3^{(1)}$, all the v_{ijk} but $v_{111}, v_{113}, v_{123}$ are zero, and we discover that there are only 4 independent equations left:

$$\begin{cases} v_{123,23} = 0, \\ v_{113,23} + v_{123,13} = 0, \\ v_{123,12} = 0, \\ v_{111,23} + v_{113,12} = 0. \end{cases}$$

This proves that $M_3^{(1)}$ is not 2-acyclic and that the criterion is sufficient, but not necessary, for formal integrability.

By the general formula found in Section C, the general solution will depend only on a certain number of constants and 2 arbitrary functions of one variable. We now check this property by direct integration:

$$y_{33} = 0 \Rightarrow y = a(x^1, x^2)x^3 + b(x^1, x^2),$$

$$y_{22} = 0 \Rightarrow \begin{cases} a(x^1, x^2) = \alpha(x^1)x^2 + \beta(x^1), \\ b(x^1, x^2) = \gamma(x^1)x^2 + \delta(x^1), \end{cases}$$

and from the second-order equations we obtain

$$y = \alpha(x^1)x^2 x^3 + \beta(x^1)x^3 + \gamma(x^1)x^2 + \delta(x^1).$$

Taking into account the third-order equation $y_{112} = 0$, we find

$$\partial_{11}\alpha(x^1)x^3 + \partial_{11}\gamma(x^1) = 0 \Rightarrow \begin{cases} \partial_{11}\alpha(x^1) = 0, \\ \partial_{11}\gamma(x^1) = 0. \end{cases}$$

Hence α and γ depend together on 4 arbitrary constants while $\beta(x^1)$ and $\delta(x^1)$ are arbitrary functions of x^1.

Example 4. $n = 3$, $m = 1$, $q = 2$.

This example is among the rare ones which we have been able to find and in which there is one unknown and a symbol that is 2-acyclic but not involutive.

Consider the following second-order equation, where α is a constant and $a(x^1, x^2, x^3)$ an arbitrary function:

$$
R_2 \begin{cases}
y_{33} & = 0 \\
y_{23} - \alpha y_{11} = 0 \\
y_{22} - a(x)y_3 = 0 \\
y_{13} & = 0 \\
y_{12} & = 0
\end{cases}
\quad
\begin{array}{ccc}
1 & 2 & 3 \\
1 & 2 & \bullet \\
1 & 2 & \bullet \\
1 & \bullet & \bullet \\
1 & \bullet & \bullet
\end{array}
$$

At first sight the equations are in solved form and linearly independent. Hence R_2 is a vector bundle, with $\dim R_2 = 10 - 5 = 5$ and symbol

$$
M_2 \begin{cases}
v_{33} & = 0 \\
v_{23} - \alpha v_{11} = 0 \\
v_{22} & = 0 \\
v_{13} & = 0 \\
v_{12} & = 0
\end{cases}
\quad
\begin{array}{ccc}
1 & 2 & 3 \\
1 & 2 & \bullet \\
1 & 2 & \bullet \\
1 & \bullet & \bullet \\
1 & \bullet & \bullet
\end{array}
$$

Of course, M_2 is a vector bundle, with $\dim M_2 = 6 - 5 = 1$. However, we discover that, although M_3 is a vector bundle for all α, its dimension depends to a large extend on α. Indeed, we have:

$$
\begin{cases}
\alpha = 0 \Rightarrow \dim M_3 = 1 & \text{(all } v_{ijk} = 0, \text{ except } v_{111}, \text{ which is arbitrary)} \\
\alpha \neq 0 \Rightarrow \dim M_3 = 0 \Rightarrow & M_3 \text{ is involutive.}
\end{cases}
$$

Hence we have to distinguish these two cases carefully.

1) $\alpha = 0$. Now M_2 is involutive because $\dim M_3 = 10 - (3 + 4 + 2) = 1$ and the coordinate system is δ-regular. This gives as characters $\alpha_2^1 = 1$, $\alpha_2^2 = 0$, $\alpha_2^3 = 0$.

By the criterion for formal integrability, we have only to look for $R_2^{(1)}$ or, equivalently, to study the dot prolongations in the scheme associated to R_2, taking into account the other prolongations. The reader should verify him/herself that only two dots actually bring new, second-order, equations, these being

$$
\partial_3 a(x)y_3 = 0, \qquad \partial_1 a(x)y_3 = 0.
$$

Hence, there are two possibilities:
- $\partial_1 a(x) = 0$, $\partial_3 a(x) = 0 \Rightarrow R_2^{(1)} = R_2$ is involutive,
- $(\partial_1 a)(\partial_3 a) \neq 0 \Rightarrow R_2^{(1)} \subset R_2$.

In the second case $R_2^{(1)}$ contains the additional equation $y_3 = 0$ and has constant coefficients indeed, with $M_2^{(1)} = M_2$ involutive. It follows that $R_2^{(1)}$ is involutive, since $y_{13} = 0$, $y_{23} = 0$, $y_{33} = 0$ are among the remaining equations.

2) $\alpha \neq 0$. This is the generic and most interesting case. We shall prove that M_2 is 2-acyclic, although not involutive because of Proposition B.6 and the fact that $M_3 = 0$. It remains to prove the exactness of the sequence

$$0 \longrightarrow \wedge^2 T^* \otimes M_2 \overset{\delta}{\longrightarrow} \wedge^3 T^* \otimes T^* \otimes E,$$

i.e. the injectivity of the map δ.
In fact the kernel of δ is defined by the following three equations:

$$\begin{cases} v_{11,23} + v_{12,31} + v_{13,12} = 0, \\ v_{21,23} + v_{22,31} + v_{23,12} = 0, \\ v_{31,23} + v_{23,31} + v_{33,12} = 0. \end{cases}$$

Substitution of the equations of M_2 gives

$$v_{11,23} = 0, \quad v_{11,12} = 0, \quad v_{11,31} = 0,$$

taking into account the crucial fact that $\alpha \neq 0$. But $\dim M_2 = 1$, and the only parametric component is v_{11}. Hence δ is injective and M_2 is 2-acyclic.
We are now in a position to use the criterion with $R_2^{(1)}$ only, and do not need one more prolongation and the use of $R_3^{(1)}$ instead. Prolongating again with respect to the dots, as before, we find the two additional equations

$$\partial_1 a(x) y_3 = 0, \qquad \partial_3 a(x) y_3 = 0.$$

Hence we have two possibilities:
- $\partial_1 a(x) = 0, \partial_3 a(x) = 0 \Rightarrow a(x) = a(x^2)$.
  In this case $R_2^{(1)} = R_2$ is formally integrable.
- $(\partial_1 a)(\partial_3 a) \neq 0 \Rightarrow R_2^{(1)} \subset R_2$, with additional equation $y_3 = 0$. In this case $R_2^{(1)} \subset R_2$ is *not* formally integrable, because it contains $y_{13} = 0$, $y_{33} = 0$ but not $y_{23} = 0$.
  Nevertheless, the interesting property is that $M_2^{(1)} = M_2$ is still 2-acyclic, and we can check that $R_2^{(2)} \subset R_2^{(1)}$ contains the additional equation $y_{11} = 0$, giving thus $y_{23} = 0$ also. Hence, now only $R_2^{(2)}$ is formally integrable, as can be verified directly.

Recapitulating the above results, we obtain, as in the Introduction, an intrinsic tree incorporating all possible formally integrable systems, in dependence on the choice of the constant α and the function $a(x)$:

$$\alpha = 0 \qquad\qquad \alpha \neq 0$$

$$a(x^2) \quad \partial_1 a \partial_3 a \neq 0 \qquad a(x^2) \quad \partial_1 a \partial_3 a \neq 0$$

The method is therefore absolutely general for linear systems depending on parametric coefficients or arbitrary functions. The most generic situation is on the righthand side. In Chapter VII we shall see that this procedure can be extremely useful for applications.

We also notice that the resulting 'leaves' of the tree can be very different. To see this on the above example, we give a scheme indicating the number of equations obtained at order $1, 2, 3$, in each of the four different situations finally possible.

case	1	2	3	4
order 1	0	1	0	1
order 2	5	5	5	6
order 3	9	9	10	10

This scheme can be readily generalized to other examples, although in general the computations cannot easily be done by hand and one often needs a computer.

Example 5. $n = 4$, $m = 1$, $q = 1$.

We want to study the linear system

$$R_1 \quad \begin{cases} y_4 - x^3 y_2 - y = 0 & \boxed{\begin{matrix} 1 & 2 & 3 & 4 \end{matrix}} \\ y_3 - x^4 y_1 = 0 & \begin{matrix} 1 & 2 & 3 & \bullet \end{matrix} \end{cases}$$

R_1 is a vector bundle with $\dim R_2 = 5 - 2 = 3$, and its symbol M_1 is also a vector bundle, with $\dim M_1 = 4 - 2 = 2$:

$$M_1 \quad \begin{cases} v_4 - x^3 v_2 = 0 & \boxed{\begin{matrix} 1 & 2 & 3 & 4 \end{matrix}} \\ v_3 - x^4 v_1 = 0 & \begin{matrix} 1 & 2 & 3 & \bullet \end{matrix} \end{cases}$$

M_1 is involutive, and $\dim M_2 = 10 - (4 + 3) = 3$. Indeed, prolongation of the second equation with respect to the dot gives $v_{34} - x^4 v_{14} = 0$, while prolongation with respect to the multiplicative variables gives, respectively:

$$\begin{array}{lll} \text{first}/x^3 & \Rightarrow v_{34} - x^3 v_{23} = 0 & \quad 1 \\ \text{second}/x^2 & \Rightarrow v_{23} - x^4 v_{12} = 0 & \quad x^3 \\ \text{first}/x^1 & \Rightarrow v_{14} - x^3 v_{12} = 0 & \quad -x^4 \end{array}$$

with as sum the previous equation.

We may look for $R_1^{(1)}$. Using the scheme on the right we easily find

$$R_1^{(1)} \quad \begin{cases} y_4 - x^3 y_1 - y = 0 & \boxed{\begin{matrix} 1 & 2 & 3 & 4 \end{matrix}} \\ y_3 - x^4 y_1 = 0 & \begin{matrix} 1 & 2 & 3 & \bullet \end{matrix} \\ y_2 - y_1 = 0 & \begin{matrix} 1 & 2 & \bullet & \bullet \end{matrix} \end{cases}$$

The reader should compute the prolongations to see that it is quite accidental that $R_1^{(1)}$ is still a vector bundle. Its symbol is:

$$M_1^{(1)} \quad \begin{cases} v_4 - x^3 v_1 = 0 & \boxed{\begin{matrix} 1 & 2 & 3 & 4 \end{matrix}} \\ v_3 - x^4 v_1 = 0 & \begin{matrix} 1 & 2 & 3 & \bullet \end{matrix} \\ v_2 - v_1 = 0 & \begin{matrix} 1 & 2 & \bullet & \bullet \end{matrix} \end{cases}$$

Prolongation of the last equation with respect to x^4 (dot) gives

$$v_{24} - v_{14} = 0.$$

Prolongation with respect to the multiplicative variables gives, respectively:

first/x^2	$\Rightarrow v_{24} - x^3 v_{12} = 0$		1
third/x^1	$\Rightarrow v_{12} - v_{11} = 0$		x^3
first/x^1	$\Rightarrow v_{14} - x^3 v_{11} = 0$		-1

and we recover the previous equation. Hence $M_1^{(1)}$ is involutive, and we leave it to the reader to verify that $R_1^{(2)} = R_1^{(1)}$. It follows that $R_1^{(1)}$ is involutive.

Now, when computing R_3 by hand or by using a computer, among the 20 equations obtained when differentiating twice the two equations defining R_1, only 16 of them are independent, because they contain principal jets of order 3. The 4 remaining ones are

$$\begin{cases} x^3(y_{22} - y_{12}) + 2(y_2 - y_1) = 0, \\ y_{12} - y_{11} = 0, \\ y_{22} - y_{12} = 0, \\ y_{12} - y_{11} = 0, \end{cases}$$

and only 3 of these are linearly independent. However, first prolongation of the equations defining R_1 gives the additional first-order equation $y_2 - y_1 = 0$. Hence R_3 is defined by $2 + 8 + 2 + 16 = 28$ linearly independent equations, and we find $\dim R_3 = 35 - 28 = 7$. As a byproduct, using the short exact sequence

$$0 \longrightarrow M_3 \longrightarrow R_3 \longrightarrow R_2^{(1)} \longrightarrow 0$$

we obtain $\dim R_2^{(1)} = \dim R_3 - \dim M_3$. Since M_1 is involutive, we immediately obtain

$$\beta_1^1 = 0, \quad \beta_1^2 = 0, \quad \beta_1^3 = 1, \quad \beta_1^4 = 1,$$

from which we deduce the characters

$$\alpha_1^1 = 1, \quad \alpha_1^2 = 1, \quad \alpha_1^3 = 0, \quad \alpha_1^4 = 0.$$

We successively obtain

$$\begin{cases} \dim M_1 = \alpha_1^1 + \alpha_1^2 + \alpha_1^3 + \alpha_1^4 = 2, \\ \dim M_2 = \alpha_1^1 + 2\alpha_1^2 + 3\alpha_1^3 + 4\alpha_1^4 = 3, \\ \dim M_3 = \alpha_1^1 + 3\alpha_1^2 + 6\alpha_1^3 + 10\alpha_1^4 = 4, \end{cases}$$

and finally, $\dim R_2^{(1)} = 7 - 4 = 3$. Looking now at the scheme of $M_1^{(1)}$ and using the fact that $R_1^{(1)}$ is involutive, the number of linearly independent equations defining $\rho_1(R_1^{(1)})$ is $3 + 9 = 12$, and therefore $\dim \rho_1(R_1^{(1)}) = 15 - 12 = 3$. Accordingly, $\rho_1(R_1^{(1)}) = R_2^{(1)}$, in agreement with Theorem C.1.

This example shows that the latter theorem is not evident at first sight, since even a direct verification of it is rather delicate.

Example 6. This example will illustrate general differential elimination. The main idea involved is very closely related to the results of Section C. Indeed, let us consider, for simplicity, a linear system involving two kinds of unknowns, say y and z. If we put $z = 0$, the resulting system in y is not integrable, in general. If we can bring it to a formally integrable or involutive form, then we may be able to apply the criterion. Introducing in this procedure the parts containing z as righthand terms, we arrive at a system for which the criterion works by assumption for $z = 0$, i.e. when the righthand terms disappear. Hence, in verifying the criterion in the case these terms are present can only give a number of conditions involving z only; these *are* the desired compatibility conditions. Let us assume that the system in y is of order q, and let us introduce the numbers r, s of the algorithm (provided it exists!). This means that in fact we have to push prolongations up to order $q + r + s$, i.e. we have to prolongate $r + s$ times. Involutivity, or at least 2-acyclicity, can be checked by one more prolongation, provided that the resulting system is involutive or 2-acyclic, which introduces the compatibility conditions. If z appears already through jets of order p, the compatibility conditions will be of maximal order $p + r + s + 1$. As a byproduct, if the homogeneous system in y is already formally integrable and z appears only in righthand members of order $p = 0$, then the order of the compatibility conditions for the second members is 1 plus the number of prolongations needed to reach 2-acyclicity. Many other examples of this situation will be given in Chapter IV.

Let us consider the following linear system with 2 independent variables x^1, x^2 and two unknowns y, z.:

$$R_3 \quad \begin{cases} y_{111} + y_1 - z_1 = 0, \\ y_2 - y_1 - y = 0. \end{cases}$$

Instead of putting $z = 0$, bringing the corresponding homogeneous equation to formally integrable form and subsequently introducing again z, we retain z from the very beginning, but transfer it to the righthand side of the equations.

The coordinate system is not δ-regular, but it becomes such by exchanging x^1 and x^2. It follows that M_3 is involutive, and we may start anew with $R_3^{(1)}$:

$$R_3^{(1)} \quad \begin{cases} y_{111} + y_1 = z_1, \\ y_{22} - y_{11} - 2y_1 - y = 0, \\ y_{12} - y_{11} - y_1 = 0, \\ y_2 - y_1 - y = 0. \end{cases}$$

Since $M_3^{(1)} = M_3$, we may therefore start anew with $R_3^{(2)}$:

$$R_3^{(2)} \quad \begin{cases} y_{222} - 2y_{11} - 2y_1 - y = z_1 \\ y_{122} - 2y_{11} = z_1 \\ y_{112} - y_{11} + y_1 = z_1 \\ y_{111} + y_1 = z_1 \\ y_{22} - y_{11} - 2y_1 - y = 0 \\ y_{12} - y_{11} - y_1 = 0 \\ y_2 - y_1 - y = 0 \end{cases}$$

1	2
1	•
1	•
1	•
•	•
•	•
•	•

The coordinate system is δ-regular for $M_3^{(2)} = 0$, which is trivially involutive, and we leave it to the reader to prove that $R_3^{(2)}$ is involutive when $z = 0$. Prolongation with respect to the dots and elimination of the principal derivatives makes all parametric derivatives disappear, and we are left with the second-order equation

$$z_{12} - z_{11} - z_1 = 0.$$

In this case, although $p = 1, q = 3, r = 0$, and $s = 2$, we obtain an equation of order 2 for z, instead of an equation of order $1 + 0 + 2 + 1 = 4$. This difference becomes clear by examining the prolongations, as we had only do two prolongations of the first-order equation in y, which does not contain z.

Exchanging the roles of y and z, we may similarly eliminate z to find compatibility conditions for y. But $z_1 = 0$ is clearly involutive, and the corresponding system for z is a fibered manifold for given y with $y \neq 0$ if and only if

$$y_2 - y_1 - y = 0.$$

We therefore start from a differential correspondence between y and z until we obtain resolvent systems for y and z separately.

We shall write out the scheme in a systematic way, as follows:

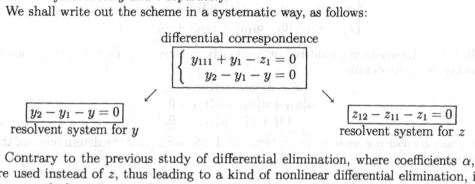

differential correspondence

$$\left\{ \begin{array}{l} y_{111} + y_1 - z_1 = 0 \\ y_2 - y_1 - y = 0 \end{array} \right.$$

$\boxed{y_2 - y_1 - y = 0}$
resolvent system for y

$\boxed{z_{12} - z_{11} - z_1 = 0}$
resolvent system for z

Contrary to the previous study of differential elimination, where coefficients α, a are used instead of z, thus leading to a kind of nonlinear differential elimination, in the case of a linear system there is no tree. We also notice that the previous technique can be extended (on one side only) to the situation of a correspondence that is linear in y with arbitrary nonlinear terms in z.

Examples from mechanics will be given in Chapter VII, and the preceding procedure will be quite important for understanding certain concepts of classical control theory.

Example 7. $n = 3, m = 1, q = 2$.

The purpose of the example is to provide a complete technical study of the δ-regularity condition for systems of coordinates. As it is a quite delicate question, we shall give all the details of the calculations involved.

Consider the following second-order system:

$$R_2 \quad \left\{ \begin{array}{l} y_{33} - y_{13} + y_3 = 0, \\ y_{23} - y_{13} + y_3 = 0, \\ y_{12} - y_{11} + y_1 = 0. \end{array} \right.$$

Its symbol is defined by the equations

$$M_2 \quad \begin{cases} v_{33} - v_{13} = 0 \\ v_{23} - v_{13} = 0 \\ v_{12} - v_{11} = 0 \end{cases} \quad \begin{array}{|ccc|} \hline 1 & 2 & 3 \\ 1 & 2 & \bullet \\ 1 & \bullet & \bullet \\ \hline \end{array}$$

Prolongation of the last equation with respect to the nonmultiplicative variable x^2 shows that M_2 is not involutive $\cdots$ in *that* coordinate system! Hence there are two possibilities:

- M_2 is not involutive or even 2-acyclic *at all*.
- M_2 is involutive by a scheme test in another (δ-regular) coordinate system.

To decide between these we effect an arbitrary change of coordinates $x^i = a^i_j \bar{x}^j$. The only '*a priori*' condition is that $\det a \neq 0$, and hence that we can introduce the inverse matrix $b = a^{-1}$. In the new coordinate system $\dim M_2 = 6 - 3 = 3$ remains the same, and we want to have minimal dimension for $(M_2)^1$, which amounts to having maximal rank for the system

$$\begin{cases} v_{33} - v_{13} = 0, & v_{11}a^1_1 + v_{12}a^2_1 + v_{13}a^3_1 = 0, \\ v_{23} - v_{13} = 0, & v_{21}a^1_1 + v_{22}a^2_1 + v_{23}a^3_1 = 0, \\ v_{12} - v_{11} = 0, & v_{31}a^1_1 + v_{32}a^2_1 + v_{33}a^3_1 = 0, \end{cases}$$

which has already been considered in Section B. Substituting the first 3 equations in the last ones, we obtain

$$\begin{aligned} (a^1_1 + a^2_1)v_{11} + a^3_1 v_{13} &= 0, \\ a^1_1 v_{11} + a^2_1 v_{22} + a^3_1 v_{13} &= 0, \\ (a^1_1 + a^2_1 + a^3_1)v_{13} &= 0. \end{aligned}$$

By sheer coincidence we have 3 equations for 3 unknowns, and the determinant of the system is

$$\det \begin{pmatrix} a^1_1 + a^2_1 & 0 & a^3_1 \\ a^1_1 & a^2_1 & a^3_1 \\ 0 & 0 & a^1_1 + a^2_1 + a^3_1 \end{pmatrix} = a^2_1(a^1_1 + a^2_1)(a^1_1 + a^2_1 + a^3_1).$$

Hence, for $\bar{x}$ to be a δ-regular coordinate system, we *need* the condition

$$a^2_1(a^1_1 + a^2_1)(a^1_1 + a^2_1 + a^3_1) \neq 0.$$

As a byproduct, if $\bar{x} = x$, we have $a^1_1 = 1$, $a^2_1 = 0$, $a^3_1 = 0$, and the condition is *not* satisfied.

Let us prove that this necessary condition is also sufficient. Indeed, if it is satisfied we obtain $v_{11} = v_{22} = v_{13} = 0$, and thus $(M_2)^1 = 0$, implying that $(M_2)^2 = 0$, while $(M_2)^3 = 0$ by definition. It follows that the maximum value of α^1_2 is 3, and in that case $\alpha^2_2 = \alpha^3_2 = 0$ are uniquely determined for any system $\bar{x} = ax$ satisfying the previous condition. In that case we can immediately verify that

$$\dim M_3 = 3 = \alpha^1_2 + 2\alpha^2_2 + 3\alpha^3_2,$$

and it follows that M_2 is involutive, with $\dim M_{2+r} = 3$ *in any coordinate system.*
E.g., let us verify that $H_2^2(M_2)$ vanishes in the original system x, which is now known
not δ-regular. We have to prove the exactness of the sequence (with $E = X \times \mathbb{R}$):

$$0 \longrightarrow M_4 \overset{\delta}{\longrightarrow} T^* \otimes M_3 \overset{\delta}{\longrightarrow} \wedge^2 T^* \otimes M_2 \overset{\delta}{\longrightarrow} \wedge^3 T^* \otimes T^* \longrightarrow 0$$

$$0 \longrightarrow \boxed{3} \longrightarrow \boxed{9} \longrightarrow \boxed{9} \longrightarrow \boxed{3} \longrightarrow 0$$

We have to study the 3 equations

$$\begin{cases} v_{11,23} + v_{12,31} + v_{13,12} = 0, \\ v_{21,23} + v_{22,31} + v_{23,12} = 0, \\ v_{31,23} + v_{32,31} + v_{33,12} = 0. \end{cases}$$

Using the 3 equations defining M_2 we obtain the 3 linearly independent equations

$$\begin{cases} v_{11,23} + v_{11,31} + v_{13,12} = 0, \\ v_{11,23} + v_{22,31} + v_{13,12} = 0, \\ v_{13,23} + v_{13,31} + v_{13,12} = 0, \end{cases}$$

which can be solved for $v_{11,31}, v_{22,31}, v_{13,31}$. Since M_2 is always 0- and 1-acyclic, the
image of the second δ has dimension $9 - 3 = 6$, while the kernel of the third δ has
also dimension $9 - 3 = 6$, and it follows that $H_2^2(M_2) = 0$.

We have looked at the δ-regularity condition from the 'α' point of view, and now we
shall look at it from the 'β' point of view, to check the equivalence of these two points
of view. The previous condition on a_i^i implies that α_2^1 has maximum value 3, hence
$\alpha_2^1 + \alpha_2^2$ has also maximum value 3. Since $\alpha_2^1 + \alpha_2^2 + \alpha_2^3 = \dim M_2 = 3$, we find that
α_2^3 has minimum value 0, and hence β_2^3 has maximum value 1, because $\alpha_2^3 + \beta_2^3 = 1$.
The converse need not be true, as can be seen by the change of coordinates , $\bar{x}^1 = x^1$,
$\bar{x}^2 = x^3$, $\bar{x}^3 = x^2$, which leads to the system

$$\begin{cases} v_{23} - v_{12} = 0, \\ v_{22} - v_{12} = 0, \\ v_{13} - v_{11} = 0, \end{cases}$$

where $\beta_2^3 = 0$. One should also notice that in the initial system x we have $\beta_2^3 = 1$,
although this system is not δ-regular.

We now have $\bar{x}^j = b_i^j x^i$, and obtain in succession

$$\frac{\partial y}{\partial x^i} = \frac{\partial y}{\partial \bar{x}^n} b_i^n + \dots,$$

$$\frac{\partial^2 y}{\partial x^i \partial x^j} = \frac{\partial^2 y}{\partial \bar{x}^n \bar{x}^n} b_i^n b_j^n + \dots.$$

For a general linear system $A_k^{\tau\mu}(x) y_k^k = 0$, in the new coordinate system we obtain
$A_k^{\tau\mu}(x) \bar{y}_{(0,\dots,q)}^k \chi_\mu + (\text{class} < n) = 0$, where we have set $\chi_i = b_i^n$. Annihilating all jets of
class $1, \dots, n - 1$, we are left with equations of the form $A_k^{\tau\mu}(x) \bar{y}_{(0,\dots,q)}^k \chi_\mu = 0$, where

$\chi_\mu = (\chi_1)^{\mu_1} \ldots (\chi_n)^{\mu_n}$ and summation is over all different multi-indices. Accordingly, we obtain the important formula

$$\beta_q^n = \max_\chi \mathrm{rk}(A_k^{\tau\mu}(x)\chi_\mu).$$

This is the only way to look for the number of arbitrary functions in n variables in the formal power series, or for the number of arbitrary unknowns, $\cdots$ when the system is formally integrable!

In our case we have $\beta_2^3 = 1$, unless

$$(\chi_3)^2 - \chi_1\chi_3 = 0, \quad \chi_2\chi_3 - \chi_1\chi_3 = 0, \quad \chi_1\chi_2 - (\chi_1)^2 = 0.$$

An easy verification leads to the following three possibilities

1) $\chi_3 = 0$, $\chi_1 = 0$, χ_2 arbitrary, nonzero.
2) $\chi_3 = 0$, $\chi_2 = \chi_1$ arbitrary, nonzero.
3) $\chi_1 = \chi_2 = \chi_3$ arbitrary, nonzero.

At first sight it may seem that these conditions on b for computing β_2^3 have nothing to do with the conditions on a for computing α_2^1. A tricky piece of algebra will prove that they are in fact direct consequences, i.e. the conditions on a imply those on b for maximal rank. Let us prove this, i.e. let us prove that we cannot have the preceding solutions for χ. Indeed, since $b = a^{-1}$ we have

$$\begin{cases} a_1^1 = \dfrac{1}{\det b}(b_2^2 b_3^3 - b_3^2 b_2^3) \approx b_2^2 \chi_3 - b_3^2 \chi_2, \\[2mm] a_1^2 = \dfrac{1}{\det b}(b_3^2 b_1^3 - b_1^2 b_3^3) \approx b_3^2 \chi_1 - b_1^2 \chi_3, \\[2mm] a_1^3 = \dfrac{1}{\det b}(b_1^2 b_2^3 - b_2^1 b_1^3) \approx b_1^2 \chi_2 - b_2^1 \chi_1, \end{cases}$$

and we successively have

$$\begin{cases} \chi = (0,1,0) & \Rightarrow a_1^2 = 0, \\ \chi = (1,1,0) & \Rightarrow a_1^1 + a_1^2 = 0, \\ \chi = (1,1,1) & \Rightarrow a_1^1 + a_1^2 + a_1^3 = 0, \end{cases}$$

i.e. $\alpha_2^1 = 3 \Rightarrow \alpha_2^3 = 0$, $\beta_2^3 = 1$.

Now we know that M_2 is involutive, and we may, directly or using a computer, verify that, *in any coordinate system*, $R_2^{(1)} = R_2$. Accordingly, R_2 is involutive. We shall explicitly give this verification for the δ-regular coordinate system obtained from the cyclic permutation $(1,2,3) \rightarrow (3,1,2)$, and we obtain the new system in solved form:

$$R_2 \quad \begin{cases} \Phi^3 \equiv y_{33} - y_{13} - y_3 = 0 \\ \Phi^2 \equiv y_{23} - y_{12} - y_2 = 0 \\ \Phi^1 \equiv y_{22} - y_{12} \quad\quad = 0 \end{cases} \quad \boxed{\begin{array}{ccc} 1 & 2 & 3 \\ 1 & 2 & \bullet \\ 1 & 2 & \bullet \end{array}}$$

We check prolongation with respect to the upper dot:

$$d_3\Phi^2 \equiv y_{233} - y_{123} - y_{23} = 0,$$

$$-d_2\Phi^3 \equiv -y_{233} + y_{123} + y_{23} = 0,$$

prolongation with respect to the lower dot:

$$d_3\Phi^1 \equiv y_{223} - y_{123} = 0,$$

$$-d_2\Phi^2 \equiv -y_{223} + y_{122} + y_{22} = 0,$$

$$d_1\Phi^2 \equiv y_{123} - y_{112} - y_{12} = 0,$$

$$-d_1\Phi^1 \equiv -y_{122} + y_{112} = 0,$$

$$-\Phi^1 \equiv -y_{22} + y_{12} = 0,$$

and thus the compatibility conditions:

$$\begin{cases} d_3\Phi^2 - d_2\Phi^3 = 0, \\ d_3\Phi^1 - d_2\Phi^2 + d_1\Phi^2 - d_1\Phi^1 - \Phi^1 = 0, \end{cases}$$

which are already in solved δ-regular form!

We make a final comment on this exercise, and use the last system to illustrate how to use differential methods in algebra for solving the 'belonging' problem.

Indeed, according to the comment made in the Introduction, consider the following three polynomials in χ_1, χ_2, χ_3:

$$P_1 \equiv (\chi_2)^2 - \chi_1\chi_1, \quad P_2 \equiv \chi_2\chi_3 - \chi_1\chi_2 - \chi_2, \quad P_3 \equiv (\chi_3)^2 - \chi_1\chi_3 - \chi_3.$$

We want to know if the polynomial

$$Q \equiv (\chi_2)^2(\chi_3)^2 + (\chi_1)^3\chi_2 - \alpha\chi_1\chi_2$$

belongs to the ideal generated by (P_1, P_2, P_3) for a certain value of α.

For this we transfer P_1, P_2, P_3 to Φ^1, Φ^2, Φ^3, respectively, and Q to $\Omega \equiv y_{2233} + y_{1112} - \alpha y_{12} = 0$. Using the fact that R_2 is involutive, we have a unique way of computing the principal derivatives, but *the final meaning is intrinsic*. Namely, let us adjoin to R_4 the equation $\Omega = 0$, to obtain a system $R'_4(\alpha) \subseteq R_4$. Then $Q \in (P_1, P_2, P_3) \Leftrightarrow R'_4(\alpha) = R_4$. This last problem, however, is a problem in pure linear algebra. We obtain in succession:

$$y_{2233} - y_{1223} - y_{223} = 0,$$

$$-y_{1223} + y_{1122} + y_{122} = 0,$$

$$y_{223} - y_{122} - y_{22} = 0,$$

$$-y_{1122} + y_{1112} = 0,$$

$$y_{22} - y_{12} = 0.$$

Adding up we obtain:

$$y_{2233} + y_{1112} - y_{12} = 0,$$

and the only possible value is $\alpha = 1$, for otherwise $y_{12} = 0$. As we have seen the importance of δ-regular coordinates, for checking involution, and of involution, for checking formal integrability, it does not seem possible that the 'belonging' problem can be solved without randomly choosing a matrix in the algorithm. This remark is independent of *any* progress on the complexity of the algorithm, provided we want an intrinsic procedure. Finally, since for $\alpha = 1$:

$$\Omega = d_{22}\Phi^3 - d_{12}\Phi^2 + d_2\Phi^2 - d_{11}\Phi^1 + \Phi^1,$$

we immediately obtain

$$Q = (\chi_2)^2 P_3 - \chi_1\chi_2 P_2 + \chi_2 P_2 - (\chi_1)^2 P_1 + P_1.$$

A similar comment can be made regarding the question whether or not a polynomial Q has a zero in common with polynomials $P_1, \ldots, P_r$. Indeed, Hilbert's theorem states that we can find polynomials $A_1, \ldots, A_r$ and B such that

$$A_1 P_1 + \cdots + A_r P_r + BQ = 1.$$

It follows that by adjoining the equation defining Ω we obtain a formally integrable system that should be of the simple form $y = 0$.

The next two examples are adaptations of the famous jacobian conjecture.

Example 8. $n = 2$, $m = 1$, $q = 3$.
We shall prove that the three polynomials

$$\begin{cases} P_1 \equiv (\chi_1)^3 + \chi_1(\chi_2)^2 - \chi_2, \\ P_2 \equiv (\chi_1)^2\chi_2 + (\chi_2)^3 + \chi_1, \\ P_3 \equiv \chi_1\chi_2 - 1 \end{cases}$$

cannot have a common zero.

Equivalently, this means that the linear system

$$\mathrm{R}_3 \quad \begin{cases} y_{222} + y_{112} + y_1 = 0 \\ y_{122} + y_{111} - y_2 = 0 \\ y_{12} - y \quad\quad = 0 \end{cases} \quad \begin{array}{|cc|} \hline 1 & 2 \\ 1 & \bullet \\ \bullet & \bullet \\ \hline \end{array}$$

has solution $y = 0$ only.

Its symbol M_3,

$$M_3 \quad \begin{cases} v_{222} + v_{112} = 0 \\ v_{122} + v_{111} = 0 \end{cases} \quad \begin{array}{|cc|} \hline 1 & 2 \\ 1 & \bullet \\ \hline \end{array}$$

is easily seen to be involutive. Hence we may start anew with $\mathrm{R}_3^{(1)}$:

$$\mathrm{R}_3^{(1)} \quad \begin{cases} y_{222} + 2y_1 = 0, \\ y_{122} \quad\quad = 0, \\ y_{112} - y_1 = 0, \\ y_{111} \quad\quad = 0, \\ y_{22} + y_{11} = 0, \\ y_{12} - y \quad = 0, \end{cases}$$

The reader will notice that in this situation we have to take into account not only prolongations of the second-order equation, but also the crossed prolongations of the third-order equations. This situation is therefore the most general and complicated one.

The symbol $M_3^{(1)} = 0$ of $R_3^{(1)}$ is clearly involutive, being of finite type, and we may start anew with $R_3^{(2)}$:

$$R_3^{(2)} \quad \begin{cases} y_{222} + 2y_1 &= 0, \\ y_{122} &= 0, \\ y_{112} - y_1 &= 0, \\ y_{111} &= 0, \\ y_{22} &= 0, \\ y_{12} &= 0, \\ y_{11} &= 0, \\ y &= 0. \end{cases}$$

Therefore the system $R_3^{(2)}$ is not formally integrable, but $R_3^{(3)} = 0$ is clearly involutive, with single solution $y = 0$.

As a final remark, since the maximum number of prolongations for getting $R_{q+r}^{(s)}$ from R_q is $r + s$, it follows that the polynomials $A_1, \ldots, A_r, B$ have degree $r + s$ at most. In our case $r = 0$, $s = 3$, but we obtained $y = 0$ after 2 prolongations. We leave it to the reader to modify P_3 such that 3 prolongations are needed.

Example 9. $n = 2$, $m = 1$, $q = 4$.
The polynomial transformation of the plane

$$\begin{cases} X = x - 2y(y^2 + x) - (y^2 + x)^2, \\ Y = y + (y^2 + x) \end{cases}$$

has constant jacobian

$$\det \begin{pmatrix} 1 - 2y - 2x - 2y^2 & -6y^2 - 2x - 4y^3 - 4xy \\ 1 & 1 + 2y \end{pmatrix} = 1.$$

Surprisingly, it admits a polynomial inverse:

$$\begin{cases} x = X + 2Y(Y^2 + X) - (Y^2 + X)^2, \\ y = Y - (Y^2 + X). \end{cases}$$

The 'jacobian conjecture' states that this is a general fact, for any number of unknowns. As of yet it is neither proved nor disproved!
According to the above result, we have

$$(x = 0, y = 0) \Longleftrightarrow (X = 0, Y = 0).$$

Keeping in mind the previous example, we see that the system

$$R_4 \quad \begin{cases} y_{2222} + 2y_{122} + 2y_{222} + 2y_{12} + y_{12} + y_{11} - y_1 = 0, \\ y_{22} + y_2 + y_1 = 0, \end{cases}$$

can only have a solution $y = $ const, i.e. it amounts to $y_1 = 0$, $y_2 = 0$. As before, we could adjoin the polynomial $Z = xy - 1$ and the corresponding equation $y_{12} - y = 0$, so that the resulting system has unique solution $y = 0$.

Twice prolongation of the second-order equation gives

$$y_{2222} = -y_{222} - y_{122},$$
$$y_{222} = -y_{22} - y_{12} = -y_{12} + y_2 + y_1,$$
$$y_{122} = -y_{12} - y_{11}.$$

By substitution in the first fourth-order equation we arrive at $y_2 = 0$ only. By another prolongation we arrive at $y_{22} = 0$, and thus $y_1 = 0$. Therefore $y_1 = 0$, $y_2 = 0$ are among the equations defining $R_4^{(3)}$.

We can use this latter property of the jacobian to construct a very particular type of system. Let us regard the matrix as a matrix of differential equations with constant coefficients and apply it to 2 unknowns. We obtain the third-order system of 2 equations

$$\begin{cases} 4y_{222}^2 + 6y_{22}^2 + 4y_{12}^1 + 2y_{22}^1 + 2y_1^2 + 2y_2^1 + 2y_1^1 - y^1 = 0, \\ 2y_2^2 + y^2 + y^1 = 0. \end{cases}$$

Of course, this system is not formally integrable, and we leave it to the reader to verify that its only solution is $y^1 = 0$, $y^2 = 0$ (apply the inverse matrix of operators obtained from the fact that the inverse of the polynomial matrix given above is also a polynomial matrix, since the determinant equals 1).

Example 10. If x and y are indeterminates, we can verify that

$$\det \begin{pmatrix} x^2 & xy - 1 \\ xy + 1 & y^2 \end{pmatrix} = 1.$$

Accordingly, the system

$$R_2 \quad \begin{cases} y_{11}^1 + y_{12}^2 - y^2 = 0, \\ y_{12}^1 + y_{22}^2 + y^1 = 0, \end{cases}$$

with $n = 2$, $m = 2$, $q = 2$, has only the solution $y^1 = 0$, $y^2 = 0$. In fact, we leave it to the reader to verify that R_2 has an involutive symbol,

$$M_2 \quad \begin{cases} v_{22}^2 + v_{12}^1 = 0 \quad \boxed{1 \ \ 2} \\ v_{12}^2 + v_{11}^1 = 0 \quad \boxed{1 \ \ \bullet} \end{cases}$$

and that $R_2^{(4)} = 0$, although $R_0^{(4)} = \pi_0^2(R_2^{(2)}) = 0$.

Example 11. $n = 2$, $m = 3$, $q = 1$.
Let us consider the following first-order system:

$$R_1 \quad \begin{cases} \Phi^3 \equiv y_2^2 + y_2^3 - y_1^3 - y_1^2 = 0 & \boxed{1 \ \ 2} \\ \Phi^2 \equiv y_2^1 - y_2^3 - y_1^3 - y_1^2 = 0 & \boxed{1 \ \ 2} \\ \Phi^1 \equiv y_1^1 - 2y_1^3 - y_1^2 \quad\ \ = 0 & \boxed{1 \ \ \bullet} \end{cases}$$

Since it is a linear homogeneous system with constant coefficients, the involutivity of the symbol M_1 implies the involutivity of the system, with one compatibility condition:

$$d_2\Phi^1 - d_1\Phi^2 + d_1\Phi^3 \equiv 0,$$

which can be regarded as the curvature map. Let us compute the characters. We recall that $(M_1)^0 = M_1$, $(M_1)^1$ is determined by the two equations

$$v_2^2 + v_2^3 = 0, \qquad v_2^1 - v_2^3 = 0,$$

and $(M_1)^2 = 0$. Accordingly we have

$$\begin{cases} \alpha_1^1 = \dim(M_1)^0 - \dim(M_1)^1 = (6-3) - (3-2) = 2, \\ \alpha_1^2 = \dim(M_1)^1 - \dim(M_1)^2 = (3-2) - 0 = 1. \end{cases}$$

Hence, although this system has as many equations as number of unknowns, one of the unknowns can be given arbitrarily. Since $\dim M_1 = \alpha_1^1 + \alpha_1^2 = 3$ while $\dim F_0 = \beta_1^1 + \beta_1^2 = 3$, we may as well look for β_1^2 to determine the characters. Introducing the covector $\chi = \chi_1 \, dx^1 + \chi_2 \, dx^2$ as in Example 1, we immediately obtain

$$\det \begin{pmatrix} 0 & \chi_2 - \chi_1 & \chi_2 - \chi_1 \\ \chi_2 & -\chi_1 & -\chi_1 - \chi_2 \\ \chi_1 & \chi_1 & -2\chi_1 \end{pmatrix} = 0,$$

and the generic rank of the matrix is 2, provided $\chi_2 - \chi_1 \neq 0$. It follows that $\beta_1^2 = 2$ and therefore $\alpha_1^2 = m - \beta_1^2 = 3 - 2 = 1$. In the actual coordinate system we employ, we have $\chi_1 = 0$, $\chi_2 = 1$, and this coordinate system is δ-regular, a fact already revealed by the involutivity of the symbol. In that case we may look for solutions $y = f(x)$, by providing

$$f^1(0,0) = 1, \quad f^2(x^1,0) = (x^1)^2, \quad f^3(x^1,x^2) = x^1 + x^2.$$

Using $\Phi^1 = 0$ for $x^2 = 0$, we immediately find

$$f^1(x^1,0) = (1 + x^1)^2.$$

Using now the Cauchy–Kowalewskaya theorem for $\Phi^2 = 0$, $\Phi^3 = 0$, we discover that the unique analytic solution of this analytic system with the given conditions is of polynomial type:

$$f^1(x^1,x^2) = (1 + x^1 + x^2)^2, \quad f^2(x^1,x^2) = (x^1 + x^2)^2, \quad f^3(x^1,x^2) = x^1 + x^2.$$

The compatibility condition already obtained for Φ^1, Φ^2, Φ^3 then ensures that $\Phi^1 = 0$ identically for all x^2, whenever it holds for $x^2 = 0$.

Example 12. $n = 3$, $m = 3$, $q = 1$.

A more complicated example is as follows (with the same remark as in the previous example):

$$
R_1 \begin{cases}
\Phi^6 \equiv y_3^3 - 2y_2^1 - 6y_1^2 - 5y_1^1 = 0 \\
\Phi^5 \equiv y_3^2 + 2y_2^1 - 2y_1^2 - 3y_1^1 = 0 \\
\Phi^4 \equiv y_3^1 - 2y_2^1 - 2y_1^2 - y_1^1 = 0 \\
\Phi^3 \equiv y_2^3 - y_2^1 - y_1^2 - y_1^1 = 0 \\
\Phi^2 \equiv y_2^2 + y_2^1 - y_1^2 - y_1^1 = 0 \\
\Phi^1 \equiv y_1^3 - y_1^2 - 2y_1^1 = 0
\end{cases}
\quad
\begin{array}{ccc}
1 & 2 & 3 \\
1 & 2 & 3 \\
1 & 2 & 3 \\
1 & 2 & \bullet \\
1 & 2 & \bullet \\
1 & \bullet & \bullet
\end{array}
$$

One may check, by hand or using a computer, that the coordinate system is δ-regular and that the system R_1 is involutive, with characters:

$$
\begin{cases}
\alpha_1^1 = \dim(M_1)^0 - \dim(M_1)^1 = (9-6) - (6-5) = 2, \\
\alpha_1^2 = \dim(M_1)^1 - \dim(M_1)^2 = (6-5) - (3-3) = 1, \\
\alpha_1^3 = \dim(M_1)^2 - \dim(M_1)^3 = (3-3) - 0 = 0.
\end{cases}
$$

The four compatibility conditions will be studied in examples of the next Chapter.
  Starting with the initial conditions

$$
f^1(x^1, x^2, 0) = x^1 + x^2, \quad f^2(x^1, 0, 0) = (x^1)^2, \quad f^3(0, 0, 0) = 1,
$$

we may use Φ^1 to obtain

$$
f^3(x^1, 0, 0) = (1 + x^1)^2,
$$

and similarly $\Phi^2 = 0$, $\Phi^3 = 0$ to obtain

$$
f^2(x^1, x^2, 0) = (x^1 + x^2)^2, \quad f^2(x^1, x^2, 0) = (1 + x^1 + x^2)^2.
$$

Finally we may use $\Phi^4 = 0$, $\Phi^5 = 0$ and $\Phi^6 = 0$ to achieve the integration. However, we will no longer find polynomial expressions.

Example 13. $n = 2$, $m + 1$, $q = 3$.
We give this example to show that higher-order equations can be treated by an absolutely similar method. Consider the following third-order system:

$$
R_3 \begin{cases}
\Phi^3 \equiv y_{222} - y_{111} - 3y_{11} - y_2 - 2y_1 & = 0, \\
\Phi^2 \equiv y_{122} - y_{111} - 2y_{11} - y_2 & = 0, \\
\Phi^1 \equiv y_{112} - y_{111} - y_{11} & = 0.
\end{cases}
$$

We start with studying the symbol

$$
M_3 \begin{cases}
v_{222} - v_{111} = 0 \\
v_{122} - v_{111} = 0 \\
v_{112} - v_{111} = 0
\end{cases}
\quad
\begin{array}{cc}
1 & 2 \\
1 & \bullet \\
1 & \bullet
\end{array}
$$

and leave it to the reader to verify that it is involutive. Since $\dim M_3 = \dim(M_3)^0 = 4 - 3 = 1$ and $(M_3)^1$ is determined by $v_{222} = 0$, we have $\dim(M_3)^1 = 0$. Accordingly, $\alpha_3^1 = 1 - 0 = 1$, $\alpha_3^2 = 0 - 0 = 0$.

Again we leave it to the reader to verify that the system R_3 is involutive, by giving the following two compatibility conditions of order one:

$$\begin{cases} d_2\Phi^2 - d_1\Phi^3 + d_1\Phi^1 + 2\Phi^1 = 0, \\ d_2\Phi^1 - d_1\Phi^2 + d_1\Phi^1 + \Phi^1 = 0, \end{cases}$$

which are differentially independent.

We shall now deal with linear systems having degenerate symbols. There is no general rule, but we shall provide two examples concerning the kind of solution that can be found.

Example 14. $n = 1, m = 1, q = 1$.
Let us start with the ODE

$$R_1 \quad \{xy_x + y = 2x + 1.$$

Its symbol

$$M_1 \quad \{xv_x = 0$$

is not a vector bundle, since $x = 0$ is a critical point. Away from $x = 0$ the system is involutive, and the general solution is of the form

$$y = \frac{c}{x} + x + 1,$$

where c is a constant. This solution cannot be expanded around the origin as an analytic function, unless $c = 0$. Conversely, we may look for an analytic solution at the origin in the form of a series

$$y = \sum_\alpha a_\alpha x^\alpha.$$

By substitution we obtain

$$a_0 = 1, \quad a_1 = 1, \quad (\alpha + 1)a_\alpha = 0, \quad \forall \alpha \geq 2,$$

and thus $a_\alpha = 0$ for all $\alpha \geq 2$. This leads to the single solution $y = x + 1$, defined at the origin.

Example 15. $n = 4, m = 1, q = 2$.
The following example, taken from [86] is much more involved:

$$R_2 \quad \begin{cases} y_{44} + (x^2)^2 y_{22} + x^2 y_2 = 0, \\ y_{34} + x^1 x^2 y_{12} = 0, \\ y_{33} + (x^1)^2 y_{11} + x^1 y_1 = 0, \end{cases}$$

with symbol

$$M_2 \quad \begin{cases} v_{44} + (x^2)^2 v_{22} = 0 & \boxed{1 \ 2 \ 3 \ 4} \\ v_{34} + x^1 x^2 v_{12} = 0 & 1 \ 2 \ 3 \ \bullet \\ v_{33} + (x^1)^2 v_{11} = 0 & 1 \ 2 \ 3 \ \bullet \end{cases}$$

M_2 is not involutive and, even worse, M_3 is not a vector bundle unless $x^1 x^2 \neq 0$. Under this condition R_2 is formally integrable and R_3 is even involutive. We give the elementary solution $y = x^4 \log x^1$.

As in the previous example we can inquire about analytic solutions defined at $(x^1, x^2, x^3, x^4) = (0, 0, x^3, x^4)$, of the form

$$y = \sum_\alpha a_\alpha(x^3, x^4)(x^1)^{\alpha_1}(x^2)^{\alpha_2},$$

where $\alpha = (\alpha_1, \alpha_2)$ is a multi-index. By substitution we obtain immediately the homogeneity condition

$$\begin{cases} \partial_{44}a_\alpha + (\alpha_2)^2 a_\alpha = 0 & \boxed{3 \quad 4} \\ \partial_{34}a_\alpha + \alpha_1\alpha_2 a_\alpha = 0 & 3 \quad \bullet \\ \partial_{33}a_\alpha + (\alpha_1)^2 a_\alpha = 0 & 3 \quad \bullet \end{cases}$$

This gives an infinite number of systems, in two variables, for the various coefficients of the series. The new symbol is involutive because it is of finite type. We can immediately verify that the system is not formally integrable, unless one takes into account the additional equation

$$\alpha_1 \partial_4 a_\alpha - \alpha_2 \partial_3 a_\alpha = 0, \qquad \forall \alpha_1, \alpha_2 \geq 0.$$

Returning to the original system, we find to our surprise that any solution analytic in (x^1, x^2) at the origin $(0, 0)$ must satisfy, besides R_2, the new equation

$$x^1 y_{14} - x^2 y_{23} = 0.$$

We immediately see that, of course, $y = x^4 \log x^1$ does not satisfy this equation.

We now turn to the study of nonlinear systems. The examples given are sufficiently elementary to enable the reader to mature the new concepts. More complicated examples, dealing with applications, will be given in the subsequent Chapters.

Example 16. $n = 2$, $m = 1$, $q = 2$.

A very instructive example, which has already been considered in the Introduction, is provided by the following second-order system:

$$\mathcal{R}_2 \quad \begin{cases} y_{22} - \frac{1}{2}(y_{11})^2 = 0, \\ y_{12} - y_{11} = 0. \end{cases}$$

We first notice that $\mathcal{R}_2$ is a fibered manifold over $J_1(\mathcal{E})$, since it is already in parametrized form. Accordingly, the symbol

$$M_2 \quad \begin{cases} v_{22} - y_{11}v_{11} = 0 & \boxed{1 \quad 2} \\ v_{12} - v_{11} = 0 & 1 \quad \bullet \end{cases}$$

is a vector bundle over $\mathcal{R}_2$. However, its first prolongation,

$$M_3 \quad \begin{cases} v_{222} - y_{11}v_{111} = 0, \\ v_{122} - y_{11}v_{111} = 0, \\ v_{112} - v_{111} = 0, \\ (y_{11} - 1)v_{111} = 0 \end{cases}$$

is *not* a vector bundle over $\mathcal{R}_2$. As a byproduct, for computational purposes we have to distinguish carefully between the case $y_{11} - 1 = 0$ (an equation for the solutions) and the case $y_{11} - 1 \neq 0$ (an inequality for the jets).

In the first case the system

$$\begin{cases} y_{22} - \frac{1}{2} = 0 & \boxed{1 \quad 2} \\ y_{12} - 1 = 0 & 1 \quad \bullet \\ y_{11} - 1 = 0 & 1 \quad \bullet \end{cases}$$

is involutive, and the general solution is of the form

$$y = \frac{1}{2}(x^1)^2 + x^1 x^2 + \frac{1}{4}(x^2)^2 + \alpha x^1 + \beta x^2 + \gamma.$$

In the second case $M_3 = 0$ is trivially involutive , and we may start anew with the system

$$\mathcal{R}_3 \begin{cases} y_{222} = 0 & \boxed{1 \quad 2} \\ y_{122} = 0 & 1 \quad \bullet \\ y_{112} = 0 & 1 \quad \bullet \\ y_{111} = 0 & 1 \quad \bullet \\ y_{22} - \frac{1}{2}(y_{11})^2 = 0 & \bullet \quad \bullet \\ y_{12} - y_{11} = 0 & \bullet \quad \bullet \end{cases}$$

which is easily seen to be involutive.

Example 17. $n = 2$, $m = 1$, $q = 2$.

Contrary to what one could believe at first sight, the following second-order non-linear system, which looks more complicated than that of the preceding example, behaves surprisingly differently. Let

$$\mathcal{R}_2 \begin{cases} \Phi^2 \equiv y_{22} - \frac{1}{3}(y_{11})^3 = 0, \\ \Phi^1 \equiv y_{12} - \frac{1}{2}(y_{11})^2 = 0. \end{cases}$$

It too is a fibered manifold in parametric form, and its symbol is

$$M_2 \begin{cases} v_{22} - (y_{11})^2 v_{11} = 0 & \boxed{1 \quad 2} \\ v_{12} - y_{11} v_{11} = 0 & 1 \quad \bullet \end{cases}$$

It is a vector bundle over $\mathcal{R}_2$ with $\dim M_2 = 1$. Nevertheless, the first prolongation

$$M_3 \begin{cases} v_{222} - (y_{11})^3 v_{111} = 0 & \boxed{1 \quad 2} \\ v_{122} - (y_{11})^2 v_{111} = 0 & 1 \quad \bullet \\ v_{112} - y_{11} v_{111} = 0 & 1 \quad \bullet \end{cases}$$

is also a vector bundle over $\mathcal{R}_2$, contrary to what happened in the previous example. Since the equations of M_2 are solved for v_{22}, it follows that $\beta_2^2 = 1$, and we leave it to the reader to verify that M_2 is involutive with characters $\alpha_2^1 = 1$, $\alpha_2^2 = 0$. This implies that all M_{2+r} are vector bundles over $\mathcal{R}_2$, with $\dim M_{2+r} = 1$ for all $r \geq 0$, in agreement with Proposition B.13.

For the construction of the curvature map we introduce the exact sequence

$$0 \longrightarrow M_{q+1} \longrightarrow S_{q+1}T^* \otimes E \longrightarrow T^* \otimes F_0 \longrightarrow F_1 \longrightarrow 0$$

with $q = 2$, and we find the dimensions:

$$0 \longrightarrow \boxed{1} \longrightarrow \boxed{4} \longrightarrow \boxed{4} \longrightarrow \boxed{1} \longrightarrow 0.$$

It follows that $\dim F_1 = 1$, and the curvature map is zero, because

$$d_2\Phi^1 - d_1\Phi^2 + y_{11}d_1\Phi^1 \equiv 0,$$

which is just a manner of verifying the surjectivity of $\pi_2^3 \colon \mathcal{R}_3 \to \mathcal{R}_2$ via dot prolongation.

Example 18. $n = 2$, $m = 1$, $q = 2$.

We modify the preceding example by introducing an arbitrary function $u(x^1, x^2)$ in the coefficients:

$$\mathcal{R}_2 \quad \begin{cases} y_{22} - \frac{1}{3}u(y_{11})^3 = 0, \\ y_{12} - \frac{1}{2}(y_{11})^2 = 0. \end{cases}$$

Of course, for $u = 1$ we recover Example 17, which is involutive, and the question is now to study the dependence of formal integrability on u. First of all, $\mathcal{R}_2$ is a fibered manifold for any u, and M_2 is a vector bundle over $\mathcal{R}_2$ with $\dim M_2 = 1$. Nevertheless, M_3 is not necessarily a vector bundle over $\mathcal{R}_2$ for any u, because the equations defining M_3 contain

$$(1 - u)(y_{11})^2 v_{111} = 0.$$

- For $y_{11} = 0$ we obtain the involutive system

$$y_{22} = 0, \quad y_{12} = 0, \quad y_{11} = 0,$$

with solution $y = ax^1 + bx^2 + c$, where $a, b, c = \text{const.}$
- For $y_{11} \neq 0$ we have to distinguish the case $u = 1$ from the case $u \neq 1$.
    - For $u = 1$ we obtain the involutive system of Example 17.
    - For $u \neq 1$ we have $v_{111} = 0$, thus $\dim M_3 = 0$. Hence M_3 is trivially involutive and M_2 is no longer involutive. So, we have to study the prolongation $\mathcal{R}_3$ of $\mathcal{R}_2$. Among the third-order equations we have

$$\begin{cases} y_{222} + \dfrac{1}{3}\dfrac{\partial_1 u}{u-1}(y_{11})^4 - \dfrac{1}{3}\partial_2 u(y_{11})^3 = 0 & \boxed{\begin{matrix} 1 & 2 \end{matrix}} \\[2ex] y_{122} + \dfrac{1}{3}\dfrac{\partial_1 u}{u-1}(y_{11})^3 = 0 & \boxed{\begin{matrix} 1 & \bullet \end{matrix}} \\[2ex] y_{112} + \dfrac{1}{3}\dfrac{\partial_1 u}{u-1}(y_{11})^2 = 0 & \boxed{\begin{matrix} 1 & \bullet \end{matrix}} \\[2ex] y_{111} + \dfrac{1}{3}\dfrac{\partial_1 u}{u-1}y_{11} = 0 & \boxed{\begin{matrix} 1 & \bullet \end{matrix}} \end{cases}$$

Prolongation with respect to the last dot gives the following additional second-order equation for $\mathcal{R}_3^{(1)}$:

$$\left[\partial_1\left(\frac{\partial_1 u}{u-1}\right) - \frac{1}{3}\left(\frac{\partial_1 u}{u-1}\right)^2\right] y_{11} - \partial_2\left(\frac{\partial_1 u}{u-1}\right) = 0.$$

Since $u \neq 1$, we may assume that $u > 1$ and set $v = \log(u-1)$. We obtain the simplification

$$\left[\partial_{11} v - \frac{1}{3}(\partial_1 v)^2\right] y_{11} - \partial_{12} v = 0.$$

* If $\partial_{11} v - \frac{1}{3}(\partial_1 v)^2 = 0$, then $\partial_{12} v = 0$, and we must have

$$u = 1 + a(x^1 + b)^{-3}, \qquad a \neq 0, \quad a, b = \text{const}.$$

It follows that $\mathcal{R}_3^{(1)} = \mathcal{R}_3$ is involutive.
* If $\partial_{11} v - \frac{1}{3}(\partial_1 v)^2 \neq 0$, we may set

$$w = \frac{\partial_{12} v}{\partial_{11} v - \frac{1}{3}(\partial_1 v)^2},$$

and we find the following equations among the equations defining $\mathcal{R}_3^{(1)}$:

$$y_{11} - w = 0, \quad y_{111} + \frac{1}{3} w \partial_1 v = 0, \quad y_{112} + \frac{1}{3}(w)^2 \partial_1 v = 0.$$

We must now study $\pi_3^4(\rho_1(\mathcal{R}_3^{(1)})) = \pi_3^4(\mathcal{R}_4^{(1)}) = \mathcal{R}_3^{(2)}$, which need not be a fibered manifold, unless

$$\partial_1 w + \frac{1}{3} w \partial_1 v = 0, \quad \partial_2 w + \frac{1}{3}(w)^2 \partial_1 v = 0.$$

Using crossed derivatives, the reader can easily verify that this system in v is involutive. If these conditions for v, and hence u, hold, then $\mathcal{R}_3^{(2)} = \mathcal{R}_3^{(1)}$ is involutive.

The reader will notice that the situation $u = 1$ is *absolutely nongeneric*, contrary to what one could believe.

Below, we draw the tree obtained, framing the part that depends on u only. Again, the most generic situation is on the righthand side.

$$
\begin{array}{cc}
y_{11} = 0 & y_{11} \neq 0
\end{array}
$$

$$
\begin{array}{cc}
u = 1 & u \neq 1
\end{array}
$$

$$
\begin{cases} \partial_{11} v - \frac{1}{3}(\partial_1 v)^2 = 0 \\ \partial_{12} v = 0 \end{cases}
\qquad
\begin{array}{l} \partial_{11} v - \frac{1}{3}(\partial_1 v)^2 \neq 0 \\ \begin{cases} \partial_{122} v + \cdots = 0 \\ \partial_{112} v + \cdots = 0 \end{cases} \end{array}
$$

Example 19. $n = 1$, $m = 1$, $q = 1$.

This example is taken from control theory, but we use the previously introduced notation in it. The problem is to study the dependence on $u(x)$ of the system

$$\mathcal{R}_1 \quad \begin{cases} y_x = y, \\ y^2 + y = u. \end{cases}$$

The symbol $M_1 = 0$ is trivially involutive, and $\mathcal{R}_1$ is a fibered manifold if and only if $2y + 1 \neq 0$, or $y \neq -\frac{1}{2}$, using the jacobian matrix of $y^2 + y$. Hence $\mathcal{R}_1$ is a fibered manifold if and only if $u \neq -\frac{1}{4}$. In that case we have to study $\mathcal{R}_1^{(1)}$, which is defined by the single additional equation $2y^2 + y = \partial_x u$. Using linear combinations of the various equations, we obtain

$$\mathcal{R}_1^{(1)} \quad \begin{cases} y_x = y, \\ y = 2u - \partial_x u, \\ (2u - \partial_x u)^2 + u - \partial_x u = 0. \end{cases}$$

The symbol $M_1^{(1)} = 0$ is trivially involutive, and the rank with respect to y is equal to 1; *however*, $\mathcal{R}_1^{(1)}$ is a fibered manifold if and only if the last equation is satisfied by u. In that case $\mathcal{R}_1^{(1)}$ is defined by the first two equations and is a fibered manifold indeed. We can start anew with $\mathcal{R}_1^{(1)}$, which need not be involutive. We have

$$\pi_1^2(\rho_1(\mathcal{R}_1^{(1)})) = \pi_1^2(\mathcal{R}_2^{(1)}) = \pi_1^3(\mathcal{R}_3) = \mathcal{R}_1^{(2)} \subseteq \mathcal{R}_1^{(1)}.$$

In fact, $\mathcal{R}_1^{(2)}$ is a fibered manifold if and only if

$$\partial_{xx} u - 3 \partial_x u + 2u = 0.$$

In that case $\mathcal{R}_1^{(2)} = \mathcal{R}_1^{(1)}$ is involutive, and the study is complete. One should notice that we have indeed obtained 2 equations for u. However, differentiating the first-order equation, we arrive at

$$(2\partial_x u - 4u - 1)\partial_{xx} u - (4\partial_x u - 8u - 1)\partial_x u = 0,$$

and, after some work (!), the equation

$$(2\partial_x u - 4u - 1)(\partial_{xx} u - 3\partial_x u + 2u) = 0,$$

which is automatically satisfied whenever the preceding second-order equation for u is satisfied. Again, after some work (!), the reader will be able to verify that $2\partial_x u - 4u - 1$ vanishes if and only if $u = -\frac{1}{4}$, which is a 'forbidden' value. This result, which may seem quite unexpected, will be explained in terms of the differential algebraic appproach, when we revisit this example in Chapter VI. The above intrinsic procedure should be compared to and explains the corresponding problems encountered in the Gröbner basis approach of [66].

Example 20. Among the PDE describing the behavior of functions on space-time, a particularly important class is made up by the so-called *evolution equations*. For them people tend to believe that Cauchy data are required, by fixing the unknowns as functions of space when time has a given initial value, say zero. *This is often* *wrong* and we give a simple example of this. This example will be taken up again in Chapter VII for another purpose.

Consider the linearized *Euler equations* for an incompresible fluid with speed $\vec{v}$ and pressure p:

$$\vec{\nabla}\cdot\vec{v} = 0, \quad \partial_t\vec{v} + \vec{\nabla}p = 0.$$

With n space dimensions and with time $t = x^{n+1}$, this system need not be formally integrable, but if we find $\beta_1^{n+1} = n+1$, then this (maximum) value cannot be exceeded, even not by adding new first-order equations. Introducing $\chi = \chi_i\,dx^i + \chi_t\,dt$, we have to look for the rank of the system of equations

$$\chi_i v^i = 0, \quad \chi_t v^i + \chi_i p = 0,$$

where we do not care too much about upper or lower placement of the index i. A similar result will be obtained in Chapter VII for the nonlinear Euler equations, but here we only want to point out the problem involved.

As a byproduct, we have to look for the rank of the $(n+1) \times (n+1)$-matrix of the system:

$$\begin{pmatrix} \chi_1 & \cdots & \chi_n & 0 \\ \chi_t & & 0 & \chi_1 \\ & 0 & \ddots & \vdots \\ & & \chi_t & \chi_n \end{pmatrix} \begin{pmatrix} v^1 \\ \vdots \\ v^n \\ p \end{pmatrix} = \begin{pmatrix} 0 \\ \vdots \\ 0 \\ 0 \end{pmatrix}$$

Let us multiply by $-\chi_i/\chi_t$ the column with label i, and add to the last column all the resulting columns for $i = 1,\ldots,n$. We are left with $-[(\chi_1)^2 + \cdots + (\chi_n)^2]/\chi_t$ in the upper right corner, and therefore the determinant is

$$(-1)^{n-1}(\chi_t)^{n-1}[(\chi_1)^2 + \cdots + (\chi_n)^2].$$

Hence the generic rank of the matrix is $n + 1$, and $\beta_1^{n+1} = n + 1$, so that $\alpha_1^{n+1} = 0$. However, in the case of the original space-time coordinate system we have $\chi_1 = \cdots = \chi_n = 0$, $\chi_t = 1$. Hence this *space-time coordinate system is not δ-regular*, contrary to the intuitive belief! In Chapter VI we shall explain how this result can give a new approach to turbulence.

CHAPTER IV

Linear Systems

As already said in the Introduction, most of mathematical physics is based on a confusion between two differential sequences. The purpose of this Chapter is to construct these two sequences, namely the Janet sequence and the Spencer sequence, for linear systems of PDE. At the same time, the key result of this Chapter will be a fundamental commutative diagram linking these two sequences. We will give many examples illustrating the content of this Chapter. In contrast to our previous books [146], we shall here insist on the effective point of view in constructing the differential sequences. We ask the reader to carefully read this Chapter, which is certainly the most difficult of the whole book. It is particularly striking that such a difficult tool is *absolutely necessary* for revisiting the foundations of continuum mechanics and thermodynamics, $\cdots$ although at first sight there does not seem to exist any link with applications!

A. Linear Sequences

Before presenting the details, we recall from the Introduction that, in essence, we have to answer many questions at the same time:

- When can we construct differential sequences?
- What differential sequences can be constructed?
- How to construct differential sequences?
- Are some differential sequences better than others?
- How to connect differential sequences?
- How to use differential sequences?

Before answering these delicate questions, we would like to warn the reader. Most computer algebra approaches to PDE stop at solving the problems proposed in the preceding Chapter. Of course, the computations can become very intricate and one could hardly do without a computer. Nevertheless, everybody can understand them. Now we study a quite different situation. Indeed, the importance of jet theory and homological algebra lies in the fact that they allow us to prove results that cannot be discovered by hand or on a computer. In particular, we could say that this Chapter starts at the point where the preceding Chapter ended, namely with an involutive linear system.

Let E be a vector bundle over X with $\dim E = m$, and let $R_q \subset J_q(E)$ be an involutive system of order q on E. We thus assume that the characters $\alpha_q^1, \ldots, \alpha_q^n$ of the involutive symbol $M_q \subset S_q T^* \otimes E$ are already known. It is essential to notice that, even though involutivity must be checked in a δ-regular coordinate system, once we know by pure linear algebra that the symbol has this property, all the constructions in this Chapter can be done in an *arbitrary* coordinate system. As already said, we can define $F_0 = J_q(E) / R_q$ and consider the canonical projection $\Phi: J_q(E) \to F_0$, or we can start with such an epimorphism and define $R_q = \ker \Phi$. With some abuse of language we shall say that the operator $\mathcal{D} = \Phi \circ j_q$ is *involutive*, and we have the left beginning of a sequence that will become the *Janet sequence*:

$$0 \longrightarrow \Theta \longrightarrow E \stackrel{\mathcal{D}}{\longrightarrow} F_0,$$

where Θ is the set of solutions of $\mathcal{D}$, i.e. the set of sections ξ of E such that $\mathcal{D}\xi = 0$.

For constructing the Janet sequence we can use two methods:

- '*step by step*';
- '*as a whole*'.

We shall see that each method has its advantages and disadvantages, and we will describe these. The '*step by step*' method will make it possible to inductively construct operators $\mathcal{D}_1, \dots, \mathcal{D}_n$. In fact, we shall construct from $\mathcal{D}$ a first-order operator $\mathcal{D}_1$, prove that it is involutive, and start anew with $\mathcal{D}_1$ as if we could forget $\mathcal{D}$. Nevertheless, all formal information concerning $\mathcal{D}_1$, such as characters, dimensions, etc., comes from $\mathcal{D}$, and this will provide the reason for stopping at $\mathcal{D}_n$ while still convincing the reader that the sequence can be constructed '*as a whole*'.

This second construction, which is much more sophisticated, will prove that each operator $\mathcal{D}_r$ can be constructed as easy as the others, although the first construction might give the feeling that the difficulty of constructing them increases with r.

The scheme of the step by step method is based on a successive construction of 3-row diagrams, each central row of one becoming the bottom row of the next, as in the following picture:

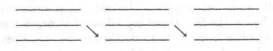

The top rows will depend on the symbol M_q only, and its involutivity will make them exact. This property will be used in a chase, to prove inductively that the central row is exact; the reader immediately understands the importance of the results presented in Chapter I. At the same time we shall recover Corollary III.C.4 by a simpler chase.

The starting diagram is as follows:

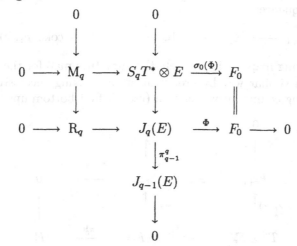

where $\sigma_0(\Phi) = \sigma(\Phi)$ is the restriction of Φ to $S_q T^* \otimes E$. Sometimes, the diagram can be achieved in a better way, but usually $\sigma(\Phi)$ need not be even surjective. This diagram makes it possible to define the symbol and to verify involutivity by computing the characters. We recall the numerical relations

$$\dim M_q = \alpha_q^1 + \cdots + \alpha_q^n,$$
$$\dim R_q = \dim J_q(E) - \dim F_0.$$

The second diagram is as follows:

$$
\begin{array}{ccccccccc}
 & & 0 & & 0 & & 0 & & 0 \\
 & & \downarrow & & \downarrow & & \downarrow & & \downarrow \\
0 \longrightarrow & M_{q+1} & \longrightarrow & S_{q+1}T^* \otimes E & \xrightarrow{\sigma_1(\Phi)} & T^* \otimes F_0 & \xrightarrow{\sigma_0(\Psi_1)} & F_1 & \longrightarrow 0 \\
 & \downarrow & & \downarrow & & \downarrow & & \| & \\
0 \longrightarrow & R_{q+1} & \longrightarrow & J_{q+1}(E) & \xrightarrow{\rho_1(\Phi)} & J_1(F_0) & \xrightarrow{\Psi_1} & F_1 & \longrightarrow 0 \\
 & \downarrow & & \downarrow{\scriptstyle \pi_q^{q+1}} & & \downarrow{\scriptstyle \pi_0^1} & & \downarrow & \\
0 \longrightarrow & R_q & \longrightarrow & J_q(E) & \xrightarrow{\Phi} & F_0 & \longrightarrow & 0 & \\
 & \downarrow & & \downarrow & & \downarrow & & & \\
 & 0 & & 0 & & 0 & & &
\end{array}
$$

In this diagram F_1 is defined by the top row, as the cokernel of $\sigma_1(\Phi)$, and it can be identified, after a chase, with the cokernel of $\rho_1(\Phi)$, since both are defined up to isomorphism. We let Ψ_1 be the canonical projection $J_1(F_0) \to F_1$, and $\sigma_0(\Psi_1)$ its restriction to $T^* \otimes F_0$. In the above identification we use, in a crucial way, the fact that $\pi_q^{q+1}: R_{q+1} \to R_q$ is an epimorphism. The snake Theorem (Chapter I) gives the following exact sequence:

$$0 \longrightarrow M_{q+1} \longrightarrow R_{q+1} \longrightarrow R_q \longrightarrow F_1 \longrightarrow \operatorname{coker} \rho_1(\Phi) \longrightarrow 0,$$

where the connecting map $\cdots$ is precisely the curvature map for the general case.

Continuing in a similar way by prolongation and taking successive cokernels, we obtain the following commutative diagram (read it first bottom upwards!):

$$
\begin{array}{ccccc}
& 0 & & 0 & \\
& \uparrow & & \uparrow & \\
0 \longrightarrow & F_{r+1} & = & F_{r+1} & \longrightarrow 0 \\
& {\scriptstyle \sigma_0(\Psi_{r+1})}\uparrow & & {\scriptstyle \Psi_{r+1}}\uparrow & \uparrow \\
0 \longrightarrow & T^* \otimes F_r & \longrightarrow & J_1(F_r) \xrightarrow{\pi_0^1} F_r & \longrightarrow 0 \\
& {\scriptstyle \sigma_1(\Psi_r)}\uparrow & & {\scriptstyle \rho_1(\Psi_r)}\uparrow \quad {\scriptstyle \Psi_r}\uparrow & \\
& \vdots & & \vdots \qquad \vdots & \\
& {\scriptstyle \sigma_{r-1}(\Psi_2)}\uparrow & & {\scriptstyle \rho_{r-1}(\Psi_2)}\uparrow \quad {\scriptstyle \rho_{r-2}(\Psi_2)}\uparrow & \\
0 \longrightarrow & S_r T^* \otimes F_1 & \longrightarrow & J_r(F_1) \xrightarrow{\pi_{r-1}^r} J_{r-1}(F_1) & \longrightarrow 0 \\
& {\scriptstyle \sigma_r(\Psi_1)}\uparrow & & {\scriptstyle \rho_r(\Psi_1)}\uparrow \quad {\scriptstyle \rho_{r-1}(\Psi_1)}\uparrow & \\
0 \longrightarrow & S_{r+1} T^* \otimes F_0 & \longrightarrow & J_{r+1}(F_0) \xrightarrow{\pi_r^{r+1}} J_r(F_0) & \longrightarrow 0 \\
\cdots \quad {\scriptstyle \sigma_{r+1}(\Phi)}\uparrow \cdots & & \cdots \; {\scriptstyle \rho_{r+1}(\Phi)}\uparrow \cdots & \cdots \; {\scriptstyle \rho_r(\Phi)}\uparrow \cdots & \cdots \cdots \text{ cut} \\
0 \longrightarrow & S_{q+r+1} T^* \otimes E & \longrightarrow & J_{q+r+1}(E) \xrightarrow{\pi_{q+r}^{q+r+1}} J_{q+r}(E) & \longrightarrow 0 \\
& \uparrow & & \uparrow \qquad \uparrow & \\
0 \longrightarrow & M_{q+r+1} & \longrightarrow & R_{q+r+1} \longrightarrow R_{q+r} & \longrightarrow 0 \\
& \uparrow & & \uparrow \qquad \uparrow & \\
& 0 & & 0 \qquad 0 &
\end{array}
$$

Applying the Spencer δ-map to the left column and chasing as in Proposition III.B.13, we see that the involutivity of M_q inductively implies the exactness of the left column, which, also inductively, implies the exactness of the central row. As a byproduct we see that, if R_q is involutive, then this diagram is commutative *and* exact. Otherwise, the 2-acyclicity of M_q implies the exactness of the left column at $S_{r+1}T^* \otimes F_0$ *only*, and Theorem I.8 inductively implies Corollary III.C.4.

Accordingly, we may cut it along the dotted line, and define

$$N_{r+1} = \ker \sigma_r(\Psi_1) = \operatorname{im} \sigma_{r+1}(\Phi) = \rho_r(N_1),$$
$$B_{r+1} = \ker \rho_r(\Psi_1) = \operatorname{im} \rho_{r+1}(\Phi) = \rho_r(B_1),$$

in order to obtain the commutative diagram

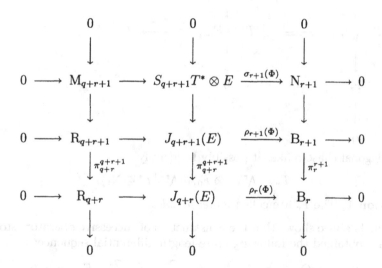

Defining $\mathcal{D}_r = \Psi_r \circ j_1$, our step-by-step construction is complete if we can prove that $\mathcal{D}_1$ is involutive. Indeed, by the above diagram, B_1 is formally integrable. Applying the Spencer δ-sequence to the top row of this diagram, we see that N_r is a vector bundle over X for all $r \geq 1$, and also that N_1 is involutive (in fact, s-acyclic whenever M_q is $(s+1)$-acyclic). Accordingly, we may forget about $\mathcal{D}$ and start anew with $\mathcal{D}_1$, while noting that $F_0, F_1, \ldots, F_r$ are linked by the following commutative and exact diagram:

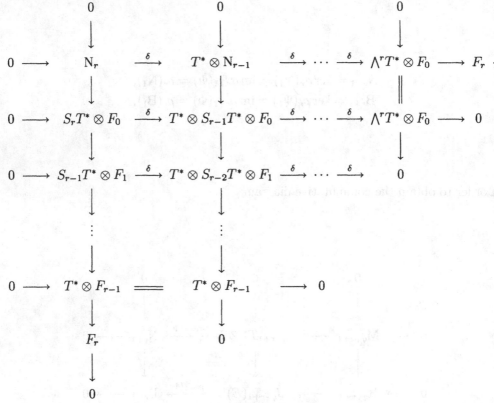

where a diagonal chase makes it possibly to identify

$$F_r = \wedge^r T^* \otimes F_0 / \delta(\wedge^{r-1} T^* \otimes N_1).$$

Definition 1. The F_r are called *Janet bundles*.

The formula above shows that the construction of successive operators stops at $\mathcal{D}_n$, and we have obtained the following finite-length differential sequence:

$$0 \longrightarrow \Theta \longrightarrow E \xrightarrow{\mathcal{D}} F_0 \xrightarrow{\mathcal{D}_1} \cdots \xrightarrow{\mathcal{D}_n} F_n \longrightarrow 0.$$

Definition 2. This sequence is called the *Janet sequence*.

In this sequence, we notice that $\mathcal{D}$ is involutive of order q and $\mathcal{D}_1, \ldots, \mathcal{D}_n$ are involutive of order 1. In actual practice it is extremely difficult to determine the images and the kernels of the δ-maps without the use of a computer, and we hope that in the future software packages will be created for this purpose. Notice that the Janet sequence is *formally exact*, i.e. it is exact at the level of prolonged morphisms. It follows that it is the best sequence we can hope for, in the sense that each $\mathcal{D}_{r+1}$ represents effectively *all* the compatibility conditions for the preceding $\mathcal{D}_r$. Indeed, if we want to solve $\mathcal{D}\xi = \eta$, we must have $\mathcal{D}_1\eta = 0$. However, by formal exactness,

$B_{r+1} = \rho_r(B_1)$, which means that *all compatibility conditions are obtained from first-order compatibility conditions.*

Remark 3. In view of its use and importance in classical differential geometry, which will be revealed in the next Chapter, in [144] we called this sequence a *P-sequence*, with 'P' for '*physical*' and not 'Pommaret'! Therefore, it is quite astonishing to discover that in applications to mathematical physics another differential sequence is even more important!

It remains to express the characters of $\mathcal{D}_1$ in terms of those of $\mathcal{D}$, since we can then find the characters of $\mathcal{D}_r$ by induction. For this we note that, by the definition of N_{r+1},

$$\dim N_{r+1} = \dim S_{q+r+1}T^* \otimes E - \dim M_{q+r+1}$$
$$= \sum_{i=1}^{n} \frac{(r+i)!}{(r+1)!\,(i-1)!} \left(m\frac{(q+n-i-1)!}{(q-1)!\,(n-i)!} - \alpha_q^i \right)$$
$$= \sum_{i=1}^{n} \frac{(r+i)!}{(r+1)!\,(i-1)!} \beta_q^i.$$

If $\alpha_q^1, \ldots, \alpha_q^n$ are the characters of M_q, Proposition III.B.9 2) implies that

$$\dim(M_{q+1})^{i-1} - \dim(M_{q+1})^i = \dim(M_q)^{i-1},$$

i.e.

$$\alpha_{q+1}^i = \alpha_q^i + \cdots + \alpha_q^n.$$

To compute the characters of N_1 we use a formal trick. Indeed, the previous computation implies that the characters of 'N_0' should be $\beta_q^1, \ldots, \beta_q^n$. Accordingly, the characters of N_1 are

$$\beta_q^1 + \cdots + \beta_q^n, \ldots, \beta_q^i + \cdots + \beta_q^n, \ldots, \beta_q^n,$$

and we check that $\dim N_1 = \sum_{i=1}^{n} i\beta_q^i$. Thus, there is a simple interpretation of the scheme already produced:

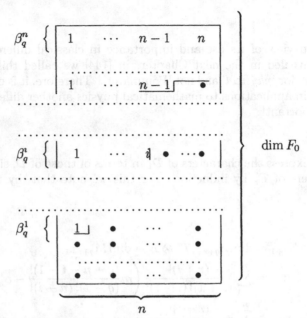

Once we have the scheme for $\mathcal{D}$, that of $\mathcal{D}_1$ is uniquely determined by the following two rules:

- $\dim F_1$ = number of dots in the scheme of $\mathcal{D}$;
- the principal jets of class i are described by the dots in column i.

In Section C we shall present many examples of this inductive combinatorial construction. In particular, we shall see that the Janet sequence of $\mathcal{D}$ stops at F_r, with r the maximum number of dots in a row. Hence, from a simple inspection of the scheme of $\mathcal{D}$ we can easily determine the dimensions of the various Janet bundles in the Janet sequence for $\mathcal{D}$.

Among the linear differential operators of order q that are involutive, the simplest one is j_q itself, which has zero symbol. Accordingly, we introduce the following Janet sequence for it:

$$0 \longrightarrow E \xrightarrow{\ j_q\ } C_0(E) \xrightarrow{\ D_1\ } C_1(E) \xrightarrow{\ D_2\ } \cdots \xrightarrow{\ D_n\ } C_n(E) \longrightarrow 0.$$

It depends on q and E only. In this sequence we have

$$C_r(E) = C_{q,r}(E) = \wedge^r T^* \otimes J_q(E) / \delta(\wedge^{r-1} T^* \otimes S_{q+1} T^* \otimes E),$$

because $N_1 \simeq S_{q+1} T^* \otimes E$ (recall that j_q has zero symbol), and $C_0(E) = J_q(E)$. Such a sequence always terminates at $C_n(E)$.

We shall now construct the preceding sequence '*as a whole*'. For this we first extend the Spencer operator $D: J_{q+1}(E) \to T^* \otimes J_q(E): \xi_{q+1} \to j_1(\xi_q) - \xi_{q+1}$, to an operator

$D: \wedge^s T^* \otimes J_{q+1}(E) \rightarrow \wedge^{s+1} T^* \otimes J_q(E)$ by setting

$$D(\omega \otimes \xi_{q+1}) = d\omega \otimes \xi_q + (-1)^s \omega \wedge D\xi_{q+1},$$

for all $\omega \in \wedge^s T^*$ and $\xi_{q+1} \in J_{q+1}(E)$ projecting to $\xi_q \in J_q(E)$. Recall that, since D is an operator, we must deal with sections.

Proposition 4. *There is a commutative diagram*

$$
\begin{array}{ccc}
\wedge^s T^* \otimes J_{q+1}(E) & \xrightarrow{\rho_1(\Phi)} & \wedge^s T^* \otimes J_1(F_0) \\
\downarrow{\scriptstyle D} & & \downarrow{\scriptstyle D} \\
\wedge^{s+1} T^* \otimes J_q(E) & \xrightarrow{\Phi} & \wedge^s T^* \otimes F_0
\end{array}
$$

PROOF. In local coordinates we may define Φ and $\rho_1(\Phi)$ on sections as follows:

$$
\rho_1(\Phi) \quad
\begin{cases}
\Phi & \left\{ A_k^{\tau\mu}(x)\xi_{\mu,I}^k(x) = \eta_{,I}^\tau(x), \right. \\
& A_k^{\tau\mu}(x)\xi_{\mu+1_i,I}^k(x) + \partial_i A_k^{\tau\mu}(x)\xi_{\mu,I}^k(x) = \eta_{i,I}^\tau(x).
\end{cases}
$$

Differentiating the first equations with respect to x^i and subtracting the second equations, we obtain

$$A_k^{\tau\mu}(x)\left(\partial_i \xi_{\mu,I}^k(x) - \xi_{\mu+1_i,I}^k(x) \right) = \partial_i \eta_{,I}^\tau(x) - \eta_{i,I}^\tau(x).$$

Multiplication by $dx^i \wedge dx^I$ finishes the proof. $\square$

Corollary 5. *There is a restriction*

$$D: \wedge^s T^* \otimes R_{q+1} \rightarrow \wedge^{s+1} T^* \otimes R_q.$$

PROOF. We only have to introduce the left kernels and consider the restriction of D. $\square$

Lemma 6. $D \circ D = 0.$

PROOF. Setting $\omega = (\omega_\mu^k) = (\xi_{\mu,I}^k(x)\, dx^I)$ for the bundle-valued forms involved, we immediately obtain

$$(D\omega)_\mu^k = (\partial_i \xi_{\mu,I}^k(x) - \xi_{\mu+1_i,I}^k(x))\, dx^i \wedge dx^I.$$

Hence,

$$
\begin{aligned}
(D \circ D\omega)_\mu^k &= dx^i \wedge [\partial_i (D\omega)_\mu^k - (D\omega)_{\mu+1_i}^k] \\
&= dx^i \wedge dx^j \wedge (\partial_{ij}\omega_\mu^k - \partial_i\omega_{\mu+1_j}^k - \partial_j\omega_{\mu+1_i}^k + \omega_{\mu+1_i+1_j}^k) \\
&= -dx^i \wedge dx^j \wedge (\partial_i\omega_{\mu+1_j}^k + \partial_j\omega_{\mu+1_i}^k) \\
&= 0. \qquad \square
\end{aligned}
$$

We may therefore consider the sequence

$$0 \longrightarrow \Theta \xrightarrow{j_{q+r+n}} R_{q+r+n} \xrightarrow{D} T^* \otimes R_{q+r+n-1} \xrightarrow{D} \cdots \xrightarrow{D} \wedge^n T^* \otimes R_{q+r} \longrightarrow 0.$$

Definition 7. This sequence is called the *first Spencer sequence*.

This sequence does not have very nice properties, as we shall now prove. The first problem is that it does not stabilize in the order of jets, i.e. if we want to have R_{q+r} at the end, we need to start with R_{q+r+n}. The second problem is that each D involved does not describe *all* the compatibility conditions of the preceding one, or of j_{q+r+n}. In Example II.B.9 we have already considered the situation for the first D. We leave it to the reader, as an exercise in local coordinates, to similarly treat the second D.

Remark 8. We can prove in a direct way that the sequence

$$0 \longrightarrow E \xrightarrow{\ j_{q+1}\ } J_{q+1}(E) \xrightarrow{\ D\ } T^* \otimes J_q(E)$$

is not formally exact. Indeed, on the level of the zero-th prolongation we have

$$0 \longrightarrow J_{q+2}(E) \longrightarrow J_1(J_{q+1}(E)) \longrightarrow T^* \otimes J_q(E) \longrightarrow 0,$$

because $j_1 \circ j_{q+1} = j_{q+2}$. Counting the dimensions we obtain:

$$\dim J_1(J_{q+1}(E)) - \dim T^* \otimes J_q(E) - \dim J_{q+2}(E)$$
$$= m(n+1)\frac{(q+n+1)!}{(q+1)!\,n!} - mn\frac{(q+n)!}{q!\,n!} - m\frac{(q+n+2)!}{(q+2)!\,n!}$$
$$= m(q+1)\frac{(q+n)!}{(q+2)!\,(n-2)!} > 0.$$

Nevertheless, the first Spencer sequence does have an interesting property:

Proposition 9. *The cohomology of the first Spencer sequence stabilizes at sufficiently high order.*

PROOF. Indeed, we have the following commutative diagram:

$$
\begin{array}{ccccccc}
 & & 0 & & 0 & & \\
 & & \downarrow & & \downarrow & & \\
0 & \longrightarrow & M_{q+r+1} & \xrightarrow{-\delta} & T^* \otimes M_{q+r} & \xrightarrow{-\delta} & \cdots \\
 & & \downarrow & & \downarrow & & \downarrow \\
0 \longrightarrow \Theta & \xrightarrow{j_{q+r+1}} & R_{q+r+1} & \xrightarrow{\ D\ } & T^* \otimes R_{q+r} & \xrightarrow{\ D\ } & \cdots \\
 \| & & \downarrow {\scriptstyle \pi^{q+r+1}_{q+r}} & & \downarrow {\scriptstyle \pi^{q+r}_{q+r-1}} & & \\
0 \longrightarrow \Theta & \xrightarrow{j_{q+r}} & R_{q+r} & \xrightarrow{\ D\ } & T^* \otimes R_{q+r-1} & \xrightarrow{\ D\ } & \cdots \\
 & & \downarrow & & \downarrow & & \downarrow \\
 & & 0 & & 0 & & 0
\end{array}
$$

where the restriction of D to the symbol level is easily seen to be $-\delta$ in local coordinates. Hence, since M_{q+r} becomes involutive for r large enough, the upper Spencer δ-sequence becomes exact, and the cohomology of the first Spencer sequence stabilizes, though it need not always vanish [72, 172]. $\square$

The main use of this Proposition lies in the following important corollary:

Corollary 10. *The Spencer sequences*

$$0 \longrightarrow E \xrightarrow{\ j_{q+r+1}\ } J_{q+r+1}(E) \xrightarrow{\ D\ } T^* \otimes J_{q+r}(E) \xrightarrow{\ D\ } \cdots$$

are locally exact, i.e. any local section killed by one D is the image of a local section under D.

PROOF. All δ-sequences

$$0 \longrightarrow S_{q+r+1}T^* \otimes E \xrightarrow{\ \delta\ } T^* \otimes S_{q+r}T^* \otimes E \xrightarrow{\ \delta\ } \cdots$$

are exact, by Proposition III.B.3. The proof then follows by induction from the case $q = r = 0$, as in the beginning of this Section. Indeed, in that case we have the following commutative and exact diagram:

$$
\begin{array}{ccccccccc}
 & & & & 0 & & 0 & & \\
 & & & & \downarrow & & \downarrow & & \\
 & & 0 & \longrightarrow & T^* \otimes E & \xrightarrow{-\delta} & T^* \otimes E & \longrightarrow & 0 \\
 & & & & \downarrow & & \downarrow & \| & \\
0 & \longrightarrow & E & \xrightarrow{j_1} & J_1(E) & \xrightarrow{D} & T^* \otimes E & \longrightarrow & 0 \\
 & & \| & & \downarrow{\pi_0^1} & & \| & & \\
0 & \longrightarrow & E & = & E & \longrightarrow & 0 & & \\
 & & \downarrow & & \downarrow & & & & \\
 & & 0 & & 0 & & & &
\end{array}
$$

To aid the reader we provide the next diagram, which can be useful in applications to understand how the central row can become the bottom row.

$$
\begin{array}{ccccccccccc}
 & & 0 & & 0 & & 0 & & & & \\
 & & \downarrow & & \downarrow & & \downarrow & & & & \\
0 & \longrightarrow & S_2 T^* \otimes E & \xrightarrow{-\delta} & T^* \otimes T^* \otimes E & \xrightarrow{-\delta} & \wedge^2 T^* \otimes E & \longrightarrow & 0 & & \\
 & & \downarrow & & \downarrow & & \downarrow & \| & & & \\
0 & \longrightarrow & E & \xrightarrow{j_2} & J_2(E) & \xrightarrow{D} & T^* \otimes J_1(E) & \xrightarrow{D} & \wedge^2 T^* \otimes E & \longrightarrow & 0 \\
 & & \| & & \downarrow{\pi_1^2} & & \downarrow{\pi_0^1} & & \downarrow & & \\
0 & \longrightarrow & E & \xrightarrow{j_1} & J_1(E) & \xrightarrow{D} & T^* \otimes E & \longrightarrow & 0 & & \\
 & & \downarrow & & \downarrow & & \downarrow & & & & \\
 & & 0 & & 0 & & 0 & & & &
\end{array}
$$

etc. $\square$

Now we introduce the vector bundles

$$C_r = \bigwedge^r T^* \otimes R_q \,/\delta(\bigwedge^{r-1} T^* \otimes M_{q+1})$$

and state

Definition 11. The C_r are called *Spencer bundles*.

The importance of these bundles lies in the following proposition.

Proposition 12. *There is a differential sequence*

$$0 \longrightarrow \Theta \xrightarrow{\ j_q\ } C_0 \xrightarrow{\ D_1\ } \cdots \xrightarrow{\ D_n\ } C_n \longrightarrow 0,$$

which is called the second Spencer sequence.

PROOF. We have the following commutative diagram [144, 172]:

$$
\begin{array}{ccccc}
& & \bigwedge^{r-2}T^* \otimes R_{q+2} & \xrightarrow{\pi^{q+2}_{q+1}} & \bigwedge^{r-2}T^* \otimes R_{q+1} \longrightarrow 0 \\
& & \downarrow{\scriptstyle D} & & \downarrow{\scriptstyle D} \\
0 \longrightarrow \bigwedge^{r-1}T^* \otimes M_{q+1} \longrightarrow & & \bigwedge^{r-1}T^* \otimes R_{q+1} & \xrightarrow{\pi^{q+1}_q} & \bigwedge^{r-1}T^* \otimes R_q \longrightarrow 0 \\
\downarrow{\scriptstyle -\delta} & & \downarrow{\scriptstyle D} & & \quad D'\downarrow \;\searrow^{\;C_{r-1}\,\to\,0}_{\;\;\nearrow\,D_r} \\
0 \longrightarrow \delta(\bigwedge^{r-1}T^* \otimes M_{q+1}) \longrightarrow & & \bigwedge^r T^* \otimes R_q & \longrightarrow & C_r \longrightarrow 0
\end{array}
$$

To construct D' we take a section $a_q \in \bigwedge^{r-1}T^* \otimes R_q$, lift it to $a_{q+1} \in \bigwedge^{r-1}T^* \otimes R_{q+1}$, then apply D to obtain a section $Da_{q+1} \in \bigwedge^r T^* \otimes R_q$, which we finally project to a section $D'a_q \in C_r$. This last projection is well defined by the definition of C_r.

If $a_q = -\delta m_{q+1}$ with $m_{q+1} \in \bigwedge^{r-2}T^* \otimes M_{q+1}$, we may regard m_{q+1} as a section of $\bigwedge^{r-2}T^* \otimes R_{q+1}$, and lift it to $m_{q+2} \in \bigwedge^{r-2}T^* \otimes R_{q+2}$. It follows that $D'a_q = D' \circ Dm_{q+1} = D' \circ D \circ \pi^{q+2}_{q+1}(m_{q+2})$ is the projection on C_r of $D \circ Dm_{q+2} = 0$. Hence D' factors through $D_r \colon C_{r-1} \to C_r$ and $D \circ D = 0 \Rightarrow D' \circ D = 0 \Rightarrow D_r \circ D_{r-1} = 0$. $\square$

We see that this sequence stabilizes the order, since only R_q and R_{q+1} are needed to construct it. Later we shall see that it is also formally exact.

Corollary 13. *The sequence*

$$0 \longrightarrow E \xrightarrow{\ j_q\ } C_0(E) \xrightarrow{\ D_1\ } \cdots \xrightarrow{\ D_n\ } C_n(E) \longrightarrow 0$$

is locally exact.

PROOF. We may use the previous diagram, with $J_q(E)$ instead of R_q. Any section of C_{r-1} killed by D_r can be lifted to a section $a_q \in \bigwedge^{r-1}T^* \otimes J_q(E)$ killed by D'. Then a_q can be lifted to $a_{q+1} \in \bigwedge^{r-1}T^* \otimes J_{q+1}(E)$ such that $D' \circ \pi^{q+1}_q(a_{q+1}) = 0 \Rightarrow Da_{q+1} = -\delta m_{q+1}$ for some section $m_{q+1} \in \bigwedge^{r-1}T^* \otimes S_{q+1}T^* \otimes E$. Setting $a'_{q+1} = a_{q+1} + m_{q+1}$, we obtain $Da'_{q+1} = 0$. Hence, by Corollary 10, we can find a $b_{q+2} \in \bigwedge^{r-2}T^* \otimes R_{q+2}$ such that $a'_{q+1} = Db_{q+2}$. Now $\pi^{q+1}_q(a'_{q+1}) = \pi^{q+1}_q(a_{q+1}) = a_q$, and if $b_{q+1} \in \bigwedge^{r-2}T^* \otimes J_{q+1}(E)$

is the projection of b_{q+2}, we have $Db_{q+1} = D \circ \pi_{q+1}^{q+2}(b_{q+2}) = \pi_q^{q+1} \circ Db_{q+2} = \pi_q^{q+1}(a'_{q+1}) = a_q$, and by projecting we can deduce local exactness. $\square$

Remark 14. Using the same diagram again, but now with F_0 instead of R_q, $B_1 \subset J_1(F_0)$ instead of $R_{q+1} \subset J_1(R_q)$, and $N_1 \subset T^* \otimes F_0$ instead of $M_{q+1} \subset S_{q+1}T^* \otimes E$, we find a manner of constructing $\mathcal{D}_r \colon F_{r-1} \to F_r$ 'as a whole'.

A relation between Janet bundles and Spencer bundles is provided by the following proposition.

Proposition 15. *We have the short exact sequences*

$$0 \longrightarrow C_r \longrightarrow C_r(E) \xrightarrow{\Phi_r} F_r \longrightarrow 0,$$

where the epimorphisms Φ_r are induced by $\Phi = \Phi_0$.

PROOF. We may apply the δ-sequence to the short exact sequence

$$0 \longrightarrow M_{q+r} \longrightarrow S_{q+r}T^* \otimes E \xrightarrow{\sigma_r(\Phi)} N_r \longrightarrow 0$$

to obtain the following commutative and exact diagram:

$$
\begin{array}{ccccccc}
& 0 & & 0 & & 0 & \\
& \downarrow & & \downarrow & & \downarrow & \\
0 \longrightarrow & M_{q+r} & \longrightarrow & S_{q+r}T^* \otimes E & \xrightarrow{\sigma_r(\Phi)} & N_r & \longrightarrow 0 \\
& \downarrow{\scriptstyle\delta} & & \downarrow{\scriptstyle\delta} & & \downarrow{\scriptstyle\delta} & \\
0 \longrightarrow & T^* \otimes M_{q+r-1} & \longrightarrow & T^* \otimes S_{q+r-1}T^* \otimes E & \xrightarrow{\sigma_{r-1}(\Phi)} & T^* \otimes N_{r-1} & \longrightarrow 0 \\
& \downarrow{\scriptstyle\delta} & & \downarrow{\scriptstyle\delta} & & \downarrow{\scriptstyle\delta} & \\
& \vdots & & \vdots & & \vdots & \\
& \downarrow{\scriptstyle\delta} & & \downarrow{\scriptstyle\delta} & & \downarrow{\scriptstyle\delta} & \\
0 \longrightarrow & \textstyle\bigwedge^{r-1}T^* \otimes M_{q+1} & \longrightarrow & \textstyle\bigwedge^{r-1}T^* \otimes S_{q+1}T^* \otimes E & \xrightarrow{\sigma_1(\Phi)} & \textstyle\bigwedge^{r-1}T^* \otimes N_1 & \longrightarrow 0 \\
& \downarrow{\scriptstyle\delta} & & \downarrow{\scriptstyle\delta} & & \downarrow{\scriptstyle\delta} & \\
0 \longrightarrow & \delta(\textstyle\bigwedge^{r-1}T^* \otimes M_{q+1}) & \longrightarrow & \delta(\textstyle\bigwedge^{r-1}T^* \otimes S_{q+1}T^* \otimes E) & \xrightarrow{\sigma(\Phi)} & \delta(\textstyle\bigwedge^{r-1}T^* \otimes N_1) & \longrightarrow 0 \\
& \downarrow & & \downarrow & & \downarrow & \\
& 0 & & 0 & & 0 &
\end{array}
$$

where we notice that the involutivity of M_{q+1} is used in a crucial manner in the chase proving the exactness of the right column. Accordingly, we have the following commutative and exact diagram:

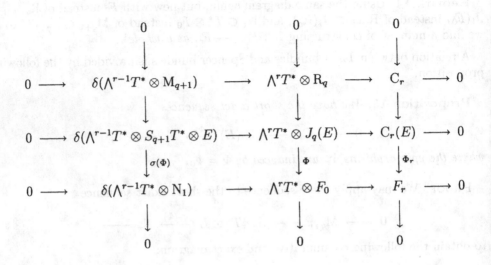

allowing us to induce Φ_r from $\Phi\colon J_q(E) \to F_0$. The left epimorphism has been denoted by $\sigma(\Phi)$, though it is induced by $\sigma(\Phi)\colon \bigwedge^r T^* \otimes S_q T^* \otimes E \to \bigwedge^r T^* \otimes F_0$, which is itself induced by $\sigma(\Phi)\colon S_q T^* \otimes E \to F_0 \cdots$ which, in general, is *not* an epimorphism! $\square$

We hope that the reader begins to understand that, in view of the difficulty of chasing in local coordinates, if the former methods can be applied to mathematical physics or to engineering sciences, they are bound to give quite unexpected results!

We now have everything at hand for relating the Janet sequence and the (second) Spencer sequence in a single commutative diagram, which will be essential for revisiting the foundations of modern physics, when its connection with group theory will have been established. This diagram was first obtained in [144, p. 183], but it took ten years before it was discovered, in [146], that it can be used in applications.

Theorem 16. *The Janet sequence for j_q projects onto the Janet sequence for $\mathcal{D}$ and the kernel of this projection is the second Spencer sequence, in the following* *formally exact commutative diagram*

$$
\begin{array}{ccccccccccc}
& & 0 & & 0 & & & & 0 & & \\
& & \downarrow & & \downarrow & & & & \downarrow & & \\
0 \longrightarrow & \Theta & \xrightarrow{\ j_q\ } & C_0 & \xrightarrow{\ D_1\ } & C_1 & \xrightarrow{\ D_2\ } & \cdots \xrightarrow{\ D_n\ } & C_n & \longrightarrow & 0 \\
& & & \downarrow & & \downarrow & & & \downarrow & & \\
0 \longrightarrow & E & \xrightarrow{\ j_q\ } & C_0(E) & \xrightarrow{\ D_1\ } & C_1(E) & \xrightarrow{\ D_2\ } & \cdots \xrightarrow{\ D_n\ } & C_n(E) & \longrightarrow & 0 \\
& \| & & \downarrow{\scriptstyle \Phi_0} & & \downarrow{\scriptstyle \Phi_1} & & & \downarrow{\scriptstyle \Phi_n} & & \\
0 \longrightarrow & \Theta \longrightarrow E & \xrightarrow{\ \mathcal{D}\ } & F_0 & \xrightarrow{\ D_1\ } & F_1 & \xrightarrow{\ D_2\ } & \cdots \xrightarrow{\ D_n\ } & F_n & \longrightarrow & 0 \\
& & & \downarrow & & \downarrow & & & \downarrow & & \\
& & & 0 & & 0 & & & 0 & &
\end{array}
$$

PROOF. The proof proceeds by an induction that is completely similar to the induction used for constructing the Janet sequence '*step-by-step*'. Keeping in mind that $C_0 = R_q$ and $C_0(E) = J_q(E)$, the first commutative '*cell*' on the lower left just expresses the definition of $\mathcal{D} = \Phi_0 \circ j_q$. The first prolongation of this cell gives rise to the following commutative and exact diagram

$$
\begin{array}{ccccccccc}
& & & 0 & & 0 & & & \\
& & & \downarrow & & \downarrow & & & \\
0 \longrightarrow & R_{q+1} & \longrightarrow & J_1(R_q) & \longrightarrow & C_1 & \longrightarrow & 0 & \\
& & & \downarrow & & \downarrow & & & \\
0 \longrightarrow & J_{q+1}(E) & \longrightarrow & J_1(J_q(E)) & \longrightarrow & C_1(E) & \longrightarrow & 0 & \\
& \| & & \downarrow{\scriptstyle J_1(\Phi)} & & \downarrow{\scriptstyle \Phi_1} & & & \\
0 \longrightarrow & R_{q+1} & \longrightarrow & J_{q+1}(E) & \xrightarrow{\ \rho_1(\Phi)\ } & J_1(F_0) & \xrightarrow{\ \Psi_1\ } & F_1 & \longrightarrow 0 \\
& & & & & \downarrow & & \downarrow & \\
& & & & & 0 & & 0 &
\end{array}
$$

which depends only on the left commutative '*cell*'. Indeed, the central and lower sequences are exact according to the cokernel construction of the Janet sequences for j_q and $\mathcal{D}$. The epimorphism Φ_1 is therefore induced by the epimorphism $J_1(\Phi)$. The kernel of Φ_1 is C_1, by Proposition 15 for $r = 1$. We are thus able to induce the morphism $J_1(R_q) \to C_1$, or $J_1(C_0) \to C_1$, defining the first-order operator D_1. A simple chase, which is left to the reader, shows that the kernel of this morphism is $J_1(R_q) \cap J_{q+1}(E) = R_{q+1}$, by definition.

We may inductively repeat the same reasoning, and successively construct the epi-morphisms $\Phi_1, \ldots, \Phi_n$, where Φ_{r+1} is induced by $J_1(\Phi_r)$. We already know that the Janet sequences for j_q and $\mathcal{D}$ are formally exact by construction, so it follows that the second Spencer sequence is also formally exact, as we wanted. $\quad\square$

Remark 17. In actual practice, and unless $M_{q+1} = 0$, it is usually much easier to construct the second Spencer sequence from the Janet sequence. Indeed, the second Spencer sequence always strictly ends at C_n, because the Janet sequence for j_q always strictly ends at $C_n(E)$, in contrast to what usually happens for the Janet sequence for $\mathcal{D}$. If $M_{q+1} = 0$, which is the most useful case in applications, the first and second Spencer sequences are isomorphic, and we need not distinguish them by their proper name, although we shall call the bundles involved 'Spencer bundles', with a slight abuse of language.

Corollary 18. *The (local) cohomology at C_{r+1} in the second Spencer sequence is equal to the (local) cohomology at F_r in the Janet sequence.*

PROOF. From Corollary 13 we know that the Janet sequence for j_q is locally exact. Accordingly, the central row of the diagram of Theorem 16 is locally exact, and the corollary follows from a chase. Since this chase is useful in applications, we will give its details. We have the following commutative diagram, with locally exact central row and exact columns (we leave out the zeros):

$$
\begin{array}{ccccc}
C_r & \xrightarrow{D_{r+1}} & C_{r+1} & \xrightarrow{D_{r+2}} & C_{r+2} \\
\downarrow & & \downarrow & & \downarrow \\
C_{r-1}(E) & \xrightarrow{D_r} & C_r(E) & \xrightarrow{D_{r+1}} & C_{r+1}(E) & \xrightarrow{D_{r+2}} & C_{r+2}(E) \\
\downarrow{\scriptstyle \Phi_{r-1}} & & \downarrow{\scriptstyle \Phi_r} & & \downarrow{\scriptstyle \Phi_{r+1}} \\
F_{r-1} & \xrightarrow{\mathcal{D}_r} & F_r & \xrightarrow{\mathcal{D}_{r+1}} & F_{r+1}
\end{array}
$$

We shall prove that there is a short exact sequence:

$$ 0 \longrightarrow B(C_{r+1}) \longrightarrow Z(C_{r+1}) \longrightarrow H(F_r) \longrightarrow 0 $$

Indeed, let $a_{r+1} \in C_{r+1}$ be such that $D_{r+2}a_{r+1} = 0$. Because of the commuting upper right cell, there is a way to consider a_{r+1} as a section $b_{r+1} \in C_{r+1}(E)$ killed by D_{r+2}. By Corollary 13, we can find $b_r \in C_r(E)$ such that $b_{r+1} = D_{r+1}b_r$. Since $\Phi_{r+1}(b_{r+1}) = 0$, we have $\mathcal{D}_{r+1}(\Phi_r(b_r)) = 0$ and $\Phi_r(b_r) \in Z(F_r)$. Now, if $b'_r \in C_r(E)$ is such that $b_{r+1} = D_{r+1}b'_r$, then $D_{r+1}(b'_r - b_r) = 0$ and, again by Corollary 13, there is $c_{r-1} \in C_{r-1}(E)$ such that $b'_r - b_r = D_r c_{r-1}$. Hence we find $\Phi_r(b'_r) - \Phi_r(b_r) = \Phi_r(b'_r - b_r) = \Phi_r(D_r(c_{r-1})) = \mathcal{D}_r(\Phi_{r-1}(c_{r-1}))$ and thus $\Phi_r(b'_r) - \Phi_r(b_r) \in B(F_r)$, a result showing that the right map of the sequence is well defined. It is also surjective. Indeed, let $f_r \in F_r$ be such that $\mathcal{D}_{r+1}f_r = 0$. We can find $b_r \in C_r(E)$ such that $f_r = \Phi_r(b_r)$ and we have $\Phi_{r+1}(D_{r+1}(b_r)) = 0$, a result showing that $D_{r+1}(b_r) = c_{r+1} \in C_{r+1}$ with $D_{r+2}(c_{r+1}) = 0$, i.e. $c_{r+1} \in Z(C_{r+1})$. Finally, if $\Phi_r(b_r) = \mathcal{D}_r(f_{r-1})$ for some $f_{r-1} \in F_{r-1}$, then we can find $b_{r-1} \in C_{r-1}(E)$ such that $f_{r-1} = \Phi_{r-1}(b_{r-1})$, and it follows that $\Phi_r(D_r(b_{r-1})) = \Phi_r(b_r)$. Accordingly, $\Phi_r(b_r - D_r(b_{r-1})) = 0$, and we can find $a_r \in C_r$ such that $b_r = a_r + D_r(b_{r-1})$. Hence $b_{r+1} = D_{r+1}(a_r) = D_{r+1}(b_r)$ and $a_{r+1} = D_{r+1}(a_r) \in \mathrm{B}(C_{r+1})$. From the previous short exact sequence we deduce that $\mathrm{H}(C_{r+1}) \simeq \mathrm{H}(F_r)$. $\square$

Remark 19. When we wish to study an inhomogeneous system $\mathcal{D}\xi = \eta$ such that $\mathcal{D}_1\eta = 0$, we in fact need to study the problem of local exactness of F_0. By Corollary 18, this problem amounts to studying the local exactness at C_1 but, most of the time, the operators D_1 and D_2 are quite different from the operators $\mathcal{D}$ and $\mathcal{D}_1$. A main part of Spencer's life has been devoted to this problem for the particular class of operators $\mathcal{D}$ described in the next Chapter. Of course, looking for local exactness resembles looking for a potential in case a field satisfies the field equations. However, at this moment we have no reason to consider a field as a section of F_r instead of as a section of the corresponding C_{r+1}.

Remark 20. We will say a few words about the lengths of the various sequences that we have introduced. As can be seen from the scheme of multiplicative and nonmultiplicative variables for $\mathcal{D}$, the Janet sequence strictly terminates at F_n if and only if rows with n dots appear in this scheme. Equivalently, this situation is encountered whenever $\pi_{q-1}^q \colon R_q \to J_{q-1}(E)$ is not surjective. E.g., for j_q we have $R_q = 0$ and, of course, the Janet sequence for j_q strictly terminates at $C_n(E)$. For the second Spencer sequence, we reason as follows. From the short exact sequence

$$0 \longrightarrow R_{q+1} \longrightarrow J_1(C_) \longrightarrow C_1 \longrightarrow 0$$

and the fact that $C_0 = R_q$ we see that this Spencer sequence is determined by looking at R_{q+1} as a first-order system on R_q. Taking into account that $\pi_q^{q+1} \colon R_{q+1} \to R_q$ is surjective, this comment shows that the second Spencer sequence is in fact the Janet sequence for the transitive system $R_{q+1} \subset J_1(R_q)$, which strictly terminates at C_n. Using Proposition II.B.6 we find, in particular,

$$J_r(R_{q+1}) \cap J_{r+1}(R_q) \subset J_r(J_{q+1}(E)) \cap J_{r+1}(J_q(E)) = J_{q+r+1}(E).$$

It follows that

$$J_r(R_{q+1}) \cap J_{r+1}(R_q) = R_{q+r+1},$$

and the analogy is completed at any order.

B. Algebraic Properties

The purpose of this Section is to study certain relations between properties of a system $R_q \subset J_q(E)$ and properties of its symbol $M_q \subset S_q T^* \otimes E$ by means of purely algebraic techniques. At the same time we greatly simplify the exposition in [144]. These results will be quite important for applications to engineering sciences, as given in Chapter VII.

Let $\mathcal{D} = \Phi \circ j_q \colon E \to F \colon \xi \to A_k^{\tau\mu}(x)\partial_\mu \xi^k(x)$ be a linear partial differential operator of order q. With any covector $\chi = \chi_i\, dx^i \in T^*$ we associate its so-called 'characteristic matrix'

$$A(x,\chi) = (A_k^\tau(x,\chi)) = (A_k^{\tau\mu}(x)\chi_\mu),$$

where $|\mu| = q$, $\chi_\mu = (\chi_1)^{\mu_1} \ldots (\chi_n)^{\mu_n}$, and we consider the resulting map

$$\sigma_\chi(\mathcal{D}): E \rightarrow F.$$

Definition 1. The map $\sigma_\chi(\mathcal{D})$ is called the *symbol* of $\mathcal{D}$ at $\chi \in T^*$.

This map must be constructed for each $x \in X$; however, for simplicity reasons we shall not indicate x explicitly. The usefulness of this definition can be seen at once by looking at the computation of the character α_q^n of an involutive operator $\mathcal{D}$ of order q. Indeed, taking into account the results of Chapter III, §§ B and D, we immediately find the relations

$$\alpha_q^n + \beta_q^n = m,$$

$$\begin{cases} \alpha_q^n = \min_\chi \dim \ker \sigma_\chi(\mathcal{D}), \\ \beta_q^n = \max_\chi \dim \operatorname{im} \sigma_\chi(\mathcal{D}) \\ \qquad = \max_\chi \operatorname{rk} \sigma_\chi(\mathcal{D}). \end{cases}$$

Finally, if we have two differential operators, $\mathcal{D}: E \rightarrow F$ of order q, with $\mathcal{D}\xi = (A_k^{\tau\mu}(x)\partial_\mu \xi^k)$, and $\mathcal{D}': F \rightarrow G$ of order r, with $\mathcal{D}'\eta = (B_\tau^{\alpha\nu}(x)\partial_\nu \eta^\tau)$, we can immediately verify that

$$(\mathcal{D}'\circ\mathcal{D})\xi = \mathcal{D}'(\mathcal{D}\xi) = B_\tau^{\alpha\nu}(x)A_k^{\tau\mu}(x)\partial_{\mu+\nu}\xi + (\text{order} < q+r),$$

where $|\mu| = q$, $|\nu| = r$. hence we have proved

Lemma 2. $\sigma_\chi(\mathcal{D}'\circ\mathcal{D}) = \sigma_\chi(\mathcal{D}')\circ\sigma_\chi(\mathcal{D})$.

All the algebraic properties of $\mathcal{D}$ will be obtained from these two results, which will be generalised to nonlinear systems in Chapter VI.

First of all, we can use the definitions in Chapter I to classify linear differential operators.

Definition 3. An operator $\mathcal{D}: E \rightarrow F$ is said to be '*underdetermined*' ('*determined*') if there is $\chi \in T^*$ such that $\sigma_\chi(\mathcal{D}): E \rightarrow F$ is surjective (bijective). Otherwise $\mathcal{D}$ is said to be '*overdetermined*'.

Underdetermined (and hence also determined) operators have very specific features.

Proposition 4. *If the operator $\mathcal{D}: E \rightarrow F$ is underdetermined, then it is involutive, and its Janet sequence has the form*

$$0 \longrightarrow \Theta \longrightarrow E \overset{\mathcal{D}}{\longrightarrow} F \longrightarrow 0.$$

PROOF. First of all, if $\mathcal{D} = \Phi \circ j_q$, then Φ is necessarily an epimorphism and $F = F_0$. By definition, we now have $\beta_q^n = \dim F$. Therefore, *by linearly changing coordinates if necessary*, the $\dim F$ given equations can be solved with respect to $y_{(0,\ldots,q)}^k$ for $k = 1, \ldots, \dim F$, and the principal jets are thus all of class n. Accordingly, there is no dot in the scheme for $\mathcal{D}$ and no compatibility condition can exist (since no crossed derivative can exist). A similar comment can be made for the symbol M_q of the corresponding system $R_q \subset J_q(E)$, and hence M_q is involutive. It follows that

$$\dim M_{q+1} = \dim S_{q+1}T^* \otimes E - \dim T^* \otimes F,$$

and thus $F_1 = 0$. This result immediately proves that R_q is formally integrable, since the curvature map is trivially zero. □

Remark 5. When $\mathcal{D}$ is determined we have

$$
\begin{aligned}
\dim R_{q+r} &= \dim J_{q+r}(E) - \dim J_r(F) \\
&= m\frac{(q+r+n)!}{(q+r)!\,n!} - m\frac{(r+n)!}{r!\,n!} \\
&= m\frac{(q+r+1)\dots(q+r+n) - (r+1)\dots(r+n)}{n!} \\
&= m\frac{nqr^{n-1} + \dots}{n!}.
\end{aligned}
$$

Accordingly, $\beta_q^1 = 0, \dots, \beta_q^{n-1} = 0, \beta_q^n = m \Rightarrow$

$$
\alpha_q^i = m\frac{(q+n-i-1)!}{(q-1)!\,(n-i)!} \quad \text{for } 1 \le i \le n-1, \qquad \alpha_q^n = 0.
$$

In the Hilbert polynomial $\dim R_{q+r}$ the coefficient at r^n is zero and the coefficient at r^{n-1} is $\alpha_q^{n-1}/(n-1)!$, i.e. $mq/(n-1)!$, as above.

Having taken care of the surjectivity of the symbol of $\mathcal{D}$, we shall now consider its injectivity.

Definition 6. A nonzero covector $\chi \in T^*$ is said to be '*characteristic*' for $\mathcal{D}$ if $\sigma_\chi(\mathcal{D})$ fails to be injective. $\mathcal{D}$ is said to be '*elliptic*' if it has no nonzero '*real*' characteristic covector (although it may have complex characteristic covectors).

When looking effectively for characteristic covectors, we see that two cases have to be distinguished:

- $\max_\chi \operatorname{rk} \sigma_\chi(\mathcal{D}) < m \Leftrightarrow \alpha_q^n > 0$. In this case any covector is characteristic, and the notion of characteristic covector is not so important.
- $\max_\chi \operatorname{rk} \sigma_\chi(\mathcal{D}) = m \Leftrightarrow \alpha_q^n = 0$. Generically, $\sigma_\chi(\mathcal{D})$ is injective, and it fails to be injective if and only if all the determinants of the $(m \times m)$-submatrices vanish. This leads to a system of algebraic equations for characteristic covectors.

In the *last* case it makes sense to give

Definition 7. For each $x \in X$ the set of covectors is an algebraic set, V_x, called the '*characteristic variety*' (with some abuse of language). We call $V = \cup_{x \in X} V_x$ the *characteristic variety* of $\mathcal{D}$.

Remark 8. When looking at the scheme of an involutive operator $\mathcal{D}$, we have the following picture:

$\beta_q^n(\mathcal{D})$ { 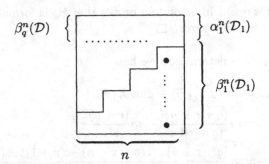 } $\alpha_1^n(\mathcal{D}_1)$

$\left.\right\} \beta_1^n(\mathcal{D}_1)$

$\underbrace{}_{n}$

It relates the characters of $\mathcal{D}$ and $\mathcal{D}_1$. Accordingly, either $\mathcal{D}_1$ does not exist (as in Proposition 4) or, if it does exist, $\alpha_1^n(\mathcal{D}_1) \geq 1 > 0$. Accordingly, all covectors are characteristic for $\mathcal{D}_1, \dots, \mathcal{D}_n$, and it follows that the definition of characteristic variety is surely not optimal for specifying covectors, even though it is the only one known in the literature. In particular, a necessary condition for an operator $\mathcal{D}_1$ to be the compatibility conditions of an operator $\mathcal{D}$ is that $V(\mathcal{D}_1) = T^*$. (See Chapter VII.)

We now give an alternative definition, which is more useful in practice than Definitions 6 and 7 and includes them both. For this we first notice that $\sigma_\chi(\mathcal{D})$ has generic rank β_q^n, except at the covectors annihilating all determinants of rank β_q^n.

Definition 9. A covector as above will be called '*systatic*', and we can introduce, as before, the '*systatic variety*' $W = \cup_{x \in X} W_x$ of $\mathcal{D}$.

We notice the following property:

$$\begin{cases} \bullet\ \alpha_q^n > 0 : & W \subset V = T^*, \\ \bullet\ \alpha_q^n = 0 : & W = V \subset T^*. \end{cases}$$

In fact, the concept of systatic variety seems more important than that of characteristic variety, because of the following theorem [144].

Theorem 10. *If $\mathcal{D}$ is involutive, the symbol sequence*

$$0 \longrightarrow \ker \sigma_\chi(\mathcal{D}) \longrightarrow E \xrightarrow{\sigma_\chi(\mathcal{D})} F_0 \xrightarrow{\sigma_\chi(\mathcal{D}_1)} \cdots \xrightarrow{\sigma_\chi(\mathcal{D}_n)} F_n \longrightarrow 0$$

is exact if and only if $\chi \notin W$.

PROOF. Let us look at the scheme in Remark 8.

- If $\chi \notin W$, then

$$\dim \operatorname{im} \sigma_\chi(\mathcal{D}) = \operatorname{rk} \sigma_\chi(\mathcal{D}) = \beta_q^n(\mathcal{D}) = \alpha_1^n(\mathcal{D}_1) = \dim \ker \sigma_\chi(\mathcal{D}_1).$$

Hence the sequence is exact at F_0. The exactness for each F_r now follows from the inductive construction of the Janet sequence and the fact that $\mathcal{D}_1, \dots, \mathcal{D}_n$ are involutive whenever $\mathcal{D}$ is involutive.

- Conversely, if the sequence is exact, the definition of the n-character implies

$$\dim \operatorname{im} \sigma_\chi(\mathcal{D}) \leq \beta_q^n(\mathcal{D}) = \alpha_q^n(\mathcal{D}_1) \leq \dim \ker \sigma_\chi(\mathcal{D}_1).$$

hence equality follows and $\chi \notin W$. $\square$

Remark 11. Since j_q is involutive and of finite type, the symbol sequence of the Janet sequence for j_q is exact if and only if $\chi \neq 0$.

As a straightforward consequence of Theorem 10 and Proposition I.4 we have

Corollary 12. $\sum_{r=0}^{n}(-1)^r \dim F_r = m - \alpha_q^n = \beta_q^n$.

In particular, if $\mathcal{D}$ is elliptic, then $\alpha_q^n = 0$ and the alternate sum of dimensions of the vector bundles appearing in the Janet sequence is zero.

Finally, combining Theorems A.16 and 10 and Remark 11 we can obtain (by a chase left to the reader):

Corollary 13. *If $R_q \subset J_q(E)$ is involutive, the symbol sequence of the second Spencer sequence is exact if and only if $\chi \notin W$.*

In fact, the symbol sequence of the Janet sequence for j_q projects onto the symbol sequence of the Janet sequence for $\mathcal{D}$, and the kernel is the symbol sequence of the second Spencer sequence. However, here we only have maps in the corresponding commutative and exact diagram of vector bundles. In particular, $\ker \sigma_\chi(\mathcal{D}) \simeq \ker \sigma_\chi(D_1)$, and so D_1 is elliptic if and only if $\mathcal{D}$ is elliptic. In the next Chapter we shall see that this is indeed Spencer's main reason for his lifelong systematic use of the second Spencer sequence.

Nevertheless, as will be seen in the examples in the next Section, apart from certain rare cases the Spencer sequence is always more difficult to deal with than the Janet sequence and, most of the time, it may seem unrelated to any classical approach involving crossed derivatives.

We end this Section with a result concerning operators of finite type.

Theorem 14. *A formally integrable operator is of finite type if and only if its characteristic variety consists of the zero covector only.*

PROOF. If the operator is of finite type, $M_{q+r} = 0$ for r large enough, and among the equations defining M_{q+r} we have $v_\mu^k = 0$, with $\mu = (q + r, 0, \ldots, 0), \ldots, \mu = (0, \ldots, 0, q + r)$. Accordingly, among the polynomial equations defining the characteristic variety we have $(\chi_1)^{q+r} = 0, \ldots, (\chi_n)^{q+r} = 0$, leading unambiguously to $\chi_1 = \cdots = \chi_n = 0$.

The converse is more tricky and more useful, since it can be used as a test for checking the property of having finite type.

First of all, we must have $\alpha_q^n = 0$, and the matrix $A(x, \chi)$ must have maximum generic rank equal to m. The characteristic variety is therefore obtained by equating to zero the determinants of all $(m \times m)$-submatrices. The heart of the proof is Hilbert's theorem of zeros, stating that if a polynomial vanishes at all zeros of an ideal, then a power of this polynomial belongs to the ideal. In our case there is, for all $i = 1, \ldots, m$, a sufficiently large a such that $(\chi_i)^a$ belongs to the ideal generated by the previous polynomial determinants. Developing each determinant with respect to one of its rows, we deduce from Hilbert's theorem of zeros that linear combinations of the rows of $A(x, X)$ with polynomial coefficients are equal to χ_μ times the identity $(m \times m)$-matrix, for μ with $|\mu| = q + r$ sufficiently large. Transforming these polynomial coefficients into corresponding operators, as described in the Introduction, we see that certain well-defined prolongations make it possible to determine all jets of order $q + r$. We thus have the property of being of finite type. □

Remark 15. The reader is warned that the previous Theorem allows us to test whether an operator has a finite-dimensional space of solutions or not only in case the operator is formally integrable. A good counterexample is provided by the Janet example, for which the characteristic variety is 1-dimensional while the space of solutions is 12-dimensional.

C. Examples

The purpose of this Section is to familiarise the reader with the constructions of various differential sequences presented in the previous sections. Most examples are adapted from the examples in the final Section of Chapter III, while examples related to group theory are postponed to the next Chapter, as they are much more difficult to handle.

Example 1. $n = 3$, m is arbitrary, $q = 1, 2$.

For subsequent use in many of the examples to follow, we start by constructing the Janet sequence for j_q when $q = 1, 2$. The reader will understand the difficulty of extending these computations to the general case without using a computer. Also, the Janet sequence for j_2 does project onto the Janet sequence for j_1 by taking successive cokernels of prolongations of the initial 'projection cell', namely:

$$
\begin{array}{ccc}
0 \longrightarrow J_{q+r+1}(E) & \longrightarrow & J_r(J_{q+1}(E)) \\
\Big\downarrow \pi^{q+r+1}_{q+r} & & \Big\downarrow J_r(\pi^{q+1}_q) \\
0 \longrightarrow J_{q+r}(E) & \longrightarrow & J_r(J_q(E)) \\
\Big\downarrow & & \Big\downarrow \\
0 & & 0
\end{array}
$$

Nevertheless, the splitting of the Spencer bundles which is used in practice does not project in a similar way. This is a delicate point, and we give more details. First of all, by the definition of the Spencer bundles and the δ-map, there is a commutative and exact diagram:

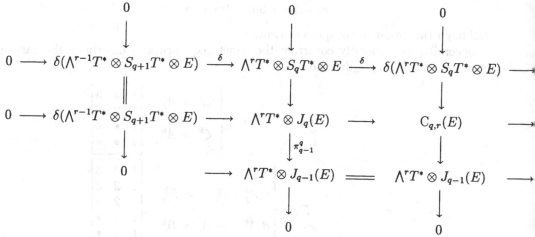

hence we have a splitting

$$C_{q,r}(E) \simeq \wedge^r T^* \otimes J_{q-1}(E) \times_X \delta(\wedge^r T^* \otimes S_q T^* \otimes E).$$

Replacing q by $q+1$, we see that *the first factor does project, while the second does not*. A similar comment can be made regarding any involutive system $R_q \subset J_q(E)$ and the corresponding Spencer bundles C_r, to obtain the splittings

$$C_r \simeq \wedge^r T^* \otimes R_{q-1} \times_X \delta(\wedge^{r-1} T^* \otimes M_q),$$

where the projection R_{q-1} of R_q on $J_{q-1}(E)$ is a vector bundle, by Remark III.B.14.

Now, exactly as we have proved that $C_{q,r}(E)$ projected onto $\wedge^r T^* \otimes J_{q-1}(E)$, we can prove that $C_{q+1,r}(E)$ projects onto $\wedge^r T^* \otimes J_q(E)$, and thus onto $C_{q,r}(E)$, as required. Hence, it remains to explain how the preceding splitting is automatically performed in practice.

We first treat the case $q = 1$.

From the theoretical point of view, *taking into account the fact that $n = 3$*, we have

$$\begin{cases} C_{1,0}(E) = J_1(E), & 4m, \\ C_{1,1}(E) = T^* \otimes E \times_X \wedge^2 T^* \otimes E, & 6m, \\ C_{1,2}(E) = \wedge^2 T^* \otimes E \times_X \wedge^3 T^* \otimes E, & 4m, \\ C_{1,3}(E) = \wedge^3 T^* \otimes E, & m, \end{cases}$$

where we have taken into account the exactness of the sequences

$$0 \longrightarrow S_3 T^* \xrightarrow{\delta} T^* \otimes S_2 T^* \xrightarrow{\delta} \wedge^2 T^* \otimes T^* \xrightarrow{\delta} \wedge^3 T^* \longrightarrow 0,$$
$$0 \longrightarrow S_4 T^* \xrightarrow{\delta} T^* \otimes S_3 T^* \xrightarrow{\delta} \wedge^2 T^* \otimes S_2 T^* \xrightarrow{\delta} \wedge^3 T^* \otimes T^* \longrightarrow 0,$$

when $n = 3$. The dimensions of the various Spencer bundles are indicated on the right side.

First we directly verify Corollary B.12:

$$m - 4m + 6m - 4m + m = 0,$$

although this result is not quite evident.

Secondly, we explicitly construct the Janet sequence and determine the various schemes for $D_1, \ldots, D_n$ in case $m = 1$:

$$j_1 \begin{cases} \partial_i \xi^k = A_i^k \\[2mm] \xi^k = A^k \end{cases} \quad \begin{array}{ccc} 1 & 2 & 3 \\ 1 & 2 & \bullet \\ 1 & \bullet & \bullet \\ \bullet & \bullet & \bullet \end{array}$$

$$D_1 \begin{cases} \partial_i A_j^k - \partial_j A_i^k = B_{ij}^k \\[2mm] \partial_i A^k \quad - A_i^k = B_i^k \end{cases} \quad \begin{array}{ccc} 1 & 2 & 3 \\ 1 & 2 & 3 \\ 1 & 2 & 3 \\ 1 & 2 & \bullet \\ 1 & 2 & \bullet \\ 1 & \bullet & \bullet \end{array}$$

$$D_2 \begin{cases} \partial_r B_{ij}^k + \partial_i B_{jr}^k + \partial_j B_{ri}^k = C_{ijr}^k \\[2mm] \partial_i B_j^k - \partial_j B_i^k \quad + B_{ij}^k = C_{ij}^k \end{cases} \quad \begin{array}{ccc} 1 & 2 & 3 \\ 1 & 2 & 3 \\ 1 & 2 & 3 \\ 1 & 2 & \bullet \end{array}$$

$$D_3 \quad \left\{ \partial_r C_{ij}^k + \partial_i C_{jr}^k + \partial_j C_{ri}^k - C_{ijr}^k = D_{rij}^k \right. \quad \boxed{1 \quad 2 \quad 3}$$

and obtain the Janet sequence for j_1 in case $m = 1$, $n = 3$:

$$0 \longrightarrow \boxed{1} \xrightarrow{j_1} \boxed{4} \xrightarrow{D_1} \boxed{6} \xrightarrow{D_2} \boxed{4} \xrightarrow{D_3} \boxed{1} \longrightarrow 0.$$

If m is arbitrary we only need multiply the previous dimensions by m to obtain the correct result.

Via these explicit computations we discover at once that the main difficulty has been to take into account the fact that D is *not* formally integrable.

We now treat the case $q = 2$, which is more involved. As before, we have:

$$\begin{cases} C_{2,0}(E) = J_2(E), & 10m, \\ C_{2,1}(E) = T^* \otimes J_1(E) \times_X \delta(T^* \otimes S_2 T^* \otimes E), & 20m, \\ C_{2,2}(E) = \wedge^2 T^* \otimes J_1(E) \times_X \wedge^3 T^* \otimes T^* \otimes E, & 15m, \\ C_{2,3}(E) = \wedge^3 T^* \otimes J_1(E), & 4m. \end{cases}$$

Again we check:

$$m - 10m + 20m - 15m + 4m = 0,$$

but now the dimensions are increasing and we no longer have only forms with values in jet bundles, and the bundle $\delta(T^* \otimes S_2 T^*) \subset \wedge^2 T^* \otimes T^*$ cannot be easily described in local coordinates.

Nevertheless, we may give j_2, D_1, D_2, D_3 in local coordinates and perceive the importance of the Spencer operator, although counting the number of linearly independent equations becomes more difficult:

$$j_2 \quad \begin{cases} \partial_{ij}\xi^k = A^k_{ij}, \\ \partial_i\xi^k = A^k_i, \\ \xi^k = A^k, \end{cases}$$

$$D_1 \quad \begin{cases} \partial_i A^k_{rj} - \partial_j A^k_{ri} = B^k_{r,ij}, \\ \partial_i A^k_j - A^k_{ij} = B^k_{j,i}, \\ \partial_i A^k - A^k_i = B^k_i, \end{cases}$$

$$D_2 \quad \begin{cases} \partial_s B^k_{r,ij} + \partial_i B^k_{r,js} + \partial_j B^k_{r,si} = C^k_{ijs,r}, \\ \partial_i B^k_{r,j} - \partial_j B^k_{r,i} + B^k_{r,ij} = C^k_{ij,r}, \\ \partial_i B^k_j - \partial_j B^k_i + B^k_{j,i} - B^k_{i,j} = C^k_{ij}, \end{cases}$$

$$D_3 \quad \begin{cases} \partial_r C^k_{ij} + \partial_i C^k_{jr} + \partial_j C^k_{ri} - C^k_{ij,r} - C^k_{jr,i} - C^k_{ri,j} = D^k_{rij}, \\ \partial_r C^k_{ij,s} + \partial_i C^k_{jr,s} + \partial_j C^k_{ri,s} - C^k_{ijr,s} = D^k_{rij,s}. \end{cases}$$

In these formulas we have used a comma to separate the various tensorial parts. Such a notation will be helpful in checking the projections $C_{2,r}(E) \to C_{1,r}(E)$:

$$\begin{cases} B^k_{ij} = B^k_{j,i} - B^k_{i,j} & \text{for } C_{2,1}(E) \to C_{1,1}(E), \\ C^k_{ijr} = C^k_{ij,r} + C^k_{jr,i} + C^k_{ri,j} & \text{for } C_{2,2}(E) \to C_{2,1}(E). \end{cases}$$

We will now use these results for constructing the Janet sequence and the Spencer sequences in specific examples.

In the following examples we shall either introduce inhomogeneous (right) terms and look for compatibility conditions, or, equivalently, consider the equations and look for linear differential dependence between them.

Example 2 (see Example III.D.1). The Janet example $R_2 \subset J_2(E)$ is not formally integrable, but $R_5^{(2)}$ is involutive with zero symbol. Accordingly, the right terms (the equations themselves) of the defining equations with second terms can be expressed by means of linear combinations of the second terms (the equations themselves) of the equations defining R_2. Of course, the expression is only defined up to possible compatibility conditions (differential identities) for these two second terms (equations). It follows that studying $R_5^{(2)}$ involves 5th order derivatives, and the final compatibility conditions (differential identities) are of order 6 at most. This is why, in the Introduction, the computer scheme given by the second student, as well as the direct computation given by the first student, are both correct. We believe there is no way to shortcut the transition via a formally integrable system.

Example 3 (see Example III.D.2). Adding a second term, we have

$$R_2 \quad \begin{cases} y_{11} + y_{22} = v, \\ x^1 y_1 + x^2 y_2 = u. \end{cases}$$

The system $R_2^{(1)}$ obtained is involutive, with zero symbol, if we do not consider formal power series at the origin. The second terms will involve the first derivatives of (u, v). Accordingly, $R_2^{(2)}$ will have second terms involving the second-order derivatives of (u, v), and hence the compatibility conditions are of order 2 only. Their number is $\dim Q_2$ in the following kernel/cokernel exact sequence:

$$0 \longrightarrow R_4 \longrightarrow J_4(E) \longrightarrow J_2(F) \longrightarrow Q_2 \longrightarrow 0.$$

We now have, in succession:

$$\dim R_3^{(1)} = \dim \rho_1(R_2^{(1)}) = \dim R_2^{(1)},$$

because $M_2^{(1)} = 0$. Counting the number of equations defining $R_2^{(1)}$, we thus have

$$\dim R_3^{(1)} = 6 - 4 = 3 - 1 = 2.$$

Now M_2 is involutive, with characters $\alpha_2^1 = 2$, $\alpha_2^2 = 0$, and $\dim M_4 = 2$. It follows that

$$\dim R_4 = \dim R_3^{(1)} + \dim M_4 = 2 + 2 = 4,$$
$$\dim Q_2 = -\dim J_4(E) + \dim J_2(F) + \dim R_4 = 15 - 12 - 2 = 1,$$

and we have found in a purely intrinsic way that there is only one (generating) compatibility condition, of order 2. A little work using a computer rapidly gives

$$\partial_{11} u + \partial_{22} u - x^1 \partial_1 v - x^2 \partial_2 v - 2v = 0,$$

which is thus the only possible one.

If we look for solutions analytic at the origin, we set

$$y = \sum_\mu a_\mu \frac{x^\mu}{\mu!},$$

with $\mu = (\mu_1, \mu_2)$ and $\mu! = \mu_1! \mu_2!$. In particular, $|\mu| a_\mu = 0$ whenever $u = v = 0$, and the only possibility is a constant. When u, v are given and analytic at the origin, the solution is uniquely determined up to a constant by only the first-order equation, provided that u and v do satisfy the compatibility condition.

Example 4 (see Example III.D.3). Adding a second term, we have

$$y_{112} = w, \qquad y_{33} = u, \qquad y_{22} = v.$$

We leave it to the reader to prove that the 3 generating compatibility conditions are:

$$\begin{cases} \partial_{22} u - \partial_{33} v = 0, \\ \partial_{11} v - \partial_2 w = 0, \\ \partial_{112} u - \partial_{33} w = 0. \end{cases}$$

Example 5 (see Example III.D.4 with $\alpha, a = \text{const} \neq 0$). This *delicate* example will make it possible to study the distinction between 2-acyclicity and involution in constructing the Janet sequence.

Let us, again, consider the formally integrable system R_2 with 2-acyclic symbol M_2:

$$R_2 \begin{cases} \Phi^5 \equiv y_{33} & = 0 \\ \Phi^4 \equiv y_{23} - \alpha y_{11} = 0 \\ \Phi^3 \equiv y_{22} - a y_3 = 0 \\ \Phi^2 \equiv y_{13} & = 0 \\ \Phi^1 \equiv y_{12} & = 0 \end{cases} \quad \begin{matrix} 1 & 2 & 3 \\ 1 & 2 & \bullet \\ 1 & 2 & \bullet \\ 1 & \bullet & \bullet \\ 1 & \bullet & \bullet \end{matrix}$$

By Corollary III.C.4, the only compatibility conditions are those obtained when checking the equality $R_2^{(1)} = R_2$, i.e.

$$\begin{cases} \Psi^5 \equiv d_3\Phi^4 - d_2\Phi^5 & + \alpha d_1\Phi^2 = 0 \\ \Psi^4 \equiv d_3\Phi^3 - d_2\Phi^4 - \alpha d_1\Phi^1 + a\Phi^5 = 0 \\ \Psi^3 \equiv d_3\Phi^2 - d_1\Phi^5 & = 0 \\ \Psi^2 \equiv d_3\Phi^1 - d_2\Phi^2 & = 0 \\ \Psi^1 \equiv d_2\Phi^1 - d_1\Phi^3 & - a\Phi^2 = 0 \end{cases} \quad \begin{matrix} 1 & 2 & 3 \\ 1 & 2 & 3 \\ 1 & 2 & 3 \\ 1 & 2 & 3 \\ 1 & 2 & \bullet \end{matrix}$$

The symbol of this new system is surely *not* involutive, since otherwise M_2 would be involutive by the general theory. Nevertheless, this system ($B_1 \subset J_1(F)$ in the general notation) must be formally integrable. Trying the test, the symbol cannot be 2-acyclic, because otherwise M_2 would be 3-acyclic, i.e. we would have an exact sequence

$$0 \longrightarrow \wedge^3 T^* \otimes M_2 \longrightarrow 0,$$

which is impossible. Since M_3 is involutive, the prolongation of the preceding system has an involutive symbol, and one can verify formal integrability, as desired. By this result, the differential identities that must be satisfied by Ψ should be, *in contrast to the construction of the Janet sequence*, of order 2. However, the system is much too complicated to be able to find the result, and in particular the number of differential identities, by hand.

Our next purpose is to find such a result in an intrinsic way, using kernel/cokernel sequences.

First of all, we have the short exact sequence

$$0 \longrightarrow R_2 \longrightarrow J_2(E) \longrightarrow F_0 \longrightarrow 0,$$
$$0 \longrightarrow \boxed{5} \longrightarrow \boxed{10} \longrightarrow \boxed{5} \longrightarrow 0.$$

Since $M_3 = 0$, we have $\dim R_5 = \dim R_4 = \dim R_3 = \dim R_2 = 5$, and we have the short exact sequence

$$0 \longrightarrow R_3 \longrightarrow J_3(E) \longrightarrow J_1(F_0) \longrightarrow F_1 \longrightarrow 0,$$
$$0 \longrightarrow \boxed{5} \longrightarrow \boxed{20} \longrightarrow \boxed{20} \longrightarrow \boxed{5} \longrightarrow 0.$$

The short exact sequence

$$0 \longrightarrow R_4 \longrightarrow J_4(E) \longrightarrow J_2(F_0) \longrightarrow J_1(F_1) \longrightarrow 0,$$

$$0 \longrightarrow \boxed{5} \longrightarrow \boxed{35} \longrightarrow \boxed{50} \longrightarrow \boxed{20} \longrightarrow 0,$$

is implied by the 2-acyclicity of M_2 (*it is not evident!*) and proves that the Ψ do not satisfy any first-order differential identity. Nevertheless, the following short exact sequence proves, by counting dimensions, that there is only 1 generating second-order differential identity, a fact which is also not evident at first sight:

$$0 \longrightarrow R_5 \longrightarrow J_5(E) \longrightarrow J_3(F_0) \longrightarrow J_2(F_1) \longrightarrow Q \longrightarrow 0,$$

$$0 \longrightarrow \boxed{5} \longrightarrow \boxed{56} \longrightarrow \boxed{100} \longrightarrow \boxed{50} \longrightarrow \boxed{1} \longrightarrow 0.$$

Since this differential identity can only come from the unique dot in the last scheme presented (all other principal derivatives are different!), a short computation gives the second-order differential identity:

$$d_{33}\Psi^1 - d_{23}\Psi^2 + d_{13}\Psi^4 - d_{22}\Psi^3 + d_{12}\Psi^5 + \alpha d_{11}\Psi^2 + a d_3 \Psi^3 = 0.$$

Hence, in general $\mathcal{D}_2$ cannot be defined as a first-order operator if only 2-acyclicity is present and not involutivity.

Example 6 (see Example III.D.5). Consider the inhomogeneous system

$$R_1 \quad \begin{cases} y_4 - x^3 y_2 - y = u, \\ y_3 - x^4 y_1 = v. \end{cases}$$

We can complete it, and obtain

$$R_1^{(1)} \quad \begin{cases} y_4 - x^3 y_2 - y = u & \boxed{\begin{array}{cccc} 1 & 2 & 3 & 4 \end{array}} \\ y_3 - x^4 y_1 = v & \boxed{\begin{array}{cccc} 1 & 2 & 3 & \bullet \end{array}} \\ y_2 - y_1 = w & \boxed{\begin{array}{cccc} 1 & 2 & \bullet & \bullet \end{array}} \end{cases}$$

According to the study already performed in Chapter III, the condition $R_1^{(2)} = R_1^{(1)}$ gives the compatibility conditions for u, v and w, namely:

$$\begin{cases} \Psi^3 \equiv \partial_4 v - \partial_3 u - x^3 \partial_2 v + x^4 \partial_1 u - v - w = 0 & \boxed{\begin{array}{cccc} 1 & 2 & 3 & 4 \end{array}} \\ \Psi^2 \equiv \quad \partial_4 w - \partial_2 u + \partial_1 u - x^3 \partial_2 w - w = 0 & \boxed{\begin{array}{cccc} 1 & 2 & 3 & 4 \end{array}} \\ \Psi^1 \equiv \quad \partial_3 w - \partial_2 v + \partial_1 v - x^4 \partial_1 w = 0 & \boxed{\begin{array}{cccc} 1 & 2 & 3 & \bullet \end{array}} \end{cases}$$

Nevertheless, in the transition from R_1 to $R_1^{(1)}$ we see that w is determined by the first equation, if u and v are known. hence we have only two second-order equations for u and v, obtained by substituting the expression for w obtained from the first equation into the second and third equations. The Ψ satisfy the following differential identity:

$$d_4 \Psi^1 - d_3 \Psi^2 + d_2 \Psi^3 - d_1 \Psi^3 + x^4 d_1 \Psi^2 - x^3 d_2 \Psi^1 - \Psi^1 \equiv 0.$$

Adding the condition $\Psi^3 = 0$, which defines w, we obtain

$$d_4 \Psi^1 - d_3 \Psi^2 + x^4 d_1 \Psi^2 - x^3 d_2 \Psi^1 - \Psi^1 \equiv 0,$$

which is the only differential identity relating the compatibility conditions for the second terms of R_1. To verify this in an intrinsic way, consider the following exact sequence:

$$0 \longrightarrow R_2 \longrightarrow J_2(E) \longrightarrow J_1(F) \longrightarrow Q_1 \longrightarrow 0.$$

We have:

$$\dim R_2 = \dim R_1^{(1)} + \dim M_2$$
$$= (5 - 3) + 3 = 5.$$

Hence

$$\dim Q_1 = \dim J_1(F) + \dim R_2 - \dim J_2(E)$$
$$= 10 + 5 - 15 = 0,$$

and there are no first-order compatibility conditions. At the next order we have

$$0 \longrightarrow R_3 \longrightarrow J_3(E) \longrightarrow J_2(F) \longrightarrow Q_2 \longrightarrow 0,$$

with $\dim R_3 = \dim M_3 + \dim R_2^{(1)}$,

$$\dim R_2^{(1)} = \dim \rho_1(R_1^{(1)}) = \dim R_1^{(1)} + \dim \rho_1(M_1^{(1)}),$$

i.e. $\dim R_3 = 4 + 2 + 1 = 7$ and

$$\dim Q_2 = \dim J_2(F) + \dim R_3 - \dim J_3(E)$$
$$= 30 + 7 - 35 = 2,$$

as desired. We leave it to the reader to perform one more prolongation and find the unique differential identity for Ψ.

Example 7 (see Example III.D.7). Consider the involutive system

$$R_2 \quad \begin{cases} \Phi^3 \equiv y_{33} - y_{13} + y_3 = 0, \\ \Phi^2 \equiv y_{23} - y_{13} + y_3 = 0, \\ \Phi^1 \equiv y_{12} - y_{11} + y_1 = 0. \end{cases}$$

Our purpose is to use the results of Example 1 for $n = 3$ to construct the Janet sequence and the Spencer sequence, *despite the fact that the coordinate system is not δ-regular*. This is one of the best examples illustrating the intrinsic procedure.

First, we have $\dim E = 1$, $\dim F_0 = 3$, and we look for F_1 by means of the exact sequence

$$0 \longrightarrow M_3 \longrightarrow S_3 T^* \otimes E \longrightarrow T^* \otimes F_0 \longrightarrow F_1 \longrightarrow 0.$$

Since M_2 is involutive, with characters $\alpha_2^1 = 3$, $\alpha_2^2 = 0$, $\alpha_2^3 = 0$, we have $\dim M_{2+r} = 3$ for all $r \geq 0$. Accordingly,

$$\dim F_1 = \dim M_3 + \dim T^* \otimes F_0 - \dim S_3 T^* \otimes E$$
$$= 3 + 9 - 10 = 2,$$

and we have indeed two differential identities:

$$\begin{cases} d_3\Phi^2 - d_2\Phi^3 + d_1\Phi^3 - d_1\Phi^2 - \Phi^3 + \Phi^2 = 0, \\ d_3\Phi^1 - d_1\Phi^2 = 0, \end{cases}$$

which make up an involutive system. Because the system is solved for $d_3\Phi^2$, $d_3\Phi^1$, these two equations are differentially linearly independent. Equivalently, one can verify that the following sequence is exact (but the reader should recall that this property derives from M_2 being involutive):

$$0 \longrightarrow M_4 \longrightarrow S_4T^* \otimes E \longrightarrow S_2T^* \otimes F_0 \longrightarrow T^* \otimes F_1 \longrightarrow 0,$$

$$0 \longrightarrow \boxed{3} \longrightarrow \boxed{15} \longrightarrow \boxed{18} \longrightarrow \boxed{6} \longrightarrow 0.$$

Accordingly, we finally obtain

$$\dim E = 1, \quad \dim F_0 = 2, \quad \dim F_1 = 2, \quad \dim F_2 = 0, \quad \dim F_3 = 0.$$

Example 1 with the Janet sequence for j_2 implies that

$$\dim C_0(E) = 10, \quad \dim C_1(E) = 20, \quad \dim C_2(E) = 15, \quad \dim C_3(E) = 4.$$

Therefore we have the following diagram, connecting the Janet sequence and the second Spencer sequence for R_2:

$$\begin{array}{ccccccccc}
& & \boxed{7} & \xrightarrow{D_1} & \boxed{18} & \xrightarrow{D_2} & \boxed{15} & \xrightarrow{D_3} & \boxed{4} & \longrightarrow & 0 \\
& & \downarrow & & \downarrow & & \| & & \| & & \\
0 \longrightarrow & \boxed{1} & \xrightarrow{j_2} & \boxed{10} & \xrightarrow{D_1} & \boxed{20} & \xrightarrow{D_2} & \boxed{15} & \xrightarrow{D_3} & \boxed{4} & \longrightarrow 0 \\
& \| & & \downarrow & & \downarrow & & \downarrow & & \\
& \boxed{1} & \xrightarrow{D} & \boxed{3} & \xrightarrow{D_1} & \boxed{2} & \longrightarrow & 0 & &
\end{array}$$

We have $\dim E - \dim F_0 + \dim F_1 = 0$, but $\mathcal{D}$ is not elliptic, because $(1, 1, 1)$ is a real nonzero characteristic covector.

To aid the reader with the details, we shall compute $C_r = \bigwedge^r T^* \otimes R_2 / \delta(\bigwedge^{r-1} T^* \otimes M_3)$ in a direct way. We have $C_0 = R_2$, and $\dim C_0 = \dim R_2 = 10 - 3 = 7$. Also, $C_1 = T^* \otimes R_2 / \delta M_3$. Since we know that M_2 is 1-acyclic, we have $M_3 \simeq \delta M_3 \subset T^* \otimes M_2$. Thus, $\dim C_1 = \dim T^* \otimes R_2 - \dim M_3 = 21 - 3 = 18$. Now we have the short exact sequence

$$0 \longrightarrow M_4 \xrightarrow{\delta} T^* \otimes M_3 \xrightarrow{\delta} \delta(T^* \otimes M_3) \longrightarrow 0,$$

and $\dim \delta(T^* \otimes M_3) = \dim T^* \otimes M_3 - \dim M_4 = 9 - 3 = 6$. Accordingly, $\dim C_2 = \dim \bigwedge^2 T^* \otimes R_2 - \dim \delta(T^* \otimes M_3) = 21 - 6 = 15$. Finally, $\dim C_3 = \dim \bigwedge^3 T^* \otimes R_2 - \dim \delta(\bigwedge^2 T^* \otimes M_3)$. Since there is an exact sequence

$$0 \longrightarrow M_5 \xrightarrow{\delta} T^* \otimes M_4 \xrightarrow{\delta} \bigwedge^2 T^* \otimes M_3 \xrightarrow{\delta} \bigwedge^3 T^* \otimes M_2 \longrightarrow 0,$$

we thus have $\dim C_3 = \dim \bigwedge^3 T^* \otimes R_2 - \dim \bigwedge^3 T^* \otimes M_2 = 7 - 3 = 4$. Of course, it is almost impossible to write out explicitly D_1, D_2 or D_3 by hand.

Since the schemes of multiplicative/nonmultiplicative variables have an intrinsic meaning, we give them here:

$$
\begin{array}{|ccc|}
\hline
1 & 2 & 3 \\
1 & 2 & \bullet \\
1 & 2 & \bullet \\
\hline
\end{array}
\qquad \longrightarrow \qquad
\begin{array}{|ccc|}
\hline
1 & 2 & 3 \\
1 & 2 & 3 \\
\hline
\end{array}
$$

scheme for $\mathcal{D}$ scheme for $\mathcal{D}_1$

Example 8 (see Example III.D.11). The system R_2 in Example III.D.11 is involutive, with characters $\alpha_1^1 = 2$, $\alpha_1^2 = 1$. The corresponding schemes are

$$
\begin{array}{|cc|}
\hline
1 & 2 \\
1 & 2 \\
1 & \bullet \\
\hline
\end{array}
\qquad \longrightarrow \qquad
\begin{array}{|cc|}
\hline
1 & 2 \\
\hline
\end{array}
$$

scheme for $\mathcal{D}$ scheme for $\mathcal{D}_1$

Accordingly, $\dim E = 3$, $\dim F_0 = 3$, $\dim F_1 = 1$, $\dim F_2 = 0$. Notice that $\dim E - \dim F_0 + \dim F_1 - \dim F_2 = \alpha_1^2 = 1$. Moreover, if $\chi = \chi_1 \, dx^1 + \chi_2 \, dx^2 \in T^*$, the characteristic matrix $A(x, \chi)$ is as follows:

$$
\begin{pmatrix}
0 & \chi_2 - \chi_1 & \chi_2 - \chi_1 \\
\chi_2 & -\chi_1 & -\chi_1 - \chi_2 \\
\chi_1 & -\chi_1 & -2\chi_1
\end{pmatrix}
$$

Expanding the determinant of this matrix with respect to the first column, we obtain

$$
- \chi_2(-2\chi_1(\chi_2 - \chi_1) + \chi_1(\chi_2 - \chi_1)) + \chi_1(-(\chi_1 + \chi_2)(\chi_2 - \chi_1) + \chi_1(\chi_2 - \chi_1))
$$
$$
= \chi_1\chi_2(\chi_2 - \chi_1) - \chi_1\chi_2(\chi_2 - \chi_1) = 0.
$$

Since the generic rank is 2, we have thus found an alternative way for computing the second character in δ-regular coordinates.

Again using Example 1, the Janet sequence for j_1 and $n = 2$ is easily found; for it

$$
\dim E = 3, \quad \dim C_0(E) = 9, \quad \dim C_1(E) = 9, \quad \dim C_2(E) = 3.
$$

Accordingly, the dimensions of the Janet and Spencer bundles are framed in the following diagram:

$$
\begin{array}{ccccccccc}
& & \boxed{6} & \xrightarrow{D_1} & \boxed{8} & \xrightarrow{D_2} & \boxed{3} & & \\
& & \downarrow & & \downarrow & & \| & & \\
0 \longrightarrow & \boxed{3} & \xrightarrow{j_1} & \boxed{9} & \xrightarrow{D_1} & \boxed{9} & \xrightarrow{D_2} & \boxed{3} & \longrightarrow 0 \\
& \| & & \downarrow & & \downarrow & & \downarrow & \\
& \boxed{3} & \xrightarrow{D} & \boxed{3} & \xrightarrow{D_1} & \boxed{1} & \longrightarrow & 0 &
\end{array}
$$

Again we can directly verify the dimension of the Spencer bundles. We have $\dim M_1 = \alpha_1^1 + \alpha_1^2 = 3$, $\dim M_2 = \alpha_1^1 + 2\alpha_1^2 = 4$, and $\dim M_3 = \alpha_1^1 + 3\alpha_1^2 = 5$.

Hence, as in Example 7 we have

$$\begin{cases} \dim C_0 = \dim R_1 = 6, \\ \dim C_1 = \dim T^* \otimes R_1 - \dim M_2 = 12 - 4 = 8, \\ \dim C_2 = \dim \wedge^2 T^* \otimes R_1 - \dim \wedge^2 T^* \otimes M_1 = 6 - 3 = 3. \end{cases}$$

Example 9 (see Example III.D.12). The system R_1 is involutive and $n = 3$. Hence we may use the results of Example 1 with $m = 3$. Accordingly, the schemes are as follows:

$$\begin{array}{ccc}
\boxed{\begin{array}{ccc} 1 & 2 & 3 \\ 1 & 2 & 3 \\ 1 & 2 & 3 \\ 1 & 2 & \bullet \\ 1 & 2 & \bullet \\ 1 & \bullet & \bullet \end{array}} & \longrightarrow & \boxed{\begin{array}{ccc} 1 & 2 & 3 \\ 1 & 2 & 3 \\ 1 & 2 & 3 \\ 1 & 2 & \bullet \end{array}} \quad \longrightarrow \quad \boxed{\begin{array}{ccc} 1 & 2 & 3 \end{array}} \\
\text{scheme for } \mathcal{D} & & \text{scheme for } \mathcal{D}_1 \qquad\qquad \text{scheme for } \mathcal{D}_2
\end{array}$$

The four compatibility conditions/differential identities are as follows:

$$\begin{cases} \Psi^4 \equiv d_3\Phi^3 - d_2\Phi^6 + d_1\Phi^5 + d_2\Phi^4 + d_1\Phi^4 - 4d_1\Phi^2 = 0, \\ \Psi^3 \equiv d_3\Phi^2 - d_2\Phi^5 + d_1\Phi^5 - d_2\Phi^4 + d_1\Phi^4 - 4d_1\Phi^2 = 0, \\ \Psi^2 \equiv d_3\Phi^1 - d_1\Phi^6 + d_1\Phi^5 + 2d_1\Phi^4 = 0, \\ \Psi^1 \equiv d_2\Phi^1 - d_1\Phi^3 + d_1\Phi^2 = 0. \end{cases}$$

They are related by the single differential identity

$$d_3\Psi^1 - d_2\Psi^2 + d_1\Psi^4 - d_1\Psi^3 = 0.$$

The dimensions of the Janet bundles in the Janet sequence for j_1 are as follows, with $n = 3$, $m = 3$:

$$\dim E = 3, \quad \dim C_0(E) = 12, \quad \dim C_1(E) = 18, \quad \dim C_2(E) = 12, \quad \dim C_3(E) = 3.$$

Accordingly, the dimensions of the Janet bundles and the Spencer bundles are linked in the following diagram:

$$\begin{array}{ccccccccc}
& \boxed{6} & \xrightarrow{D_1} & \boxed{14} & \xrightarrow{D_2} & \boxed{11} & \xrightarrow{D_3} & \boxed{3} & \longrightarrow 0 \\
& \downarrow & & \downarrow & & \downarrow & & \| & \\
0 \longrightarrow \boxed{3} & \xrightarrow{j_1} & \boxed{12} & \xrightarrow{D_1} & \boxed{18} & \xrightarrow{D_2} & \boxed{12} & \xrightarrow{D_3} & \boxed{3} \longrightarrow 0 \\
\| & & \downarrow & & \downarrow & & \downarrow & & \\
\boxed{3} & \xrightarrow{\mathcal{D}} & \boxed{6} & \xrightarrow{D_1} & \boxed{4} & \xrightarrow{D_2} & \boxed{1} & \longrightarrow & 0
\end{array}$$

Again we can directly check the dimensions of the Spencer bundles. We have

$$\begin{cases} \dim C_0 = \dim R_1 = 6, \\ \dim C_1 = \dim T^* \otimes R_1 - \dim M_2 = 18 - 4 = 14, \\ \dim C_2 = \dim \wedge^2 T^* \otimes R_1 - \dim \delta(T^* \otimes M_2) \\ \qquad = \dim \wedge^2 T^* \otimes R_1 - \dim T^* \otimes M_2 + \dim M_3 = 18 - 12 + 5 = 11, \\ \dim C_3 = \dim \wedge^3 T^* \otimes R_1 - \dim \delta(\wedge^2 T^* \otimes M_2) \\ \qquad = \dim \wedge^3 T^* \otimes R_1 - \dim \wedge^3 T^* \otimes M_1 = \dim \wedge^3 T^* \otimes E = 3. \end{cases}$$

The last equality is obtained because the system is transitive, and we have used the characters $\alpha_1^1 = 2$, $\alpha_1^2 = 1$, $\alpha_1^3 = 0$.

This example shows that, even for simple systems, it is almost impossible to write down Spencer sequences.

Example 10 (see Example III.D.13). The system R_3 under consideration is involutive. The characters of its symbol M_3 are $\alpha_3^1 = 1$, $\alpha_3^2 = 0$, and hence $\dim M_{3+r} = 1$ for all $r \geq 0$. The schemes are as follows:

$$\begin{array}{ccc} \begin{array}{cc} 1 & 2 \\ 1 & \bullet \\ 1 & \bullet \end{array} & \longrightarrow & \begin{array}{cc} 1 & 2 \\ 1 & 2 \end{array} \\ \text{scheme for } \mathcal{D} & & \text{scheme for } \mathcal{D}_1 \end{array}$$

Accordingly, $\dim E = 1$, $\dim F_0 = 3$, $\dim F_1 = 2$, $\dim F_2 = 0$. Since we have not yet considered the Janet sequence for j_3, we shall look directly for the dimensions of the Spencer bundles, because all prolongations of the symbol M_3 have the same dimension (namely, 1):

$$\begin{cases} \dim C_0 = \dim R_3 = \dim J_3(E) - \dim F_0 = 10 - 3 = 7, \\ \dim C_1 = \dim T^* \otimes R_3 - \dim M_4 = 14 - 1 = 13, \\ \dim C_2 = \dim \wedge^2 T^* \otimes R_3 - \dim \wedge^2 T^* \otimes M_3 = \dim \wedge^2 T^* \otimes J_2(E) = 6. \end{cases}$$

Therefore there is a diagram

$$\begin{array}{ccccccccc} & & \boxed{7} & \xrightarrow{D_1} & \boxed{13} & \xrightarrow{D_2} & \boxed{6} & \longrightarrow & 0 \\ & & \downarrow & & \downarrow & & \| & & \\ 0 \longrightarrow & \boxed{1} & \xrightarrow{j_3} \boxed{10} & \xrightarrow{D_1} & \boxed{15} & \xrightarrow{D_2} & \boxed{6} & \longrightarrow & 0 \\ & \| & \downarrow & & \downarrow & & \downarrow & & \\ & \boxed{1} & \xrightarrow{D} \boxed{3} & \xrightarrow{D_1} & \boxed{2} & \longrightarrow & 0 & & \end{array}$$

The characteristic variety is not zero, because the symbol is not of finite type. It is defined by the three polynomial equations:

$$\begin{cases} P_1 \equiv (\chi_2)^3 - (\chi_1)^3 = 0, \\ P_2 \equiv \chi_1(\chi_2)^2 - (\chi_1)^3 = 0, \\ P_3 \equiv (\chi_1)^2 \chi_2 - (\chi_1)^3 = 0. \end{cases}$$

Notice the relation

$$(\chi_2 - \chi_1)^3 \equiv P_1 + 3P_3 - 3P_2.$$

Hence the characteristic covectors are defined by the condition $\chi_1 = \chi_2$, and $\mathcal{D}$ is not elliptic. When $\chi_1 \neq \chi_2$, $\sigma_\chi(\mathcal{D})$ is injective, and we have $\dim \operatorname{im} \sigma_\chi(\mathcal{D}) = 1$.
We also have

$$\sigma_\chi(\mathcal{D}_1) = \begin{pmatrix} \chi_1 & \chi_2 & -\chi_1 \\ \chi_1 + \chi_2 & -\chi_1 & 0 \end{pmatrix}.$$

The generic rank is 2, unless $\chi_1 - \chi_2 = 0$. So, when $\chi_1 \neq \chi_2$, $\sigma_\chi(\mathcal{D}_1)$ is surjective and

$$\dim \operatorname{im} \sigma_\chi(\mathcal{D}) = \dim \ker \sigma_\chi(\mathcal{D}_1) = 1,$$

as predicted by Section B.

All the previous Examples could eventually be produced and solved by a symbolic computation. In contrast with this, the following examples, involving arbitrary dimensions, cannot be solved without appealing to the general procedure.

Example 11 (Poincaré sequence). We shall recover the Poincaré sequence from the general procedure, and compute the corresponding second Spencer sequence, which has a particularly simple structure. We start with the operator $\wedge^0 T^* \to \wedge^1 T^*$: $f(x) \to \partial_i f(x)$, which is a *gradient* operator if $n = 3$. It is an involutive operator with corresponding system R_1 and symbol $M_1 = 0$. Since R_1 is, moreover, transitive, we have $R_1 \simeq E = \wedge^0 T^*$. Accordingly, $C_r = \wedge^r T^*$. Now $\wedge^1 T^* = T^*$, and $N_1 \simeq S_2 T^* \subset T^* \otimes T^*$. Therefore

$$F_r = \wedge^r T^* \otimes T^* / \delta(\wedge^{r-1} T^* \otimes S_2 T^*).$$

However, there are exact sequences

$$0 \longrightarrow S_{r+1} T^* \xrightarrow{\delta} T^* \otimes S_r T^* \xrightarrow{\delta} \cdots \xrightarrow{\delta} \wedge^r T^* \otimes T^* \xrightarrow{\delta} \wedge^{r+1} T^* \longrightarrow 0,$$

and thus $F_r \simeq \wedge^{r+1} T^*$, a nonevident reason for the shift in grading with respect to a usual Janet sequence.

Combining these two results we find

$$C_r(E) \simeq \wedge^r T^* \times_X \wedge^{r+1} T^*.$$

It is *absolutely a coincidence* that the Janet sequence and the second Spencer sequence for d coincide, although with a shift in grading. In the last Chapter we shall see that this result has led physicists to misleading concepts.

The fact that the second Spencer sequence is also the Poincaré sequence follows from the definition of Spencer operator and the fact that a section ξ_1 of R_1 satisfies $\xi_i = 0$. Accordingly, $(D\xi_1)_i = \partial_i \xi - \xi_i = \partial_i \xi$, and we recover the definition of exterior derivative.

The fact that the Janet sequence is also the Poincaré sequence is *much more* subtle and involves many tricks, which we will now describe in detail.

The basic diagram is as follows:

$$\bigwedge^r T^* \otimes R_1$$
$$\downarrow$$
$$\delta(\bigwedge^{r-1} T^* \otimes S_r T^*) \longrightarrow \bigwedge^r T^* \otimes J_1(\bigwedge^0 T^*)$$
$$\downarrow \qquad \searrow$$
$$\bigwedge^r T^* \otimes T^* \qquad \longrightarrow F_r$$

Since $\delta: \bigwedge^r T^* \otimes T^* \to \bigwedge^{r+1} T^*$, any $(r+1)$-form in F_r can be written as $\omega_{i,i_1\ldots i_r}\, dx^{i_1} \wedge \cdots \wedge dx^{i_r}$ with $i_1 < \cdots < i_r$. The lift to $\bigwedge^r T^* \otimes J_1(\bigwedge^0 T^*)$, taking into account the definition of R_1, will be $(\omega_{i_1\ldots i_r}, \omega_{i,i_1\ldots i_r})$, with $\omega_{i_1\ldots i_r}$ *arbitrary*. To apply D we lift again, to $\bigwedge^r T^* \otimes J_2(\bigwedge^0 T^*)$, by considering $(\omega_{i_1\ldots i_r}, \omega_{i,i_1\ldots i_r}, \omega_{ij,i_1\ldots i_r})$, with now $\omega_{i_1\ldots i_r}$ and $\omega_{ij,i_1\ldots i_r}$ arbitrary. Changing $\omega_{ij,i_1\ldots i_r}$ amounts to changing the image by an element of $\delta(\bigwedge^r T^* \otimes S_2 T^*)$, and this change is therefore killed by projecting onto F_{r+1}. So, we can choose $\omega_{ij,i_1\ldots i_r} = 0$, since the result does not depend on the chosen lift of the element. The image under D is thus:

$$\begin{cases} (\partial_i \omega_{i_1\ldots i_r} - \omega_{i,i_1\ldots i_r})\, dx^i \wedge dx^{i_1} \wedge \cdots \wedge dx^{i_r}, \\ (\partial_i \omega_{j,i_1\ldots i_r} - 0)\, dx^i \wedge dx^{i_1} \wedge \cdots \wedge dx^{i_r}. \end{cases}$$

Projection onto $\bigwedge^{r+1} T^* \otimes T^*$ preserves the second component (index j for T^*):

$$\partial_i \omega_{j,i_1\ldots i_r}\, dx^i \wedge dx^{i_1} \wedge \cdots \wedge dx^{i_r},$$

which we then have to project onto $\bigwedge^{r+2} T^* = F_{r+1}$. To this end we have only to take the exterior product with dx^j:

$$\partial_i \omega_{j,i_1\ldots i_r}\, dx^j \wedge dx^i \wedge dx^{i_1} \wedge \cdots \wedge dx^{i_r} = -\partial_i \omega_{j,i_1\ldots i_r}\, dx^i \wedge dx^j \wedge dx^{i_1} \wedge \cdots \wedge dx^{i_r},$$

and we recognise *minus* the exterior derivative of

$$\omega_{j,i_1\ldots i_r}\, dx^j \wedge dx^i \wedge dx^{i_1} \wedge \cdots \wedge dx^{i_r},$$

where all the forms have to be decomposed with respect to an ordinary basis of exterior products. The latter form is the element of $F_r = \bigwedge^{r+1} T^*$ we started with.

Regarding the symbol sequence, we have

$$\sigma_\chi(d) = \chi\wedge,$$

and it is known to be exact for any nonzero covector.

For the sake of being complete, we indicate the dimensions of the Janet and Spencer bundles in the case of classical vector geometry, $n = 3$, while taking into account the results of Example 1 concerning the Janet sequence for j_1:

$$
\begin{array}{ccccccccc}
\boxed{1} & \xrightarrow{D_1} & \boxed{3} & \xrightarrow{D_2} & \boxed{3} & \xrightarrow{D_3} & \boxed{1} & \longrightarrow & 0 \\
\downarrow & & \downarrow & & \downarrow & & \| & & \\
0 \longrightarrow \boxed{1} & \xrightarrow{j_1} & \boxed{4} & \xrightarrow{D_1} & \boxed{6} & \xrightarrow{D_2} & \boxed{4} & \xrightarrow{D_3} & \boxed{1} \longrightarrow 0 \\
\| & & \downarrow & & \downarrow & & \downarrow & & \downarrow \\
\boxed{1} & \xrightarrow{D} & \boxed{3} & \xrightarrow{D_1} & \boxed{3} & \xrightarrow{D_2} & \boxed{1} & \longrightarrow & 0
\end{array}
$$

Up to sign we successively recognise the three well-known operators *grad*, *rot* and *div*. From the previous computation we can understand why on the one hand $\mathcal{D}_1, \mathcal{D}_2$ are 'different' from $\mathcal{D}$, while on the other hand D_1, D_2, D_3 cannot be 'separated', at least conceptually.

Example 12. The case that $\mathcal{D}$ is the '*divergence*' operator will provide one example of a situation in which the Janet and Spencer sequences are so totally different (unlike in the preceding Example) that it is quite a surprise that they could have been mixed up in mathematical physics.

Set $E = T$ and $\mathcal{D} = \mathrm{div}: T \to \bigwedge^n T^*: \xi \to \mathcal{L}(\xi)\omega$, where $\omega = dx^1 \wedge \cdots \wedge dx^n$ is the volume form (we now work locally, over a coordinate domain) and $\mathcal{L}$ is the ordinary Lie derivative. The corresponding system R_1 has sections satisfying $\xi_r^r(x) = 0$. We have $\beta_1^1 = \cdots = \beta_1^{n-1} = 0$, $\beta_1^n = 1$, and thus $\alpha_1^1 = \cdots = \alpha_1^{n-1} = n$, $\alpha_1^n = n - 1$ (since $m = n$). Hence R_1 is involutive, and we have

$$
\dim M_2 = n(1 + \cdots + n - 1) + n(n-1) = \frac{n(n-1)(n+2)}{2}.
$$

Accordingly, the exact sequence

$$
0 \longrightarrow M_2 \longrightarrow S_2 T^* \otimes T^* \longrightarrow T^* \otimes \textstyle\bigwedge^n T^* \longrightarrow F_1 \longrightarrow 0
$$

implies

$$
\begin{aligned}
\dim F_1 &= \dim M_2 + \dim T^* \otimes \textstyle\bigwedge^n T^* - \dim S_2 T^* \otimes T \\
&= \frac{n(n-1)(n+2)}{2} + n - \frac{n^2(n+1)}{2} \\
&= 0.
\end{aligned}
$$

It follows that the Janet sequence ends at F_0! Using the results of Example 1 for

$m = n = 3$, we obtain the diagram:

$$\boxed{11} \xrightarrow{D_1} \boxed{18} \xrightarrow{D_2} \boxed{12} \xrightarrow{D_3} \boxed{3} \longrightarrow 0$$

$$0 \longrightarrow \boxed{3} \xrightarrow{j_1} \boxed{12} \xrightarrow{D_1} \boxed{18} \xrightarrow{D_2} \boxed{12} \xrightarrow{D_3} \boxed{3} \longrightarrow 0$$

$$\boxed{3} \xrightarrow{D} \boxed{1} \longrightarrow 0$$

Example 13. Let $E = T$ and $n = 3$ and consider the system $\hat{R}_1$ defining the conformal isometries of the Euclidean metric

$$\omega_{rj}\xi_i^r + \omega_{ir}\xi_j^r = A(x)\omega_{ij} \Rightarrow \omega_{rj}\xi_i^r + \omega_{ir}\xi_j^r - \frac{2}{n}\omega_{ij}\xi_r^r = 0,$$

where $\omega_{ij} = 0$ if $i \neq j$ and $= 1$ if $i = j$. This system is formally integrable, since it is homogeneous with constant coefficients. Although this system will be studied in great depth in Chapter V (because of its importance for applications in the final Chapter), we shall here prove that it is of finite type. We do this by looking at the characteristic matrix

$$\begin{pmatrix} \frac{4}{3}\chi_1 & -\frac{2}{3}\chi_2 & -\frac{2}{3}\chi_3 \\ -\frac{2}{3}\chi_1 & \frac{4}{3}\chi_2 & -\frac{2}{3}\chi_3 \\ -\frac{2}{3}\chi_1 & -\frac{2}{3}\chi_2 & \frac{4}{3}\chi_3 \\ 0 & \chi_3 & \chi_2 \\ \chi_3 & 0 & \chi_1 \\ \chi_2 & \chi_1 & 0 \end{pmatrix}$$

Among the various determinants that have to be equated to zero, we have (using permutations):

$$\chi_1\chi_2\chi_3 = 0,$$
$$\chi_1(\chi_2)^2 + (\chi_1)^3 = 0 \Rightarrow \chi_2(\chi_3)^2 + (\chi_2)^3 = 0 \Rightarrow \chi_3(\chi_1)^2 + (\chi_3)^3 = 0.$$

Since one of the χ_i must be zero, all others must be zero as well. Accordingly, the characteristic covector is $\chi = 0$, and the system $\hat{R}_1$ is of finite type. Nevertheless, it is not quite evident to deduce $M_3 = 0$, as we shall see.

We leave it to the reader to verify that the case $n = 2$ is totally different. (See [156] for more details.)

CHAPTER V

Group theory

The purpose of this Chapter is to set up the group-theoretical concepts needed in applications. In particular, we shall point out the many reasons why the theory of Lie groups is insufficient and has to be replaced by the formal theory of Lie pseudogroups. Since the main results have already been given in our three books [144, 145, 146], here we shall simplify and shorten the proofs whenever this is possible. We present only the very basic concepts that are absolutely necessary for understanding the general theory and the many examples that will be treated in great technical detail in the last Section. Although we have tried to make this Chapter largely independent of the two preceding ones, it is nevertheless necessary to have good knowledge of jet theory to understand the key points (compare to [117] and [162]).

A. Groups and Algebras

The purpose of this Section is to recall the basic results in the theory of Lie groups and Lie groups of transformations. In particular, the material here presented will be used in constructing a nonlinear differential sequence, the '*gauge sequence*', which is of great importance in nowadays mathematical physics.

Definition 1. A *Lie group* G is a manifold with a distinguished point e, called *identity*, and endowed with two maps:

- $G \times G \to G$: $(a, b) \to ab$, *composition*;
- $G \to G$: $a \to a^{-1}$, *inverse*,

satisfying the following relations:

$$(ab)c = a(bc) = abc, \qquad \forall a, b, c \in G,$$
$$aa^{-1} = a^{-1}a = e, \qquad \forall a \in G,$$
$$ae = ea = a, \qquad \forall a \in G.$$

A *Lie subgroup* $H \subset G$ is a subset of G for which these maps are defined by restriction.

Definition 2. A Lie group G is said to *act* on a manifold X if there is a map $X \times G \to X$: $(x, a) \to ax$ such that

$$ex = x, \qquad \forall x \in X,$$
$$(ab)x = a(bx), \qquad \forall a, b \in G, \quad \forall x \in X.$$

The set $G_x = \{a \in G | \ ax = x\}$ is called the *isotropy subgroup* of G at x. A subset $S \subset X$ is said to be *invariant* under the action of G if $aS \subset S$ for all $a \in G$. The *orbit* of $x \in X$ is the set $Gx = \{ax| \ a \in G\}$. If G acts on two manifolds X and Y, a map $f: X \to Y$ is said to be *equivariant* if $f(ax) = af(x)$ for all $a \in G$ and all $x \in X$.

To study the action of G on X it is often convenient to introduce the *graph* $X \times G \to X \times X$: $(x, a) \to (x, ax)$ of this action. In the product $X \times X$, the first factor is called the *source*, and the second the *target*.

Actions are classified by means of properties of their graphs, as follows.

Definition 3. The action is *free* if the graph is injective. The action is *transitive* if the graph is surjective. The action is *simply transitive* if the graph is an isomorphism. In this last case X is also called a *principal homogeneous space* for G. The action is *effective* if $ax = x$, $\forall x \in X$, implies $a = e$.

Remark 4. The action of G on X determines a so-called *Lie group of transformations* of X, and the transformation $x \to ax$ is often denoted in a functorial way, by $x \to y = f(x, a)$. Its *source* is x, its *target* y, and a is a parameter of it. Nevertheless, since global actions usually do not exist, one most often considers the situation when f belongs to the set $\mathrm{aut}(X)$ of local diffeomorphisms of X. For simplicity, instead of indicating the (open) source set, we shall use the global notation as a manner of specifying the corresponding local coordinates.

All formal results on Lie groups of transformations derive from the following, celebrated, *three fundamental theorems of Lie*. Of course, these will be constantly used in the sequel. We shall give very elementary proofs of these Theorems, referring to [40, 92, 144, 195] for more details. Note that these Theorems are also valid for Lie groups acting on themselves from the *right* or from the *left*.

Let $\dim G = r$.

Theorem 5 (First Fundamental Theorem). *The orbits* $x = f(x_0, a)$ *satisfy*

$$\frac{\partial x^i}{\partial a^\sigma} = \xi^i_\rho(x)\omega^\rho_\sigma(a),$$

with $\det \omega \neq 0$. *The vector fields* $\xi_\rho = \xi^i_\rho \partial_i$, *called infinitesimal generators of the action, are linearly independent over the constants, provided that the action is effective.*

PROOF. $x = f(x_0, a)$, $y = f(x, b)$ imply $y = f(x_0, c)$ with $c = ba = \varphi(b, a)$ (composition in G). Fixing x_0 and c while differentiating, we obtain

$$\begin{cases} \dfrac{\partial f(x, b)}{\partial x} \, dx + \dfrac{\partial f(x, b)}{\partial b} \, db = 0, \\[2mm] \dfrac{\partial \varphi(b, a)}{\partial a} \, da + \dfrac{\partial \varphi(b, a)}{\partial b} \, db = 0. \end{cases}$$

Differentiating the identity $y \equiv f(f(y, b^{-1}), b)$ with respect to y, where $x = f(y, b^{-1})$, and taking the determinant, we obtain $\det(\partial f/\partial x) \neq 0$, and hence also $\det(\partial \varphi/\partial b) \neq 0$. So, we can express dx in terms of db and db in terms of da, hence dx in terms of da. Setting $b = e$ we arrive at the required Frobenius-type system.

If the action is effective, we can choose sufficiently many points x_α such that $x_\alpha = f(x_\alpha, a)$ has only the solution $a = e$. Accordingly, $\mathrm{rk}(\partial f(x_\alpha)/\partial a) = r$ and thus $\mathrm{rk}(\xi_\rho^i(x_\alpha)) = r$, because $\det \omega \neq 0$. Hence there cannot exist constants λ^ρ such that $\lambda^\rho \xi_\rho = 0$, even though it is possible that $\mathrm{rk}(\xi_\rho^i(x)) \leq r$. □

A very useful concept related to vector fields is as follows.

Definition 6. If $\xi, \eta \in T$, we define their *bracket*, $[\xi, \eta] \in T$, by the local formula

$$([\xi, \eta])^i(x) = \xi^r(x)\partial_r\eta^i(x) - \eta^r(x)\partial_r\xi^i(x).$$

We leave it to the reader to verify the following properties. Note that the third property implies that the definition above does not depend on the coordinate system.

Proposition 7. *For all $\xi, \eta, \zeta \in T$ and $f\colon X \to Y$ the following relations hold:*

$$[\xi, \eta] + [\eta, \xi] = 0,$$
$$[\xi, [\eta, \zeta]] + [\eta, [\zeta, \xi]] + [\zeta, [\xi, \eta]] = 0 \qquad (\textit{Jacobi identity}),$$
$$[T(f)(\xi), T(f)(\eta)] = T(f)([\xi, \eta]).$$

The first two properties give rise to the notion of *Lie algebra* on the sections of T. The third property gives $T(f)\colon T(X) \to T(Y)$ as a morphism of Lie algebras. These properties are particularly useful in proving the following result.

Theorem 8 (Second Fundamental Theorem). *If $\xi_1, \ldots, \xi_r$ are the infinitesimal generators of the effective action of a Lie group G on X, then*

$$[\xi_\rho, \xi_\sigma] = c_{\rho\sigma}^\tau \xi_\tau,$$

where $c_{\rho\sigma}^\tau$ are constants, called the structure constants of the corresponding Lie algebra of vector fields.

PROOF. By the First Fundamental Theorem, the system defining the orbits, which is of order one on $G \times G \to G$ and has zero symbol, must be transitive because the orbits $x = f(x_0, a)$ depend on the n arbitrary constants x_0. Using the criterion for formal integrability, we have to take crossed derivatives in order to obtain the conditions:

$$([\xi_\rho, \xi_\sigma])^i(x) + \alpha_\rho^r(a)\alpha_\sigma^s(a)\left(\frac{\partial\omega_s^\tau(a)}{\partial a^r} - \frac{\partial\omega_r^\tau(a)}{\partial^s}\right)\xi_\tau^i(x) = 0,$$

where α is the inverse matrix of ω. They are of the form

$$[\xi_\rho, \xi_\sigma] = c_{\rho\sigma}^\tau(a)\xi_\tau,$$

and must be identities on $X \times G$ also. Differentiation with respect to a and taking into account the fact that the ξ_ρ are linearly independent over the constants, because the action is effective, we find $c_{\rho\sigma}^\tau(a) = c_{\rho\sigma}^\tau = \mathrm{const}$, as required. We have

$$\frac{\partial\omega_r^\tau(a)}{\partial a^s} - \frac{\partial\omega_s^\tau(a)}{\partial a^r} = c_{\rho\sigma}^\tau\omega_r^\rho(a)\omega_s^\sigma(a).$$

In the sequel, the reader should take care of the signs! □

Remark 9. By the construction in the proof of Theorem 5, the 1-forms $\omega^\tau = \omega^\tau_\sigma(a)\, da^\sigma$ are 1-forms on G with values in the *Lie algebra* $\mathcal{G} = T_e(G)$, and the latter conditions are called the *Maurer–Cartan equations* on G. For later use we notice that the vector fields $\alpha_\tau = \alpha^\sigma_\tau(a)\partial/\partial a^\sigma$ on G satisfy the 'dual' equations

$$[\alpha_\rho, \alpha_\sigma] = c^\tau_{\rho\sigma}\alpha_\tau.$$

The Lie algebra structure on $\mathcal{G}$ is defined by the structure constants. Indeed, using the two first relations of Proposition 7 and the fact that the infinitesimal generators are linearly independent over the constants (the action is effective), we obtain the algebraic conditions

$$\begin{cases} c^\tau_{\rho\sigma} + c^\tau_{\sigma\rho} = 0, \\ c^\lambda_{\mu\rho}c^\mu_{\sigma\tau} + c^\lambda_{\mu\sigma}c^\mu_{\tau\rho} + c^\lambda_{\mu\tau}c^\mu_{\rho\sigma} = 0 \quad \text{(Jacobi identities)}. \end{cases}$$

The third relation in Proposition 7 shows that the definitions above are coherent, since we can construct, for any vector in $\mathcal{G}$, a vector field on G by using the tangent map of the composition. In fact, since we may compose on the *right* or on the *left* and since the right action commutes with the left action, we leave it to the reader to prove that any corresponding *right invariant* vector field α commutes with any corresponding *left invariant* vector field β. Later on, we will obtain this result by another approach.

The following technical result will be quite useful in the applications in Chapter VIII. Once more, it is a result that seems very simple when presented abstractly, but the proof must be given in local coordinates $\cdots$ because the applications are done in local coordinates (anyway, any proof amounts to doing the same computations).

Consider a vector λ in $\mathcal{G}$. We can transfer it to $a \in G$, as a vector $\lambda^\rho\beta_\rho$, and pull it back to $\mathcal{G}$ by $\omega = \alpha^{-1}$. Accordingly, to any $a \in G$ we can associate a map, $Ad(a): \mathcal{G} \to \mathcal{G}$, called the *adjoint action* of G on $\mathcal{G}$. It is described by a matrix $M^\tau_\rho(a) = \omega^\tau_r(a)\beta^r_\rho(a)$.

Proposition 10. *The adjoint map satisfies*

$$\frac{\partial M^\tau_\mu}{\partial a^r} + c^\tau_{\rho\sigma}\omega^\rho_r M^\sigma_\mu = 0.$$

PROOF. By Remark 9 we have $[\alpha, \beta] = 0$ with $\alpha = \omega^{-1}$. Hence, successively,

$$\alpha_\tau^r \frac{\partial \beta_\mu^\sigma}{\partial a^r} - \beta_\mu^s \frac{\partial \alpha_\tau^\sigma}{\partial a^s} = 0,$$

$$\frac{\partial \beta_\mu^\sigma}{\partial a^r} - \beta_\mu^s \omega_r^\tau \frac{\partial \alpha_\tau^\sigma}{\partial a^s} = 0,$$

$$\frac{\partial \beta_\mu^\sigma}{\partial a^r} + \alpha_\tau^\sigma \beta_\mu^s \frac{\partial \omega_r^\tau}{\partial a^s} = 0,$$

$$\omega_s^\tau \frac{\partial \beta_\mu^s}{\partial a^r} + \beta_\mu^s \frac{\partial \omega_r^\tau}{\partial a^s} = 0,$$

$$\frac{\partial M_\mu^\tau}{\partial a^r} + \beta_\mu^s \left(\frac{\partial \omega_r^\tau}{\partial a^s} - \frac{\partial \omega_s^\tau}{\partial a^r} \right) = 0,$$

$$\frac{\partial M_\mu^\tau}{\partial a^r} + c_{\rho\sigma}^\tau \omega_r^\rho \omega_s^\sigma \beta_\mu^s = 0,$$

by using the Maurer–Cartan equations. $\square$

Theorem 11 (Third Fundamental Theorem). *For any Lie algebra $\mathcal{G}$ one can construct an analytic group G such that $\mathcal{G} = T_e(G)$.*

PROOF. In practice we want to recover G from given constants satisfying the conditions of Remark 9. For this we write the Maurer–Cartan equations as

$$d\omega^\tau + \frac{1}{2} c_{\rho\sigma}^\tau \omega^\rho \wedge \omega^\sigma = 0,$$

taking care of the sign as in the proof of Theorem 8. Closing the system by applying the exterior derivative once more amounts to verifying that the system is involutive using the criterion. We recall that the exterior derivative is an involutive operator, because *the Poincaré sequence is a Janet sequence.* After a short computation we are left with the fact that the Maurer–Cartan equations are involutive if and only if the structure constants satisfy the Jacobi relations, whenever $\det \omega \neq 0$. Under the conditions of the Theorem we therefore have an involutive system, and we can apply the Cartan–Kähler Theorem III.C.10 to find analytic Maurer–Cartan forms $\omega(a)$. The action of the group on itself is then defined by the analytic involutive system $\partial b / \partial a = \alpha(b) \omega(a)$, which leads to the invariance of the Maurer–Cartan forms under the group action. Again we should use the Cartan–Kähler Theorem for this system, to find an analytic group G. The fact that the latter system is involutive derives from the Maurer–Cartan equations *and* the commutation relations of Remark 9. In Sections B, C we shall find a new interpretation of this result. $\square$

Now we have at our disposal all the material needed to construct the key piece of machinery making it possible to apply Lie group theory in physics; namely, the construction of a differential sequence which we call the *gauge sequence*.

The starting point is a manifold X and a Lie group G. *We do not assume that G acts on X.* Let us consider the picture:

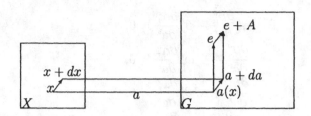

Whenever there is a map $a\colon X \to G$, we can start with a tangent vector to X at x, apply $T(a)$ to obtain a tangent vector to G at $a(x)$, and project this to a tangent vector to G at e by applying the tangent map of the composition law (on the left or right). Hence, we find, under any such a, from a vector on X a vector in $\mathcal{G}$ or, equivalently, a 1-form on X with values in $\mathcal{G}$, while using first derivatives only. Another way to describe this construction is to write the local description

$$\xi^i(x) \to \omega_\sigma^\tau(a(x))\xi^i(x)\partial_i a^\sigma(x)$$

for any $a\colon (x^i) \to (a^\sigma(x))$. Equivalently, we may define

$$\omega_\sigma^\tau(a(x))\partial_i a^\sigma(x) = A_i^\tau(x),$$

and consider the family of 1-forms $A^\tau = A_i^\tau(x)\,dx^i$, which is also called a *Yang-Mills potential*. Finally, a very compact way of using the picture is to simply set $a^{-1}\,da = A$ (we could have used $da\,a^{-1}$ instead, depending on the way we pass from $T_a(G)$ to $T_e(G) = \mathcal{G}$: by composition on the left or on the right).

We use the results in Chapter II.A to be able to regard a as a section of the trivial fibered manifold $X \times G \to X$ and, as usual, we need consider local sections only, although we will write down the concepts as if they were global (to avoid the specification of the open sets that have to be used).

We have

$$\partial_i A_j^\tau(x) - \partial_j A_i^\tau(x) = \left(\frac{\partial\omega_\sigma^\tau(a(x))}{\partial a^\rho} - \frac{\partial\omega_\rho^\tau(a(x))}{\partial a^\sigma}\right)\partial_i a^\rho(x)\partial_j a^\sigma(x)$$
$$= c_{\sigma\rho}^\tau \omega_r^\rho(a(x))\omega_s^\sigma(a(x))\partial_i a^r(x)\partial_j a^s(x)$$
$$= c_{\sigma\rho}^\tau A_i^\rho(x)A_j^\sigma(x).$$

Therefore, setting

$$\partial_i A_j^\tau(x) - \partial_j A_i^\tau(x) + c_{\rho\sigma}^\tau A_i^\rho(x)A_j^\sigma(x) = F_{ij}^\tau(x),$$

we may consider the family of 2-forms $F^\tau = \frac{1}{2}F_{ij}^\tau(x)\,dx^i \wedge dx^j$, which is also called a *Yang-Mills field*. We have $a^{-1}\,da = A \Rightarrow F = 0$. The converse is also true, locally, since the system $da = \omega(a)A(x)$ is involutive with zero symbol because $\det\omega \neq 0$ and $F = 0$. Therefore we can locally integrate this system by means of Frobenius' Theorem for involutive distributions of vector fields.

Collecting the above results, we obtain:

Theorem 12. *There exists a locally exact differential sequence*

$$X \times G \longrightarrow T^* \otimes \mathcal{G} \xrightarrow{\text{MC}} \wedge^2 T^* \otimes \mathcal{G},$$
$$a \longrightarrow a^{-1} da = A \longrightarrow dA + [A, A] = F,$$

called gauge sequence.

In describing this sequence we have used the evident notation

$$[A, A](\xi, \eta) = [A(\xi), A(\eta)] \in \mathcal{G}, \qquad \forall \xi, \eta \in T.$$

Also, with a slight abuse of language, the first-order nonlinear operator $A \to dA + [A, A]$ has been called the *Maurer-Cartan operator*.

We may now compose the image of a and $a \circ b$ under the first operator, setting $a \circ b(x) = a(x)b(x)$ in G. We successively obtain, in symbolic notation:

$$(ab)^{-1} d(ab) = b^{-1} a^{-1} (da\, b + a\, db)$$
$$= b^{-1}(a^{-1} da)b + b^{-1} db,$$

and we can state

Definition 13. The transformation

$$A \to Ad(b)\, A + b^{-1} db$$

is called a *gauge transformation*.

We now use Proposition 10 to prove a few technical, but useful, properties of the gauge transformation.

First of all, we use a^{-1} instead of a, and we obtain

$$a^{-1} a = e \Rightarrow a^{-1} da + da^{-1} a = 0,$$

i.e. $da^{-1} a = -a^{-1} da = da^{-1} (a^{-1})^{-1}$. Also, using the gauge transformation with $b = a^{-1}$, we find

$$a\, da^{-1} = (a^{-1})^{-1} da^{-1} = -Ad(a^{-1})\, (a^{-1} da).$$

The next Proposition expresses a more delicate property.

Proposition 14.

$$A' = Ad(b)A + b^{-1} db \Rightarrow F' = Ad(b)F.$$

PROOF. In local coordinates we have

$$A_i'^\tau(x) = M^\tau(b(x)) A_i(x) + \omega_\sigma^\tau(b(x)) \partial_i b^\sigma(x).$$

We now have:

$$F' = dA' + [A', A']$$
$$= d(Ad(b)A) + [Ad(b)A, Ad(b)A] + [b^{-1}db, Ad(b)A] + [Ad(b)A, b^{-1}db],$$

because, setting $b^{-1}db = B$, we have

$$dB + [B, B] = 0.$$

Substituting local coordinates, we obtain

$$
\begin{aligned}
F'^\tau_{ij} &= M^\tau_\nu(\partial_i A^\nu_j - \partial_j A^\nu_i) + c^\tau_{\rho\sigma} M^\rho_\lambda M^\sigma_\mu A^\lambda_i A^\mu_j \\
&\quad + \frac{\partial M^\tau_\nu}{\partial b^\sigma}(A^\nu_j \partial_i b^\sigma - A^\nu_i \partial_j b^\sigma) \\
&\quad + c^\tau_{\rho\sigma} B^\rho_i M^\sigma_\mu A^\mu_j + c^\tau_{\rho\sigma} M^\rho_\lambda A^\lambda_i B^\sigma_j \\
&= M^\tau_\nu(\partial_i A^\nu_j - \partial_j A^\nu_i) + c^\tau_{\rho\sigma} M^\rho_\lambda M^\sigma_\mu A^\lambda_i A^\mu_j \\
&\quad + A^\mu_j\left(\frac{\partial M^\tau_\mu}{\partial b^r}\partial_i b^r + c^\tau_{\rho\sigma} B^\rho_i M^\sigma_\mu\right) \\
&\quad - A^\mu_i\left(\frac{\partial M^\tau_\mu}{\partial b^r}\partial_j b^r + c^\tau_{\rho\sigma} B^\rho_j M^\sigma_\mu\right) \\
&= M^\tau_\nu(\partial_i A^\nu_j - \partial_j A^\nu_i) + c^\tau_{\rho\sigma} M^\rho_\lambda M^\sigma_\mu A^\lambda_i A^\mu_j,
\end{aligned}
$$

having used Proposition 10 twice.

Let us now regard $c \in \wedge^2 \mathcal{G}^* \otimes \mathcal{G}$ as a tensor under the linear transformation $Ad(a): \mathcal{G} \to \mathcal{G}$. The cornerstone of the proof is to show that c is invariant under the adjoint action. To this end we set

$$
c'(a) = Ad(a)c \Leftrightarrow c'^\nu_{\lambda\mu}(a) = c^\tau_{\rho\sigma} M^\rho_\lambda(a)\, M^\sigma_\mu(a)\, M^{-1}{}^\nu_\tau(a),
$$

and use the following consequence of Proposition 10:

$$
\frac{\partial M^{-1}{}^\nu_\tau}{\partial a^r} - c^\sigma_{\rho\tau}\omega^\rho_r M^{-1}{}^\nu_\sigma = 0.
$$

Accordingly, we successively obtain

$$
\begin{aligned}
\frac{\partial c'^\nu_{\lambda\mu}(a)}{\partial a^r} &= -c^\tau_{\rho\sigma} c^\rho_{\alpha\beta}\omega^\alpha_r M^\beta_\lambda M^\sigma_\mu M^{-1}{}^\nu_\tau \\
&\quad - c^\tau_{\rho\sigma} c^\sigma_{\alpha\beta}\omega^\alpha_r M^\rho_\lambda M^\beta_\mu M^{-1}{}^\nu_\tau \\
&\quad + c^\tau_{\rho\sigma} c^\beta_{\alpha\tau}\omega^\alpha_r M^\rho_\lambda M^\sigma_\mu M^{-1}{}^\nu_\beta \\
&= -(c^\tau_{\beta\alpha} c^\beta_{\rho\sigma} + c^\tau_{\beta\rho} c^\beta_{\sigma\alpha} + c^\tau_{\beta\sigma} c^\beta_{\alpha\rho})\omega^\alpha_r M^\rho_\lambda M^\sigma_\mu M^{-1}{}^\nu_\tau \\
&= 0,
\end{aligned}
$$

where we have used the Jacobi relations in Remark 9. Since $M(e) = \mathrm{id}_\mathcal{G}$, we have $c'^\nu_{\lambda\mu}(a) = c'^\nu_{\lambda\mu}(e) = c^\nu_{\lambda\mu}$. Collecting the results we thus find

$$
F'^\tau_{ij} = M^\tau_\nu(\partial_i A^\nu_j - \partial_j A^\nu_i + c^\nu_{\lambda\mu} A^\lambda_i A^\mu_j) = M^\tau_\nu F^\nu_{ij},
$$

which finishes the proof. $\square$

If $F = 0$ we have $F' = 0$, and another way to interpret Proposition 14 is to say that *a gauge transformation permutes the solutions of the Maurer–Cartan equations.* We shall now '*linearise*' the gauge sequence. We do this in several steps.

First we notice that a *variation* of the section (map) a gives rise to a variation of A, as in the following symbolic computation:

$$\delta A = \delta(a^{-1}da) = a^{-1}\delta da + (\delta a^{-1})da$$
$$= a^{-1}d\delta a - a^{-1}\delta aa^{-1}da$$
$$= d(a^{-1}\delta a) + (a^{-1}da)(a^{-1}\delta a) - (a^{-1}\delta a)(a^{-1}da)$$
$$= d(a^{-1}\delta a) + [a^{-1}da, a^{-1}\delta a].$$

Hence, regarding $a^{-1}\delta a = \xi$ as a section of $X \times \mathcal{G}$, we obtain

$$\delta A = d\xi + [A, \xi].$$

In local coordinates we have

$$\delta A_i^\tau(x) = \omega_\sigma^\tau(a(x))\partial_i\delta a^\sigma(x) + \frac{\partial\omega_\sigma^\tau(a(x))}{\partial a^\rho}\delta a^\rho(x)\partial_i a^\sigma(x)$$
$$= \partial_i(\omega_\sigma^\tau(a(x))\delta a^\sigma(x)) + \left(\frac{\partial\omega_\sigma^\tau(a(x))}{\partial a^\rho} - \frac{\partial\omega_\rho^\tau(a(x))}{\partial a^\sigma}\right)\delta a^\rho(x)\partial_i a^\sigma(x)$$
$$= \partial_i\xi^\tau(x) + c_{\rho\sigma}^\tau\omega_r^\rho\omega_s^\sigma\partial_i a^r(x)\delta a^s(x)$$
$$= \partial_i\xi^\tau(x) + c_{\rho\sigma}^\tau A_i^\rho(x)\xi^\sigma(x).$$

Now, if $A = a^{-1}da$ we have $dA + [A, A] = 0$, so that if δA has the previous value, we obtain an *identity*: $d\delta A + [\delta A, A] + [A, \delta A] \equiv 0$. We advise the reader to verify this relation directly, it being a nice exercise with brackets. Sometimes we shall write $\delta A = \delta_\xi A$ to specify the variation.

Finally, in symbolic notation we have

$$Ad(a^{-1})\xi = a\xi a^{-1} = aa^{-1}\delta a\, a^{-1} = \delta a\, a^{-1},$$
$$d\,Ad(a^{-1})\xi = (d\delta a)a^{-1} - \delta a\, a^{-1}da\, a^{-1},$$

$$Ad(a)d\,Ad(a^{-1})\xi = a^{-1}(d\delta a\, a^{-1} - \delta a\, a^{-1}da\, a^{-1})a$$
$$= a^{-1}d\delta a - a^{-1}\delta a\, a^{-1}\, da$$
$$= \delta A,$$

which implies the *identity*:

$$d(Ad(a^{-1})\,\delta A) = 0,$$

and thus

$$Ad(a)d\,Ad(a^{-1})\,\delta A = 0.$$

Looking at the terms involving derivatives we finally obtain

$$d\delta A + [\delta A, A] + [A, \delta A] = Ad(a)d\,Ad(a^{-1})\,\delta A = 0.$$

We now introduce the linear operator

$$\nabla: \xi \to d\xi + [A, \xi],$$
$$\nabla_i\xi^\tau(x) = \partial_i\xi^\tau(x) + c_{\rho\sigma}^\tau A_i^\rho(x)\xi^\sigma(x).$$

It depends only on A and is the '*twist*' of d under the action of $Ad(a)$ on $\mathcal{G}$ whenever $A = a^{-1}da$ in the gauge sequence.

On the other hand we may extend

$$\nabla \colon \bigwedge\nolimits^0 T^* \otimes \mathcal{G} \to \bigwedge\nolimits^1 T^* \otimes \mathcal{G}$$

to

$$\nabla \colon \bigwedge\nolimits^r T^* \otimes \mathcal{G} \to \bigwedge\nolimits^{r+1} T^* \otimes \mathcal{G}$$

by setting

$$\nabla(\omega \otimes \xi) = d\omega \otimes \xi + (-1)^r \omega \wedge \nabla\xi, \qquad \forall \omega \in \bigwedge\nolimits^r T^*, \quad \xi \in \mathcal{G}.$$

As a byproduct we have:

$$\begin{cases} \nabla \circ \nabla(\omega \otimes \xi) = (-1)^r \omega \wedge (\nabla \circ \nabla\xi), \\ \nabla \circ \nabla(f\xi) = f\nabla \circ \nabla\xi. \end{cases}$$

It follows that there is a sequence

$$\bigwedge\nolimits^0 T^* \otimes \mathcal{G} \xrightarrow{\ \nabla\ } \bigwedge\nolimits^1 T^* \otimes \mathcal{G} \xrightarrow{\ \nabla\ } \cdots \xrightarrow{\ \nabla\ } \bigwedge\nolimits^n T^* \otimes \mathcal{G} \longrightarrow 0$$

because an easy computation shows

$$\begin{aligned}(\nabla \circ \nabla\xi)^\tau_{ij} &= c^\tau_{\rho\sigma}(\partial_i A^\rho_j - \partial_j A^\rho_i + c^\rho_{\lambda\mu} A^\lambda_i A^\mu_j)\xi^\sigma \\ &\quad - (c^\tau_{\lambda\rho}c^\lambda_{\sigma\mu} + c^\tau_{\lambda\sigma}c^\lambda_{\mu\rho} + c^\tau_{\lambda\mu}c^\lambda_{\rho\sigma}) A^\rho_i A^\sigma_j \xi^\mu \\ &= 0 \end{aligned}$$

if $F = dA + [A, A] = 0$, according to the jacobi identity. We have obtained:

Theorem 15. *The gauge sequence admits a locally exact linearisation, which is isomorphic to copies of the Poincaré sequence and describes infinitesimal gauge transformations.*

PROOF. We have already constructed the ∇-sequence. Since $\nabla = Ad(a) \circ d \circ Ad(a^{-1})$, the adjoint map establishes a (local) isomorphism with the following sequence:

$$\bigwedge\nolimits^0 T^* \otimes \mathcal{G} \xrightarrow{\ d\ } \bigwedge\nolimits^1 T^* \otimes \mathcal{G} \xrightarrow{\ d\ } \cdots \xrightarrow{\ d\ } \bigwedge\nolimits^n T^* \otimes \mathcal{G} \longrightarrow 0,$$

induced by the exterior derivative, which is known to be locally exact. The first and second ∇-operators have already been used for describing infinitesimal gauge transformations in terms of variation of Yang–Mills potentials and fields. $\square$

We shall now prove a result concerning '*commutation*' of variations. Later on we shall generalise this result.

Proposition 16. $[\delta_\xi, \delta_\eta] = -\delta_{[\xi,\eta]}.$

PROOF. In this formula we have set

$$[\delta_\xi, \delta_\eta] = \delta_\xi \circ \delta_\eta - \delta_\eta \circ \delta_\xi.$$

For the variation δ_η we have

$$A \to A + \delta_\eta A = A + d\eta + [A, \eta],$$

and for δ_ξ we have

$$A + \delta_\eta A \to A + \delta_\eta A + d\xi + [A + \delta_\eta A, \xi],$$

i.e.

$$A + d\xi + d\eta + [A, \xi] + [A, \eta] + [d\eta, \xi] + [[A, \eta], \xi].$$

Exchanging ξ and η and subtracting the results, we obtain

$$[d\eta, \xi] - [d\xi, \eta] + [[A, \eta], \xi] - [[A, \xi], \eta],$$

i.e.

$$-(d[\xi, \eta] + [A, [\xi, \eta]]) = -\delta_{[\xi,\eta]} A,$$

using the Jacobi identity for the bracket. $\square$

Remark 17. Setting $A = da\, a^{-1}$, we obtain $\delta A = d\xi - [A, \xi]$, and therefore the 'correct' sign in $[\delta_\xi, \delta_\eta] = \delta_{[\xi,\eta]}$, though the definition of A is a matter of taste. In this Section we have adopted the choice dictated by the formula in the First Fundamental Theorem.

Although there is still no experimental evidence of how to use a global formulation of the above results, instead of the trivial fibered product $X \times G$ one may wish to use a fibered manifold or bundle. However, in view of the specific properties of $X \times G$ this procedure is too general, and one has to rely on the following definition.

Definition 18. A *principal bundle* $\pi: \mathcal{P} \to X$ with *base space* X and *structure group* G is a bundle $\mathcal{P}$ over X such that G acts simply transitively on the fibers of $\mathcal{P}$ and local trivialisations are of the form $\pi^{-1}(U) \to U \times G$ for an open set $U \subset X$.

Remark 19. Of course, the simplest example of a principal bundle is the trivial bundle $X \times G \to X$, where G acts on the fiber (i.e. itself) on the right or on the left. Usually, if U_α and U_β are two coordinate domains, one may define maps $a_{\alpha\beta}: U_\alpha \cap U_\beta \to G$ for changing trivialisations by acting on G *on the left*, while the action of G described in the Definition is *on the right* and thus well-defined, independently of local coordinates. However, we do not dwell on these delicate technical details, as we will not need them.

The next Proposition characterises the trivial situation.

Proposition 20. *A principal bundle is trivial if and only if it has a global section.*

PROOF. Let us prove that there is a correspondence between local sections and local trivialisations. Indeed, if $f: U \to \mathcal{P}$ is a local section, then for all $p \in \mathcal{P}$ with $\pi(p) = x \in X$ there is a unique $a \in G$ such that $p = f(x)a$, and we may define the local trivialisation $\pi^{-1}(U) \to U \times G: p \to (x, a)$. Conversely, if $\pi^{-1}(U) \to U \times G$ is a local trivialisation, the local section f is determined by choosing $f(x)$ to be the inverse image of (x, e), with e the identity in G. $\square$

As in Chapter II.A, there is the following exact sequence of vector bundles over $\mathcal{P}$:

$$0 \longrightarrow V(\mathcal{P}) \longrightarrow T(\mathcal{P}) \xrightarrow{T(\pi)} T(X) \longrightarrow 0,$$

where $T(X)$ is shorthand for $T(X) \times_X \mathcal{P}$. The action of G on $\mathcal{P}$ induces an action of G on $T(\mathcal{P})$, and thus also on $V(\mathcal{P})$. Since the action of G on $\mathcal{P}$ is simply transitive on the fibers, we may take the quotient of the latter short exact sequence with respect to the action of G, and obtain a short exact sequence of vector bundles over X. This gives rise to

Definition 21. A *connection* χ is a splitting of this latter sequence:

$$0 \longrightarrow V(\mathcal{P})/G \longrightarrow T(\mathcal{P})/G \underset{T(\pi)/G}{\overset{\chi}{\rightleftarrows}} T(X) \longrightarrow 0.$$

Identifying $\mathcal{G}$ with the Lie algebra of invariant vector fields on $\mathcal{P}$ under the action of G, we obtain an isomorphism $V(\mathcal{P}) \simeq \mathcal{P} \times \mathcal{G}$, and thus $V(\mathcal{P})/G \simeq X \times \mathcal{G} = \bigwedge^0 T^* \otimes \mathcal{G}$ is a bundle of Lie algebras over X.

Definition 22. The *curvature* $\kappa \in \bigwedge^2 T^* \otimes V(\mathcal{P})/G$ of χ is defined by the formula

$$\kappa(\xi, \eta) = [\chi(\xi), \chi(\eta)] - \chi([\xi, \eta]), \qquad \forall \xi, \eta \in T.$$

A connection is said to be *flat* if it has zero curvature.

Let us establish a link between this definition and the results previously obtained with $\mathcal{P} = X \times G$. We have

$$\chi = (\delta_i^j, A_i^\tau(x)).$$

Developing the bracket on $X \times G$ while using the third formula of Proposition 7, we obtain:

$$[\chi(\xi), \chi(\eta)] = (([\xi, \eta])^i, \xi^i \partial_i (A_j^\tau \eta^j) - \eta^j \partial_j (A_i^\tau \xi^i) + c_{\rho\sigma}^\tau A_i^\rho A_j^\sigma \xi^i \eta^j)$$

$$= (([\xi, \eta])^i, A_i^\tau ([\xi, \eta])^i + (\partial_i A_j^\tau - \partial_j A_i^\tau + c_{\rho\sigma}^\tau A_i^\rho A_j^\sigma) \xi^i \eta^j),$$

$$\chi([\xi, \eta]) = (([\xi, \eta])^i, A_i^\tau ([\xi, \eta])^i).$$

By subtraction we find

$$\kappa(\xi, \eta) = (0, (\partial_i A_j^\tau - \partial_j A_i^\tau + c_{\rho\sigma}^\tau A_i^\rho A_j^\sigma) \xi^i \eta^j),$$

and we recognise a result already obtained. Also, the system $d\xi + [A, \xi] = 0$ is a first-order system of PDE for ξ, depending on A. We leave it to the reader to prove that the curvature map, which was introduced via the computation of $\nabla \circ \nabla$, is coherent with the above definition and with the definition of curvature map for nonlinear systems in

Chapter III. One should notice that both the definition of F from A and the Jacobi relations for c participate in this identification.

We now arrive at a delicate point, namely the construction of a global substitute for the gauge sequence. For this we need the following, technical, result.

Proposition 23. $J_1(\mathcal{P})/G \simeq$ *bundle of connections.*

PROOF. First of all, recall that the definition of connection as a splitting amounts to considering a connection as a section of $T^* \otimes T(\mathcal{P})/G$ projecting to the section $\mathrm{id}_T \in T^* \otimes T$. Hence, we only have to prove that

$$J_1(\mathcal{P}) \simeq \text{sections of } T^* \otimes T(\mathcal{P}) \text{ over } \mathrm{id}_T \in T^* \otimes T.$$

Indeed, let $\bar{x} = \varphi(x)$, $\bar{y} = \psi(x,y)$ be the coordinate transformations of $\mathcal{P}$. The additional equations for $J_1(\mathcal{P})$ are, with full indices:

$$\bar{y}_j^l \partial_i \varphi^j(x) = \frac{\partial \psi^l(x,y)}{\partial x^i} + y_i^k \frac{\partial \psi^l(x,y)}{\partial y^k}.$$

With coordinates (x,y,u,v) in $T(\mathcal{P})$ we have

$$\bar{v}^l = \frac{\partial \psi^l(x,y)}{\partial x^i} u^i + \frac{\partial \psi^l(x,y)}{\partial y^k} v^k.$$

Hence, for $T^* \otimes T(\mathcal{P})$ we have

$$\bar{v}_j^l \partial_i \varphi^j(x) = \frac{\partial \psi^l(x,y)}{\partial x^r} u_{,i}^r + \frac{\partial \psi^l(x,y)}{\partial y^k} v_{,i}^k.$$

Setting $u_{,i}^r = \delta_i^r$ for a connection, we obtain the previous formula by identification. $\square$

This Proposition implies

Theorem 24. *There is a 'global' gauge sequence*

$$\mathcal{P} \longrightarrow J_1(\mathcal{P})/G \xrightarrow{\ \kappa\ } \wedge^2 T^* \otimes \mathcal{G}.$$

PROOF. The first operator is a first-order operator, defined by composition:

$$\mathcal{P} \xrightarrow{\ j_1\ } J_1(\mathcal{P}) \longrightarrow J_1(\mathcal{P})/G.$$

Hence, in contrast to the gauge sequence, it cannot be defined by a *'formula'* that could be handled by a computer. Using again the third formula of Proposition 7, we see that we can work on $T(\mathcal{P})$ instead of on $T(\mathcal{P})/G$, and from the composition we obtain (starting with a section f of $\mathcal{P}$):

$$[\chi(\xi), \chi(\eta)] - \chi([\xi,\eta]) = [T(f)(\xi), T(f)(\eta)] - T(f)([\xi,\eta]) = 0.$$

Hence we have a sequence, and we have already proved that its local description is coherent with the *'local'* gauge sequence. $\square$

Although *all* of mathematical physics relies on the content of this Section, and on *nothing else*, it is rather surprising that the following two Sections will provide new differential sequences that not only completely supersede the results of this Section, but also point out that there is a deep confusion.

Independently of any physical applications, we shall end this Section by giving two hints about facts that should have shown up a long time ago: the answer to certain questions does not support the previous approach at all, at least mathematically speaking.

The two difficulties immediately raised by the previous construction concern the wish to 'finish' these sequences *on the left* and *on the right*. We separately examine these two points for the local gauge sequence only, as they are exactly the same for the global gauge sequence.

By definition, the kernel of the first operator involves so-called *locally constant* maps from X to G. At first sight, this leads to nonsense, because the parameters, i.e. the constant coordinates on G, have a meaning $\cdots$ if and only if they are indeed '*parameters*', i.e. if and only if G *does act on* X. However, not only did we not assume the existence of an action, but the very absence of this action $\cdots$ is nowadays considered the novelty of gauge theory! Hence we arrive at a contradiction, since if we want a nice mathematical interpretation/explanation of the kernel of the first operator, we have to take into consideration an action of G on X $\cdots$ and precisely this is given up in mathematical gauge theory!

Let us now look at a way of extending the gauge sequence to the right. For the moment, the problem of local exactness has only been formulated at the connection level. Indeed, if $F \equiv dA + [A, A] = 0$, we looked, at least locally, for maps $a: X \to G$ such that $a^{-1}da = A$. However, following physicists, from a mathematical point of view the (pure) filed *is* F, i.e. we want to solve $dA + [A, A] = F$ with F given. Of course, in the *abelian* case, i.e. when $c = 0$, we have $dA = F$ and hence obtain the field equations $dF = 0$, by analogy with the Maxwell equations. In the nonabelian case $c \neq 0$ we are left with a nonlinear system for which the search of compatibility conditions does not provide a unique solution but rather a *tree* of possibilities. This tree is rather complicate, even in the very simple situations considered in the examples of Chapter III. The lack of field equations contradicts any physical interpretation, since the potentials cannot be 'measured' (see electromagnetism in Chapter VIII!). Physicists (at least those who are concerned with the fact that 'there is a problem') try to escape from this by exhibiting certain differential identities for F and, as we shall see in the next Section, these are improperly called *Bianchi identities*. Indeed, taking the exterior derivative of the system $dA + [A, A] = F$, we obtain, after some substitutions,

$$\partial_i F_{jk}^\tau + \partial_j F_{ki}^\tau + \partial_k F_{ij}^\tau + c_{\rho\sigma}^\tau (A_i^\rho F_{jk}^\sigma + A_j^\rho F_{ki}^\sigma + A_k^\rho F_{ij}^\sigma) = 0,$$

which we will write symbolically as

$$dF + [A, F] = 0,$$

by introducing the corresponding forms in $\wedge^3 T^* \otimes \mathcal{G}$. Even though we shall prove that the association with Bianchi's name, stemming from tensor analysis, is completely misleading, the above identities do not present a way of extending the gauge sequence to the right. Indeed, they do not give a differential condition for F *only*, since they explicitly involve A. Hence, the only way to preserve the existence of field equations uniquely defined for *the* field is to endow the Maurer–Cartan equations with such an interpretation, $\cdots$ but again this breaks down classical gauge theory, in which, according to the electromagnetic model, the field *must* be a 2-form and *cannot* be a 1-form. *Again a contradiction!*

B. Groupoids and Algebroids

In examples in the Introduction we have seen that Lie groups of transformations are particular instances of Lie pseudogroups of transformations, i.e. groups of transformations that are solutions of systems of PDE. Accordingly, the purpose of this Section is to use the results of Chapter III to generalize to Lie pseudogroups the results of the previous Section that were valid for Lie groups only. The exposition will be as simple as possible, and we present only the concepts that are absolutely necessary for applications.

A first, practical, question is to distinguish among the systems of nonlinear PDE those defining Lie pseudogroups. Indeed, the usual functional definition, saying that whenever $f : U \to V$ and $g : V \to W$ are two local solutions, regarded as local transformations, then the composite $g \circ f : U \to W$ is again a local solution, cannot be checked by means of computer algebra, although we feel that it is a '*built in*' property of the given system of PDE.

Similarly to the identification of a map $f : X \to X$ with its graph $f : X \to X \times X$, we may identify the bundle $J_q(X, X)$ of q-jets of maps from X to X with the q-jet bundle $J_q(X \times X)$ of $X \times X$, with *source projection* α on the first factor and *target projection* β on the second. We can then introduce the open subbundle $\Pi_q(X, X)$, or simply Π_q when there is no confusion possible about the underlying manifold X, defined by the condition that $\det(y_i^k) \neq 0$ on the first-order jets. Well-known properties of the Jacobian readily imply that this definition is independent of the coordinate system. We can similarly introduce the source projection α_q and target projection β_q, using the following diagram:

$$
\begin{array}{ccc}
\Pi_q & & (x, y_q) \\
\alpha_q \swarrow \ \searrow \beta_q & & \swarrow \ \searrow \\
X \qquad X & & x \qquad y
\end{array}
$$

For simplicity, the condition on first-order jets will never be indicated, unless explicitly needed.

Let $\mathrm{aut}(X)$ be the set of all local diffeomorphisms of X. If $f, g \in \mathrm{aut}(X)$ can be composed, we can use the chain rule for derivatives, defined as follows:

$$
j_q(g \circ f) = j_q(g) \circ j_q(f).
$$

Setting $y = f(x)$, $z = g(y)$, the reader should not not forget that in $j_q(g)$ the derivatives are taken with respect to y, while in $j_q(f)$ *and* $j_q(g \circ f)$ the derivatives are taken with respect to x. Replacing derivatives by jets, as usual, we may define the composition rule $(f_q, g_q) \to g_q \circ f_q$ for sections of Π_q with convenient source and target. More precisely, we have a *pointwise composition morphism*:

$$\gamma_q \colon \Pi_q(X,X) \times_X \Pi_q(X,X) \to \Pi_q(X,X),$$

$$\left(\left(x,y,\frac{\partial y}{\partial x} \right), \left(y,z,\frac{\partial z}{\partial y} \right) \right) \to \left(x,z,\frac{\partial z}{\partial y} \cdot \frac{\partial y}{\partial x} \right),$$

where the fibered product is taken with respect to the target projection on the right and where we have used symbolic notation for first-order jets. Of course, there is a distinguished section $\mathrm{id}_q = j_q(\mathrm{id}) = j_q(\mathrm{id}_X)$ of Π_q.

Similarly, for $f \in \mathrm{aut}(X)$ we can define $j_q(f)^{-1} = j_q(f^{-1})$. Hence, replacing derivatives by jets, we can define an inversion rule $f_q \to f_q^{-1}$ for sections of Π_q by:

$$f_q \circ f_q^{-1} = f_q^{-1} \circ f_q = \mathrm{id}_q,$$

and we have a *pointwise inversion morphism*

$$\iota_q \colon \Pi_q(X,X) \to \Pi_q(X,X),$$

$$\left(x,y,\frac{\partial y}{\partial x} \right) \to \left(y,x, \left(\frac{\partial y}{\partial x} \right)^{-1} \right).$$

We are now in a position to state:

Definition 1. A fibered submanifold $\mathcal{R}_q \subset \Pi_q(X,X)$ is called a system of *finite Lie equations* of order q, or a *Lie groupoid* of order q on X if there are the following induced morphisms

$$\begin{cases} \alpha_q \colon \mathcal{R}_q \to X & \text{(source projection)}, \\ \beta_q \colon \mathcal{R}_q \to X & \text{(target projection)}, \\ \gamma_q \colon \mathcal{R}_q \times_X \mathcal{R}_q \to \mathcal{R}_q & \text{(composition)}, \\ \iota_q \colon \mathcal{R}_q \to \mathcal{R}_q & \text{(inversion)}, \\ \mathrm{id}_q \colon X \to \mathcal{R}_q & \text{(identity)}. \end{cases}$$

It is essential to notice that composition and inversion, being pointwise defined, can be tested by means of a computer whenever $\mathcal{R}_q$ is defined by a system of algebraic PDE. Hence, while in the theory of Lie groups we have *algebraic groups* (i.e. groups of unimodular or orthogonal matrices), in the theory of Lie pseudogroups we have *algebraic groupoids*, which will play a key role in the Galois theory for PDE. The set $\Gamma \subset \mathrm{aut}(X)$ of local solutions of $\mathcal{R}_q$ is called a *Lie pseudogroup* of order q, and we notice that, although $\mathcal{R}_q$ is not required to be formally integrable, it has at least the solution $y = x$. Accordingly, we have:

$$\begin{cases} \forall f, g \in \Gamma \Rightarrow g \circ f \in \Gamma & \text{if } f \text{ and } g \text{ can be composed}, \\ \forall f \in \Gamma \Rightarrow f^{-1} \in \Gamma, \\ \mathrm{id} \in \Gamma. \end{cases}$$

However, to find out about properties of Γ we must define it by a formally integrable system. We thus give

Definition 2. A Lie pseudogroup $\Gamma \subset \mathrm{aut}(X)$ is said to be *transitive* if for all $x, y \in X$ we can find $f \in \Gamma$ such that $y = f(x)$.

Of course, the only way to test this property is to verify that the corresponding formally integrable system of finite Lie equations $\mathcal{R}_q \subset \Pi_q(X, X)$ defining Γ is *transitive*, i.e. $\pi_0^q = (\alpha_q, \beta_q): \mathcal{R}_q \to X \times X$ is an epimorphism.

We may now look for properties of the prolongations.

Definition 3. By $\mathcal{R}_{q,r} \subset J_r(\mathcal{R}_q)$ we denote the open subbundle defined by $\det(y_{0,i}^k) \neq 0$.

By definition, we have

$$\mathcal{R}_{q+r} = J_r(\mathcal{R}_q) \cap \Pi_{q+r}(X, X) = \mathcal{R}_{q,r} \cap \Pi_{q+r} \subset \Pi_{q,r}.$$

We define a groupoid structure on $\mathcal{R}_{q,r} \subset \Pi_{q,r}$ by $j_r(g_q) \circ j_r(f_q) = j_r(g_q \circ f_q)$ and $j_r(f_q)^{-1} = j_r(f_q^{-1})$. Taking the intersection with the groupoid Π_{q+r}, we obtain

Proposition 4. *If $\mathcal{R}_q \subset \Pi_q$ is a Lie groupoid, then $\mathcal{R}_{q+r} \subset \Pi_{q+r}$ is also a Lie groupoid, for all $r \geq 0$.*

Finally, we define the analog of a Lie algebra. To this end we notice that vertical vectors over the diagonal $\Delta \subset X \times X$ can be identified with vector fields on X under the projection $T(\beta)$. Applying the functor J_q to the vector bundle $T = \mathrm{id}^{-1}(V(X \times X))$ just defined, we obtain $J_q(T) = \mathrm{id}_q^{-1}(V(\Pi_q))$, using Proposition II.B.5.

Accordingly, the nonlinear system $\mathcal{R}_q \subset \Pi_q$ gives rise to a linear system $\mathrm{R}_q = \mathrm{id}_q^{-1}(V(\mathcal{R}_q)) \subset J_q(T)$. In actual practice, we only have to substitute $y_\mu^k = \partial_\mu x^k + t \xi_\mu^k(x)$ into the equations, take the derivative with respect to t, and set $t = 0$.

Using local coordinates we obtain

$$\begin{array}{ll} \mathcal{R}_q & \Phi^\tau(x, y_q) = 0, \\[2mm] \mathrm{R}_q & \dfrac{\partial \Phi^\tau}{\partial y_\mu^k}(x, \mathrm{id}_q(x)) \xi_\mu^k(x) = 0 \end{array}$$

as the simplest way of defining a section ξ_q of R_q. The main difficulty is to extend the Lie algebra structure on T, defined in Proposition A.7, to $J_q(T)$ and then to restrict it to R_q. For this we need several additional concepts, generalising the left and right translation morphisms in Lie groups together with their tangent maps, which we have used to push a vector on G back to a vector at e, i.e. back to an element of $\mathcal{G} = T_e(G)$.

Taking the tangent map of the composition γ_q while fixing the left point in the fibered product reduces to fixing (x, y_q) and therefore determines a 'sharp' isomorphism:

$$\#: \mathcal{R}_q \times \mathrm{R}_q \to V(\mathcal{R}_q),$$
$$(x, y_q)(y, \eta_q(y)) \to (x, y_q; d_\mu \eta^k(y)).$$

Taking a section η_q of R_q *over the target*, we obtain the following *vertical* vector:

$$\eta^k(y)\frac{\partial}{\partial y^k} + \eta_r^k(y)y_i^r\frac{\partial}{\partial y_i^k} + (\eta_{rs}^k(y)y_i^r y_j^s + \eta_r^k(y)y_{ij}^r)\frac{\partial}{\partial y_{ij}^k} + \cdots.$$

Similarly, taking the tangent map of the composition γ_q while fixing the right point in the fibered products comes down to fixing (y, z_q), and thus determines the *'flat'* *monomorphism*

$$\flat: R_q \times_X R_q \to T(\mathcal{R}_q),$$

which cannot be defined as simple as before. Nevertheless, taking a section ξ_q of R_q *over the source*, we obtain the following *tangent* vector:

$$\xi^i(x)\frac{\partial}{\partial x^i} - y_r^k \xi_i^r(x)\frac{\partial}{\partial y_i^k} - (y_r^k \xi_{ij}^r(x) + y_{rj}^k \xi_i^r(x) + y_{ri}^k \xi_j^r(x))\frac{\partial}{\partial y_{ij}^k} + \dots.$$

Here, we have to explain the minus sign, as this is a careful point in effective computations. Indeed, R_q on the left is *not* the reciprocal image of the vertical bundle, but only the image

$$T(\iota_q): V(\mathcal{R}_q) \to T(\mathcal{R}_q).$$

After evaluation at id_q we obtain

$$(\xi^k(x), \xi_i^k(x), \xi_{ij}^k(x), \dots) \to (\xi^k(x), -\xi_i^k(x), -\xi_{ij}^k(x), \dots),$$

a result explaining the sign.

Before proceeding further, we recall a few facts concerning the so-called *Lie derivative*. These facts will be used in a crucial way in the sequel.

The Lie derivative with respect to a vector field $\xi \in T$ acting on an r-form is a first-order differential operator $\mathcal{L}(\xi)$ in ξ uniquely defined by the following three properties:

1) $\mathcal{L}(\xi)f = \xi \cdot f = \xi^i \partial_i f$, for all $f \in \wedge^0 T^* = C^\infty(X)$;
2) $\mathcal{L}(\xi)d = d\mathcal{L}(\xi)$;
3) $\mathcal{L}(\xi)(\alpha \wedge \beta) = (\mathcal{L}(\xi)\alpha) \wedge \beta + \alpha \wedge (\mathcal{L}(\xi)\beta)$ for all $\alpha, \beta \in \wedge T^*$.

Equivalently, introducing the *interior multiplication* $i(\xi)$ by a vector field ξ as a map $\wedge^{r+1} T^* \to \wedge^r T^*$ defined in local coordinates by the formula

$$(i(\xi)\omega)_{i_1 \dots i_r} = \xi^i \omega_{i i_1 \dots i_r},$$

we leave it to the reader to prove (see [144, 195, p. 246]) that:

$$\mathcal{L}(\xi) = i(\xi)d + di(\xi).$$

Accordingly, for any $\omega \in \wedge^r T^*$ we have

$$\mathcal{L}(\xi)\omega = (\xi \cdot \omega_I) \, dx^I + \sum_{s=1}^r \omega_I \, dx^{i_1} \wedge \dots \wedge dx^{i_{s-1}} \wedge d\xi^{i_s} \wedge dx^{i_{s+1}} \wedge \dots \wedge dx^{i_r}.$$

Now, let $\xi, \eta \in T$, and compare $[\mathcal{L}(\xi), \mathcal{L}(\eta)] = \mathcal{L}(\xi) \circ \mathcal{L}(\eta) - \mathcal{L}(\eta) \circ \mathcal{L}(\xi)$ and $\mathcal{L}([\xi, \eta])$. The properties 2) and 3) are readily seen to be satisfied individually, while the restriction of 1) to functions makes it possible to state

Lemma 5. $[\mathcal{L}(\xi), \mathcal{L}(\eta)] = \mathcal{L}([\xi, \eta])$, *for all* $\xi, \eta \in T$.

We now give the infinitesimal version of the prolongation formulas that allowed us, in Chapter II.A, to define a jet bundle of order q. Indeed, using local coordinates, we want

$$w_\mu^k \equiv dy_\mu^k - y_{\mu+1_\iota}^k\, dx^i = 0 \Rightarrow \overline{w}_\mu^k \equiv d\overline{y}_\mu^k - \overline{y}_{\mu+1_\iota}^k\, d\overline{x}^i = 0.$$

hence, if we have a *projectable vector field*

$$\theta = \xi^i(x)\frac{\partial}{\partial x^i} + \eta^k(x,y)\frac{\partial}{\partial y^k}$$

on a fibered manifold $\pi\colon \mathcal{E} \to X$, i.e. a vector field coherent with π, then we can extend it to a projectable vector field

$$\rho_q(\theta) = \xi^i(x)\frac{\partial}{\partial x^i} + \eta_\mu^k(x,y_q)\frac{\partial}{\partial y_\mu^k}$$

on $J_q(\mathcal{E})$ such that the infinitesimal transformation $\overline{x}^i = x^i + t\xi^i(x)$, $\overline{y}_\mu^k = y_\mu^k + t\eta_\mu^k(x,y_q)$ preserves the above relations. Substituting into the second and subtracting the first, the condition is easily seen to become

$$\eta_{\mu+1_\iota}^k = d_i\eta_\mu^k - y_{\mu+1_\jmath}^k \partial_i\xi^j.$$

This provides, inductively, a *unique* way of defining the components of $\rho_q(\theta)$ in terms of those of θ. We now have the technical identity

$$\mathcal{L}(\theta_q)(dy_\mu^k - y_{\mu+1_\iota}^k\, dx^i) \equiv \frac{\partial\eta_\nu^l}{\partial y_\mu^k}(dy_\mu^k - y_{\mu+1_\iota}^k\, dx^i) + (d_i\eta_\mu^k - \eta_{\mu+1_\iota}^k - y_{\mu+1_\jmath}^k \partial_i\xi^j)\, dx^i.$$

It follows that $\theta_q = \rho_q(\theta)$ if and only if the Lie derivative of the so-called *fundamental forms* w_μ^k vanish whenever these forms vanish. Accordingly, if θ_1 and θ_2 are two projectable vector fields on $\mathcal{E}$, then $[\rho_q(\theta_1), \rho_q(\theta_2)]$ and $\rho_q([\theta_1,\theta_2])$ are projectable vector fields on $J_q(\mathcal{E})$ coinciding with $[\theta_1,\theta_2]$ on $\mathcal{E}$. Lemma 5 and the preceding remark show that they coincide everywhere. Thus:

Lemma 6. $[\rho_q(\theta_1), \rho_q(\theta_2)] = \rho_q([\theta_1,\theta_2])$.

We restrict the preceding results to the situation $\mathcal{E} = X \times X$, where both 'horizontal' and 'vertical' vectors can exist individually with respect to the source and the target projections.

If $\xi = \xi^i(x)\frac{\partial}{\partial x^i}$ is a horizontal vector field, we have

$$\flat(j_q(\xi)) = \rho_q(\xi) = \xi^i(x)\frac{\partial}{\partial x^i} + \partial_\nu\xi^j A_j^\nu(q),$$

by decomposing $\rho_q(\xi)$ on $j_q(\xi)$. We thus have

$$\flat(\xi_q) = \xi^i(x)\frac{\partial}{\partial x^i} + \xi_\nu^j(x)A_j^\nu(q),$$

where the vector fields $A(q)$ depend on y_q only and have already been given for $q = 2$.

Before comparing brackets as in Section A, we give two definitions, which will be used in an essential way throughout the remainder.

Definition 7. If ξ_{q+1} and η_{q+1} are two sections of $J_{q+1}(T)$, we define the *algebraic bracket* $\{\xi_{q+1}, \eta_{q+1}\}$ with values in $J_q(T)$ by the formula

$$\{j_{q+1}(\xi), j_{q+1}(\eta)\} = j_q([\xi, \eta]), \qquad \forall \xi, \eta \in T.$$

As this is *not* the bracket of a Lie algebra in any case (it is defined on $J_{q+1}(T)$ and not on $J_q(T)$), we modify this Definition as follows:

Definition 8. On sections of $J_q(T)$ with values in $J_q(T)$ we define the *differential bracket* by

$$[\xi_q, \eta_q] = \{\xi_{q+1}, \eta_{q+1}\} + i(\xi)D\eta_{q+1} - i(\eta)D\xi_{q+1}.$$

We leave it to the reader to verify, as an easy exercise on the Spencer operator, that this Definition does not depend on the lifts ξ_{q+1} and η_{q+1} of ξ_q and η_q in $J_{q+1}(T)$. We also notice that, by Definition 7 and Remark IV.A.8, that

$$[j_q(\xi), j_q(\eta)] = j_q([\xi, \eta]).$$

This is coherent with the ordinary bracket of vector fields.

Theorem 9. *The differential bracket defines a Lie algebra structure on the sections of $J_q(T)$.*

PROOF. Let ξ_q and η_q be sections of $J_q(T)$. We can write:

$$\flat(\xi_q) = \xi^i(x)\frac{\partial}{\partial x^i} + \xi_\mu^k(x)A_k^\mu(q),$$

$$\flat(\eta_q) = \eta^j(x)\frac{\partial}{\partial x^j} + \eta_\nu^j(x)A_j^\nu(q),$$

and obtain for the bracket on $T(\Pi_q)$:

$$[\flat(\xi_q), \flat(\eta_q)] = ([\xi, \eta])^i\frac{\partial}{\partial x^i} + (\xi^i\partial_i\eta_\mu^k - \eta^j\partial_j\xi_\mu^k)A_k^\mu(q)$$
$$+ \xi_\mu^k\eta_\nu^j[A_k^\mu(q), A_j^\nu(q)],$$

and thus, by Lemma 6

$$[\flat(j_q(\xi)), \flat(j_q(\eta))] = ([\xi, \eta])^i\frac{\partial}{\partial x^i} + (\xi^i\partial_{\mu+1,}\eta^k - \eta^j\partial_{\mu+1,}\xi^k)A_k^\mu(q)$$
$$+ \partial_\mu\xi^k\partial_\nu\eta^j[A_k^\mu(q), A_j^\nu(q)] =$$
$$= ([\xi, \eta])^i\frac{\partial}{\partial x^i} + \partial_\mu([\xi, \eta])^kA_k^\mu(q).$$

It follows that

$$\xi_\mu^k\eta_\nu^j[A_k^\mu(q), A_j^\nu(q)] = \left(\left(\{\xi_{q+1}, \eta_{q+1}\}\right)_\mu^k - \xi^i\eta_{\mu+1,}^k + \eta^j\xi_{\mu+1,}^k\right)A_k^\mu(q).$$

Substitution into the first formula gives

$$[\flat(\xi_q), \flat(\eta_q)] = \flat([\xi_q, \eta_q]), \qquad \forall \xi_q, \eta_q \in J_q(T).$$

The Jacobi identity for the differential bracket now follows from that for the ordinary bracket used in $T(\Pi_q)$. $\square$

Here we have used the '*flat*' map over the *source*. We could have also used the '*sharp*' map over the *target*. We advise the reader to carry out the calculations up to order two directly, using the formulas given. The difficulty of writing down explicit developments will then become clear.

Corollary 10. *The restriction of the differential bracket defines a Lie algebra structure on the sections of* R_q.

PROOF. The proof follows immediately from the restriction of the 'flat' or 'sharp' map and Proposition A.7. If $\pi_q^{q+1}\colon R_{q+1} \to R_q$ is surjective, we can choose lifts in R_{q+1}. Since we already know the restriction $D\colon R_{q+1} \to T^* \otimes R_q$, we deduce from the differential bracket $[R_q, R_q] \subset R_q$ that, *in this case*, $\{R_{q+1}, R_{q+1}\} \subset R_q$. $\square$

Definition 11. $R_q \subset J_q(T)$ is called the *Lie algebroid* of the Lie groupoid $\mathcal{R}_q \subset \Pi_q(X, X)$.

Remark 12. Using the Remark made above Theorem 9, we can see that the set $\Theta \subset T$ of solutions of $\mathcal{R}_q$ is a Lie algebra of vector fields which is infinite dimensional, in general. We shall symbolically write $[\Theta, \Theta] \subset \Theta$, meaning that if ξ_1, ξ_2 are two (local) solutions of R_q, then $[\xi_1, \xi_2]$ is also a (local) solution of R_q. Accordingly, an operator $\mathcal{D}$ acting on T and satisfying $\mathcal{D}\xi_1 = 0, \mathcal{D}\xi_2 = 0 \Rightarrow \mathcal{D}[\xi_1, \xi_2] = 0$ is called a *Lie operator*. The *divergence* operator is the simplest such operator.

Definition 13. We define the vector bundle $J_q^0(T)$ by the short exact sequence

$$0 \longrightarrow J_q^0(T) \longrightarrow J_q(T) \xrightarrow{\pi_0^q} T \longrightarrow 0,$$

and we set $R_q^0 = R_q \cap J_q^0(T)$.

We can easily see that R_q is transitive when $\mathcal{R}_q$ is transitive. In this case a *connection* χ_q, or, more precisely, an R_q-*connection*, is a splitting of the short exact sequence

$$0 \longrightarrow R_q^0 \longrightarrow R_q \underset{\pi_0^q}{\overset{\chi_q}{\longleftrightarrow}} T \longrightarrow 0.$$

Equivalently, a connection is a section of $T^* \otimes R_q$ projecting onto $\chi_0 = \mathrm{id}_T \in T^* \otimes T$.

Looking at the differential bracket, we see that sections of R_q^0 have a null zero-order projection and the Spencer operator does not appear. Accordingly, $[R_q^0, R_q^0] \subset R_q^0$ is a bracket of Lie algebras defined fiberwise, and thus R_q^0 is a bundle of Lie algebras. As in Section A we may define the *curvature* κ_q of χ_q as the section of $\wedge^2 T^* \otimes R_q^0$ defined by

$$\kappa_q(\xi, \eta) = [\chi_q(\xi), \chi_q(\eta)] - \chi_q([\xi, \eta]), \qquad \forall \xi, \eta \in T,$$

with $0 = \kappa_0 \in \wedge^2 T^* \otimes T$.

Also, we have restricted morphisms

$$\sharp\colon \mathcal{R}_q \times_X R_q^0 \to V(\mathcal{R}_q),$$
$$\flat\colon R_q^0 \times_X \mathcal{R}_q \to V(\mathcal{R}_q).$$

If $x_0 \in X$ is a fixed point, we see that the restrictions of γ_q and ι_q induce a Lie group structure on $\mathrm{GL}_q = \pi_0^{q-1}(x_0, x_0) = \Pi_q(x_0, x_0)$ which is called the *general linear group of order* q, or the *isotropy group* of Π_q at x_0. It consists of the jets having source and target x_0. We can similarly introduce the *isotropy group* $G_q = \mathcal{R}_q(x_0, x_0) \subset \mathrm{GL}_q$ of $\mathcal{R}_q$ at x_0; the Lie algebra of G_q is $\mathrm{R}_{q;x_0}^0$. The restrictions of the morphisms $\sharp$ and $\flat$ give the left and right Maurer–Cartan forms on G_q, in accordance with the construction of the gauge sequence in Section A. It follow that the vector fields $A(q)$ are the infinitesimal generators of the action of GL_q on itself. Their commutation relation can be found from the proof of Theorem 9. We can also use the relation

$$\left[x^\mu \frac{\partial}{\partial x^i}, x^\nu \frac{\partial}{\partial x^j} \right] = \nu_i x^{\mu+\nu-1_i} \frac{\partial}{\partial x^j} - \mu_j x^{\mu+\nu-1_j} \frac{\partial}{\partial x^i},$$

with the obvious notation, and Lemma 6 to obtain

$$[A_i^\mu(q), A_j^\nu(q)] = \begin{cases} \nu_i \dfrac{(\mu+\nu-1_i)!}{\mu!\,\nu!} A_j^{\mu+\nu-1_i}(q) - \mu_j \dfrac{(\mu+\nu-1_j)!}{\mu!\,\nu!} A_i^{\mu+\nu-1_j}(q), \\ \quad\text{if } 2 \leq |\mu| + |\nu| \leq q+1, \\ 0 \quad\text{otherwise.} \end{cases}$$

However, we shall not use these commutation relations. We advise the reader to treat the case $q = 2$ in some detail, and to deal with $\sharp$ similarly as we have done with $\flat$, in order to become familiar with these technical results.

The merit of the previous Definition lies in the following Proposition:

Proposition 14. *If* $\mathcal{R}_q$ *is a transitive Lie groupoid of order* q, *then there are explicit isomorphisms* $(\forall x, y \in X)$:

$$\mathrm{R}_{q;x}^0 \simeq \mathrm{R}_{q;y}^0, \qquad \mathrm{M}_{q+r;x} \simeq \mathrm{M}_{q+r;y}, \qquad \mathrm{H}_{q+r;x}^s \simeq \mathrm{H}_{q+r;y}^s,$$

showing the vector bundle structure of all these families of vector spaces over X.

PROOF. Since $\mathcal{R}_q$ is transitive for all $(x, y) \in X \times X$, we can find $(x, y_q) \in \mathcal{R}_q$ over (x, y). We can use the restrictions of $\sharp$ and $\flat$ to R_q^0 at this point of $\mathcal{R}_q$ to establish an isomorphism between each of $\mathrm{R}_{q;x}^0$ and $\mathrm{R}_{q;y}^0$ and the space of tangent vectors on $\mathcal{R}_q$ at (x, y_q) that are killed by $T(\pi_0^q)$, i.e. having projections $u^i = 0$, $v^k = 0$ on $T(X \times X)$. Since, again, we shall not use these results in the sequel, we leave it to the reader to establish the isomorphism in case $q = 1, 2$ before having a look at the examples in the last Section of this Chapter.

The restriction of the previous isomorphism to the symbol, and its prolongation, is much easier to establish, and we give full details. Looking at the induction formula for prolongation (just above Lemma 6) and taking into account the Remark concerning the sign (above the introduction of the Lie derivative), we have

$$\begin{cases} \sharp: \eta_{r_1,\ldots,r_q}^k(y) \to \eta_{r_1,\ldots,r_q}^k(y) y_{i_1}^{r_1} \cdots y_{i_q}^{r_q} \dfrac{\partial}{\partial y_{i_1 \ldots i_q}^k}, \\ \flat: \xi_{i_1 \ldots i_q}^r(x) \to -y_r^k \xi_{i_1 \ldots i_q}^r(x) \dfrac{\partial}{\partial y_{i_1 \ldots i_q}^k}. \end{cases}$$

The isomorphism $M_{q;x} \simeq M_{q;y}$ is therefore defined, in a purely technical way, by the formula

$$y_r^k \xi_{i_1 \dots i_q}^\tau(x) = \eta_{r_1, \dots, r_q}^k(y) y_{i_1}^{\tau_1} \cdots y_{i_q}^{\tau_q}$$

for tensors in $S_q T^* \otimes T$. Since the definition of δ-map is intrinsic, the isomorphism extends to the prolongation of the symbol and to the cohomology groups. It follows that $R_q^0, M_{q+r}, H_{q+r}^s(M_q)$ are vector bundles over X for all $r, s \geq 0$. $\square$

We conclude this Section by proving that all the results given above are coherent with those obtained in the previous Section.

Indeed, let $X \times G \to X \times X$ be a Lie group of transformations of X with $\dim X = n$, $\dim G = r$. We may extend the graph of the action to a morphism

$$X \times G \to \Pi_q(X, X),$$
$$(x, a) \to (x, y_\mu^k = \partial_\mu f^k(x, a)).$$

Since $\dim G$ equals the number of parameters (and is thus finite), we can choose q sufficiently large in order to eliminate the r parameters. This amounts to looking for the image $\mathcal{R}_q \subset \Pi_q(X, X)$ of the previous morphism. The canonical projection π_q^{q+1} induces an epimorphism $\mathcal{R}_{q+1} \to \mathcal{R}_q$. Also, since $a = $ const and we consider the various prolongations of $y = f(x, a)$, the resulting equations, which no longer involve a, are also prolongated, and so $\mathcal{R}_{q+1} \subset \rho_1(\mathcal{R}_q)$, with:

$$\dim \mathcal{R}_q \leq \dim \mathcal{R}_{q+1} \leq n + r.$$

In particular, $\mathcal{R}_q$ is transitive if and only if the action is transitive.

Proposition 15. *If the action of G on X is effective, there is an isomorphism $X \times \mathcal{G} \simeq R_q$ of Lie algebras, provided q is sufficiently large.*

PROOF. Using the First Fundamental Theorem, see Section A, we have

$$v_\mu^k = \frac{\partial}{\partial a^\sigma}(\partial_\mu f^k(x, a))\lambda^\sigma = \partial_\mu \left(\frac{\partial f^k}{\partial a^\sigma}(x, a)\right)\lambda^\sigma = d_\mu \xi_i^k(y)\omega_\sigma^i(a)\lambda^\sigma.$$

Evaluating at $a = e$, i.e. $y = x$, while using a basis of infinitesimal generators, we thus obtain an epimorphism

$$X \times \mathcal{G} \to R_q,$$
$$(x, \lambda^\tau(x)) \to (x, \xi_\mu^k(x) = \lambda^\tau(x)\partial_\mu \xi_\tau^k(x)),$$

which is described for sections. Hence it remains to prove that this epimorphism is also a monomorphism if q is sufficiently large. Indeed, assume that

$$\mathrm{rk}(q) = \mathrm{rk}(q+1) = s < r.$$

Permuting, if necessary, the infinitesimal generators, we can find nonzero *functions* $A^l(x)$ such that

$$\begin{cases} \partial_\mu \xi_\sigma^k(x) = \sum_{l=1}^{s} A^l(x) \partial_\mu \xi_l^k(x), & 0 \le |\mu| \le q, \\ \partial_{\mu+1_i} \xi_\sigma^k(x) = \sum_{l=1}^{s} A^l(x) \partial_{\mu+1_i} \xi_l^k(x), & \sigma = s+1, \ldots, r. \end{cases}$$

Differentiating the first relations with respect to x^i and subtracting the second relations, we obtain

$$\sum_{l=1}^{s} \partial_i A^l(x) \partial_\mu \xi_l^k(x) = 0 \Rightarrow A^l(x) = A^l = \text{const}.$$

Since the infinitesimal generators are independent over the constants (the action is assumed to be effective), we arrive at a contradiction. So, $\text{rk}(q) < \text{rk}(q+1) \le r$. Consequently, the rank becomes maximal (equal to r) for q sufficiently large.

This result, which is not evident at first sight, will be clearly illustrated in the last Section. It remains to prove that *the latter isomorphism is an isomorphism of Lie algebras*. To this end we note that

$$\partial_i \xi_\mu^k(x) - \xi_{\mu+1_i}^k(x) = \partial_i(\lambda^\tau(x) \partial_\mu \xi_\tau^k(x)) - \lambda^\tau(x) \partial_{\mu+1_i} \xi_\tau^k(x)$$
$$= \partial_i \lambda^\tau(x) \partial_\mu \xi_\tau^k(x).$$

Accordingly, if λ and λ' are two sections of $X \times \mathcal{G}$, we can introduce corresponding sections ξ_q and ξ_q' of R_q, with bracket

$$[\xi_q, \xi_q'] = \{\xi_{q+1}, \xi_{q+1}'\} + i(\xi)D\xi_{q+1}' - i(\xi')D\xi_{q+1}.$$

By substitution and bilinearity of the algebraic bracket we find

$$([\xi_q, \xi_q'])_\mu^k = \lambda^\rho \lambda'^\sigma \partial_\mu([\xi_\rho, \xi_\sigma])^k + ((\lambda^\rho \xi_\rho^i)\partial_i \lambda'^\tau - (\lambda'^\rho \xi_\rho^i)\partial_i \lambda^\tau)\partial_\mu \xi_\tau^k$$
$$= (c_{\rho\sigma}^\tau \lambda^\rho \lambda'^\sigma + \lambda^\rho(\xi_\rho \cdot \lambda'^\tau) - \lambda'^\sigma(\xi_\sigma \cdot \lambda^\tau))\partial_\mu \xi_\tau^k.$$

The diffferential bracket for the sections of $X \times \mathcal{G}$ is thus defined by the 'ordinary' bracket $[\xi, \xi']$ on T and is thus a bracket of Lie algebras *depending on the action* by means of the infinitesimal generators. Surprisingly, this differs from the 'ordinary' nondifferential bracket on $\mathcal{G}$ expressed by the first term in the development. □

It seems that these results are not well known, since they crucially depend on the use of the Spencer operator.

Corollary 16. *If the action of G on X is effective, there is an isomorphism $X \times G \simeq \mathcal{R}_q$ for q sufficiently large.*

PROOF. According to the proof of the Proposition, we have to prove that the matrix

$$\left(\frac{\partial y_\mu^k}{\partial a^\sigma}\right) = (d_\mu \xi_\rho^k(y) \omega_\sigma^\rho(a))$$

has maximal rank equal to r. Since $\det \omega \neq 0$, this amounts to proving that the matrix

$$(d_\mu \xi^k_\rho(y))$$

has maximal rank equal to r. This can be done as before, but working on the target instead of the source, while taking into account that the parameters are constant. Of course, we have to stress that a global action need not exist and that we use our notations to simplify any explicit calculations (see also the examples in Section E). $\square$

Remark 17 (Important). By the previous isomorphism, the gauging procedure in Section A, i.e. the use of maps from X to G, amounts to the introduction of sections of $\mathcal{R}_q$ that need not be jets of sections of $X \times X$. Hence a section f_q of $\mathcal{R}_q$ is of the form $j_q(f)$ if and only if $y = f(x, a)$ with $a = \text{const}$. Indeed, let $a \colon X \to G \colon (x) \to (a(x))$, be a map from X to G (equivalently: a section of $X \times G$). The corresponding section of $\mathcal{R}_q$ is given by the formulas

$$y^k_\mu = f^k_\mu(x) = \partial_\mu f^k(x, a(x)),$$

in which we have to understand that first partial derivatives with respect to x are taken and *then* $a = a(x)$ is inserted. In particular, $y^k = f^k(x) = f^k(x, a(x))$, although the notation is somewhat confusing. Using the Spencer operator we immediately obtain

$$\partial_i f^k_\mu(x) - f^k_{\mu+1_i}(x) = \partial_\mu \left(\frac{\partial f^k(x, a(x))}{\partial a^\sigma} \right) \partial_i a^\sigma(x)$$

$$= d_\mu \xi^k_\rho(y) \omega^\rho_\sigma(a(x)) \partial_i a^\sigma(x).$$

By Corollary 16, the image of the Spencer operator is zero if and only if $\omega^\rho_\sigma(a(x)) \partial_i a^\sigma(x) = 0$, i.e. $\partial_i a^\sigma(x) = 0$ (since $\det \omega \neq 0$). Hence $a \colon X \to G$ is the 'constant' map $(x) \to (a)$ looked for at the end of Section A. But the interpretation of a gauge in terms of a section of $\mathcal{R}_q$, although it may seem natural, is in absolute contradiction with the gauge framework from which it arose $\cdots$ since now we need an action of G on X in order to construct $\mathcal{R}_q$. The complete explanation of this contradiction will be given in Section D.

Remark 18. As in the previous Remark we may introduce the so-called constant section $e \colon X \to X \times G \colon (x) \to (x, e)$, and we have, in succession:

$$e^{-1}(V(X \times G)) = X \times T_e(G) = X \times \mathcal{G} \simeq \text{id}_q^{-1}(V(\mathcal{R}_q)) = R_q.$$

This completes the interpretation of Section A. In particular, introducing a section of $X \times \mathcal{G}$ comes down to transforming the constant components λ^τ of a vector in $\mathcal{G}$ into functions $\lambda^\tau(x)$ on X, to which we can associate a section of R_q as in Proposition 15. The following useful picture should be kept in mind:

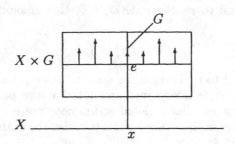

It describes the preceding interpretation.

We end this Section by recapitulating the corresponding concepts described in it:

Lie group	.....	Lie groupoid
gauging of the group G	.....	section of the groupoid $\mathcal{R}_q$
gauging of the Lie algebra $\mathcal{G}$	.....	section of the algebroid R_q

C. Differential Invariants

The purpose of this Section is to study the particular features of systems of PDE defining Lie groupoids and algebroids. All results were sketched by Vessiot in 1903 [191], and they are so useful in practice that it is almost unbelievable that they have never appeared in Spencer's work [101].

First we will recapitulate in compact form the main results of the previous Section, in a way similar to that adopted for Lie groups in Section A. We deal with transitive Lie groupoids only.

Lie groupoid: $\mathcal{R}_q$:

$$\begin{cases} \forall f_q, g_q \in \mathcal{R}_q \Rightarrow g_q \circ f_q \in \mathcal{R}_q \\ \quad \text{if } f_q \text{ and } g_q \text{ can be composed,} \\ \forall f_q \in \mathcal{R}_q \Rightarrow f_q^{-1} \in \mathcal{R}_q, \\ \text{id}_q \in \mathcal{R}_q. \end{cases}$$

Lie algebroid: R_q:

$$\forall \xi_q, \eta_q \in R_q \Rightarrow [\xi_q, \eta_q] \in R_q.$$

Now, if $x_0 \in X$ is a given point, we may introduce the isotropy Lie groups $G_q = \mathcal{R}_q(x_0, x_0)$ and $\text{GL}_q = \Pi_q(x_0, x_0)$, consisting of the jets with source and target x_0. We have $G_q \subset \text{GL}_q$.

The subbundle of Π_q consisting of the jets of transformations with arbitrary source and fixed target x_0 is a *principal bundle*, $\Pi_q(X, x_0)$, with structure group GL_q. The action is defined by composition at the target with jets having source and target x_0. Therefore we can define the bundle $\mathcal{F} = \Pi_q(X, x_0)/G_q$ of homogeneous spaces with typical fiber GL_q/G_q. The image $\mathcal{R}_q(X, x_0)/G_q$ of $\mathcal{R}_q(X, x_0) \subset \Pi_q(X, x_0)$ under the canonical projection $\Phi \colon \Pi_q(X, x_0) \to \mathcal{F}$ makes sense, thanks to the Lie groupoid structure of $\mathcal{R}_q$. Moreover, $\mathcal{R}_q$ being transitive, this image defines a section ω of the bundle $\mathcal{F} \to X$. The dimension of the fibers of $\mathcal{F}$ is equal to $\dim \mathrm{GL}_q - \dim G_q = \dim J_q(T) - \dim \mathrm{R}_q$.

Now we shall study some features of $\mathcal{F}$ which were first discovered by Vessiot.

To this end we notice that groupoid composition at the source commutes with groupoid composition at the target. Hence, a groupoid action of $\Pi_q(X, X)$ can be defined on $\Pi_q(X, x_0)$ by composition at the source. This action factors through a groupoid action of $\Pi_q(X, X)$ on $\mathcal{F}$, which induces a *natural bundle* structure of order q over X. We shall say that $\mathcal{F}$ becomes *associated* with $\Pi_q(X, X)$ under the *natural action*:

$$\lambda: \mathcal{F} \times_X \Pi_q(X, X) \to \mathcal{F}, \qquad \lambda^{-1}: \Pi_q(X, X) \times_X \mathcal{F} \to \mathcal{F},$$

which is compatible with the following commutative diagram:

$$
\begin{array}{ccc}
\mathcal{F} \times_X \Pi_q & \xrightarrow{\ \lambda\ } & \mathcal{F} \\
\downarrow & & \downarrow{\scriptstyle \pi} \\
\Pi_q & \xrightarrow{\ \beta_q\ } & X
\end{array}
$$

This simple definition in terms of a groupoid action should be compared with the sophisticated definition of bundle associated with a principal bundle present in the literature [1, 27, 51]. Let us compare the situation with that of a man having, respectively, no leg, one leg, and two legs:

group   bundle   groupoid

It is beyond doubt that a man having two legs can do much more than with only one leg, or even with no leg! This is exactly the philosophy of the application to mathematical physics.

The preceding property of $\mathcal{F}$ allows us to *lift* any $f \in \mathrm{aut}(X)$ to a local automorphism of $\mathcal{F}$ fibered over f by defining an action on the sections of $\mathcal{F}$ of *either* jet sections f_q or jets of maps $j_q(f)$. The action of f_q or $j_q(f)$ on a section ω of $\mathcal{F}$ is a new section $\bar{\omega} = f_q(\omega)$ of $\mathcal{F}$, defined by the following commutative diagram:

$$
\begin{array}{ccc}
\mathcal{F} \times_X \Pi_q & \longrightarrow & \mathcal{F} \\
{\scriptstyle (\omega, f_q)}\big\uparrow\big\uparrow & & {\scriptstyle \pi}\big\uparrow\big\uparrow{\scriptstyle \bar{\omega}} \\
X & \xrightarrow{\ f\ } & X
\end{array}
$$

In particular, we have $\bar{\omega}(f(x)) = \lambda(\omega(x), f_q(x))$ and for any $\varphi \in \text{aut}(X)$ we have the following *natural transformations*, which can be used (cf., e.g., the tensorial case) for constructing the bundle $\mathcal{F}$ by patching together adapted local coordinates (x, u), where $u = (u^1, \ldots, u^m)$ are local coordinates in the fiber of $\mathcal{F}$:

$$\mathcal{F} \quad \begin{cases} \bar{u} = \lambda(u, j_q(\varphi)(x)), \\ \bar{x} = \varphi(x). \end{cases}$$

Using now the *special* section ω already constructed at the beginning of this Section, we can extend $\Phi \colon \Pi_q(X, x_0) \to \mathcal{F}$ to $\Phi_\omega \colon \Pi_q(X, X) \to \mathcal{F} \colon f_q \to f_q^{-1}(\omega)$.

Hence we can give the equivalent definition:

Definition 1.

$$\mathcal{R}_q = \{f_q \in \Pi_q(X, X) \mid f_q(\omega) = \omega\},$$
$$\Gamma = \{f \in \text{aut}(X) \mid j_q(f)(\omega) = \omega\}.$$

To make this Definition effective, we need to determine the groupoid action λ, or, equivalently, determine the canonical projection $\Pi_q(X, x_0) \to \mathcal{F}$ by constructing the orbit space. It is for this purpose that we introduce the concept of differential invariants, following the way first proposed by Lie [109].

Definition 2. A *differential invariant* of order q is a function on Π_q that is invariant under the groupoid action of $\mathcal{R}_q$ at the target.

We have $\Phi(g_q \circ f_q) = \Phi(f_q)$, $\forall f_q \in \Pi_q$, $\forall g_q \in \mathcal{R}_q$, and thus $\Phi(j_q(g \circ f)) = \Phi(j_q(g) \circ j_q(f)) = \Phi(j_q(f))$, $\forall f \in \text{aut}(X)$, $\forall g \in \Gamma$, by the classical definition. In particular, $\Phi_\omega(f_q) = f_q^{-1}(\omega)$ is a differential invariant, because we successively have:

$$\Phi_\omega(g_q \circ f_q) = (g_q \circ f_q)^{-1}(\omega) = (f_q^{-1} \circ g_q^{-1})(\omega) = f_q^{-1}(g_q^{-1}(\omega)) = f_q^{-1}(\omega) = \Phi_\omega(f_q),$$

by Definition 1, whenever $g_q \in \mathcal{R}_q$ can be composed with $f_q \in \Pi_q$. Moreover,

$$\Phi_\omega(y_q) = \lambda^{-1}(\omega(y), y_\mu^k), \qquad 1 \leq |\mu| \leq q.$$

This formula clearly exhibits the dependence of Φ_ω on ω.

However, in actual practice the way to construct differential invariants is not to use the transformation laws of the bundle $\mathcal{F}$, but, on the contrary, to deduce these laws from knowledge of the differential invariants. Also, unless $\mathcal{R}_q$ is defined by algebraic PDE (see Chapter VI), one cannot look for differential invariants without integrating an involutive distribution of vector fields by means of Frobenius' theorem.

Indeed, the infinitesimal counterpart of Definition 2 is to introduce the distribution $\sharp(R_q)$ on Π_q. This distribution is involutive by Theorem B.9 and the remark ending the proof of this Theorem, since $[\sharp(\xi_q), \sharp(\eta_q)] = \sharp([\xi_q, \eta_q])$ and we can find a *fundamental set* of differential invariants $\Phi^\tau(y_q)$, i.e. a maximum set of functionally independent functions on Π_q killed by $\sharp(\xi_q)$, $\forall \xi_q \in R_q$. It follows that $\mathcal{R}_q$ (more precisely, the connected component of the identity; we will not dwell on this since it is not very useful in applications) is defined as the orbit through $id_q(x)$, $\forall x \in X$. Setting $\Phi^\tau(id_q(x)) = \omega^\tau(x)$, we obtain the *Lie form*:

$$\mathcal{R}_q \qquad \Phi^\tau(y_q) = \omega^\tau(x),$$

where we have separated the source from the target. The crucial question is then to establish a link between the right members $\omega(x)$ and a section ω of $\mathcal{F}$. Following Vessiot, to this end we notice that changes of source commutes with changes of target. Hence, changes of source permute the differential invariants, since we have a maximal set. We see that a fundamental set of differential invariants allows us to *explicitly* determine a section ω of $\mathcal{F}$ (cf. the comment concerning $J_1(\mathcal{P})/G$ at the end of Section A!) and, at the same time, determine $\mathcal{F}$ as a natural bundle. By analogy with tensor bundles, we state:

Definition 3. $\mathcal{F}$ is called a *natural bundle*, and a section ω of it determines a *geometric object*, or *structure*, on X.

In Section E we shall see that tensors are 'exceptional', i.e. particularly simple, geometric objects, since in that Section we shall see examples of nonlinear geometric objects of higher order.

Passing to the infinitesimal point of view, we may define a *formal Lie derivative* $L(\xi_q)\omega$ of ω with respect to the section ξ_q of $J_q(T)$ and with value in the vector bundle

$$F_0 = \omega^{-1}(V(\mathcal{F})) = J_q(T)/\mathrm{R}_q,$$

by *varying* the section ω. Since we work at the source, we can introduce $f_{q,t} = \exp(t b(\xi_q) \circ id_q$, and define:

$$L(\xi_q)\omega = \frac{d}{dt}(f_{q,t}^{-1}(\omega))\Big|_{t=0} \in F_0.$$

Another way is by defining the *ordinary Lie derivative* $\mathcal{L}(\xi)\omega$ of ω with respect to a vector field $\xi \in T$ as follows: set $f_t = \exp(t\xi) \in \mathrm{aut}(X)$, and then

$$\mathcal{L}(\xi)\omega = \frac{d}{dt}(j_q(f_t)^{-1}(\omega))\Big|_{t=0},$$

and substituting jet sections instead of jets of vector fields.

Considering finite changes of coordinates on $\mathcal{F}$ and passing to the infinitesimal limit, we obtain

$$A_k^\mu(q) \cdot \Phi^\tau \equiv L_k^{\tau\mu}(\Phi),$$

and thus

$$\begin{cases} \overline{u}^\tau = u^\tau + t\partial_\mu \xi^k(x) L_k^{\tau\mu}(u) + \dots, \\ \overline{x} = x + t\xi(x) + \dots, \end{cases}$$

using the First Fundamental Theorem (Section A) for the action of GL_q at the source. Accordingly we have, in succession,

$$(\mathcal{L}(\xi)\omega)^\tau = -L_k^{\tau\mu}(\omega(x))\partial_\mu \xi^k + \xi^r \partial_r \omega^\tau(x),$$
$$(L(\xi_q)\omega)^\tau = -L_k^{\tau\mu}(\omega(x))\xi_\mu^k + \xi^r \partial_r \omega^\tau(x).$$

Therefore we can give

Definition 4.

$$R_q = \{\xi_q \in J_q(T)|\ L(\xi_q)\omega = 0\},$$
$$\Theta = \{\xi \in T|\ \mathcal{L}(\xi)\omega = 0\}.$$

It remains to study the relation between the determination of differential invariants and the procedure for prolongating them. The main formula is given in the following proposition:

Proposition 5. *For any function* Φ *on* Π_q *we have*

$$\sharp(\eta_{q+1})d_i\Phi = d_i(\sharp(\eta_q) \cdot \Phi) - y_i^k \sharp\left(D\eta_{q+1}\left(\frac{\partial}{\partial y^k}\right)\right) \cdot \Phi.$$

PROOF. The main difficulty lies in dealing with sections of jet bundles and not with jets of sections. We have

$$d_i\Phi = \partial_i\Phi + y_{\mu+1_i}^k \frac{\partial\Phi}{\partial y_\mu^k},$$

and therefore, with $\sharp(\eta_q) = \zeta_\mu^k(y_q)\frac{\partial}{\partial y_\mu^k}$, we obtain

$$\sharp(\eta_{q+1})d_i\Phi = \partial_i(\sharp(\eta_q)\Phi) + \zeta_{\mu+1_i}^k(y_{q+1})\frac{\partial\Phi}{\partial y_\mu^k} + y_{\mu+1_i}^k \zeta_\nu^l(y_q)\frac{\partial^2\Phi}{\partial y_\mu^k \partial y_\nu^l}$$

$$= d_i(\sharp(\eta_q)\Phi) + (\zeta_{\mu+1_i}^k - d_i\zeta_\mu^k)\frac{\partial\Phi}{\partial y_\mu^k}.$$

Finally, we only have to notice that differentiation of sections leads to higher-order sections to see that

$$d_i\zeta_\mu^k - \zeta_{\mu+1_i}^k = y_i^k \sharp\left(D\eta_{q+1}\left(\frac{\partial}{\partial y^k}\right)\right) \cdot y_\mu^k,$$

using the Spencer operator at the target. $\square$

It immediately follows that, if Φ is a differential invariant of order q and $\eta_{q+1} \in R_{q+1}$, then $D\eta_{q+1} \in T^* \otimes R_q$, and the right hand side of the preceding formula vanishes identically. Accordingly, $d_i\Phi$ is a differential invariant of order $q + 1$.

We conclude this section by making a few comments about associated bundles. Indeed, we can extend the definition of natural bundle of order q to that of a bundle associated with a Lie groupoid of order q, by taking $\mathcal{R}_q$ instead of Π_q in the corresponding action morphism and diagram.

Notice that $J_q(T)$ is associated with Π_{q+1}, although $J_q^0(T)$ is associated with Π_q (an exercise). Similarly, R_q is associated with $\mathcal{R}_{q+1}$, although R_q^0 is associated with $\mathcal{R}_q$. Since $S_q T^* \otimes T$ is a tensor bundle, it is associated with Π_1, hence with Π_q and so with $\mathcal{R}_q$. Accordingly, the symbol M_q and its various prolongations are associated with $\mathcal{R}_q$ (by Proposition B.14), and so N_1 is associated with $\mathcal{R}_q$. Now $F_0 = J_q(T)/R_q = J_q^0(T)/R_q^0$ whenever R_q is transitive (an exercise). It follows that

F_0 and, more generally, the Janet bundles F_r are associated with $\mathcal{R}_q$. This is in contrast to what happens for the Spencer bundles, which are associated with $\mathcal{R}_{q+1}$ only, because $C_0 = \mathrm{R}_q$.

Finally, one can prove (exercise) that if $\mathcal{F}$ is associated with $\mathcal{R}_q$, then $V(\mathcal{F})$ is associated with $\mathcal{R}_q$ while $J_r(\mathcal{F})$ becomes associated with $\mathcal{R}_{q+r}$. These results are used mainly in the deformation theory of structures on manifolds [144].

D. Nonlinear Sequences

This Section is the 'hardest' one, and it is of great importance for applications to mathematical physics. In Chapter IV we have constructed the full Janet and Spencer sequences for a linear differential operator. Here we shall construct a nonlinear version of a part of these sequences, for a given involutive system of finite Lie equations. A very delicate point is to relate concepts introduced in these two sequences, since it is almost impossible to describe certain results in local coordinates.

We start with the nonlinear Janet sequence.

As a basic motivation for constructing such a sequence we now sketch the *equivalence problem* for structures on X. The key idea is to fix the bundle $\mathcal{F}$ and to vary the section. Indeed, if ω and $\overline{\omega}$ are given sections of $\mathcal{F}$, we may look for $f \in \mathrm{aut}(X)$ such that $j_q(f)^{-1}(\omega) = \overline{\omega}$. The solution of this problem is related to a number of compatibility conditions, usually depending on ω, to be satisfied by $\overline{\omega}$. Once these formal conditions are fulfilled, the *local solvability* of the problem depends on a delicate analysis (there are counterexamples, [144, 173]). Here we shall concern ourselves with the formal problem only.

Suppose that the system $\mathcal{R}_q$ of finite Lie equations has a 2-acyclic symbol. In fact, by Proposition B.14, this property can be immediately checked on R_q. To apply the test for formal integrability, we only need to check that $\pi_q^{q+1} : \mathcal{R}_{q+1} \to \mathcal{R}_q$ is surjective or, equivalently, see that the curvature map vanishes identically. We treat this problem by considering the Lie form

$$\mathcal{R}_q \qquad \Phi^\tau(y_q) = \omega^\tau(x).$$

Consider the maximal number of formal derivatives $d_i\Phi^\tau$ that are linearly independent with respect to the jet coordinates of strict order $q+1$ that appear linearly. Any other $d_j\Phi^\sigma$ is such that we can find a linear combination $d_j\Phi^\sigma + A_{j\tau}^{i\sigma}(y_q)d_i\Phi^\tau$ not containing jet coordinates of strict order $q+1$. Using the comment following Proposition C.5 while applying $\sharp(\mathrm{R}_{q+1})$, we see that $(\sharp(\mathrm{R}_q)A_{j\tau}^{i\sigma}(y_q))d_i\Phi^\tau$ does not contain jet coordinates of strict order $q + 1$, i.e. we obtain a contradiction unless $\sharp(\mathrm{R}_q)A_{j\tau}^{i\sigma}(y_q) = 0$. So, we can use linearly independent combinations of the form $A_\tau^{\alpha i}(\Phi)d_i\Phi$ to eliminate the jet coordinates of strict order $q + 1$. According to the Lie form, in which the y_q are contained in the differential invariants only, and the fact that the previous expressions are killed by $\sharp(\mathrm{R}_q)$, it follows that $\pi_q^{q+1} : \mathcal{R}_{q+1} \to \mathcal{R}_q$ is surjective if and only if we obtain differential *identities* of the form

$$A_\tau^{\alpha i}(\Phi)d_i\Phi^\tau + B^\alpha(\Phi) \equiv 0,$$

which describe the curvature map. Therefore we *must have*

$$A_\tau^{\alpha i}(\omega(x))\partial_i\omega^\tau(x) + B^\alpha(\omega(x)) = 0,$$

and for solving the equivalence problem

$$\Phi^\tau(j_q(f)(x)) = \overline{\omega}(x)$$

we *must also have*

$$A_\tau^{\alpha i}(\overline{\omega}(x))\partial_i\overline{\omega}^\tau(x) + B^\alpha(\overline{\omega}(x)) = 0.$$

Summarising the results obtained thus far, the equivalence problem amounts to looking for $f \in \text{aut}(X)$ such that $j_q(f)^{-1}(\omega) = \overline{\omega}$, and a necessary formal solvability condition is that the two sections ω and $\overline{\omega}$ of $\mathcal{F}$ must satisfy the *same* compatibility conditions. As a byproduct we notice that these compatibility conditions are independent of the coordinate system, since any change of source can be lifted to sections and the resulting section satisfies the same compatibility conditions, if regarded as a system $\mathcal{B}_1 \subset J_1(\mathcal{F})$ on $\mathcal{F}$ of the form

$$I(u_1) \equiv A(u)u_x + B(u) = 0.$$

We combine the previous results in a nonlinear differential sequence, called a nonlinear Janet sequence. To this end we first exhibit a number of natural bundles of order q; these will play the role of the Janet bundles in the linear situation with $E = T$.

Recall that $J_r(\mathcal{F})$ is a natural bundle of order $q + r$ when $\mathcal{F}$ is a natural bundle of order q. In particular, the infinitesimal rules of $J_r(\mathcal{F})$ are, for $r = 1$,

$$\xi^i(x)\frac{\partial}{\partial x^i} + \partial_\mu\xi^k(x)L_k^{\tau\mu}(u)\frac{\partial}{\partial u^\tau}$$

$$+ \left(\partial_{\mu+1_i}\xi^k(x)L_k^{\tau\mu}(u) + \partial_\mu\xi^k(x)\frac{\partial L_k^{\tau\mu}(u)}{\partial u^\sigma}u_i^\sigma - u_r^\tau\partial_i\xi^r(x)\right)\frac{\partial}{\partial u_i^\tau},$$

the first two components being those of $\mathcal{F}$.

By Definition C.4 [118], the equations for R_q, in so-called *Medolaghi form*, are:

$$\Omega^\tau \equiv -L_k^{\tau\mu}(\omega(x))\xi_\mu^k + \xi^r\partial_r\omega^\tau(x) = 0,$$

where (ω, Ω) is a section of $V(\mathcal{F})$, i.e. Ω is a *perturbation* of ω. By the above, the equations for R_{q+1} are obtained from those of R_q by adjoining the equations

$$\Omega_i^\tau \equiv -L_k^{\tau\mu}(\omega(x))\xi_{\mu+1_i}^k - \frac{\partial L_k^{\tau\mu}(\omega(x))}{\partial u^\sigma}\partial_i\omega^\sigma(x)\xi_\mu^k + \partial_r\omega^\tau(x)\xi_i^r + \xi^r\partial_r(\partial_i\omega^\tau(x)) = 0.$$

By analogy with special and general relativity, we call the section of $\mathcal{F}$ with which we started *special* and an arbitrary section of $\mathcal{F}$ *general*, taking into account that GL_q acts transitively along the fibers of $\mathcal{F}$ since it acts transitively along the fibers of the principal bundle $\Pi_q(X, x_0)$. Thus, we may consider the family $R_q(\omega)$ of systems obtained by keeping $\mathcal{F}$ fixed and by varying the section ω. A prolongation of it will be denoted by $R_{q+1}(j_1(\omega))$, depending on the choice of the section $j_1(\omega)$ of $J_1(\mathcal{F})$. In the finite framework we were able to define $\mathcal{R}_q(\omega)$ and $\mathcal{R}_{q+1}(j_1(\omega))$ similarly. In particular, we notice that R_q^0, M_{q+r} and H_{q+r}^s, which depend only on ω and *not* on $j_1(\omega)$, are vector bundles by Proposition B.14.

$\mathcal{B}_1$ and $J_1(\mathcal{F})$ are affine bundles over $\mathcal{F}$; they are also natural. The symbol $\mathcal{N}_1 \subset T^* \otimes V(\mathcal{F})$ of $\mathcal{B}_1$ is thus a natural bundle of order q, and we can introduce the natural Janet bundles

$$\mathcal{F}_r = \wedge^r T^* \otimes V(\mathcal{F})/\delta(\wedge^{r-1} T^* \otimes \mathcal{N}_1)$$

as a family of natural vector bundles over $\mathcal{F}$. The usual Janet bundles are therefore $F_r = \omega^{-1}(\mathcal{F}_r)$ and are associated with $\mathcal{R}_q$ only, since $f_q \in \Pi_q$ must preserve ω. Corollary II.A.22 implies that the quotient $J_1(\mathcal{F})/\mathcal{B}_1$ of affine bundles over $\mathcal{F}$ can be identified with $\mathcal{F}_1$ by projection:

$$(u_1) \to (u, v^\alpha = A_\tau^{\alpha i}(u)u_i^\tau + B(u)),$$

adopting local coordinates (u, v) on $\mathcal{F}_1$ with infinitesimal transformation rules of the form

$$\xi^i \frac{\partial}{\partial x^i} + \partial_\mu \xi^k(x) \left(L_k^{\tau\mu} \frac{\partial}{\partial u^\tau} + M_{\beta,k}^{\alpha,\mu}(u)v^\beta \frac{\partial}{\partial v^\alpha} \right).$$

This shows that $\mathcal{F}_1$ is a vector bundle over $\mathcal{F}$. For future applications it is essential to notice that setting $v^\alpha = A_\tau^{\alpha i}(u)u_i^\tau$ alone is not sufficient to obtain natural *vector bundles* over $\mathcal{F}$, if $q > 1$. *This is a very delicate point* and only an understanding of the many examples in Section E will allow the reader to become familiar with the new language of natural bundles and geometric objects.

The *nonlinear Janet sequence* associated with a Lie form is as follows:

$$0 \longrightarrow \Gamma \longrightarrow \operatorname{aut}(X) \underset{\omega \circ \alpha}{\overset{\Phi \circ j_q}{\rightrightarrows}} \mathcal{F} \underset{0}{\overset{I \circ j_1}{\rightrightarrows}} \mathcal{F}_1,$$

$$f \to j_q(f)^{-1}(\omega),$$

where $\alpha \circ f(x) = x$ whenever $x \in \operatorname{dom} f$. Of course, its infinitesimal counterpart is the *linear Janet sequence*

$$0 \longrightarrow \Theta \longrightarrow T \overset{\mathcal{D}}{\longrightarrow} F_0 \overset{\mathcal{D}_1}{\longrightarrow} \cdots \overset{\mathcal{D}_n}{\longrightarrow} F_n \longrightarrow 0,$$

where $\mathcal{D}\xi = \mathcal{L}(\xi)\omega$ and $[\Theta, \Theta] \subset \Theta$ (with the ordinary bracket of vector fields).

We now come to the difficult part of this Section, namely the study of $\mathcal{R}_q(\omega)$ or $R_q(\omega)$ under the condition that we now suppose this system to be formally integrable and to have 2-acyclic symbol for a special section.

First of all, by the above-said, the symbol $M_q(\omega)$ remains 2-acyclic, and we only have to study the surjectivity of $\mathcal{R}_{q+1} \to \mathcal{R}_q$ or of $R_{q+1} \to R_q$ (where we leave out ω for simplicity). The reader should notice that often we introduce inequalities to define $\mathcal{F}$ (see the examples in Section E, where $\det \omega \neq 0$ often appears).

Passing to the prolongations, we have

$$\mathcal{R}_{q+r} = \{f_{q+r} \in \Pi_{q+r} | \ f_{q+r}(j_r(\omega)) = j_r(\omega)\},$$
$$R_{q+r} = \{\xi_{q+r} \in J_{q+r}(T) | \ L(\xi_{q+r})j_r(\omega) = 0\}.$$

The following important Theorem holds.

Theorem 1. $\pi_q^{q+1}\colon \mathcal{R}_{q+1} \to \mathcal{R}_q$ *is an epimorphism if and only if there is an equivariant section* $c\colon \mathcal{F} \to \mathcal{F}_1\colon (x, u) \to (x, u, v = c(u))$ *such that* $I(j_1(\omega)) = c(\omega)$.

PROOF. For every section $f_q \in \mathcal{R}_q$ we have $f_q^{-1}(\omega) = \omega$. If we lift such a section to an arbitrary section $f_{q+1} \in \Pi_{q+1}$, then the sections $f_{q+1}^{-1}(j_1(\omega))$ and $j_1(\omega)$ of $J_1(\mathcal{F})$ are, in general, different although they have the same projection ω on $\mathcal{F}$. Thus, the curvature of f_q is the projection of the difference $f_{q+1}^{-1}(j_1(\omega)) - j_1(\omega)$ to $\mathcal{F}_1$, i.e. $\kappa(f_q) = f_q^{-1}(I(j_1(\omega))) - I(j_1(\omega))$ because I is quasi-linear in the first-order jets. So, $\kappa(f_q) = 0$ if and only if $f_q(I(j_1(\omega))) = I(j_1(\omega))$ whenever $f_q(\omega) = \omega$. In this case we can modify f_{q+1} by an element of $S_{q+1}T^* \otimes T$ to obtain exactly $f_{q+1}^{-1}(j_1(\omega)) = j_1(\omega)$ and $f_{q+1} \in \mathcal{R}_{q+1}$. Now, similarly as in the special case, we must have a differential identity of the form $I(j_1(\omega)) = c(\omega)$ for a section $c \colon \mathcal{F} \to \mathcal{F}_1$ that is equivariant for all $f_q \in \Pi_q$ when ω varies. So, it remains to determine such equivariant sections.

Passing to the infinitesimal point of view, we must have $L(\xi_q)(I(j_1(\omega))) = 0$ if $L(\xi_q)\omega = 0$. In particular, there is a commutative and exact diagram

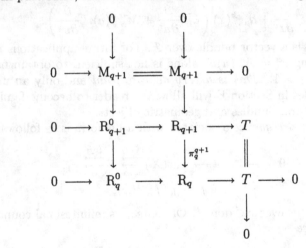

and the surjectivity of $R_{q+1} \to R_q$ splits into that of $R_{q+1}^0 \to R_q^0$ *and* that of $R_{q+1} \to T$. To study the surjectivity of the isotropy part we notice that

$$L_k^{\tau\mu}(\omega(x))\xi_{\mu+1_\tau}^k + \frac{\partial L_k^{\tau\mu}(\omega(x))}{\partial u^\sigma}\partial_i\omega^\sigma(x)\xi_\mu^k - \partial_r\omega^\tau(x)\xi_i^r = 0$$

should not imply conditions of order q other than

$$L_k^{\tau\mu}(\omega(x))\xi_\mu^k = 0$$

whenever $1 \leq |\mu| \leq q$. This is a linear algebra problem, depending only on $j_1(\omega)$, and certain conditions

$$I_*(j_1(\omega)) = 0,$$

called *integrability conditions of the first kind*, must be satisfied. Equivalently, we must have $L(\xi_q^0)(I(j_1(\omega))) = 0$ whenever $L(\xi_q^0)\omega = 0$, i.e. for any section $\xi_q^0 \in R_q^0$. Accordingly, we must have

$$M_{\beta,k}^{\alpha,\mu}(\omega(x))v^\beta(x)\xi_\mu^k = 0, \qquad 1 \leq |\mu| \leq q,$$

with $v(x) = I(j_1(\omega)(x))$, whenever

$$L_k^{\tau\mu}(\omega(x))\xi_\mu^k = 0, \qquad 1 \le |\mu| \le q.$$

Hence, the integrability conditions of the first kind factorise through *linear conditions* among the v, with coefficients depending on the u, and thus define a natural vector subbundle $\mathcal{E}_1$ of $\mathcal{F}_1$ which is again a vector bundle over $\mathcal{F}$. For this subbundle we have the rank condition

$$\mathrm{rk}(L(u), M(u)v) = \mathrm{rk}(L(u)) = \dim F_0.$$

Using a symbolic notation, the equivariance of c finally amounts to saying that the submanifold $\mathcal{E}_1 \subseteq \mathcal{F}_1$ defined by $v - c(u) = 0$ is invariant under the infinitesimal generators $L(u)\frac{\partial}{\partial u} + M(u)v\frac{\partial}{\partial v}$, and we must have

$$L(u)\frac{\partial c(u)}{\partial u} = M(u)c(u).$$

Using the rank condition *for the subbundle* $\mathcal{E}_1$, we integrate a linear system of the form

$$\frac{\partial c(u)}{\partial u} = N(u)c(u).$$

The space of solutions of this system (which has zero symbol) is a vector space over the 'constants' of dimension $\dim E_1 \le \dim F_1$, where $E_1 = \omega^{-1}(\mathcal{E}_1)$. In fact, these so-called constants may be functions of x, and we have to prove that they are 'absolute constants'. Indeed, we can express each such 'function' in the form $I_{**} = a(u)v$. The restrictions of these forms to the subbundle $\mathcal{E}_1$ of $\mathcal{F}_1$ are killed by the infinitesimal generators $L(u)\frac{\partial}{\partial u} + M(u)v\frac{\partial}{\partial v}$. Using a connection $\chi_q : T \to R_q$, the equivariance of c also implies that the $a(u)v$ are killed by ∂_i when we substitute $j_1(\omega)$, for otherwise $R_{q+1} \to T$ is not surjective and we obtain differential conditions of the form

$$I_{**}(j_1(\omega)) = c,$$

called *integrability conditions of the second kind*, which amount to *checking the surjectivity of* $R_{q+1} \to T$ *whenever* $R_{q+1}^0 \to R_q^0$ *is surjective*. A change of connection does not change the result, since if $\chi_q' : T \to R_q$ is another connection, then $\chi_q' - \chi_q \in T^* \otimes R_q^0$, while all results are already R_q^0-invariants. $\square$

We advise the reader to follow this proof on an example, because it, again, is quite technical and involves specific properties of geometric objects and natural bundles.

Remark 2. By looking at the equations defining R_{q+1} and R_q, we see that a straightforward application of linear algebra techniques in the study of the surjectivity of $R_{q+1} \to R_q$ should, in general, imply second-order differential conditions on ω, namely $I_* = 0$, $d_i I_{**} = 0$. The existence of integrating factors, which, even in simple examples, is not evident at first sight, has been one of the major discoveries of Vessiot, who followed the lines set out by Lie. It is quite surprising that these results cannot be found in the literature on this subject, and are also not acknowledged, despite their importance [126, 173].

Remark 3. By construction, the integrability conditions of the first and second kinds are invariant under lifting an arbitrary change of coordinates, and only those of the first kind are separately invariant. The total set of integrability conditions therefore depends on an equivariant section c and determines an affine natural bundle $\mathcal{B}_1(c) \subset J_1(\mathcal{F})$. The special section may correspond to the case $c = 0$.

Definition 4. The constants introduced above are called *structure constants* and characterise the equivariant sections $c \colon \mathcal{F} \to \mathcal{F}_1$ in the transitive case.

The system $\mathcal{B}_1(c)$, which is known to be formally integrable for $c = 0$, need not be formally integrable for any equivariant section c.

Theorem 5. *The system $\mathcal{B}_1(c)$ is formally integrable if and only if the structure constants satisfy polynomial conditions $J(c) = 0$ of degree ≤ 2.*

These polynomial conditions are called *Jacobi conditions*.

PROOF. Using the natural projection $T^* \otimes \mathcal{F}_1 \to \mathcal{F}_2 \colon v_x \to \gamma(u)v_x$, and omitting indices (for simplicity), we have

$$d_x I \equiv A(u)u_{xx} + \partial_u A(u)u_x u_x + \partial_u B(u)u_x,$$

and thus

$$\gamma(u)d_x I \equiv a(u)u_x u_x + b(u)u_x.$$

Since $\mathcal{F}_2$ is a natural bundle of order q, we may factor the dependence on u_x through a dependence on v and obtain expressions of the form

$$\gamma(u)d_x I \equiv \alpha(u)II + \beta(u)I,$$

because $\mathcal{B}_1(0)$ is formally integrable for $v = 0$. Using the specific affine combinations (I_*, I_{**}) in u_x and substituting $I_* = 0$, $I_{**} = c$, we obtain polynomial conditions in c of degree ≤ 2 and with coefficients depending on u. Since the action of GL_q is transitive on the fibers of $\mathcal{F}$, we may choose any point $(x_0, u_0) \in \mathcal{F}$ and obtain the required result.

If $q = 1$, then I is linear in u_x and $B(u), b(u), \beta(u)$ are absent. This leads to homogeneous Jacobi conditions of degree 2. $\square$

Remark 6 (Important Remark). It is quite a surprise that the only situations of classical differential geometry in which differential conditions for geometric objects are *known* to involve certain 'constants' are the constant Riemannian curvature for a metric, involving one constant c, and the Maurer–Cartan conditions for the Maurer–Cartan forms, involving the classical structure constants $c_{\rho\sigma}^\tau$ of Section A. By looking at Section E, the reader will notice that *there is no conceptual difference at all* between the constant Riemannian curvature condition

$$\rho_{lij}^k \equiv \partial_i \gamma_{lj}^k - \partial_j \gamma_{li}^k + \gamma_{ri}^k \gamma_{lj}^r - \gamma_{rj}^k \gamma_{li}^r$$
$$= c(\delta_j^k \omega_{il} - \delta_i^k \omega_{lj})$$

and the Maurer–Cartan conditions

$$\partial_j \omega_i^\tau - \partial_i \omega_j^\tau = c_{\rho\sigma}^\tau \omega_i^\rho \omega_j^\sigma.$$

In particular, the quadratic terms $\gamma\gamma - \gamma\gamma$ are *not at all* the analog of the quadratic terms $c\omega\omega$, contrary to common belief nowadays.

We shall now deal with the nonlinear Spencer sequence.

We start with a few remarks concerning the association of $J_q(T)$ with Π_{q+1}. A way to exhibit the proper formula is to start with the tangent map for vector fields,

$$\eta^k(f(x)) = \partial_r f^k(x)\xi^r(x),$$

differentiate μ times in the usual manner, and then substitute jets for derivatives. E.g., at order zero the preceding formula expresses that $\eta = j_1(f)(\xi)$, while the formula

$$\eta^k(f(x)) = f_r^k(x)\xi^r(x)$$

expresses that $\eta = f_1(\xi)$. Similarly, at order two we obtain the additional relations

$$\begin{cases} \eta_s^k(f(x))\partial_i f^s(x) = \partial_r f^k(x)\xi_i^r(x) + \partial_{ir} f^k(x)\xi^r(x), \\ \eta_{st}^k(f(x))\partial_i f^s(x)\partial_j f^t(x) + \eta_s^k(f(x))\partial_{ij} f^s(x) = \partial_r f^k(x)\xi_{ij}^r(x) + \\ \qquad \partial_{jr} f^k(x)\xi_i^r(x) + \partial_{ir} f^k(x)\xi_j^r(x) + \partial_{ijr} f^k(x)\xi^r(x), \end{cases}$$

which expresses that $\eta_2 = j_3(f)(\xi_2)$, while the additional formulas

$$\eta_s^k(f(x))f_i^s(x) = f_r^k(x)\xi_i^r(x) + f_{ir}^k(x)\xi^r(x),$$

$$\eta_{st}^k(f(x))f_i^s(x)f_j^t(x) + \eta_s^k(f(x))f_{ij}^s(x) = f_r^k(x)\xi_{ij}^r(x) + f_{rj}^k(x)\xi_i^r(x) +$$
$$f_{ri}^k(x)\xi_j^r(x) + f_{rij}^k(x)\xi^r(x)$$

express that $\eta_2 = f_3(\xi_2)$. More generally, using the Leibniz rule for derivatives, we have

Lemma 7. *The association of $J_q(T)$ is described by the formula*

$$f_\mu^r \eta_r^k + \cdots = f_r^k \xi_\mu^r + \cdots + f_{\mu+1,r}^k \xi^r.$$

We will repeatedly use this formula. We also notice that $J_q^0(T)$ is associated with Π_q, since we have to take out the factor $f_{\mu+1,r}^k$ with $|\mu| = q$ whenever $\xi^r = 0$ in the formula above.

We are now ready for a central result, describing the *first nonlinear Spencer sequence.*

Theorem 8. *There is a nonlinear differential sequence*

$$0 \longrightarrow \mathrm{aut}(X) \xrightarrow{j_{q+1}} \Pi_{q+1}(X,X) \xrightarrow{\overline{D}} T^* \otimes J_q(T) \xrightarrow{\overline{D}'} \wedge^2 T^* \otimes J_{q-1}(T),$$

with $\overline{D}f_{q+1} \equiv f_{q+1}^{-1} \circ j_1(f_q) - \mathrm{id}_{q+1} = \chi_q$ *and* $(\overline{D}'\chi_q)(\xi,\eta) \equiv (D\chi_q)(\xi,\eta) - \{\chi_q(\xi), \chi_q(\eta)\} = 0.$

PROOF. First of all we notice that, in contrast to the Janet sequence, there is no need for double arrows, because we have vector bundles on the right and the kernels of $\overline{D}, \overline{D}'$ are automatically taken with respect to the zero section, as usual. Also, the reader should convince him/herself that it is almost impossible to write down the preceding formulas for $q \geq 2$, although they look simple.

Let us deal with $\overline{D}$.

There are inclusions $\Pi_{q+1} \subset \Pi_{q,1} \subset J_1(\Pi_q)$, and the composition $f_{q+1}^{-1} \circ j_1(f_q)$ is well defined in $J_1(\Pi_q)$ over $f_q^{-1} \circ f_q = \mathrm{id}_q \in \Pi_q$. But id_{q+1} is also a section of Π_{q+1}, and thus of $J_1(\Pi_q)$ over id_q. Since $J_1(\Pi_q)$ is an affine bundle over Π_q (Proposition II.B.3), it follows that $f_{q+1}^{-1} \circ j_1(f_q) - \mathrm{id}_{q+1}$ is a section of $T^* \otimes V(\Pi_q)$ over $\mathrm{id}_q \in \Pi_q$, and $\mathrm{id}_q^{-1}(V(\Pi_q)) = J_q(T)$. For $q = 1$ we successively obtain

$$\begin{cases} \chi_{,i}^k = g_l^k \partial_i f^l - \delta_i^k = A_i^k - \delta_i^k, \\ \chi_{r,i}^k = g_l^k(\partial_i f_r^l - A_i^s f_{rs}^l), \end{cases}$$

where we have set $g_l^k(x) f_i^l(x) = \delta_i^k$ in Π_1. More generally, in a symbolic way we can obtain

$$f_{q+1}(\chi_q) = j_1(f_q) - f_{q+1} = D f_{q+1}$$

and the formula from Lemma 7:

$$f_r^k \chi_{\mu,i}^r + \cdots + f_{\mu+1_r}^k \chi_{,i}^r = \partial_i f_\mu^k - f_{\mu+1_i}^k,$$

allowing us to determine χ_q inductively.

(*At this point the reader should be quite surprised by the fact that these formulas describe the foundations of engineering physics!*)

Of course, we have

$$\overline{D} f_{q+1} = 0 \Leftrightarrow j_1(f_q) - f_{q+1} = 0 \Leftrightarrow f_{q+1} = j_{q+1}(f),$$

and the left part of the sequence is well defined.

Let us now deal with $\overline{D}'$.

Setting $\partial_i = \partial/\partial x^i$ we have

$$\flat(\chi_q(\partial_i)) = \chi_{,i}^r \partial_r + \zeta_{\mu,i}^k \frac{\partial}{\partial y_\mu^k},$$

with

$$\begin{aligned} \zeta_{\mu,i}^k &= -(f_r^k \chi_{\mu,i}^r + \cdots + f_{\mu+1_r}^k \chi_{,i}^r) + f_{\mu+1_r}^k \chi_{,i}^r \\ &= -(\partial_i f_\mu^k - f_{\mu+1_i}^k) + f_{\mu+1_r}^k \chi_{,i}^r \\ &= -\partial_i f_\mu^k + A_i^r f_{\mu+1_r}^k. \end{aligned}$$

It follows that

$$\begin{aligned} \flat(\chi_q(\partial_i))(y_\mu^k - f_\mu^k(x)) &= \zeta_{\mu,i}^k - \chi_{,i}^r \partial_r f_\mu^k(x) \\ &= -A_i^r(\partial_r f_\mu^k - f_{\mu+1_r}^k), \end{aligned}$$

and the vector fields

$$A_i^r d_r - \flat(\chi_q(\partial_i)) = \partial_i + \cdots$$

are hence tangent to the manifold $f_q(X) \subset \Pi_q$. By restriction, and using Proposition A.7, we obtain

$$[A_i^r d_r - \flat(\chi_q(\partial_i)), A_j^s d_s - \flat(\chi_q(\partial_j))] = 0.$$

Developing this bracket, using Lemma B.6 and Definition B.8, we obtain

$$\flat((\overline{D}'\chi_q)(\partial_i, \partial_j)) = (\partial_i A_j^r - \partial_j A_i^r - A_i^s \chi_{s,j}^r + A_j^s \chi_{s,i}^r)\, dr$$

(the substitutions involved are left to the reader as an exercise).

A direct verification shows that the right hand expression is identically zero, and the construction of $\overline{D}'$ follows from the fact that $\flat$ is an isomorphism. $\square$

Using Corollary B.10 we immediately obtain

Corollary 9. *There is a restricted first nonlinear Spencer sequence*

$$0 \longrightarrow \Gamma \xrightarrow{j_{q+1}} \mathcal{R}_{q+1} \xrightarrow{\overline{D}} T^* \otimes \mathrm{R}_q \xrightarrow{\overline{D}'} \wedge^2 T^* \otimes J_{q-1}(T).$$

Remark 10 (Important Remark). Local expressions of $\overline{D}'$ for $q = 2$ are:

$$\begin{cases} \partial_i \chi_{,j}^k - \partial_j \chi_{,i}^k - \chi_{i,j}^k + \chi_{j,i}^k - (\chi_{,i}^r \chi_{r,j}^k - \chi_{,j}^r \chi_{r,i}^k) = 0, \\ \partial_i \chi_{l,j}^k - \partial_j \chi_{l,i}^k - \chi_{li,j}^k + \chi_{lj,i}^k - (\chi_{,i}^r \chi_{lr,j}^k + \chi_{l,i}^r \chi_{r,j}^k - \chi_{l,j}^r \chi_{r,i}^k - \chi_{,j}^r \chi_{lr,i}^k) = 0. \end{cases}$$

Setting $\chi_{,i}^k = A_i^k - \delta_i^k$ we obtain

$$\begin{cases} \partial_i A_j^k - \partial_j A_i^k - A_i^r \chi_{r,j}^k + A_j^r \chi_{r,i}^k = 0, \\ \partial_i \chi_{l,j}^k - \partial_j \chi_{l,i}^k - \chi_{l,i}^r \chi_{r,j}^k + \chi_{l,j}^r \chi_{r,i}^k - A_i^r \chi_{lr,j}^k + A_j^r \chi_{lr,i}^k = 0. \end{cases}$$

Here we immediately discover, with minor variation of notation, the so-called *torsion* and *curvature* of Cartan [33, 34]. So, all of present mathematical physics is based on the following, deep, confusion: *curvature alone is an operator in the Janet sequence, while torsion+curvature is an operator in the Spencer sequence.* In particular, *any attempt at identifying these operators will be unsuccessful*, according to Theorem IV.A.16 and Remark 6. Later in this Section we shall explain how this confusion arises from early classical gauge theory! We also notice that the additional terms in the curvature depend on second-order jets and cannot be provided by a classical approach, although they are as important as the usual quadratic terms. Indeed, in Section E we shall see that for Riemannian structures *only the quadratic terms remain* in the curvature, because there are no second-order jets, while for complex-analytic structures *the quadratic terms disappear* and we are left with the additional terms only!

Proposition 11. *The nonlinear Spencer sequence of Theorem 8 is locally exact if* $\det A \neq 0$.

PROOF. As in the previous Remark, we have

$$\partial_i \chi_{\mu,j}^k - \partial_j \chi_{\mu,i}^k - A_i^r \chi_{\mu+1r,j}^k + A_j^r \chi_{\mu+1r,i}^k + \cdots = 0.$$

Let $\chi_{q+1} \in T^* \otimes J_{q+1}(T)$ be such that $\overline{D}'\chi_{q+1} = 0$, and suppose, by induction, that $\overline{D}'\chi_q = 0 \Leftrightarrow \chi = \overline{D}f_{q+1}$. We have $\pi_q^{q+1}(\chi_{q+1}) = \overline{D}(\pi_{q+1}^{q+2} \circ f_{q+2}) = \pi_q^{q+1}(\overline{D}f_{q+2})$ for any lift f_{q+2} of f_{q+1}. Hence $\chi_{q+1} = \overline{D}f_{q+2} + M_{q+1}$, with $M_{q+1} \in T^* \otimes S_{q+1}T^* \otimes T$. From the previous formula we deduce that $A_i^r M_{\mu+1r,j}^k - A_j^r M_{\mu+1r,i}^k = 0$ with $|\mu| = q$. Since we can set $M_{\mu,i}^k = A_i^r N_{\mu,r}^k$, because $\det A \neq 0$, we obtain $N_{\mu+1,j}^k - N_{\mu+1,i}^k = 0$, i.e. $\delta N_{q+1} = 0$, and we can find $U_{q+2} \in S_{q+2}T^* \otimes T$ such that $N_{q+1} = \delta U_{q+2}$ (see Proposition III.B.3). Hence,

$$\chi_{\mu,\imath}^k = -g_l^k A_i^r f_{\mu+1_r}^l + A_i^r U_{\mu+1_r}^k + \text{terms of order} \leq q + 1$$

when $|\mu| = q+1$. Setting $U_{\mu+1_r}^k = g_l^k V_{\mu+1_r}^l$ and $f_{\mu+1_r}^{l} = f_{\mu+1_r}^l + V_{\mu+1_r}^l$ when $|\mu| = q = 1$, we finally obtain $\chi_{q+1} = \overline{D} f_{q+2}'$. The proof of the Proposition now proceeds by induction. $\square$

Remark 12. In the proof of the Proposition above, the condition $\det A = \det(\partial_r f^k)/\det(f_i^k) \neq 0$ has been used in a crucial way. It reduces to $\det(\partial_i f^k) \neq 0$, because f_{q+1} is a section of Π_{q+1}, and thus $\det(f_i^k) \neq 0$. Since this will be essential for applications in Chapter VIII, we now prove that this requirement is not compatible with the use of connections in mathematical physics. Indeed, we have $\chi_0 = A - \text{id}_T$, and $-\chi_q$ is a connection if and only if $-\chi_0 = \text{id}_T$ or, equivalently, $A = 0$. Since $\det(f_i^k) \neq 0$, the only possibility is that $\partial_i f^k = 0$, so that f^k is a constant map $x \rightarrow x_0$. Accordingly, the condition $\det A \neq 0$ *makes it impossible to use principal bundles or connections!* We should also notice that the condition $A = 0$ is the only possibility for making Definition A.22 (of the curvature of a connection) compatible with the definition of $\overline{D}'$, given that we use $\chi_q' = -\chi_q$ instead of χ_q. Indeed, when $A = 0$ we have

$$
\begin{aligned}
(\overline{D}'\chi_{q+1})(\xi,\eta) &= (D\chi_{q+1})(\xi,\eta) - \{\chi_{q+1}(\xi),\chi_{q+1}(\eta)\} \\
&= (D\chi_{q+1})(\xi,\eta) - ([\chi_q(\xi),\chi_q(\eta)] - i(\chi_0(\xi))D(\chi_{q+1}(\eta)) + \\
&\quad\ i(\chi_0(\eta))D(\chi_{q+1}(\xi))) \\
&= (D\chi_{q+1})(\xi,\eta) - i(\xi)D(\chi_{q+1}(\eta)) + i(\eta)D(\chi_{q+1}(\xi)) - [\chi_q(\xi),\chi_q(\eta)] \\
&= -\chi_q([\xi,\eta]) - [\chi_q(\xi),\chi_q(\eta)],
\end{aligned}
$$

and thus

$$(\overline{D}'\chi_{q+1}')(\xi,\eta) = [\chi_q'(\xi),\chi_q'(\eta)] - \chi_q'([\xi,\eta]).$$

It is essential to notice that $\overline{D}'\chi_{q+1}$ depends only on χ_q if and only if $A = 0$. The reader should check by him/herself that the extreme-order jet terms disappear when $A = 0$.

This remark explains the additional conceptual confusion between 'torsion' in classical gauge theory and its tentative use in general relativity.

Let now $f_{q+1}, g_{q+1} \in \Pi_{q+1}$ be such that $g_{q+1} \circ f_{q+1}$ makes sense. We obtain

$$
\begin{aligned}
\overline{D}(g_{q+1} \circ f_{q+1}) &= (g_{q+1} \circ f_{q+1})^{-1} \circ j_1(g_q \circ f_q) \\
&= f_{q+1}^{-1} \circ g_{q+1}^{-1} \circ j_1(g_q) \circ j_1(f_q) \\
&= f_{q+1}^{-1} \circ \overline{D} g_{q+1} \circ j_1(f_q) + \overline{D} f_{q+1}
\end{aligned}
$$

By restriction we may state:

Definition 13 (Important Definition). For an invertible section $f_{q+1} \in \mathcal{R}_{q+1}$, the transformation

$$\chi_q \rightarrow f_{q+1}^{-1} \circ \chi_q \circ j_1(f_q) + \overline{D} f_{q+1}$$

is called a *gauge transformation*.

In this formula, $T^* \otimes R_q$ is associated with $\mathcal{R}_{q+1,1}$, while R_q is associated with $\mathcal{R}_{q+1}$. Hence, f_{q+1}^{-1} acts on R_q, while $j_1(f_q)$, and thus $j_1(f)$, acts on T^*. It follows that the gauge transformation is invertible if and only if we use invertible sections of $\mathcal{R}_{q+1}$, i.e. sections with nonzero jacobian determinant, as in the previous Remark.

Remark 14. Let $f_{q+1}, f'_{q+1} \in \mathcal{R}_{q+1}$ be such that $\chi_q = \overline{D} f_{q+1} = \overline{D} f'_{q+1}$. We may set $f'_{q+1} = g_{q+1} \circ f_{q+1}$, and obtain $\overline{D} g_{q+1} = 0$, i.e. $g_{q+1} = j_{q+1}(g)$ with $g \in \Gamma$.

We now pass to the infinitesimal point of view. By linearisation at the identity, the first nonlinear Spencer sequence becomes the initial part of the first linear Spencer sequence of Chapter IV.A.

The infinitesimal gauge transformations involve quite new and powerful, though sometimes rather technical, tools.

Passing to the limit in Definition 13 and using the Remark below it, defining the corresponding formal Lie derivative, we obtain

Definition 15. An *infinitesimal gauge transformation* has the form

$$\delta \chi_q = D \xi_{q+1} + L(j_1(\xi_{q+1})) \chi_q.$$

For practical computations we note that

$$\delta \chi_q(\zeta) = i(\zeta) D \xi_{q+1} + L(\xi_{q+1})(\chi_q(\zeta)) - \chi_q([\xi, \zeta]),$$

and it remains to define the formal Lie derivative on $J_q(T)$ for the second term only. Passing to the limit in the formula of Lemma 7, while taking into account that $\eta_q = \eta_q(f(x))$, we obtain

$$L(\xi_{q+1}) \eta_q = \{\xi_{q+1}, \eta_{q+1}\} + i(\xi) D \eta_{q+1}$$
$$= [\xi_q, \eta_q] + i(\eta) D \xi_{q+1}.$$

We leave it to the reader to prove that this definition is coherent with that of formal Lie derivative. Instead of changing the source, we can also change the target, and use the formula before Definition 13 to *pass to the target*:

$$\delta \chi_q = f_{q+1}^{-1} \circ D \eta_{q+1} \circ j_1(f)^{-1}$$

whenever $\chi_q = \overline{D} f_{q+1}$.

The *delicate* point is to relate the source and target points of view. This is done in the following, difficult, Proposition.

Proposition 16. *The same variation* $\delta \chi_q$ *is obtained if* $\eta_{q+1} = f_{q+2}(\xi_{q+1} + \chi_{q+1}(\xi))$ *and* $\chi_{q+1} = \overline{D} f_{q+2}$.

PROOF. Let $f_{q+1}, g_{q+1}, h_{q+1} \in \mathcal{R}_{q+1}$ be such that $g_{q+1} \circ f_{q+1} = f_{q+1} \circ h_{q+1}$ and pass to the limit with $g_{q+1} = \mathrm{id}_{q+1} + t \eta_{q+1} + \dots$ over the target and $h_{q+1} = \mathrm{id}_{q+1} + t \xi_{q+1} +$

... over the source. Using the chain rule for derivatives and substituting jets, we successively obtain

$$\eta^k = \xi^r \partial_r f^k,$$
$$\eta^k_r f^r_i = \xi^r \partial_r f^k_i + f^k_r \xi^r_i,$$

and, more generally,

$$d_\mu \eta^k = \eta^k_r f^r_\mu + \cdots = \xi^i(\partial_i f^k_\mu - f^k_{\mu+1_i}) + f^k_{\mu+1_r}\xi^r + \cdots + f^k_r \xi^r_\mu.$$

From the formula in the proof of Theorem 8 we obtain

$$f^k_r \chi^r_{\mu,i}\xi^i + \cdots + f^k_{\mu+1_r}\chi^r_{,i}\xi^i = \xi^i(\partial_i f^k_\mu - f^k_{\mu+1_i})$$

and by substitution we obtain $\eta_{q+1} = f_{q+2}(\bar{\xi}_{q+1})$, with $\bar{\xi}_{q+1} = \xi_{q+1} + \chi_{q+1}(\xi)$, a result not evident at first sight. $\square$

Corollary 17. $\delta\chi_q = D\bar{\xi}_{q+1} - \{\chi_{q+1}(\cdot), \bar{\xi}_{q+1}\}.$

PROOF.

$$\delta\chi_q(\zeta) = i(\zeta)D\bar{\xi}_{q+1} - \{\chi_{q+1}(\zeta), \bar{\xi}_{q+1}\}$$
$$= i(\zeta)D\xi_{q+1} + i(\zeta)D\chi_{q+1}(\xi) - \{\chi_{q+1}(\zeta), \xi_{q+1}\} - \{\chi_{q+1}(\zeta), \chi_{q+1}(\xi)\}$$
$$= i(\zeta)D\xi_{q+1} + \{\xi_{q+1}, \chi_{q+1}(\zeta)\} + i(\zeta)D\chi_{q+1}(\zeta) - \chi_q([\xi,\zeta]) +$$
$$(D\chi_{q+1})(\zeta,\xi) - \{\chi_{q+1}(\zeta), \chi_{q+1}(\xi)\}$$
$$= i(\zeta)D\xi_{q+1} + L(\xi_{q+1})(\chi_q(\zeta)) - \chi_q([\xi,\zeta]),$$

using the compatibility conditions for χ_{q+1}. Notice that this variational formula is the analog of the one already found in Section A (see Remark A.17, in particular.). $\square$

Remark 18. For future application we list a few explicit variations:

$$\delta\chi^k_{,i} = (\partial_i\xi^k - \xi^k_i) + (\xi^r \partial_r\chi^k_{,i} + \chi^k_{,r}\partial_i\xi^r - \chi^r_{,i}\xi^k_r)$$
$$= (\partial_i\bar{\xi}^k - \bar{\xi}^k_i) + (\chi^k_{r,i}\bar{\xi}^r - \chi^r_{,i}\bar{\xi}^k_r),$$
$$\delta\chi^k_{j,i} = (\partial_i\xi^k_j - \xi^k_{ij}) + (\xi^r \partial_r\chi^k_{j,i} + \chi^k_{j,r}\partial_i\xi^r + \chi^k_{r,i}\xi^r_j - \chi^r_{j,i}\xi^k_r - \chi^r_{,i}\xi^k_{jr})$$
$$= (\partial_i\bar{\xi}^k_j - \bar{\xi}^k_{ij}) + (\chi^k_{rj,i}\bar{\xi}^r + \chi^k_{k,i}\bar{\xi}^r_j - \chi^r_{j,i}\bar{\xi}^k_r - \chi^r_{,i}\bar{\xi}^k_{jr})$$

In Chapter VIII we shall see that these variations are *exactly* the ones looked for, in vain, by H. Weyl in his famous reference [196].

Corollary 19. $J_q(T)$ *is associated with* $\Pi_{q,1}$.

PROOF. The formula

$$\eta_q = f_{q+1}(\xi_q + \chi_q(\xi)) \qquad \text{with } \chi_q = \overline{D}f_{q+1}$$

depends on $j_1(f_q)$ only. If we have

$$\zeta_q = g_{q+1}(\eta_q + \sigma_q(\eta)) \qquad \text{with } \sigma_q = \overline{D}g_{q+1},$$

then Definition 13 successively gives

$$\zeta_q = (g_{q+1} \circ f_{q+1})(\xi_q) + g_{q+1}(\sigma_q(\eta)) + (g_{q+1} \circ f_{q+1})(\chi_q(\xi))$$
$$= (g_{q+1} \circ f_{q+1})(\xi_q + f_{q+1}^{-1}(\sigma_q(\eta)) + \chi_q(\xi))$$
$$= (g_{q+1} \circ f_{q+1})(\xi_q + i(\xi)\overline{D}(g_{q+1} \circ f_{q+1})),$$

because $\eta^k = \xi^r \partial_r f^k$. □

Remark 20. $\xi_q \to \bar{\xi}_q = \xi_q + \chi_q(\xi)$ is an isomorphism if and only if $\xi \to \bar{\xi} = \xi + \chi_0(\xi) = A(\xi)$ is an isomorphism, i.e. if and only if $\det A \neq 0$.

Remark 21. If $\eta_{q+1} = f_{q+2}(\xi_{q+1})$, it follows from Definition 15, Proposition 16 and Corollary 17 that

$$D\eta_{q+1} = f_{q+1} \circ (D\xi_{q+1} - \{\overline{D}f_{q+2}, \xi_{q+1}\}) \circ j_1(f)^{-1}.$$

This formula gives the transformation law of the Spencer operator and is quite useful in practice. In particular, for $\zeta_{q+1} = g_{q+2}(\eta_{q+1})$ we have

$$D\zeta_{q+1} = (g_{q+1} \circ f_{q+1}) \circ (D\xi_{q+1} - \{\overline{D}(g_{q+2} \circ f_{q+2}), \xi_{q+1}\}) \circ j_1(f)^{-1} \circ j_1(g)^{-1}$$
$$= g_{q+1} \circ (D\eta_{q+1} - f_{q+1} \circ \{f_{q+2}^{-1}\overline{D}g_{q+2}, \xi_{q+1}\}) \circ j_1(g)^{-1}$$
$$= g_{q+1} \circ (D\eta_{q+1} - \{\overline{D}g_{q+2}, \eta_{q+1}\}) \circ j_1(g)^{-1}.$$

Therefore we obtain the formula

$$\{f_{q+2}(\xi_{q+1})f_{q+2}(\eta_{q+1})\} = f_{q+1}(\{\xi_{q+1}, \eta_{q+1}\})$$

for the transformation of the algebraic bracket.

We now give a difficult Proposition which is the analog of Proposition A.14 when taking into account Remark A.17.

Proposition 22. If $\sigma_{q-1} = \overline{D}'\chi_q$, then

$$\chi_q' = f_{q+1}^{-1} \circ \chi_q \circ j_1(f) + \overline{D}f_{q+1} \Rightarrow \sigma_{q-1}' = f_q^{-1} \circ \sigma_q \circ j_1(f),$$

where f_q^{-1} acts on $J_{q-1}(T)$ and $j_1(f)$ on $\bigwedge^2 T^*$.

PROOF. Using Proposition 16 and Remark 21, concerning the transformation rule for the Spencer operator, while taking into account that the exterior derivative commutes with arbitrary changes of coordinates, we successively obtain:

$$\sigma'_{q-1} = D\chi'_q - \{\chi'_q, \chi'_q\}$$
$$= D\overline{D}f_{q+1} - \{\overline{D}f_{q+1}, \overline{D}f_{q+1}\} - \{f_{q+1}^{-1} \circ \chi_q \circ j_1(f), f_{q+1}^{-1} \circ \chi_q \circ j_1(f)\}$$
$$+ D(f_{q+1}^{-1} \circ \chi_q \circ j_1(f)) - \{f_{q+1}^{-1} \circ \chi_q \circ j_1(f), \overline{D}f_{q+1}\} - \{\overline{D}f_{q+1}, f_{q+1}^{-1} \circ \chi_q \circ j_1(f)\}$$
$$= D(f_{q+1}^{-1} \circ \chi_q \circ j_1(f)) + f_q^{-1} \circ (\{\chi_q, \overline{D}f_{q+1}^{-1}\} + \{\overline{D}f_{q+1}^{-1}, \chi_q\}) \circ j_1(f)$$
$$- f_q^{-1} \circ \{\chi_q, \chi_q\} \circ j_1(f)$$
$$= f_q^{-1} \circ D\chi_q \circ j_1(f) - f_q^{-1} \circ \{\chi_q, \chi_q\} \circ j_1(f)$$
$$= f_q^{-1} \circ (D\chi_q - \{\chi_q, \chi_q\}) \circ j_1(f)$$
$$= f_q^{-1} \circ \sigma_{q-1} \circ j_1(f).$$

The essential step has been to extend the transformation formula for $D: J_{q+1}(T) \to T^* \otimes J_q(T)$ to $D: T^* \otimes J_q(T) \to \wedge^2 T^* \otimes J_{q-1}(T)$. $\square$

As a matter of curiosity, we now give the analog of Proposition A.16.

Proposition 23. $[\delta_{\xi_{q+1}}, \delta_{\eta_{q+1}}] = \delta_{[\xi_{q+1}, \eta_{q+1}]}.$

PROOF. Since we will not use this Proposition, we will only give some hints for proving it, [156]. In fact, we first have to consider

$$\chi_q \to \chi'_q = \chi_q + \delta_{\eta_{q+1}}\chi_q = \chi_q + D\eta_{q+1} + L(j_1(\eta_{q+1}))\chi_q,$$

and then

$$\chi'_q \to \chi'_q + \delta_{\xi_{q+1}}\chi'_q = \chi'_q + D\xi_{q+1} + L(j_1(\xi_{q+1}))\chi'_q,$$

i.e. finally

$$\chi_q \to \chi_q + D\xi_{q+1} + D\eta_{q+1} + L(j_1(\xi_{q+1}))\chi_q + L(j_1(\eta_{q+1}))\chi_q$$
$$+ L(j_1(\xi_{q+1}))(D\eta_{q+1} + L(j_1(\eta_{q+1}))\chi_q).$$

Exchanging ξ and η and subtracting the resulting equations, we find

$$L(j_1(\xi_{q+1}))(D\eta_{q+1} + L(j_1(\eta_{q+1}))\chi_q) - (\xi \leftrightarrow \eta).$$

By some manipulation, using the formula

$$i(\zeta)D\{\xi_{q+1}, \eta_{q+1}\} = \{i(\zeta)D\xi_{q+1}, \eta_q\} + \{\xi_q, i(\zeta)D\eta_{q+1}\}$$

and the definition of formal Lie derivative (below Definition 15), we finally obtain

$$D[\xi_{q+1}, \eta_{q+1}] + L(j_1([\xi_{q+1}, \eta_{q+1}]))\chi_q,$$

i.e.

$$\delta_{[\xi_{q+1}, \eta_{q+1}]}\chi_q. \qquad \square$$

We end our comments concerning the nonlinear Spencer sequence with the following Proposition, which is the analog of the Bianchi identities in Section A and whose proof is left to the reader as an exercise, since we will not use this Proposition. (In the formula, the notation $\mathcal{C}(\ )$ denotes the cyclic sum.)

Proposition 24. *The following so-called Bianchi identities hold:*

$$D\sigma_{q-1}(\xi,\eta,\zeta) + C(\xi,\eta,\zeta)\{\sigma_{q-1}(\xi,\eta),\chi_{q-1}(\zeta)\} \equiv 0, \qquad \forall \xi,\eta,\zeta \in T.$$

Remark 25. As in Remark 10 we stress the confusion in gauge theory between the classical Bianchi identities for the Riemann curvature in tensor analysis and the Bianchi identities above. Indeed, in the linearised framework the classical Bianchi identities correspond to the operator $\mathcal{D}_2$ in the Janet sequence, while the gauge Bianchi identities above correspond to the operator D_3 in the Spencer sequence. As a byproduct, any attempt at identification is useless again, and we see once more that the above identities do not provide an operator for σ_{q-1}, since they explicitly contain χ_{q-1}.

As in Chapter IV.A we can prove that the first nonlinear Spencer sequence is not the 'best' sequence, since $\overline{D}$ and $\overline{D}'$ are not formally integrable, and we have to introduce a second nonlinear Spencer sequence. Since the matter is even more delicate and computational than for the first sequence, we will give some hints only and quote some basic results. These are given in full detail in [146]. The key idea is to notice that the transition from the first to the second nonlinear Spencer sequence depends in an essential manner on the condition $\det A \neq 0$. We set $B = A^{-1}$.

Setting $\chi_{\mu,\imath}^k(x) = A_\imath^r \tau_{\mu,r}^k(x)$, we find, from the proof of Proposition 11:

$$\tau_{\mu,\imath}^k = -g_\imath^k f_{\mu+1_\imath}^l + \text{terms of order} \leq |\mu|.$$

Hence, modulo the isomorphism of T^* induced by A, the operator $\overline{D}: \Pi_{q+1} \to T^* \otimes J_q(T)$ induces an operator $\overline{D}_1: \Pi_q \to C_1(T)$. Introducing the operator $\overline{D}_2: T^* \otimes J_{q+1}(T) \to \wedge^2 T^* \otimes J_q(T)$ by the formula

$$\overline{D}_2\tau_{q+1}(\xi,\eta) = D\tau_{q+1}(\xi,\eta) - [\tau_q(\xi),\tau_q(\eta)] + \tau_q(A([B(\xi),B(\eta)]) - [\xi,\eta]),$$

we notice that the components of strict order $q+1$ appear only in $D\tau_{q+1}$. Therefore $\overline{D}_2$ induces an operator $\overline{D}_2: C_1(T) \to C_2(T)$. Since $\overline{D}_2$ is induced by $\overline{D}'$ modulo the 'twist' of T^* and $\wedge^2 T^*$ by A, we finally obtain:

Theorem 26. *There is a nonlinear differential sequence*

$$0 \longrightarrow \text{aut}(X) \xrightarrow{j_q} \Pi_q \xrightarrow{\overline{D}_1} C_1(T) \xrightarrow{\overline{D}_2} C_2(T),$$

called the second nonlinear Spencer sequence, in which $\overline{D}_1$ and $\overline{D}_2$ are involutive.

Corollary 27. *There is a restricted second nonlinear Spencer sequence:*

$$0 \longrightarrow \Gamma \xrightarrow{j_q} \mathcal{R}_q \xrightarrow{\overline{D}_1} C_1 \xrightarrow{\overline{D}_2} C_2,$$

where $\overline{D}_1$ and $\overline{D}_2$ are involutive whenever $\mathcal{R}_q$ is involutive.

It remains to establish a link between the nonlinear Janet sequence and the nonlinear Spencer sequences. Of course, we can easily understand that, in the nonlinear framework, the diagram produced in Theorem IV.A.16 does not exist anymore. Solving this dilemma will surely constitute the hardest, though most striking, result of this Section.

Let $\overline{\omega}$ be a section of $\mathcal{F}$ satisfying the same compatibility/integrability conditions $\mathcal{B}_1 \subset J_1(\mathcal{F})$ as ω. Since Π_q acts transitively in the groupoid sense on $\mathcal{F}$, which is a quotient of Π_q, we can find $f_q \in \Pi_q$ such that

$$\Phi \circ f_q = f_q^{-1}(\omega) = \overline{\omega}.$$

Since $\mathcal{F}$ is a natural bundle of order q, $J_1(\mathcal{F})$ is a natural bundle of order $q+1$. Thus, we can find $f_{q+1} \in \Pi_{q+1}$ over f_q such that

$$\rho_1(\Phi) \circ f_{q+1} = f_{q+1}^{-1}(j_1(\omega)) = j_1(\overline{\omega}),$$

although we also have

$$\rho_1(\Phi) \circ j_1(f_q) = j_1(f_q)^{-1}(j_1(\omega)) = j_1(\overline{\omega}).$$

In the last two equations we can eliminate either ω or $\overline{\omega}$.

- *Elimination of ω gives* $f_{q+1}(j_1(\overline{\omega})) = j_1(f_q)(j_1(\overline{\omega}))$.
- *Elimination of $\overline{\omega}$ gives* $f_{q+1}^{-1}(j_1(\omega)) = j_1(f_q)^{-1}(j_1(\omega))$.

We examine these two situations in succession. From the first we deduce

$$(f_{q+1}^{-1} \circ j_1(f_q))j_1(\overline{\omega}) - j_1(\overline{\omega}) = -(f_{q+1}^{-1} \circ j_1(f_q))[(f_{q+1}^{-1} \circ j_1(f_q))^{-1}(j_1(\overline{\omega})) - j_1(\overline{\omega})]$$
$$= -(f_{q+1}^{-1} \circ j_1(f_q))L(\chi_q)\overline{\omega},$$

because $f_{q+1}^{-1} \circ j_1(f_q) = \mathrm{id}_{q+1} + \chi_q$. Now we have $L(\chi_q)\overline{\omega} \in T^* \otimes F_0$, and in local coordinates we obtain *over the source:*

$$\overline{\Omega}_i^\tau \equiv -L_k^{\tau\mu}(\overline{\omega}(x))\chi_{\mu,i}^k + \chi_{,i}^\tau \partial_\tau \overline{\omega}^\tau(x)$$

with $\chi_q = \overline{D}f_{q+1}$.

However, from the proof of Proposition C.5 we know that F_0 is associated with $\mathcal{R}_q$ in the transitive case. Thus, F_0 is not affected by $f_{q+1}^{-1} \circ j_1(f_q)$, which projects onto $f_q^{-1} \circ f_q = \mathrm{id}_q$. Hence, only T^* is affected by $f_1^{-1} \circ j_1(f) = A$, and *over the source* we obtain

$$(f_{q+1}^{-1} \circ j_1(f_q))(j_1(\overline{\omega})) - j_1(\overline{\omega}) \equiv -BL(\chi_q)\overline{\omega} = 0,$$

because T^* is a covariant tensor bundle. It follows that $\chi_q \in T^* \otimes R_q(\overline{\omega})$ and $\overline{D}'\chi_q = 0$ in the first nonlinear Spencer sequence for $\overline{\omega}$.

From the second relation we obtain, *over the target:*

$$(f_{q+1}^{-1} \circ j_1(f_q))^{-1}(j_1(\omega)) - j_1(\omega) \equiv L(\sigma_q)\omega = 0,$$

with $\sigma_q = \overline{D}f_{q+1}^{-1}$. At this point we notice that

$$f_{q+1} \circ f_{q+1}^{-1} = \mathrm{id}_{q+1} \Rightarrow f_{q+1} \circ \overline{D}f_{q+1} \circ j_1(f)^{-1} + \overline{D}f_{q+1}^{-1} = 0,$$

and thus $\sigma_q = -f_{q+1} \circ \chi_q \circ j_1(f)^{-1}$. It follows that $\sigma_q \in T^* \otimes R_q(\omega)$ and $\overline{D}'\sigma_q = 0$ in the first nonlinear Spencer sequence for ω.

Thus, in the nonlinear Janet sequence for ω we have related cocycles at $\mathcal{F}$ with cocycles at $T^* \otimes R_q(\overline{\omega})$ in the first nonlinear Spencer sequence for $\overline{\omega}$ *over the source/* or with cocycles at $T^* \otimes R_q(\omega)$ in the first nonlinear Spencer sequence for ω *over the target*. It is essential to notice the transition from the source to the target in the latter correspondence.

Finally, let $f_{q+1}, f'_{q+1} \in \Pi_{q+1}$ be such that $f_{q+1}^{-1}(j_1(\omega)) = f'^{-1}_{q+1}(j_1(\omega)) = j_1(\overline{\omega})$. From this equality we deduce

$$(f_{q+1}^{-1} \circ f'_{q+1})(j_1(\overline{\omega})) = j_1(\overline{\omega}),$$

and thus we can find $g_{q+1} \in \mathcal{R}_{q+1}$ such that $f'_{q+1} = f_{q+1} \circ g_{q+1}$. The new $\chi'_q = \overline{D} f'_{q+1}$ thus obtained differs from the original $\chi_q = \overline{D} f_{q+1}$ by a gauge transformation. However, we can also deduce that

$$(f'_{q+1} \circ f_{q+1}^{-1})(j_1(\omega)) = j_1(\omega),$$

and therefore we can find $g_{q+1} \in \mathcal{R}_{q+1}(j_1(\omega))$ such that

$$f'_{q+1} = g_{q+1} \circ f_{q+1} \Rightarrow f'^{-1}_{q+1} = f_{q+1}^{-1} \circ g_{q+1}^{-1}.$$

Hence, the new $\sigma'_q = \overline{D} f'^{-1}_{q+1}$ thus obtained again differs from the original $\sigma_q = \overline{D} f_{q+1}^{-1}$ by a gauge transformation.

Conversely, let $f_{q+1}, f'_{q+1} \in \Pi_{q=1}$ be such that $\overline{D} f'^{-1}_{q+1} = \overline{D} f_{q+1}^{-1}$. This implies $\overline{D}(f_{q+1}^{-1} \circ f'_{q+1}) = 0$. Proposition 11 now implies that there is a $g \in \mathrm{aut}(X)$ such that $f'_{q+1} = f_{q+1} \circ j_{q+1}(g)$, and we obtain

$$\overline{\omega}' = f'^{-1}_q(\omega) = (f_q \circ j_q(g))^{-1}(\omega) = j_q(g)^{-1}(f_q^{-1}(\omega)) = j_q(g)^{-1}(\overline{\omega}).$$

We summarise the results above in the following, important, Theorem.

Theorem 28. *Natural transformations of $\mathcal{F}$ in the nonlinear Janet sequence correspond to gauge transformations at $T^* \otimes R_q$ in the first nonlinear Spencer sequence.*

These results can be extended to the second nonlinear Spencer sequence by introducing gauge transformations at C_1 which only depend on invertible sections of $\mathcal{R}_q$. The proof, however, is much more delicate [146].

Remark 29 (Important Remark). Besides the previous Theorem, there is another fundamental distinction between the Janet sequence and the Spencer sequence that will prove to be crucial in applications. Indeed, the *larger* Γ, the *smaller* the number of differential invariants at order q in the Janet sequence, but the *larger* the dimensions of the bundles appearing in the Spencer sequence. This fact suggests a specific role of Spencer sequences in mathematical physics.

Remark 30. When $q = 1$, the differential invariants are functions of $j_1(f)$, and thus of $j_1(f_1)$ with $f_1 \in \Pi_1$, that are invariant under the action of Γ at the target. However, if $f_{q+1}, f'_{q+1} \in \mathcal{R}_{q+1}$ are such that $\overline{D} f'_{q+1} = \overline{D} f_{q+1}$ if and only if $f'_{q+1} = j_{q+1}(g) \circ f_{q+1}$ with $g \in \Gamma$, it follows that the differential invariants of order one are functions of χ_0, and thus of Λ.

We conclude this Section, like the preceding ones, by examining the results in Section A for Lie groups of transformations in the new framework, under the assumption that the action is transitive.

Surprisingly, there does no seem to exist any general behavior of the nonlinear Janet sequence, since $\mathcal{F}$ can be a natural bundle of high order q and there is no 'relation' between the 'structure constants' of $\mathcal{G}$ and the 'structure constants' appearing in the integrability conditions for the general section.

Things are quite different for nonlinear Spencer sequences. First of all, since $\mathcal{R}_{q+1} \simeq \mathcal{R}_q$ if q is large enough, and $M_{q+1} = 0$, we have $C_r = \wedge^r T^* \otimes \mathcal{R}_q$, and we need not distinguish between the two sequences.

The following, rather striking, Theorem has only local meaning, and allows us to establish a link between the *nonlinear gauge sequence* of Section A and the *nonlinear Spencer sequence* of Section D. It will also clarify the existing confusion between 'classical curvature', originating from the nonlinear Janet sequence, and 'gauge curvature', originating from the nonlinear Spencer sequence.

Theorem 31. *There is a commutative and exact diagram*

$$
\begin{array}{ccccc}
X \times G & \longrightarrow & T^* \otimes \mathcal{G} & \xrightarrow{\ MC\ } & \wedge^2 T^* \otimes \mathcal{G} \\
\Big\downarrow & & \Big\downarrow & & \Big\downarrow \\
0 \longrightarrow \Gamma & \xrightarrow{\ j_q\ } & \mathcal{R}_q & \xrightarrow{\ \overline{D}\ } T^* \otimes R_q & \xrightarrow{\ \overline{D}'\ } \wedge^2 T^* \otimes R_q
\end{array}
$$

where the vertical arrows are isomorphisms.

PROOF. With $b = a^{-1}$ and $y = f(x, a)$ we have

$$x = f(y, b), \qquad x = g(y, a), \qquad y = g(x, b),$$

and thus $x \equiv f(g(x, b), b)$. This leads to

$$\frac{\partial x}{\partial y} \frac{\partial g}{\partial b} = -\frac{\partial f}{\partial b} = -\xi(x)\omega(b),$$

by the First Fundamental Theorem of Section A. The correspondence between the section a of $X \times G$ and the section f_q of $\mathcal{R}_q$ is described by the formula:

$$f_\mu^k(x) = \partial_\mu f^k(x, a(x)).$$

Therefore we obtain, in succession:

$$
\begin{aligned}
\partial_i f_\mu^k - f_{\mu+1_i}^k &= d_i(\partial_\mu f^k(x, a(x))) - \partial_{\mu+1_i} f^k(x, a(x)) \\
&= d_i(\partial_\mu g^k(x, b(x))) - \partial_{\mu+1_i} g^k(x, b(x)) \\
&= \partial_\mu \left(\frac{\partial g^k}{\partial b^\sigma} \right) \partial_i b^\sigma(x) \\
&= -\partial_\mu \left(\frac{\partial f^k(x, a(x))}{\partial x^r} \xi_\tau^r(x) \right) \omega_\sigma^\tau(b(x)) \partial_i b^\sigma(x).
\end{aligned}
$$

Thus,

$$\chi_{\mu,i}^k(x) = -\partial_\mu \xi_\tau^k(x) \omega_\sigma^\tau(b(x)) \partial_i b^\sigma(x),$$

by Lemma 7 and Theorem 8.

The commutativity of the left square thus follows from Proposition B.15 and Remark A.17, by noting that

$$ab = e \ \Rightarrow\ -b^{-1}db = -adb = dab = daa^{-1}.$$

Regarding the commutativity of the right square, we have

$$\chi^k_{\mu,\imath}(x) = A^\tau_\imath(x)\partial_\mu \xi^k_\tau(x),$$

and we successively obtain

$$\partial_\imath \chi^k_{\mu,\jmath} - \partial_\jmath \chi^k_{\mu,\imath} - \chi^k_{\mu+1,\jmath} + \chi^k_{\mu+1,\imath} = (\partial_\imath A^\tau_\jmath - \partial_\jmath A^\tau_\imath)\partial_\mu \xi^k_\tau,$$

$$(\{\chi_{q+1}(\partial_\imath), \chi_{q+1}(\partial_\jmath)\})^k_\mu = A^\rho_\imath A^\sigma_\jmath \partial_\mu([\xi_\rho, \xi_\sigma])^k = c^\tau_{\rho\sigma} A^\rho_\imath A^\sigma_\jmath \partial_\mu \xi^k_\tau.$$

The commutativity of the right square now follows from the definition of $\overline{D}'$ and the sign condition provided by Remark A.17.

It follows that both sequences are locally exact. □

Remark 32 (Important Remark). *By a 'miracle', the upper sequence no longer depends on the action.*

However, though the proof of this fact may seem simple, the reader may have noticed that it depends on many difficult results in this Chapter.

E. Examples

In this Section we give a few explicit examples, starting from low-dimensional simple cases and ending with complicated cases of arbitrary dimension. As a rule, we have selected examples in such a way that each sheds some light on a delicate theoretical point, or on the confusion sketched in the Introduction and culminating in Theorems D.1, D.8, D.26, D.28, D.31 or Remarks D.6, D.32.

Example 1. $X = \mathbb{R}$, $\dim G = 2$.

We consider the affine Lie group of transformations

$$\Gamma \qquad y = a^1 x + a^2.$$

Inverting and setting $a = (a^1, a^2)$, $b = a^{-1}$, we get

$$x = \frac{1}{a^1}y - \frac{a^2}{a^1} \Rightarrow b^1 = \frac{1}{a^1}, \quad b^2 = -\frac{a^2}{a^1},$$

$$\frac{\partial y}{\partial a^1} = x = \frac{1}{a^1}y - \frac{a^2}{a^1}, \qquad \frac{\partial y}{\partial a^2} = 1.$$

Thus, $\xi_1 = \partial/\partial x$, $\xi_2 = x\partial/\partial x$ are the infinitesimal generators over the source, and the Maurer–Cartan forms are:

$$\begin{cases} w^1 = -\dfrac{a^2}{a^1}\, da^1 + da^2 = -\dfrac{db^2}{b^1}, \\[2mm] w^2 = \dfrac{1}{a^1}\, da^1 = -\dfrac{1}{b^1}\, db^1. \end{cases}$$

On one hand we have $[\xi_1, \xi_2] = \xi_1$, i.e. $c^1_{12} = 1$, $c^2_{12} = 0$, and we can check that

$$dw^1 + w^1 \wedge w^2 = 0, \qquad dw^2 = 0.$$

Now we notice that the action is transitive and effective, and consider this Lie group of transformations as a Lie pseudogroup Γ. We obtain, in succession:

$$y = a^1 x + a^2, \qquad y_x = a^1, \qquad y_{xx} = 0.$$

Hence, the defining system is

$$\mathcal{R}_2 \qquad y_{xx} = 0 \qquad (y_x \neq 0).$$

The corresponding Lie algebroid is defined by

$$R_2 \qquad \xi_{xx} = 0.$$

Looking for differential invariants, we have

$$\overline{y} = a^1 y + a^2 \Rightarrow \overline{y}_x = a^1 y_x, \quad \overline{y}_{xx} = a^1 y_{xx}.$$

Hence, the Lie form is

$$\Phi(y_2) \equiv \frac{y_{xx}}{y_x} = 0.$$

Regarding the natural bundle $\mathcal{F}$ we notice that

$$y_x = y_{\overline{x}} \partial_x \varphi, \qquad y_{xx} = y_{\overline{x}\overline{x}} (\partial_x \varphi)^2 + y_{\overline{x}} \partial_{xx} \varphi.$$

It follows that the transition laws of $\mathcal{F}$ are:

$$\begin{cases} u = \overline{u} \partial_x \varphi + \dfrac{\partial_{xx} \varphi}{\partial_x \varphi}, \\ \overline{x} = \varphi(x). \end{cases}$$

The infinitesimal laws are generated by the vector fields

$$\xi(x) \frac{\partial}{\partial x} - (\partial_{xx} \xi(x) + \partial_x \xi(x) u) \frac{\partial}{\partial u}.$$

Accordingly, the general Lie equations are

$$\mathcal{R}_2(\omega) \qquad \frac{y_{xx}}{y_x} + \omega(y) y_x = \omega(x),$$

$$R_2(\omega) \qquad \xi_{xx} + \omega(x) \xi_x + \xi \partial_x \omega(x) = 0,$$

with the special section $\omega(x) = 0$.

Since $\mathcal{R}_2 \simeq \Pi_1$, $R_2 \simeq J_1(T)$, an affine connection may be defined as a section of $T^* \otimes J_1(T)$ projecting onto $\mathrm{id}_T \subset T^* \otimes T$.

The association law of $J_1(T)$ is easily seen to be defined by the following formulas:

$$\begin{cases} \eta(f(x)) = f_x(x) \xi(x), \\ \eta_y(f(x)) f_x(x) = f_x(x) \xi_x(x) + f_{xx}(x) \xi(x). \end{cases}$$

Accordingly, sections of $T^* \otimes J_1(T)$ transform along the following laws:

$$\begin{cases} \eta_{,y}(f(x)) f_x(x) = f_x(x) \xi_{,x}(x), \\ \eta_{y,y}(f(x))(f_x(x))^2 = f_x(x) \xi_{x,x}(x) + f_{xx}(x) \xi_{,x}(x), \end{cases}$$

i.e.

$$\eta_{,y}(f(x)) = \xi_{,x}(x), \qquad \eta_{y,y}(f(x)) = \frac{1}{f_x(x)}\xi_{x,x}(x) + \frac{f_{xx}(x)}{(f(x))^2}\xi_{,x}(x).$$

Setting $\eta_{,y} = \xi_{,x} = 1$, we obtain

$$\eta_{y,y}(f(x)) = \frac{1}{f_x(x)}\xi_{x,x}(x) + \frac{f_{xx}(x)}{(f_x(x))^2},$$

i.e. *exactly* the transition rules of $\mathcal{F}$, up to sign. *This is sheer coincidence* and has led to wrong models in mathematical physics. Indeed, in general a connection is a section of $T^* \otimes R_q$ projecting onto $\mathrm{id}_T \in T^* \otimes T$, and it is thus almost never a geometric object.

The correspondence between the sections of $X \times G$ and those of $\mathcal{R}_2$ is as follows:

$$f(x) \equiv a^1(x)x + a^2(x), \qquad f_x(x) \equiv a^1(x), \qquad f_{xx}(x) = 0.$$

Conversely,

$$a^1(x) = f_x(x), \qquad a^2(x) = f(x) - xf_x(x).$$

Regarding the definition of $\overline{D}$, we obtain

$$\chi_{,x} = \frac{\partial_x f(x)}{f_x(x)} - 1 = \frac{\partial_x a^1(x)}{a^1(x)}x + \frac{\partial_x a^2(x)}{a^1(x)}, \qquad \chi_{x,x} = \frac{\partial_x f_x(x)}{f_x(x)} = \frac{\partial_x a^1(x)}{a^1(x)}.$$

Example 2. In a completely similar way we shall now treat the projective Lie group of transformations

$$\Gamma \qquad y = \frac{a^1x + a^2}{a^3x + a^4} \Leftrightarrow x = \frac{a^2 - a^4y}{a^3y - a^1}.$$

We first treat it as a group with 4 parameters acting in a noneffective way, and we show that the corresponding infinitesimal generators are linearly independent:

$$\begin{cases} \dfrac{\partial y}{\partial a^1} = \dfrac{x}{a^3x + a^4} = \dfrac{a^2 - a^4y}{a^2a^3 - a^1a^4}, \\[2mm] \dfrac{\partial y}{\partial a^2} = \dfrac{1}{a^3x + a^4} = \dfrac{a^3y - a^1}{a^2a^3 - a^1a^4}, \\[2mm] \dfrac{\partial y}{\partial a^3} = -\dfrac{(a^1x + a^2)x}{(a^3x + a^4)^2} = \dfrac{a^4y^2 - a^2y}{a^2a^3 - a^1a^4}, \\[2mm] \dfrac{\partial y}{\partial a^4} = -\dfrac{(a^1x + a^2)x}{(a^3x + a^4)^2} = \dfrac{a^3y^2 - a^1y}{a^2a^3 - a^1a^4}. \end{cases}$$

Therefore we have 3 linearly independent infinitesimal generators in the source:

$$\xi_1 = \frac{\partial}{\partial x}, \qquad \xi_2 = x\frac{\partial}{\partial x}, \qquad \xi_3 = x^2\frac{\partial}{\partial x},$$

$$[\xi_1, \xi_2] = \xi_1, \qquad [\xi_1, \xi_3] = 2\xi_2, \qquad [\xi_2, \xi_3] = \xi_3,$$

and the structure constants are:

$$\begin{cases} c_{12}^1 = 1, & c_{12}^2 = 0, & c_{12}^3 = 0, \\ c_{13}^1 = 0, & c_{13}^2 = 2, & c_{13}^3 = 0, \\ c_{23}^1 = 0, & c_{23}^2 = 0, & c_{23}^3 = 1. \end{cases}$$

We do not give the corresponding Maurer–Cartan forms, as we shall not need them.
We obtain, in succession,

$$\begin{cases} y = \dfrac{a^1 x + a^2}{a^3 x + a^4}, & y_x = \dfrac{a^1 a^4 - a^2 a^3}{(a^3 x + a^4)^2}, \\ y_{xx} = \dfrac{-2a^3(a^1 a^4 - a^2 a^3)}{(a^3 x + a^4)^3}, & y_{xxx} = \dfrac{6(a^3)^2(a^1 a^4 - a^2 a^3)}{(a^3 x + a^4)^4}. \end{cases}$$

The system defining the Lie groupoid is thus

$$R_3 \qquad \frac{y_{xxx}}{y_x} - \frac{3}{2}\left(\frac{y_{xx}}{y_x}\right)^2 = 0.$$

The corresponding Lie algebroid is defined by

$$R_3 \qquad \xi_{xxx} = 0.$$

We can readily verify that ξ_1, ξ_2, ξ_3 are 3 linearly independent solutions of R_3. Let us prove that any solution is a linear combination with constant coefficients. To this end we consider the section

$$\begin{cases} \xi(x) = & \lambda^1(x) + x\lambda^2(x) + x^2\lambda^3(x), \\ \xi_x(x) = & \lambda^2(x) + 2x\lambda^3(x), \\ \xi_{xx}(x) = & 2\lambda^3(x), \\ \xi_{xxx}(x) = & 0. \end{cases}$$

We know that $\xi(x)$ is a solution if and only if the following components of the Spencer sequence vanish:

$$\partial_x\xi(x) - \xi_x(x) = 0, \qquad \partial_x\xi_x(x) - \xi_{xx}(x) = 0, \qquad \partial_x\xi_{xx}(x) = 0.$$

We are led to three equations in three unknowns:

$$\begin{cases} \partial_x\lambda^1 + x\partial_x\lambda^2 + x^2\partial_x\lambda^3 = 0, \\ \partial_x\lambda^2 + 2x\partial_x\lambda^3 = 0, \\ 2\partial_x\lambda^3 = 0. \end{cases}$$

We deduce, in succession, $\partial_x\lambda^3 = 0, \partial_x\lambda^2 = 0, \partial_x\lambda^1 = 0$. We leave it to the reader to compare this result with the direct substitution of $\xi(x)$ as a solution.

Looking for the differential invariants under the target transformation

$$\bar{y} = \frac{a^1 y + a^2}{a^3 y + a^4} \Rightarrow \bar{y}_x = \frac{(a^1 a^4 - a^2 a^3) y_x}{(a^3 y + a^4)^2}, \ldots,$$

we discover, after a straightforward but painful substitution, that the Lie form is

$$\Phi(y_3) \equiv \frac{y_{xxx}}{y_x} - \frac{3}{2}\left(\frac{y_{xx}}{y_x}\right)^2 = 0.$$

We now construct a basis for $\sharp(\mathbb{R}^3)$, and check that Φ is the only invariant (up to a functorial change):

$$\sharp(\eta_3) \equiv \eta(y)\frac{\partial}{\partial y} + \eta_y(y)y_x\frac{\partial}{\partial y_x} + (\eta_{yy}(y)(y_x)^2 + \eta_y(y)y_{xx})\frac{\partial}{\partial y_{xx}}$$

$$+ (\eta_{yyy}(y)(y_x)^3 + 3\eta_{yy}(y)y_x y_{xx} + \eta_y(y)y_{xxx})\frac{\partial}{\partial y_{xxx}}.$$

Taking into account that $\eta_{yyy}(y) = 0$ if $\eta_3 \in R_3$ over the target, we can decompose this distribution with respect to the basis $(1,0,0),(0,1,0),(0,0,1)$, and we obtain the three linearly independent vector fields

$$\frac{\partial}{\partial y}, \quad y_x\frac{\partial}{\partial y_x} + y_{xx}\frac{\partial}{\partial y_{xx}} + y_{xxx}\frac{\partial}{\partial y_{xxx}}, \quad (y_x)^2\frac{\partial}{\partial y_{xx}} + 3y_x y_{xx}\frac{\partial}{\partial y_{xxx}}.$$

Since we have 4 independent variables $(y, y_x, y_{xx}, y_{xxx})$ we can have $4-3 = 1$ invariant, and we can directly check that the given one is quite convenient.

The reader will have noticed that we have not used a basis made up from the infinitesimal generators, i.e. *we need not integrate* R_3.

Using now the prolonged transformation

$$y_{xxx} = y_{\overline{x}\overline{x}\overline{x}}(\partial_x\varphi)^3 + 3y_{\overline{x}\overline{x}}\partial_x\varphi\partial_{xx}\varphi + y_{\overline{x}}\partial_{xxx}\varphi,$$

we now obtain a natural bundle $\mathcal{F}$ of order 3, with transition laws

$$\begin{cases} u = \overline{u}(\partial_x\varphi)^2 + \dfrac{\partial_{xxx}\varphi}{\partial_x\varphi} - \dfrac{3}{2}\left(\dfrac{\partial_{xx}\varphi}{\partial_x\varphi}\right)^2, \\ \overline{x} = \varphi(x). \end{cases}$$

The infinitesimal transition laws are generated by the vector field

$$\xi(x)\frac{\partial}{\partial x} - (2u\partial_x\xi(x) + \partial_{xxx}\xi(x))\frac{\partial}{\partial u}.$$

Accordingly, the general Lie equations are:

$$R_3(\omega) \quad \frac{y_{xxx}}{y_x} - \frac{3}{2}\left(\frac{y_{xx}}{y_x}\right)^2 + \omega(y)y_x^2 = \omega(x),$$

$$R_3(\omega) \quad \xi_{xxx} + 2\omega(x)\xi_x + \xi\partial_x\omega(x) = 0.$$

To study connections we only have to notice that $R_3 \simeq \Pi_2$, $R_3 \simeq J_2(T)$, and use the formulas of the previous Exercise after prolongation at order 3. We finally obtain:

$$\begin{cases} \eta_{y,y} = \dfrac{1}{f_x(x)}\xi_{x,x} + \dfrac{f_{xx}(x)}{(f_x(x))^2}, \\ \eta_{yy,y} = \dfrac{1}{(f_x(x))^2}\xi_{xx,x} + \dfrac{f_{xx}(x)}{(f_x(x))^3}\xi_{x,x} + \dfrac{f_{xxx}(x)}{(f_x(x))^3} - \dfrac{(f_{xx}(x))^2}{(f_x(x))^4}. \end{cases}$$

The first formula permutes the first-order components, while the second formula mixes all components. Therefore it is *sheer coincidence* that by setting

$$-\xi_{x,x} = u, \qquad -\xi_{xx,x} + \frac{1}{2}(\xi_{x,x})^2 = v$$

the above transformation laws *split* into

$$\begin{cases} f_x(x)\overline{u} = u - \dfrac{f_{xx}(x)}{f_x(x)}, \\[2mm] (f_x(x))^2\overline{v} = v - \left(\dfrac{f_{xxx}(x)}{f_x(x)} - \dfrac{3}{2}\left(\dfrac{f_{xx}(x)}{f_x(x)} \right)^2 \right), \end{cases}$$

which are *precisely* the transition rules of the natural bundles of affine and projective structures.

We stress once more that a connection is far from being a geometric object, since it is related to the Spencer sequence and *not* to the Janet sequence. This confusion arose, in the first place, with the Christoffel symbols, which are, as we shall see later, at the same time a geometric object *and* a connection.

Another confusion is related to the following computation. Let us prolongate the transition laws of the second-order affine geometric object:

$$J_1(\mathcal{F}) \quad \begin{cases} (\partial_x\varphi)^2\overline{u}_{\overline{x}} + \partial_{xx}\varphi\overline{u} = u_x - \dfrac{\partial_{xxx}\varphi}{\partial_x\varphi} + \left(\dfrac{\partial_{xx}\varphi}{\partial_x\varphi} \right)^2, \\[2mm] \mathcal{F} \quad \begin{cases} \partial_x\varphi\overline{u} = u - \dfrac{\partial_{xx}\varphi}{\partial_x\varphi}, \\[2mm] \overline{x} = \varphi(x). \end{cases} \end{cases}$$

Of course, the upper formula mixes u and u_x. However, setting $v = u_x - \frac{1}{2}u^2$ we obtain

$$(\partial_x\varphi)^2\overline{v} = v - \left(\dfrac{\partial_{xxx}\varphi}{\partial_x\varphi} - \dfrac{3}{2}\left(\dfrac{\partial_{xx}\varphi}{\partial_x\varphi} \right)^2 \right),$$

and the formulas for $J_1(\mathcal{F})$ split into two blocks in such a way that

$$J_1(\mathcal{F}_{\text{aff}}) \simeq \mathcal{F}_{\text{aff}} \times_X \mathcal{F}_{\text{proj}},$$

with evident notations. Looking to the Introduction, with metric ω and Christoffel symbols γ, we see that this *purely accidental* splitting is similar to the well-known isomorphism

$$j_1(\omega) \simeq (\omega, \gamma)$$

in Riemannian geometry.

Example 3. $n = 2$, $q = 1$.
This Example has first been treated by Vessiot [191].
Consider the Lie pseudogroup

$$\Gamma \qquad \left\{ y^1 = f(x^1), \quad y^2 = \frac{x^2}{\partial_1 f(x^1)} \right\}.$$

The defining involutive systems are

$$\begin{array}{llll}
\mathcal{R}_1 & y^2 y^1 - x^2 = 0, & y_2^1 = 0, & x^2 y_2^2 - y^2 = 0, \\
R_1 & x^2 \xi_1^1 + \xi^2 = 0, & \xi_2^1 = 0, & x^2 \xi_2^2 - \xi^2 = 0.
\end{array}$$

In particular, we obtain $y_1^1 y_2^2 - y_2^1 y_1^2 = 1$, and Γ is volume preserving. We easily obtain

$$\sharp(R_1) = \left\{ \frac{\partial}{\partial y_1}, \frac{\partial}{\partial y_2} + \frac{1}{\partial y^2}\left(y_1^2 \frac{\partial}{\partial y_1^2} + y_2^2 \frac{\partial}{\partial y_2^2} - y_1^1 \frac{\partial}{\partial y_1^1} - y_2^1 \frac{\partial}{\partial y_2^1} \right), y_1^1 \frac{\partial}{\partial y_1^2} + y_2^1 \frac{\partial}{\partial y_2^2} \right\}.$$

We have 6 independent variables $(y^1, y^2, y_1^1, y_1^2, y_2^1, y_2^2)$ and 3 linearly independent vector fields, so that there are 3 functorially independent differential invariants. A fundamental set could be

$$\Phi^1(y_1) \equiv y^2 y_1^1, \qquad \Phi^2(y_1) \equiv y^2 y_2^1, \qquad \Phi^3(y_1) \equiv y_1^1 y_2^2 - y_2^1 y_1^2,$$

and we have the differential identity

$$d_2 \Phi^1 - d_1 \Phi^2 \equiv \Phi^3.$$

A simple computation shows that $\mathcal{F} = T^* \times_X \wedge^2 T^*$, with special section $\omega = (\alpha, \beta) = (x^2 \, dx^1, dx^1 \wedge dx^2)$. We can use local coordinates for the 1-form α and the 2-form β:

$$\alpha = \omega^1(x) \, dx^1 + \omega^2(x) \, dx^2, \qquad \beta = \omega^3(x) \, dx^1 \wedge dx^2,$$

and obtain the special section $\omega = (x^2, 0, 1)$ of $\mathcal{F}$. We have $\mathcal{F}_1 = \mathcal{F} \times_X \wedge^2 T^*$, and we may set $v = u_2^1 - u_1^2$. Accordingly, v/u^3 transforms like a scalar, and the integrability condition of the general equations

$$\mathcal{R}_1(\omega) \quad \begin{cases} \omega^1(y) y_1^1 + \omega^2(y) y_1^2 = \omega^1(x), \\ \omega^1(y) y_2^1 + \omega^2(y) y_2^2 = \omega^2(x), \\ (y_1^1 y_2^2 - y_2^1 y_1^2) \omega^3(y) = \omega^3(x), \end{cases}$$

$$R_1(\omega) \quad \begin{cases} \omega^1(x) \xi_1^1 + \omega^2(x) \xi_1^2 + \xi^r \partial_r \omega^1(x) = 0, \\ \omega^1(x) \xi_2^1 + \omega^2(x) \xi_2^2 + \xi^r \partial_r \omega^2(x) = 0, \\ \omega^3(x)(\xi_1^1 + \xi_2^2) + \xi^r \partial_r \omega^3(x) = 0, \end{cases}$$

is the system

$$\mathcal{B}_1(c) \subset J_1(\mathcal{F}) \qquad u_2^1 - u_1^2 = c u^3.$$

For the special section we have $c = 1$.

These results allow us to construct the nonlinear Janet sequence at once. For the first nonlinear Spencer sequence we obtain the following two components of the so-called 'torsion' condition

$$\begin{cases} \partial_1 A_2^1 - \partial_2 A_1^1 - A_1^1 \chi_{1,2}^1 - A_2^2 \chi_{2,2}^1 + A_2^1 \chi_{1,1}^1 + A_2^2 \chi_{2,1}^1 = 0, \\ \partial_1 A_2^2 - \partial_2 A_1^2 - A_1^1 \chi_{1,2}^2 - A_1^2 \chi_{2,2}^2 + A_2^1 \chi_{1,1}^2 + A_2^2 \chi_{2,1}^2 = 0. \end{cases}$$

By linearisation we obtain the following diagram, linking the linear Janet sequence with the linear second Spencer sequence (*exercise*: check that R_1 is involutive):

$$
\begin{array}{ccccccc}
& & 0 & & 0 & & 0 \\
& & \downarrow & & \downarrow & & \downarrow \\
0 \longrightarrow \Theta \xrightarrow{j_1} & & C_0 \xrightarrow{D_1} & & C_1 \xrightarrow{D_2} & & C_2 \longrightarrow 0 \\
& & \downarrow & & \downarrow & & \| \\
0 \longrightarrow T \xrightarrow{j_1} & & C_0(T) \xrightarrow{D_1} & & C_1(T) \xrightarrow{D_2} & & C_2(T) \longrightarrow 0 \\
& \| & & \downarrow & & \downarrow & \downarrow \\
0 \longrightarrow \Theta \longrightarrow T \xrightarrow{D} & & F_0 \xrightarrow{D_1} & & F_1 \longrightarrow & & 0 \\
& & \downarrow & & \downarrow & & \\
& & 0 & & 0 & &
\end{array}
$$

with respective dimensions

$$
\begin{array}{ccccccc}
& & \boxed{3} \xrightarrow{D_1} & & \boxed{5} \xrightarrow{D_2} & & \boxed{2} \\
& & \downarrow & & \downarrow & & \| \\
0 \longrightarrow \boxed{2} \xrightarrow{j_1} & & \boxed{6} \xrightarrow{D_1} & & \boxed{6} \xrightarrow{D_2} & & \boxed{2} \longrightarrow 0 \\
& \| & & \downarrow & & \downarrow & \\
& \boxed{2} \xrightarrow{D} & & \boxed{3} \xrightarrow{D_1} & & \boxed{1} \longrightarrow & 0
\end{array}
$$

Finally, we consider the equivalence problem. The new section $\overline{\omega} = (1,0,1)$ of $\mathcal{F}$ gives rise to the Lie pseudogroup

$$\overline{\Gamma} \qquad \{y^1 = x^1 + a, \quad y^2 = x^2 + h(x^1)\}$$

and to the constant $\overline{c} = 0$. Hence there cannot exist $f \in \mathrm{aut}(X)$ such that $j_1(f)^{-1}(\omega) = \overline{\omega}$.

Example 4. $n = 3$, $q = 1$

In the previous Example we had the situation that only one compatibility conditions exists. Now we investigate a case in which only one compatibility condition of the first kind exists. This Example has first been treated by Vessiot [191].

Consider

$$\Gamma \qquad \{y^1 = f^1(x^1), \quad y^2 = f^2(x^1, x^2, x^3), \quad y^3 = f^3(x^1, x^2, x^3)\},$$

$$\mathcal{R}_1 \qquad y^1_2 = 0, \qquad y^1_3 = 0,$$

$$R_1 \qquad \xi^1_2 = 0, \qquad \xi^1_3 = 0.$$

We see that R_1 and $\mathcal{R}_1$ are involutive, with scheme

$$\begin{array}{|ccc|}\hline 1 & 2 & 3 \\ 1 & 2 & \bullet \\ \hline \end{array}$$

It follows that the diagram linking the linear Janet sequence to the linear second Spencer sequence is:

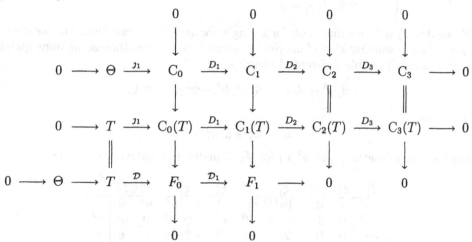

with respective dimensions

$$\boxed{10} \longrightarrow \boxed{17} \longrightarrow \boxed{12} \longrightarrow \boxed{3} \longrightarrow 0$$

$$0 \longrightarrow \boxed{3} \longrightarrow \boxed{12} \longrightarrow \boxed{18} \longrightarrow \boxed{12} \longrightarrow \boxed{3} \longrightarrow 0$$

$$\boxed{3} \longrightarrow \boxed{2} \longrightarrow \boxed{1} \qquad 0 \qquad 0$$

Again we can check that the dimensions of the Spencer bundles are much larger than those of the Janet bundles.

A possible Lie form for $\mathcal{R}_1$ is

$$\Phi^1 \equiv \frac{y_2^1}{y_1^1} = 0, \qquad \Phi^2 \equiv \frac{y_3^1}{y_1^1} = 0,$$

and the natural bundle $\mathcal{F}$ is defined by the laws

$$\mathcal{F} \quad \begin{cases} \dfrac{\partial_3 \varphi^1 + \overline{u}^1 \partial_3 \varphi^2 + \overline{u}^2 \partial_3 \varphi^3}{\partial_1 \varphi^1 + \overline{u}^1 \partial_1 \varphi^2 + \overline{u}^2 \partial_1 \varphi^3} = u^2, \\ \dfrac{\partial_2 \varphi^1 + \overline{u}^1 \partial_2 \varphi^2 + \overline{u}^2 \partial_2 \varphi^3}{\partial_1 \varphi^1 + \overline{u}^1 \partial_1 \varphi^2 + \overline{u}^2 \partial_1 \varphi^3} = u^1, \\ \overline{x} = \varphi(x). \end{cases}$$

The special section is $\omega^1 = 0, \omega^2 = 0$.

Accordingly, the general finite system has a similar form, and we obtain

$$R_1(\omega) \quad \begin{cases} \Omega^2 \equiv \xi_3^1 + \omega^1(x)\xi_3^2 + \omega^2(x)\xi_3^3 - \omega^2(x)(\xi_1^1 + \omega^1(x)\xi_1^2 + \omega^2(x)\xi_1^3) + \\ \quad \xi^r \partial_r \omega^2(x) = 0, \\ \Omega^1 \equiv \xi_2^1 + \omega^1(x)\xi_2^2 + \omega^2(x)\xi_2^3 - \omega^1(x)(\xi_1^1 + \omega^1(x)\xi_1^2 + \omega^2(x)\xi_1^3) + \\ \quad \xi^r \partial_r \omega^1(x) = 0. \end{cases}$$

Since $\dim F_1 = 1$, we may look for a single compatibility condition, but *we cannot 'a priori' know whether it is of the first or second kind*. Nevertheless, we immediately obtain the only possible differential identity

$$d_3 \Phi^1 - d_2 \Phi^2 + \Phi^1 d_1 \Phi^2 - \Phi^2 d_1 \Phi^1 \equiv 0.$$

We introduce

$$v = u_3^1 - u_2^2 + u^1 u_1^2 - u^2 u_1^1,$$

using local coordinates (x, u^1, u^2, v) for $\mathcal{F}_1$. Finally, the matrix $(L(u), M(u)v)$ is:

ξ_1^1	ξ_2^1	ξ_3^1	ξ_1^2	ξ_2^2	ξ_3^2	ξ_1^3	ξ_2^3	ξ_3^3	
$-u^1$	1	0	$-(u^1)^2$	u^1	0	$-u^1 u^2$	u^2	0	u^1
$-u^2$	0	1	$-u^1 u^2$	0	u^1	$-(u^2)^2$	0	u^2	u^2
$-v$	0	0	$-2u^1 v$	v	0	$-2u^2 v$	0	v	v

A practical way to obtain the third line is to consider the 'linearisation' of v, namely:

$$d_3 \Omega^1 - d_2 \Omega^2 + \omega^1 d_1 \Omega^2 + (\partial_1 \omega^2)\Omega^1 - \omega^2 d_1 \Omega^1 - (\partial_1 \omega^1)\Omega^2,$$

which does not contain jets of order 2 anymore. The fact that the rank must be 2 implies that the first (3×3)-matrix on the left gives $v = 0$. Hence there is only one compatibility condition of the first kind, and we have obtained

$$\mathcal{B}_1 \subset J_1(\mathcal{F}) \qquad u_3^1 - u_2^2 + u^1 u_1^2 - u^2 u_1^1 = 0.$$

There is no other formal obstruction to the equivalence problem.

Example 5. $n = 3, q = 1$.

In this Example there are two compatibility conditions of the first kind and two compatibility conditions of the second kind, with one degenerate jacobi condition.

Consider the situation when

$$\Gamma = \{f \in \mathrm{aut}(X) |\, j_1(f)(\alpha) = \alpha, \quad j_1(f)(\beta) = \beta\},$$

with $\alpha = dx^1, \beta = dx^2 \wedge dx^3$,

$$\mathcal{R}_1 \quad \begin{cases} y_1^1 = 1, \qquad y_2^1 = 0, \qquad y_3^1 = 0, \\ \dfrac{\partial(y^2, y^3)}{\partial(x^2, x^3)} = 1, \qquad \dfrac{\partial(y^2, y^3)}{\partial(x^3, x^1)} = 0, \qquad \dfrac{\partial(y^2, y^3)}{\partial(x^1, x^2)} = 0. \end{cases}$$

In fact, because of the third equation, the last two equations imply $y_1^2 = 0$, $y_1^3 = 0$, and we obtain

$$\Gamma = \left\{ f \in \mathrm{aut}(X) \,\middle|\, y^1 = x^1 + a, \quad y^2 = f^2(x^2, x^3), \quad y^3 = f^3(x^2, x^3), \quad \frac{\partial(f^2, f^3)}{\partial(x^2, x^3)} = 1 \right\}.$$

Accordingly, we have

$$\text{R}_1 \qquad \xi_1^1 = 0, \quad \xi_2^1 = 0, \quad \xi_3^1 = 0, \quad \xi_2^2 + \xi_3^3 = 0, \quad \xi_1^2 = 0, \quad \xi_1^3 = 0,$$

and we leave it to the reader to verify that this system, which is already formally integrable (being homogeneous with constant coefficients), is also involutive, with Janet sequence

$$0 \longrightarrow \Theta \longrightarrow T \xrightarrow{\;D\;} F_0 \xrightarrow{\;D_1\;} F_1 \xrightarrow{\;D_2\;} F_2 \longrightarrow 0$$
$$\boxed{3} \qquad\quad \boxed{6} \qquad\quad \boxed{4} \qquad\quad \boxed{1}$$

Of course, we immediately have $\mathcal{F} = T^* \times_X \wedge^2 T^*$, with the inequality $\alpha \wedge \beta \neq 0$ in $\wedge^3 T^*$. Therefore, the only problem left is that of studying the finite and infinitesimal Lie equations obtained by introducing $\omega = (\alpha, \beta)$ and adopting local coordinates

$$u^1 = \alpha_1, \quad u^2 = \alpha_2, \quad u^3 = \alpha_3, \quad u^4 = \beta_{23}, \quad u^5 = \beta_{31}, \quad u^6 = \beta_{12}$$

for $\mathcal{F}$ with the condition $u^1 u^4 + u^2 u^5 + u^3 u^6 \neq 0$.

Of course, since $d\alpha$ and $d\beta$ are again forms, we immediately obtain

$$\mathcal{F}_1 = \mathcal{F} \times_X \wedge^2 T^* \times_X \wedge^3 T^*.$$

Hence $\mathcal{F}_1$ is a vector bundle over $\mathcal{F}$ with fiber dimension $3 + 1 = 4$, and we can adopt local coordinates

$$v^1 = u_2^3 - u_3^2, \quad v^2 = u_3^1 - u_1^3, \quad v^3 = u_1^2 - u_2^1, \quad w = u_1^4 + u_2^5 + u_3^6$$

for the fibers.

The matrix $L(u)$ is as follows:

ξ_1^1	ξ_2^1	ξ_3^1	ξ_1^2	ξ_2^2	ξ_3^2	ξ_1^3	ξ_2^3	ξ_3^3	
u^1	0	0	u^2	0	0	u^3	0	0	u^1
0	u^1	0	0	u^2	0	0	u^3	0	u^2
0	0	u^1	0	0	u^2	0	0	u^3	u^3
0	$-u^5$	$-u^6$	0	u^4	0	0	0	u^4	u^4
u^5	0	0	$-u^4$	0	$-u^6$	0	0	u^5	u^5
u^6	0	0	0	u^6	0	$-u^4$	$-u^5$	0	u^6

The matrix $(M(u)v, N(u)w)$ does not depend on u, since (v^1, v^2, v^3) and w separately transform as tensors. Consider the row corresponding to v^1:

$$0 \quad -v^2 \quad -v^3 \quad 0 \quad v^1 \quad 0 \quad 0 \quad 0 \quad v^1.$$

We want to express it as a linear combination of the 6 rows of the matrix $L(u)$ with respective coefficients $(\lambda^1, \lambda^2, \lambda^3, \lambda^4, \lambda^5, \lambda^6)$. We can use the 5 columns $\xi_1^1, \xi_1^2, \xi_3^2, \xi_1^3, \xi_2^3$ to determine these coefficients. The resulting (5×5)-matrix is:

$$\begin{pmatrix} u^1 & u^2 & 0 & u^3 & 0 \\ 0 & 0 & 0 & 0 & u^3 \\ 0 & 0 & u^2 & 0 & 0 \\ u^5 & -u^4 & -u^6 & 0 & 0 \\ u^6 & 0 & 0 & -u^4 & -u^5 \end{pmatrix}$$

A straightforward computation shows that the determinant of this matrix is

$$u^2 u^3 u^4 (u^1 u^4 + u^1 u^5 + u^3 u^6).$$

Since GL_1 acts transitively on the fibers of $\mathcal{F}$, we may assume that $(u^1, u^2, u^3) \neq (0, 0, 0)$, because otherwise the inequality $u^1 u^4 + u^2 u^5 + u^3 u^6 \neq 0$ cannot be satisfied. A similar comment holds for $(u^4, u^5, u^6) \neq (0, 0, 0)$. Accordingly, the determinant of the matrix above is generically nonzero, and the only possibility is that $\lambda^1 = \lambda^2 = \lambda^3 = \lambda^4 = \lambda^6 = 0$. Setting $\lambda^4 = \lambda$, we obtain

$$v^1 = \lambda u^4, \quad v^2 = \lambda u^5, \quad v^3 = \lambda u^6.$$

Elimination of λ gives 2 compatibility conditions of the first kind. In fact, we have $u^1 v^1 + u^2 v^2 + u^3 v^3 = \lambda(u^1 u^4 + u^2 u^5 + u^3 u^6)$ for the determination of $\lambda(x)$ as a purely scalar function, since the left and right terms are in $\wedge^3 T^*$. The projection $R_2 \to T$ is thus surjective, under the condition that $\lambda = c = \text{const}$.

Finally, w and $u^1 u^4 + u^2 u^5 + u^3 u^6$ are in $\wedge^3 T^*$, and for the same reason we must have $w = \mu(x)(u^1 u^4 + u^2 u^5 + u^3 u^6)$ with $\mu(x) = c' = \text{const}$.

We recapitulate the results. The integrability conditions are

$$d\alpha = c\beta, \qquad d\beta = c'\alpha \wedge \beta,$$

with two structure constants, c and c'. In particular, we notice that c is uniquely determined by the differential condition $\alpha \wedge d\alpha = c\alpha \wedge \beta$.

Closing this system, we obtain

$$0 = dd\alpha = c \, d\beta = cc'\alpha \wedge \beta \Rightarrow cc' = 0.$$

The special section considered above is:

$$\omega^1(x) = 1, \quad \omega^2(x) = \omega^3(x) = 0, \quad \omega^4(x) = 1, \quad \omega^5(x) = \omega^6(x) = 0 \Rightarrow c = c' = 0,$$

i.e. the critical point of the algebraic set determined by the jacobi condition.

Consider now the contact 1-form

$$\alpha = dx^1 - x^3 \, dx^2.$$

Any transformation preserving α also preserves $d\alpha = dx^2 \wedge dx^3$, and thus also $\alpha \wedge d\alpha = dx^1 \wedge dx^2 \wedge dx^3$. Accordingly, the pseudogroup of transformations preserving α also preserves the volume form. In this case, setting

$$\alpha = dx^1 - x^3 \, dx^2, \qquad \beta = dx^2 \wedge dx^3$$

we immediately have

$$da = \beta, \qquad d\beta = 0,$$

and thus $c = 1$, $c' = 0$, i.e. a regular point of the algebraic set determined by the jacobi condition. In Chapter VI we shall see how to use this Lie pseudogroup in analytical mechanics.

Example 6. $n = 3$, $q = 1$.
We introduce the 1-form $w = dx^1 - x^3\,dx^2$, and define

$$\Gamma = \{f \in \mathrm{aut}(X)\,|\, j_1(f)(w) = \rho w\}.$$

We may study the system obtained by eliminating the factor ρ from the three equations thus obtained, namely:

$$y_2^1 - y^3 y_2^2 = -x^3(y_1^1 - y^3 y_1^2), \qquad y_3^1 - y^3 y_3^2 = 0.$$

Prolongation of the first equation with respect to x^3 and of the second with respect to x^2 and subtraction of the results gives the additional first-order equation

$$y_2^2 y_3^3 - y_3^2 y_2^3 - x^3(y_1^3 y_3^2 - y_1^2 y_3^3) - y_1^1 - y^3 y_1^2 = 0.$$

Hence the preceding system is *not* formally integrable, and the 'object' that has to be used to describe the corresponding involutive system (an exercise) thus obtained cannot be a 1-form. The key idea in order to preserve the role of each coordinate is the observation that

$$j_1(f)(dw) = d(j_1(f)(w)) = \rho\,dw + d\rho \wedge w,$$
$$j_1(f)(w \wedge dw) = \rho^2(w \wedge dw).$$

Accordingly, at least formally and symbolically, we have to consider instead of the 1-form w a tensor density ω, a section of the natural bundle $\mathcal{F}$ with local coordinates (x, u^1, u^2, u^3), and transformation laws

$$\frac{\bar{u}^1 \partial_i \varphi^1 + \bar{u}^2 \partial_i \varphi^2 + \bar{u}^3 \partial_i \varphi^3}{\sqrt{\partial(\varphi^1, \varphi^2, \varphi^3)/\partial(x^1, x^2, x^3)}} = u^i.$$

Accordingly, the general finite Lie equations are

$$\mathcal{R}_3(\omega) \qquad \frac{\omega^1(y)y_i^1 + \omega^2(y)y_i^2 + \omega^3(y)y_i^3}{\sqrt{\partial(y^1, y^2, y^3)/\partial(x^1, x^2, x^3)}} = \omega^i(x),$$

and the corresponding infinitesimal Lie equations are:

$$R_1(\omega) \qquad \Omega^i \equiv \omega^1(x)\xi_i^1 + \omega^2(x)\xi_i^2 + \omega^3(x)\xi_i^3 - \frac{1}{2}\omega^i(x)(\xi_1^1 + \xi_2^2 + \xi_3^3) + \xi^r \partial_r \omega^i(x) = 0.$$

The upper indices for u are used on purpose, to convince the reader that natural bundles can be used even if we do not know about tensors. For the special section $\omega = (1, -x^3, 0)$ we obtain the system

$$\begin{cases} \xi_2^1 - x^3\xi_2^2 + x^3\xi_1^1 - (x^3)^2\xi_1^3 - \xi^3 = 0, \\ \xi_3^1 - x^3\xi_3^2 = 0, \end{cases}$$

which is not formally integrable. However, by a direct substitution we can check that its general solution can be written as

$$\xi^1(x) = x^3 \partial_3 W - W, \quad \xi^2(x) = \partial_3 W, \quad \xi^3(x) = -\partial_2 W - x^3 \partial_1 W,$$

where W is called a 'generating' function or a 'potential'. However, these results cannot be extended to the general situation, which has to be studied by itself. To this end we notice that $w \wedge dw$ is a 3-form and thus, using the same symbolic notation, $\omega \wedge d\omega$ is a scalar.

Now the involutive system defining the Lie algebra Θ of Γ is

$$R_1 \quad \begin{cases} \xi^3_3 + 2x^3 \xi^2_1 - \xi^1_1 = 0 & \boxed{1 \ \ 2 \ \ 3} \\ \xi^1_3 - x^3 \xi^2_3 = 0 & \boxed{1 \ \ 2 \ \ 3} \\ \xi^1_2 - x^3 \xi^2_2 + x^3 \xi^1_1 - (x^3)^2 \xi^2_1 - \xi^3 = 0 & \boxed{1 \ \ 2 \ \ \bullet} \end{cases}$$

Accordingly, the Janet sequence is

$$0 \longrightarrow \Theta \longrightarrow \underset{\boxed{3}}{T} \overset{\mathcal{D}}{\longrightarrow} \underset{\boxed{3}}{F_0} \overset{\mathcal{D}_1}{\longrightarrow} \underset{\boxed{1}}{F_1} \longrightarrow 0$$

and we notice that $3 - 3 + 1 = 1$, because of the existence of the generating function W. Hence there is only one integrability condition of the second kind, namely:

$$\omega^1(\partial_2 \omega^3 - \partial_3 \omega^2) + \omega^2(\partial_3 \omega^1 - \partial_1 \omega^3) + \omega^3(\partial_1 \omega^2 - \partial_2 \omega^1) = c,$$

where the reader should check that c is a covariant density.

For the special section $\omega = (1, -x^3, 0)$ we have $c = 1$. For the other special section $\overline{\omega} = (1, 0, 0)$ we have $\overline{c} = 0$. The new pseudogroup is *not*

$$\overline{\Gamma} = \{f \in \text{aut}(X)| \, j_1(f)(dx^1) = \rho \, dx^1\},$$

which is only defined by the involutive system $y^1_2 = 0, y^1_3 = 0$, but it is

$$\overline{\Gamma} = \{f \in \text{aut}(X)| \, j_1(f)\omega = \omega\},$$

which is defined by the involutive system

$$y^1_2 = 0, \quad y^1_3 = 0, \quad \frac{\partial(y^1, y^2, y^3)}{\partial(x^1, x^2, x^3)} = (y^1_1)^2.$$

The third equation can also be written as

$$y^2_2 y^3_3 - y^3_2 y^2_3 = y^1_1.$$

Of course, since $c \neq \overline{c}$, we cannot find $f \in \text{aut}(X)$ such that $j_1(f)(\omega) = \overline{\omega}$, because of formal reasons.

This result can be generalised to the situation $n > 3$, n odd. The existence of one structure constant has, to the best of our knowledge, never been noticed before, although the reader may discover that there is no conceptual difference with the constant Riemannian curvature situation.

Example 7. $n = 2$, $q = 1$.

In 1878, Clifford introduced abstract numbers of the form $x^1 + \epsilon x^2$ with $\epsilon^2 = 0$ in order to study helicoidal movement, or 'screw', in the mechanics of rigid bodies. The question of constructing functions of such numbers can be asked, and we can study the existence of differentials.

Let $f(x^1 + \epsilon x^2) = f^1(x^1, x^2) + \epsilon f^2(x^1, x^2)$, then

$$
\begin{aligned}
df &= (A + \epsilon B)(dx^1 + \epsilon\, dx^2) \\
&= A dx^1 + \epsilon(B\, dx^1 + A dx^2) \\
&= df^1 + \epsilon\, df^2.
\end{aligned}
$$

It follows that the corresponding map

$$f : \mathbb{R}^2 \to \mathbb{R}^2,$$
$$(x^1, x^2) \to (y^1 = f^1(x^1, x^2), y^2 = f^2(x^1, x^2))$$

must satisfy the involutive system of finite Lie equations

$$\mathcal{R}_1 \qquad y_2^1 = 0, \qquad y_2^2 - y_1^1 = 0.$$

The corresponding system of infinitesimal Lie equations is:

$$R_1 \qquad \xi_2^1 = 0, \qquad \xi_2^2 - \xi_1^1 = 0,$$

and the diagram linking the linear Janet sequence with the linear second Spencer sequence is

$$
\begin{array}{ccccccccc}
& & 0 & & 0 & & 0 & & \\
& & \downarrow & & \downarrow & & \downarrow & & \\
0 \longrightarrow & \Theta & \xrightarrow{j_1} & C_0 & \xrightarrow{D_1} & C_1 & \xrightarrow{D_2} & C_2 & \longrightarrow 0 \\
& & & \downarrow & & \| & & \| & \\
0 \longrightarrow & T & \xrightarrow{j_1} & C_0(T) & \xrightarrow{D_1} & C_1(T) & \xrightarrow{D_2} & C_2(T) & \longrightarrow 0 \\
& & & \| & & \downarrow & & \downarrow & \\
0 \longrightarrow & \Theta & \longrightarrow & T & \xrightarrow{D} & F_0 & \longrightarrow & 0 & \quad 0 \\
& & & & & \downarrow & & & \\
& & & & & 0 & & &
\end{array}
$$

with dimensions

$$
\begin{array}{ccccccc}
& & \boxed{4} & \xrightarrow{D_1} & \boxed{6} & \xrightarrow{D_2} & \boxed{2} & \longrightarrow & 0 \\
& & \downarrow & & \| & & \| \\
0 & \longrightarrow & \boxed{2} & \xrightarrow{j_1} & \boxed{6} & \xrightarrow{D_1} & \boxed{6} & \xrightarrow{D_2} & \boxed{2} & \longrightarrow & 0 \\
& & \| & & \downarrow & & \downarrow \\
& & \boxed{2} & \xrightarrow{D} & \boxed{2} & \longrightarrow & 0
\end{array}
$$

Accordingly, there are no compatibility conditions for $\mathcal{D}$.

The Lie pseudogroup corresponding to such a structure, which could be called a *screw structure*, is

$$
\Gamma \qquad \{y^1 = f^1(x^1), \quad y^2 = x^2 \partial_1 f(x^1) + g(x^1)\}.
$$

We notice that the jacobian matrices have the very particular property that they behave like a commutative subgroup of GL_1:

$$
\begin{pmatrix} A & 0 \\ B & A \end{pmatrix} \begin{pmatrix} C & 0 \\ D & C \end{pmatrix} = \begin{pmatrix} AC & 0 \\ BC + AD & AC \end{pmatrix} = \begin{pmatrix} C & 0 \\ D & C \end{pmatrix} \begin{pmatrix} A & 0 \\ B & A \end{pmatrix}.
$$

Looking for the natural bundle $\mathcal{F}$, we immediately discover that we should not try to integrate $\sharp(R_1)$, since no evident differential invariant, apart from the obvious y_2^1/y_1^1, appears.

With an eye towards the theory of complex functions, we should look for (2×2)-matrices ϵ such that $\epsilon^2 = 0$. We immediately see that

$$
\begin{pmatrix} 0 & 0 \\ 1 & 0 \end{pmatrix} \begin{pmatrix} y_1^1 & y_2^1 \\ y_1^2 & y_2^2 \end{pmatrix} = \begin{pmatrix} 0 & 0 \\ y_1^1 & y_2^1 \end{pmatrix} = \begin{pmatrix} y_2^1 & 0 \\ y_2^2 & 0 \end{pmatrix} = \begin{pmatrix} y_1^1 & y_2^1 \\ y_1^2 & y_2^2 \end{pmatrix} \begin{pmatrix} 0 & 0 \\ 1 & 0 \end{pmatrix}.
$$

Hence we obtain a basis of differential invariants by considering

$$
\frac{1}{y_1^1 y_2^2 - y_2^1 y_1^2} \begin{pmatrix} y_2^2 & -y_2^1 \\ -y_1^2 & y_1^1 \end{pmatrix} \begin{pmatrix} 0 & 0 \\ 1 & 0 \end{pmatrix} \begin{pmatrix} y_1^1 & y_2^1 \\ y_1^2 & y_2^2 \end{pmatrix} = \frac{1}{y_1^1 y_2^2 - y_2^1 y_1^2} \begin{pmatrix} -y_1^2 y_1^1 & -(y_2^1)^2 \\ (y_1^1)^2 & y_1^1 y_2^1 \end{pmatrix}.
$$

As a byproduct, we may introduce the following fundamental set of differential invariants:

$$
\Phi^1 \equiv \frac{y_2^1}{y_1^1} = 0, \qquad \Phi^2 \equiv \frac{(y_1^1)^2}{y_1^1 y_2^2 - y_2^1 y_1^2} = 1.
$$

There is another procedure which we can use in this Example, since it concerns differential polynomials! In Chapter VI we shall describe this procedure for a more general situation. The basic idea is to notice that the defining system $\mathcal{R}_1$ is, by definition, invariant under the groupoid action. Using the chain rule we obtain

$$
\frac{\partial \bar{y}^1}{\partial y^1} y_2^1 + \frac{\partial \bar{y}^1}{\partial y^2} y_2^2 = 0, \frac{\partial \bar{y}^2}{\partial y^1} y_2^1 + \frac{\partial \bar{y}^2}{\partial y^2} y_2^2 - \frac{\partial \bar{y}^1}{\partial y^1} y_1^1 - \frac{\partial \bar{y}^1}{\partial y^2} y_1^2 = 0.
$$

By a straightforward elimination we obtain

$$\begin{cases} y_2^1 + \dfrac{\frac{\partial \bar{y}^1}{\partial y^2}}{\frac{\partial \bar{y}^1}{\partial y^1}} y_2^2 = 0, \\[3ex] y_2^2 - \dfrac{\left(\frac{\partial \bar{y}^1}{\partial y^1}\right)^2}{\frac{\partial \bar{y}^1}{\partial y^1}\frac{\partial \bar{y}^2}{\partial y^2} - \frac{\partial \bar{y}^1}{\partial y^2}\frac{\partial \bar{y}^2}{\partial y^1}} y_1^1 - \dfrac{\frac{\partial \bar{y}^1}{\partial y^1}\frac{\partial \bar{y}^1}{\partial y^2}}{\frac{\partial \bar{y}^1}{\partial y^1}\frac{\partial \bar{y}^2}{\partial y^2} - \frac{\partial \bar{y}^1}{\partial y^2}\frac{\partial \bar{y}^2}{\partial y^1}} y_1^2 = 0. \end{cases}$$

Comparing with the defining equations of $\mathcal{R}_1$, we again find the same fundamental set as before.

This method is sufficiently general and quite similar to the one used for algebraic groups [82]. Of course, we have to deal with differential polynomials and, for this reason, we call the corresponding Lie pseudogroups *algebraic pseudogroups*, similarly to the consideration of algebraic groups in the theory of Lie groups.

We are now ready for determining the transition rules of $\mathcal{F}$:

$$\mathcal{F} \quad \begin{cases} u^1 = \dfrac{\partial_2 \varphi^1 + \bar{u}^1 \partial_2 \varphi^2}{\partial_1 \varphi^1 + \bar{u}^1 \partial_1 \varphi^2}, \\[3ex] u^2 = \bar{u}^2 \dfrac{(\partial_1 \varphi^1 + \bar{u}^1 \partial_1 \varphi^2)^2}{\partial(\varphi^1, \varphi^2)/\partial(x^1, x^2)}, \\[2ex] \bar{x} = \varphi(x). \end{cases}$$

There is no integrability condition.

Finally, the '*torsion*' part of $\overline{D}'$ is similar to that in Example 3, because $n = 2$. However, in the '*curvature*' part, the quadratic terms disappear, because of the commutation relation already noticed above, and we obtain

$$\partial_1 \chi_{l,2}^k - \partial_2 \chi_{l,1}^k - A_1^r \chi_{lr,2}^k + A_2^r \chi_{lr,1}^k = 0.$$

We have only 2 linearly independent such relations, corresponding to $(k,l) = (1,1)$ or $(2,1)$. We notice that the terms involving the second-order jets are not known in the literature, although they are essential in the curvature because quadratic terms do not appear.

Example 8. $n = 2$, $q = 1$.

In a similar way we shall treat the case of a complex-analytic structure, closely following [158].

Introducing a so-called imaginary quantity i for which $i^2 = -1$, we look for functions of the form

$$y^1 + iy^2 = f(x^1 + ix^2) = f^1(x^1, x^2) + if^2(x^1, x^2).$$

If derivatives exist, we must have

$$\begin{aligned} df &= (A + iB)(dx^1 + i\,dx^2) \\ &= (A\,dx^1 - B\,dx^2) + i(B\,dx^1 + A\,dx^2) \\ &= df^1 + i\,df^2. \end{aligned}$$

Accordingly, the transformations of $\mathbb{R}^2$:

$$(x^1, x^2) \rightarrow (y^1 = f^1(x^1, x^2), y^2 = f^2(x^1, x^2)),$$

which are conveniently called *holomorphic*, must be solutions of the Cauchy–Riemann system of finite Lie equations

$$\mathcal{R}_1 \qquad y_2^2 - y_1^1 = 0, \qquad y_2^1 + y_1^2 = 0.$$

The corresponding system of infinitesimal Lie equations is

$$R_1 \qquad \xi_2^2 - \xi_1^1 = 0, \qquad \xi_2^1 + \xi_1^2 = 0.$$

Both systems are involutive and, as in the previous Example, there is a commutation relation

$$\begin{pmatrix} A & -B \\ B & A \end{pmatrix} \begin{pmatrix} M & -N \\ N & M \end{pmatrix} = \begin{pmatrix} AM - BN & -(AN + BM) \\ AN + BM & AM - BN \end{pmatrix} = \begin{pmatrix} M & -N \\ N & M \end{pmatrix} \begin{pmatrix} A & -B \\ B & A \end{pmatrix},$$

where A, B, M, N are real numbers.

The reader can easily check that the characters are $\alpha_1^1 = 2$, $\alpha_1^2 = 0$, and therefore we obtain $\dim M_{1+r} = 2$ for all $r \geq 0$. We shall see that this purely accidental result should be no reason for introducing an infinite sequence of complex numbers indexed by r, as is common nowadays.

Let us look for the geometric objects preserved by holomorphic transformations. We could use the same method as in the previous Example in order to find a fundamental set of differential invariants. However, it is well known that the sections of the natural bundle $\mathcal{F}$ can be identified with mixed terms $\omega \in T^* \otimes T$ such that $\omega^2 = -\mathrm{id}_T$. In general, such a tensor is defined by one half of the matrix, i.e. by $n^2/2$ components, and thus by 2 components for $n = 2$. In fact, the relation

$$\begin{pmatrix} a & b \\ c & d \end{pmatrix} \begin{pmatrix} a & b \\ c & d \end{pmatrix} = \begin{pmatrix} a^2 + bc & b(a + d) \\ c(a + d) & bc + d^2 \end{pmatrix} = \begin{pmatrix} -1 & 0 \\ 0 & -1 \end{pmatrix}$$

implies that $b \neq 0$, since otherwise we would have $d^2 = -1$, while a, b, c, d are real numbers. Hence, we must have $a + d = 0$, $a^2 + bc = -1$, and $(a, b) \Rightarrow (c, d)$. The special section ω_0 considered in our Example is

$$\omega_0 = \begin{pmatrix} 0 & 1 \\ -1 & 0 \end{pmatrix},$$

and, as in the previous Example, we can verify that

$$\begin{pmatrix} 0 & 1 \\ -1 & 0 \end{pmatrix} \begin{pmatrix} y_1^1 & y_2^1 \\ y_1^2 & y_2^2 \end{pmatrix} = \begin{pmatrix} y_1^2 & y_2^2 \\ -y_1^1 & -y_2^1 \end{pmatrix} = \begin{pmatrix} -y_2^1 & y_1^1 \\ -y_2^2 & y_1^2 \end{pmatrix} = \begin{pmatrix} y_1^1 & y_2^1 \\ y_1^2 & y_2^2 \end{pmatrix} \begin{pmatrix} 0 & 1 \\ -1 & 0 \end{pmatrix}.$$

Using an obvious notation, we obtain the following Lie form for the general finite Lie equations:

$$\mathcal{R}_1(\omega) \qquad \omega_l^k(y) \frac{\partial y^l}{\partial x^j} \frac{\partial x^i}{\partial y^k} = \omega_j^i(x),$$

and we have for sections:

$$\omega_l^k(f(x)) f_j^l(x) g_k^i(x) = \omega_j^i(x).$$

Accordingly, the general infinitesimal Lie equations are:

$$R_1(\omega) \qquad \omega_r^i(x)\xi_j^r - \omega_j^r(x)\xi_r^i + \xi^r \partial_r \omega_j^i(x) = 0,$$

where we have to take into account that always

$$\omega_r^i(x)\omega_j^r(x) = -\delta_j^i.$$

To adapt the notation to the one commonly used in the literature, we shall introduce the usual *complex source coordinates* ($z = x^1 + ix^2$, $\overline{z} = x^1 - ix^2$) with $\partial = \partial_z$, $\overline{\partial} = \partial_{\overline{z}}$ and *complex target coordinates* ($Z = y^1 + iy^2$, $\overline{Z} = y^1 - iy^2$). In this new system of coordinates we have

$$\begin{array}{ll} \Gamma & Z = f(z), \\ \mathcal{R}_1 & Z_{\overline{z}} = 0, \end{array}$$

and the prolongations are particularly simple. The equivalence problem may be stated by using the following Lie form:

$$\Phi \equiv \frac{Z_{\overline{z}}}{Z_z} = \mu(z, \overline{z}),$$

and again we see that a complex-analytic structure is determined by an object with two components. We shall use these simple presentations to study various compatibility conditions for substructures relative to various Lie subpseudogroups, using the results of Examples 1, 2.

linear case: $\dfrac{\overline{\partial}Z}{\partial Z} = \mu$, $\partial Z = u$.

We immediately obtain the compatibility conditions

$$\overline{\partial}u - \mu\partial u - u\partial\mu = 0.$$

affine case: $\dfrac{\overline{\partial}Z}{\partial Z} = \mu$, $\dfrac{\overline{\partial}^2 Z}{\partial Z} = v$.

The system $\overline{\partial}Z - \mu\partial Z = 0$, $\partial^2 Z - v\partial Z = 0$ is not formally integrable. We may add:

$$\partial\overline{\partial}Z - \mu\partial\partial Z - \partial\mu\partial Z = 0,$$
$$\overline{\partial}\overline{\partial}Z - \mu\partial\overline{\partial}Z - \overline{\partial}\mu\partial Z = 0,$$

i.e.

$$\partial\overline{\partial}Z - (v\mu + \partial\mu)\partial Z = 0,$$
$$\overline{\partial}\overline{\partial}Z - (\mu^2 v + \overline{\partial}\mu + \mu\partial\mu)\partial Z = 0,$$

and we have a system of finite type. Testing the criterion gives:

$$(\partial^2\mu + \mu\partial v + v\partial\mu - \overline{\partial}v)\partial Z = 0.$$

Since we cannot have $\partial Z = 0$, because the denominator would then be undefined, we must have the compatibility condition

$$\partial^2\mu + \mu\partial v + v\partial\mu - \overline{\partial}v = 0.$$

Another proof is to notice that

$$\partial^2 \mu + \mu \partial v - \overline{\partial} v$$

is a *new* differential invariant and, in fact, the only one *not* containing derivatives of Z of order > 2. Hence it must be a rational function of $(\mu, \partial \mu, \overline{\partial} \mu, v)$, which happens to be $-v \partial \mu$ in this case.

projective case: $\dfrac{\overline{\partial} Z}{\partial Z} = \mu, \quad \dfrac{\partial^3 Z}{\partial Z} - \dfrac{3}{2} \left(\dfrac{\partial^2 Z}{\partial Z} \right)^2 = w.$

We could proceed as before, and find the compatibility condition

$$\partial^3 \mu + 2w \partial \mu + \mu \partial w - \overline{\partial} w = 0,$$

but we are now dealing with a nonlinear system.

Since there is no compatibility condition, when linearising the diagram linking the linear Janet sequence with the linear second Spencer sequence is

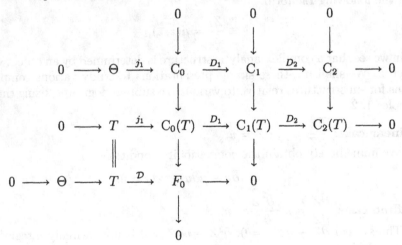

with dimensions

We have complete knowledge of the nonlinear Janet sequence, and we now study the nonlinear first Spencer sequence.

As in the previous Example, the 'torsion' term of $\overline{D}'$ is similar, and the 'curvature' term of $\overline{D}'$ does not contain the quadratic terms in order one because of the commutation relation. On the other side, $\overline{D}'$ has a coordinate-free expression, and we may use local complex coordinates to obtain

$$\begin{cases} \partial\chi^z_{,\bar{z}} - \bar{\partial}\chi^z_{,z} - \chi^z_{z,\bar{z}} - \chi^z_{,z}\chi^z_{z,\bar{z}} + \chi^z_{,\bar{z}}\chi^z_{z,z} = 0, \\ \partial\chi^z_{z,\bar{z}} - \bar{\partial}\chi^z_{z,z} - \chi^z_{zz,\bar{z}} - \chi^z_{,z}\chi^z_{zz,\bar{z}} + \chi^z_{,\bar{z}}\chi^z_{zz,z} = 0, \\ \partial\chi^z_{zz,z} - \bar{\partial}\chi^z_{zz,z} - \chi^z_{zzz,\bar{z}} - \chi^z_{,z}\chi^z_{zzz,\bar{z}} + \chi^z_{,\bar{z}}\chi^z_{zzz,z} - \\ \qquad \chi^z_{z,z}\chi^z_{zz,\bar{z}} + \chi^z_{z,\bar{z}} - \chi^z_{zz,z} = 0. \end{cases}$$

We conclude this Example with a specific computation of the algebraic Lie bracket in order to express it in a very simple way.

First we recall the general formula for the algebraic bracket on $J_{q+1}(T)$:

$$(\{\xi_{q+1},\eta_{q+1}\})^k_\mu = \sum_{|\lambda|=0}^{|\mu|}\sum_{|\nu|=0}^{|\mu|}\frac{(|\lambda|+|\nu|)!}{|\lambda|!\,|\nu|!}(\xi^r_\lambda\eta^k_{\nu+1_r} - \eta^r_\lambda\xi^\lambda_{\nu+1_r})\delta^\mu_{\lambda+\nu}.$$

In complex coordinates we have

$$(\{\xi_{q+1},\eta_{q+1}\})^z_\alpha = \sum_{|\lambda|=0}^{|\alpha|}\sum_{|\nu|=0}^{|\alpha|}\frac{(|\lambda|+|\nu|)!}{|\lambda|!\,|\nu|!}(\xi^z_\lambda\eta^z_{\nu+1_z} - \eta^z_\lambda\xi^z_{\nu+1_z})\delta^\alpha_{\lambda+\nu}.$$

We make the changes

$$\xi^z_\lambda \equiv |\lambda|!\,\hat{\xi}^z_\lambda, \qquad 0 \le |\lambda| \le q+1,$$

$$(\{\xi_{q+1},\eta_{q+1}\})^z_\alpha \equiv |\alpha|!\,(\{\hat{\xi}_{q+1},\hat{\eta}_{q+1}\})^z_\alpha.$$

The last definition of the bracket implies

$$(\{\hat{\xi}_{q+1},\hat{\eta}_{q+1}\})^z_\alpha = \sum_{|\lambda|=0}^{|\alpha|}\sum_{|\nu|=0}^{|\alpha|}(|\nu|+1)(\hat{\xi}^z_\lambda\eta^z_{\nu+1_z} - \hat{\eta}^z_\lambda\hat{\xi}^z_{\nu+1_z})\delta^\alpha_{\lambda+\nu}.$$

Interchanging the roles of $\hat{\xi}$ and $\hat{\eta}$, we find

$$(\{\hat{\xi}_{q+1},\hat{\eta}_{q+1}\})^z_\alpha = (|\alpha|+1)(\bar{\xi}^z\hat{\eta}^z_{\alpha+1_z} - \hat{\eta}^z\hat{\xi}^z_{\alpha+1_z})$$

$$\sum_{|\lambda|=1}^{|\mu|}\sum_{|\nu|=0}^{|\mu|-1}\frac{1}{2}(|\nu|+1-|\lambda|)(\hat{\xi}^z_\lambda\eta^z_{\nu+1_z} - \hat{\eta}^z_\lambda\hat{\xi}^z_{\nu+1_z})\delta^\alpha_{\lambda+\nu}.$$

Setting, for positive integers m,

$$L_m = (z,\bar{z}) \rightarrow (z,\bar{z},0,\ldots,0,-\hat{\xi}^z_{m+1}(z,\bar{z}),0,\ldots,0),$$

the bracket now reads:

$$\{L_m, L_n\}_\alpha = -(m-n)\hat{\xi}^z_{m+1}\hat{\eta}^z_{n+1}\delta^\alpha_{m+n+1} = (m-n)L_{m+n}.$$

Passing to the projective limit, we obtain

$$[L_m, L_n] = \{L_m, L_n\}_{m+n+1} = (m-n)L_{m+n}$$

for $m, n \ge -1$, and we recognise the comutation relations of the maximal Lie subalgebra w_2 of the Virasoro algebra [20].

It can be seen from this example that certain physicists try to use complex-analytic structures in mathematical physics via the Janet sequence, while others use them, for other purposes, via the Spencer sequence [20], since they do want to copy the existing formalism of gauge theory. We hope that we have succeeded in convincing the reader that the existing link between these two approaches cannot be discovered without the results in this Chapter.

Example 9. Here we shall treat in a neat manner the example, already encountered in the Introduction, of a Lie group G acting simply transitively on a manifold X.

The simplest example of such an action is the '*special*' translation case:

$$\Gamma \qquad y^i = x^i + a^i, \qquad a = \text{const}.$$

We can immediately see that

$$\mathcal{R}_1 \qquad \Phi_i(y_1) \equiv y_i^k = \delta_i^k,$$
$$\mathrm{R}_1 \qquad \xi_i^k = 0.$$

It immediately follows that

$$\mathcal{F} = \underbrace{T^* \times_X \cdots \times_X T^*}_{n \text{ times}},$$

and we may choose local coordinates (x, u_i^τ) satisfying $\det(u_i^\tau) \neq 0$.

Accordingly the general Lie equations are:

$$\mathcal{R}_1(\omega) \qquad \Phi_i^\tau(y_1) \equiv \omega_k^\tau(y)y_i^k = \omega_i^\tau(x),$$
$$\mathrm{R}_1(\omega) \qquad \omega_r^\tau(x)\xi_i^r + \xi^r \partial_r \omega_i^\tau(x) = 0.$$

We can proceed as in the Introduction on either the finite or the infinitesimal Lie equations, to find the following integrability conditions of the second kind:

$$\partial_i \omega_j^\tau(x) - \partial_j \omega_i^\tau(x) = c_{\rho\sigma}^\tau \omega_i^\rho(x)\omega_j^\sigma(x).$$

Of course, closing this system does provide the usual jacobi relations, which are determined by homogeneous polynomials of degree 2, because the Lie equations are of order 1, in agreement with the general theory. Of course, the Bianchi identities are the same as in standard gauge theory, although the reader should not forget that we are dealing with the nonlinear Janet sequence, and *not* with the nonlinear Spencer sequence, despite the isomorphism expressed in Theorem D.31.

The confusion in mathematical physics is therefore based on a few purely accidental facts.

Since $q = 1$ there is an exact sequence

$$0 \longrightarrow \wedge^{r+1}T^* \otimes T \longrightarrow C_r(T) \longrightarrow \wedge^r T^* \otimes T \longrightarrow 0,$$

and we need to look for a splitting of this sequence. The study of this problem will lead to a striking phenomenon that no classical approach could possibly discover. Since the symbol of R_1 is zero, there is an isomorphism $\mathrm{R}_1 \simeq T$, allowing us to lift T

to $R_1 \subset J_1(T)$. Hence, we may lift $\bigwedge^r T^* \otimes T$ to $\bigwedge^r T^* \otimes R_1 \subset \bigwedge^r T^* \otimes J_1(T)$ and thus to $C_r(T)$ by projection. Accordingly, we obtain an isomorphism

$$C_r(T) \simeq \bigwedge^r T^* \otimes T \times_X \bigwedge^{r+1} T^* \otimes T.$$

Surprisingly, the projection induced by the preceding sequence is *not* that of $C_r(T)$ onto $F_r \cdots$ but that onto $\bigwedge^{r+1} T^* \otimes T$ induced by the latter isomorphism. Indeed, the Janet sequence can be seen as the tensor product by either $\mathcal{G}$ of the Poincaré sequence with $F_r = \bigwedge^{r+1} T^* \otimes \mathcal{G}$, or by T of the Poincaré sequence with $F_r = \bigwedge^{r+1} T^* \otimes T$, because $\det \omega \neq 0$. Indeed, introducing the inverse matrix $\alpha = \omega^{-1}$, we may define Θ via a connection ∇, by setting

$$\partial_i \xi^k + \xi^r \alpha_r^k(x) \omega_i^\tau(x) = 0.$$

An easy computation proves that ∇ has zero curvature and gives rise to a ∇-sequence which is a convenient Janet sequence. (The reader should not forget that the bundles are always defined up to an isomorphism.)

Finally, we notice that $C_r = \bigwedge^r T^* \otimes R_1 \simeq \bigwedge^r T^* \otimes T$ and the transition from the Spencer sequence to the Janet sequence is made by the following short exact sequences:

$$0 \longrightarrow \bigwedge^r T^* \otimes T \longrightarrow C_r(T) \xrightarrow{\Phi_r} \bigwedge^{r+1} T^* \otimes T \longrightarrow 0,$$

which are *precisely* the opposite of the previous sequences (which did *not* depend on the Lie equations but only on T and $q = 1$).

It is therefore *purely accidental*, in this case, that the Janet sequence and the Spencer sequence both coincide with the ∇-sequence

$$0 \longrightarrow \Theta \longrightarrow T \xrightarrow{\nabla} T^* \otimes T \xrightarrow{\nabla} \cdots \xrightarrow{\nabla} \bigwedge^n T^* \otimes T \longrightarrow 0,$$

although the *shift* produced by the formulas $C_r = \bigwedge^r T^* \otimes T$, $F_r = \bigwedge^{r+1} T^* \otimes T$ have led to the repeatedly mentioned confusion.

Using this Example, we shall now review the links between the nonlinear Janet sequence and the nonlinear Spencer sequence, studying in particular the transition from the source to the target. All other Examples presented in this Section could be similarly treated; however, this is the simplest case.

First of all we relate χ_0 to the differential invariants. Indeed, we have

$$\chi_{,i}^k = g_r^k \partial_i f^r - \delta_i^k = A_i^k - \delta_i^k.$$

Now, working over the source, we have:

$$\omega_r^\tau(x) A_i^r = \omega_r^\tau(x) g_k^r \partial_i f^k = \omega_k^\tau(y) \partial_i f^k.$$

In fact, the symbolic computation is as follows:

$$j_1(f)^{-1}(\omega) = j_1(f)^{-1}(f_1(\omega)) = (j_1(f)^{-1} \circ f_1)(\omega) = (f_1^{-1} \circ j_1(f))^{-1}(\omega).$$

Recall the general symbolic computation:

$$f_q^{-1}(\omega) = \overline{\omega},$$
$$j_1(f_q)^{-1}(j_1(\omega)) = j_1(\overline{\omega}),$$
$$f_{q+1}^{-1}(j_1(\omega)) \neq j_1(\overline{\omega}),$$
$$j_1(\omega) = j_1(f_q)(j_1(\overline{\omega})),$$
$$f_{q+1}^{-1} \circ j_1(f_q)(j_1(\overline{\omega})) - j_1(\overline{\omega}) = -BL(\chi_q)\overline{\omega},$$

and apply it to the present Example *over the source*:

$$\omega_k^\tau(f(x))f_i^k(x) = \overline{\omega}_i^\tau(x),$$

$$\frac{\partial \omega_k^\tau}{\partial y^l}(f(x))\partial_r f^l(x)f_i^k(x) + \omega_k^\tau(f(x))\partial_r f_i^k(x) = \partial_r \overline{\omega}_i^\tau(x),$$

$$\frac{\partial \omega_k^\tau}{\partial y^l}(f(x))f_r^l(x)f_i^k(x) + \omega_k^\tau(f(x))f_{ri}^k(x) \neq \partial_r \overline{\omega}_i^\tau(x),$$

$$\begin{cases} \dfrac{\partial \omega_k^\tau}{\partial y^l} = g_k^i \dfrac{\partial x^r}{\partial y^l}\partial_r \overline{\omega}_i^\tau(x) - g_k^i \dfrac{\partial x^r}{\partial y^l}\partial_r f_i^m(x)g_m^s \overline{\omega}_s^\tau(x), \\ \omega_k^\tau(f(x)) = g_k^i \overline{\omega}_i^\tau(x), \end{cases}$$

$$\left[g_k^j \frac{\partial x^s}{\partial y^l}\partial_s \overline{\omega}_j^\tau(x) - g_k^j \frac{\partial x^s}{\partial y^l}\partial_s f_j^m \omega_m^\tau(f(x)) \right] f_r^l f_i^k + \omega_k^\tau(f(x))f_{ri}^k - \partial_r \overline{\omega}_i^\tau(x)$$
$$= f_r^l \frac{\partial x^s}{\partial y^l}\partial_s \overline{\omega}_i^\tau(x) - f_r^l \frac{\partial x^s}{\partial y^l}\partial_s f_i^m \omega_m^\tau(f(x)) + \omega_k^\tau(f(x))f_{ri}^k - \partial_r \overline{\omega}_i^\tau(x)$$
$$= B_r^s \partial_s \overline{\omega}_i^\tau(x) - B_r^s \partial_s f_i^k \omega_k^\tau(f(x)) + \omega_k^\tau(f(x))f_{ri}^k - \partial_r \overline{\omega}_i^\tau(x)$$
$$= B_r^s(\partial_s \overline{\omega}_i^\tau(x) - A_s^u \partial_u \overline{\omega}_i^\tau) - B_r^s(\partial_s f_i^k - A_s^u f_{ui}^k)\omega_k^\tau(f(x))$$
$$= -B_r^s \chi_{,s}^u \partial_u \overline{\omega}_i^\tau - B_r^s \chi_{i,s}^u f_u^k \omega_k^\tau(f(x))$$
$$= -B_r^s \chi_{,s}^u \partial_u \overline{\omega}_i^\tau - B_r^s \chi_{i,s}^u \overline{\omega}_u^\tau(x)$$
$$= -B_r^s(\overline{\omega}_k^\tau(x)\chi_{i,s}^k + \chi_{,s}^r \partial_r \overline{\omega}_i^\tau(x)).$$

Now we shall illustrate the formula

$$(f_{q+1}^{-1} \circ j_1(f_q))^{-1}(j_1(\overline{\omega})) - j_1(\overline{\omega}) = L(\chi_q)\overline{\omega}.$$

Indeed, from the preceding formulas we have

$$[g_k^j g_l^s \partial_s \overline{\omega}_j^\tau(x) - g_k^j g_l^s f_{sj}^m \omega_m^\tau(f(x))]\partial_r f^l f_i^k + \omega_k^\tau(f(x))\partial_r f_i^k - \partial_r \overline{\omega}_i^\tau(x)$$
$$= g_l^s \partial_r f^l \partial_s \overline{\omega}_i^\tau - g_s^l \partial_r f^l f_{si}^m \omega_m^\tau(f(x)) + \omega_k^\tau(f(x))\partial_r f_i^k - \partial_r \overline{\omega}_i^\tau(x)$$
$$= \overline{\omega}_k^\tau(x)\chi_{i,r}^k + \chi_{,r}^s \partial_s \overline{\omega}_i^\tau(x).$$

Finally, we illustrate the formula

$$(f_{q+1} \circ j_1(f_q)^{-1})^{-1}(j_1(\omega)) - j_1(\omega) = L(\sigma_q)\omega,$$

with $\sigma_q = \overline{D} f_{q+1}^{-1}$ defined *over the target*. Again, from the preceding formulas we get

$$\frac{\partial x^r}{\partial y^l}\left(\frac{\partial \omega_u^\tau(f(x))}{\partial y^v} f_r^v f_i^u + \omega_u^\tau(f(x)) f_{ri}^u - \omega_u^\tau(f(x)) \partial_r f_i^u\right) - \frac{\partial \omega_k^\tau(f(x))}{\partial y^l}$$

$$= \frac{\partial \omega_k^\tau(f(x))}{\partial y^v}\left(\frac{\partial x^r}{\partial y^l} f_r^v - \delta_l^v\right) + \omega_u^\tau(f(x))\left(\frac{\partial x^r}{\partial y^l} g_k^i f_{ri}^u - g_k^i \frac{\partial f_i^u}{\partial y^l}\right)$$

$$= \omega_u^\tau(f(x)) \sigma_{k,l}^u(f(x)) + \sigma_{,l}^v(f(x)) \frac{\partial \omega_k^\tau(f(x))}{\partial y^v}.$$

The above, delicate, computations need no further comment! They explain why the link between the Janet sequence and the Spencer sequence has never been found.

Example 10 (Riemann structure). In contrast to the Examples above, here we start with the natural bundle $\mathcal{F} = S_2 T^*$, and we shall recover *all* the classical concepts of Riemannian geometry in a new framework. The particular case $n = 3$ will be treated in more detail because of its application to mechanics in Chapter VIII.

The general finite and infinitesimal Lie equations are

$$\mathcal{R}_1(\omega) \qquad \Phi_{ij}(y_1) \equiv \omega_{kl}(y) y_i^k y_j^l = \omega_{ij}(x) \qquad \text{(Lie form)},$$

$$\mathrm{R}_1(\omega) \qquad \Omega_{ij} \equiv \omega_{rj}(x)\xi_i^r + \omega_{ir}(x)\xi_j^r + \xi^r \partial_r \omega_{ij}(x) = 0 \qquad \text{(Medolaghi form)}.$$

In particular, for an $\xi_1 \in J_1(T)$ we have the variation $L(\xi_1)\omega = \Omega \in S_2 T^*$ of ω.

We check on this Example how the symbol of $\mathcal{R}_1$ is determined by that of R_1. We have:

$$\omega_{kl}(y) y_i^k v_j^l + \omega_{kl}(y) y_j^l v_i^k = 0.$$

Setting $v_i^k = y_r^k w_i^r$, we obtain

$$\omega_{rj}(x) w_i^r + \omega_{ir}(x) w_j^r = 0,$$

and we recognise the equations defining the symbol of R_1. Setting now $\xi_{j,i} = \omega_{rj}(x)\xi_i^r$, we obtain $\xi_{i,j} + \xi_{j,i} = 0$ and $M_1 \simeq \wedge^2 T^*$. Accordingly, if $\det \omega \neq 0$, we have $\dim M_1 = n(n-1)/2$, and M_1 is surely a vector bundle. However, in accordance with the general theory we have to look at M_2. We successively obtain

$$\mathrm{M}_1 \qquad \omega_{rj}(x)\xi_i^r + \omega_{ir}(x)\xi_j^r = 0,$$

$$\mathrm{M}_2 \qquad \omega_{rj}(x)\xi_{is}^r + \omega_{ir}(x)\xi_{js}^r = 0.$$

Using a convenient linear combination, we obtain

$$\begin{array}{r|l} +1 & \omega_{rs}\xi_{ji}^r + \omega_{jr}\xi_{si}^r = 0 \\ +1 & \omega_{rs}\xi_{ij}^r + \omega_{ir}\xi_{sj}^r = 0 \\ -1 & \omega_{rj}\xi_{is}^r + \omega_{ir}\xi_{js}^r = 0 \\ \hline & 2\omega_{rs}\xi_{ij}^r = 0 \end{array}$$

If $\det \omega \neq 0$, we have $M_2 = 0$, and so M_1 is of finite type while M_2 is trivially involutive. According to the general theory, M_1 cannot be involutive. However, it could be 2-acyclic, and so it is worthwhile to determine $H_1^2(M_1)$. Starting with the defining morphism $\Phi\colon J_1(T) \to S_2 T^*\colon \xi_1 \to L(\xi_1)\omega$, we successively obtain the two following commutative diagrams, with dimensions boxed in the corresponding diagrams below. In the second diagram, the exactness of the top row implies the surjectivity of $\pi_1^2 \colon R_2 \to R_1$ *and* the surjectivity of $\rho_1(\Phi)$ (an exercise in chasing).

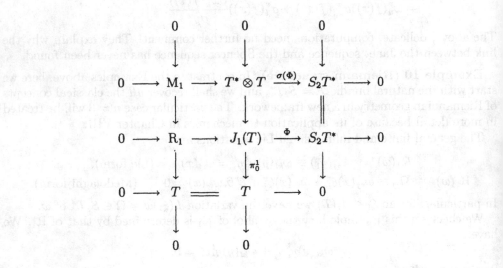

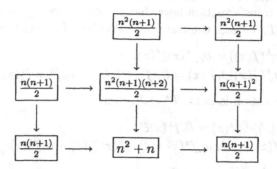

The trick now is to describe the prolongation $\rho_1(\Phi)$ by introducing the so-called Christoffel symbols

$$\gamma_{ij}^k = \frac{1}{2}\omega^{kr}(\partial_i\omega_{rj} + \partial_j\omega_{ri} - \partial_r\omega_{ij}),$$

where we introduce the inverse of ω as a section of S_2T defined by the conditions $\omega^{ir}(x)\omega_{rj}(x) = \delta_j^i$. Again, in this situation a confusion has emerged between the Janet sequence and the Spencer sequence, but nobody knows about it because in the literature the Lie equations are almost never introduced.

Indeed, on the one hand γ is a geometric object, and the condition $\det\omega \neq 0$, allowing us to define γ, also allows us to introduce an isomorphism $j_1(\omega) \simeq (\omega,\gamma)$. To see this, we could look directly at the transformation laws of γ, which can be found in any textbook on tensor calculus. However, we prefer to express the conditions for a section $f_2 \in \Pi_2$ to become a section of $\mathcal{R}_2(\omega,\gamma)$ by preserving (ω,γ): It suffices to add to the equations of $\mathcal{R}_1(\omega)$ already written down the following equations:

$$g_l^k(x)(f_{ij}^l(x) + \gamma_{rs}^l(f(x))f_i^r(x)f_j^s(x)) = \gamma_{ij}^k(x),$$

describing the relation $f_2^{-1}(\gamma) = \gamma$ over the source. From this relation we deduce that γ is a section of a natural bundle of order 2 which is an affine bundle over X modeled on $S_2T^* \otimes T$.

However, on the other hand γ is usually called a '*connection*' because of the existence of a so-called *covariant derivative* ∇ such that

$$\nabla_i\xi^k = \partial_i\xi^k + \gamma_{ri}^k(x)\xi^r$$

behave like a tensor of $T^* \otimes T$, although $\partial_i\xi^k$ does not behave at all like a tensor. We shall now explain the background of this definition. To start with, recall that a covariant derivative can be defined whenever we have a section χ of $T^* \otimes J_1(T)$ projecting onto $\mathrm{id}_T \in T^* \otimes T$, because $j_1(\xi)$ and $\chi(\xi)$ are two sections of $J_1(T)$ over the section ξ of T, and so the difference $j_1(\xi) - \chi(\xi)$ makes sense as a section of $T^* \otimes T$, because $J_1(T)$ is an affine bundle over X modeled on $T^* \otimes T$. Of course, we could use an R_1-connection as well, because $R_1 \subset J_1(T)$, by requiring

$$\omega_{rj}(x)\chi_{i,s}^r + \omega_{ir}(x)\chi_{j,s}^r + \partial_s\omega_{ij}(x) = 0.$$

Any two such R_1-connections differ by a section of $T^* \otimes M_1$. However, we deduce from the corresponding transformation laws that the difference $\chi^k_{i,j} - \chi^k_{j,i}$ is a well-defined section of $\wedge^2 T^* \otimes T$. Indeed, the transformation laws for $J_1(T)$ are:

$$\begin{cases} \eta^k(f(x)) = \partial_r f^k(x)\xi^r(x), \\ \eta^k_i(f(x))\partial_j f^l(x) = \partial_{jr} f^k(x)\xi^r(x) + \partial_r f^k(x)\xi^r_j(x). \end{cases}$$

Hence the transformation laws for $T^* \otimes J_1(T)$ are:

$$\begin{cases} \eta^k_{,s}(f(x))\partial_i f^s(x) = \partial_r f^k(x)\xi^r_{,i}(x), \\ \eta^k_{l,s}(f(x))\partial_j f^l(x)\partial_i f^s(x) = \partial_{jr} f^k(x)\xi^r_{,i}(x) + \partial_r f^k(x)\xi^r_{j,i}(x). \end{cases}$$

Setting $\xi^r_{,i} = \delta^r_i \Rightarrow \eta^k_{,s} = \delta^k_s$, we obtain

$$\eta^k_{l,s}(f(x))\partial_j f^l(x)\partial_i f^s(x) = \partial_{ij} f^k(x) + \partial_r f^k(x)\xi^r_{j,i}(x),$$

and thus

$$(\eta_{k,l}(f(x)) - \eta_{l,k}(f(x)))\partial_i f^k(x)\partial_j f^l(x) = \partial_r f^s(x)(\xi^r_{j,i}(x) - \xi^r_{i,i}(x)).$$

Hence we may look for a *symmetric connection* such that $\chi^k_{i,j} - \chi^k_{j,i} = 0$. However, we have an isomorphism

$$0 \longrightarrow T^* \otimes M_1 \overset{\delta}{\longrightarrow} \wedge^2 T^* \otimes T \longrightarrow 0.$$

Indeed, δ is injective because $H^1_1(M_1) = 0$ and $M_2 = 0$, hence surjective because both spaces have dimension $n^2(n-1)/2$. It follows that there can only be one symmetric connection, defined by

$$\omega_{rj}(x)\chi^r_{is} + \omega_{ir}(x)\chi^r_{js} + \partial_s \omega_{ij}(x) = 0.$$

A convenient linear combination gives

$$2\omega_{sk}(x)\chi^k_{ij} + \partial_i \omega_{sj}(x) + \partial_j \omega_{si}(x) - \partial_s \omega_{ij}(x) = 0,$$

i.e. $\chi^k_{ij} = -\gamma^k_{ij}$. Hence γ can be regarded as the only symmetric R_1-connection existing, and it is thus a section of $T^* \otimes R_1$, or of the first Spencer bundle C_1. Of course, an R_1-connection has no chance of being a geometric object, unless it is considered as a section of $T^* \otimes J_1(T)$. In practice, the reader should take care of the minus sign, which stems from the conventional definition of γ in tensor calculus.

However, there is an additional confusion, which is much more tricky to explain as it is deeply related to the construction of the nonlinear Spencer sequence. Indeed, recall that $\overline{D}'$ makes sense only in relation to $\overline{D}$; namely, the construction of $\overline{D}'$ makes it possible to state compatibility conditions for $\overline{D}$. However, in Cartan's formalism of 'torsion' and 'curvature', there is *no reference at all* to a previous operator, unless we refer to the gauge sequence. In that case, as the index τ for $\mathcal{G}$ and the index i for T^* in $A = (A^\tau_i) \in T^* \otimes \mathcal{G}$ are unrelated (G does not act on X), the *concept of skewsymmetrisation has no meaning at all*. Hence we *have to* use the nonlinear Spencer sequence, $\cdots$ and we arrive at a contradiction. Indeed, while the formula

$$\partial_i A^k_j - \partial_j A^k_i - A^r_i \chi^k_{r,j} + A^r_j \chi^k_{r,i} = 0$$

looks like the one of Cartan for 'torsion', the indices now range from 1 to n. The skewsymmetric term $\chi^k_{j,i} - \chi^k_{i,j}$ therefore *seems* to appear when $A^k_i = \delta^k_i$ (as everybody, Cartan, Trautman, $\cdots$, seems to believe). However, this completely contradicts the interpretation of $-\chi_q$ as a connection, provided by Remark D.12, which *implies* $A = 0$ (this is coherent with the use of principal bundles). Of course, when $A = 0$ the so-called 'torsion' part disappears completely, and terms depending on second-order jets disappear in the 'curvature' part. Hence, it remains to prove that *the desired skewsymmetric term does not come from* $\overline{D}'$, the operator corresponding to the Maurer–Cartan operator in the gauge sequence, but from the restriction of $D: T^* \otimes R_1 \to \bigwedge^2 T^* \otimes T$ to a connection obtained by keeping in mind that the restriction of D to $T^* \otimes M_1$ is just $-\delta: T^* \otimes M_1 \to \bigwedge^2 T^* \otimes T$. Indeed, by the definition of the covariant derivative ∇, we have

$$\nabla_\xi \eta = L(-\chi_1(\xi))\eta.$$

Therefore we find for the usual definition of torsion

$$\begin{aligned}
\nabla_\xi \eta - \nabla_\eta \xi - [\xi, \eta] &= L(-\chi_1(\xi))\eta - L(-\chi_1(\eta))\xi - [\xi, \eta] \\
&= [\xi, \eta] - i(\eta)D(\chi_1(\xi)) - [\eta, \xi] + i(\xi)D(\chi_1(\eta)) - [\xi, \eta] \\
&= i(\xi)D(\chi_1(\eta)) - i(\eta)D(\chi_1(\xi)) + [\xi, \eta] \\
&= (D\chi_1)(\xi, \eta).
\end{aligned}$$

This result shows that the definitions of torsion and curvature in the literature are based on a *triple confusion* between:

1) the Janet sequence and the Spencer sequence;
2) the Spencer sequence and the gauge sequence;
3) the operators $\overline{D}'$ and D.

This is one more reason for *completely* revising the concepts of 'torsion' and 'curvature' by means of their applications, in the light of this new framework.

Thus far, we have not touched upon the formal integrability of the general Killing equations. The reason is that we need one more prolongation. Accordingly, we shall need third-order conditions for the metric ω, while the discovery of an integrating factor has been done by Eisenhart in the famous reference [58]. There is the following commutative and exact diagram, in which $F_1 = \operatorname{coker} \sigma_2(\Phi)$, $Q = \operatorname{coker} \rho_2(\Phi)$:

$$
\begin{array}{ccccccccc}
& & 0 & & 0 & & & & \\
& & \downarrow & & \downarrow & & & & \\
0 \longrightarrow & S_3 T^* \otimes T & \xrightarrow{\sigma_2(\Phi)} & S_2 T^* \otimes S_2 T^* & \longrightarrow & F_1 & \longrightarrow & 0 \\
& \downarrow & & \downarrow & & \downarrow & & \downarrow & \\
0 \longrightarrow R_3 \longrightarrow & J_3(T) & \xrightarrow{\rho_2(\Phi)} & J_2(S_2 T^*) & \longrightarrow & Q & \longrightarrow & 0 \\
& \downarrow{\scriptstyle\pi_2^3} & & \downarrow{\scriptstyle\pi_1^2} & & \downarrow & & \\
0 \longrightarrow R_2 \longrightarrow & J_2(T) & \xrightarrow{\rho_1(\Phi)} & J_1(S_2 T^*) & \longrightarrow & 0 & & \\
& \downarrow & & \downarrow & & & & \\
& & 0 & & 0 & & & & \\
\end{array}
$$

So, using the Snake Theorem in Chapter I we see that the surjectivity of $\pi_2^3 \colon R_3 \to R_1$ (the *integrability conditions*) is equivalent to the injectivity of $F_1 \to Q$, and so to the isomorphism $F_1 \simeq Q$ (the *compatibility conditions*).

To study the integrability conditions for $R_2(\omega, \gamma)$, we write the second-order equations in Medolaghi form with respect to the Christoffel symbols:

$$
R_2(\omega, \gamma) \quad \left\{
\begin{array}{l}
\omega_{rj}(x)\xi_i^r + \omega_{ir}(x)\xi_j^r + \xi^r \partial_r \omega_{ij}(x) = 0 \\
\xi_{ij}^k + \gamma_{rj}^k(x)\xi_i^r + \gamma_{ir}^k(x)\xi_j^r - \gamma_{ij}^r(x)\xi_r^k + \xi^r \partial_r \gamma_{ij}^k(x) = 0.
\end{array}
\right.
$$

The second set of equations can be written as

$$
L(\xi_2)\gamma = 0,
$$

because the difference of two Christoffel symbols is a pure tensor in $S_2 T^* \otimes T$. Since $M_2 = 0$, we may start afresh the study of the criterion for formal integrability for R_2, by using $J_1(\mathcal{F})$ or, equivalently, the natural second-order bundle with section (ω, γ) which already describes the two objects individually.

Now, $q = 2$ and *the consideration of $\partial_i \gamma_{lj}^k - \partial_j \gamma_{li}^k$ only should not lead to linear terms with respect to γ*. Indeed, when prolonging the second-order equations of R_q, we should obtain *quadratic terms* in γ. In agreement with the general theory, this is the reason why, after some work at the beginning of tensor calculus, people decided to introduce the so-called *Riemann tensor*, which transforms linearly:

$$
\rho_{lij}^k \equiv \partial_i \gamma_{lj}^k - \partial_j \gamma_{li}^k + \gamma_{lj}^r \gamma_{ri}^k - \gamma_{li}^r \gamma_{rj}^k.
$$

Though it could be tempting to identify these terms with a part of $\overline{D}'$, while taking into account the change of sign already explained, the reader should not forget that *we deal with the Janet sequence and not with the Spencer sequence*. The reader should also not forget that if the identification is done, $(\delta_i^k, -\gamma_{ji}^k)$ is an R_1-connection, and we have to identify $-(\delta_i^k, -\gamma_{ji}^k) = (-\delta_i^k, \gamma_{ji}^k)$ with $(-\delta_i^k, \chi_{j,i}^k)$. This explains the choice of sign for the Riemann curvature.

Therefore, we must have

$$\rho^k_{rij}\xi^r_l + \rho^k_{lrj}\xi^r_i + \rho^k_{lir}\xi^r_j - \rho^r_{lij}\xi^k_r + \xi^r\partial_r\rho^k_{lij} = 0$$

whenever

$$\omega_{rj}\xi^r_i + \omega_{ir}\xi^r_j + \xi^r\partial_r\omega_{ij} = 0.$$

In $J^0_1(T) = T^* \otimes T$ we must therefore have

$$\rho^k_{rij}\xi^r_l + \rho^k_{lrj}\xi^r_i + \rho^k_{lir}\xi^r_j - \rho^r_{lij}\xi^k_r = 0$$

whenever

$$\omega_{rj}\xi^r_i + \omega_{ir}\xi^r_j = 0,$$

in order to ensure the surjectivity of $\pi^3_2\colon R^0_3 \to R^0_2$. We now recall some properties of the Riemann tensor that can be found in any textbook on tensor calculus:

$$\begin{cases} \rho^k_{lij} + \rho^k_{ijl} + \rho^k_{jli} = 0, \\ \omega_{rj}\rho^r_{ikl} + \omega_{ir}\rho^r_{jkl} = 0. \end{cases}$$

Setting $\rho_{ijkl} = \omega_{ir}\rho^r_{jkl}$ we can also obtain from these relations (exercise):

$$\rho_{ijkl} = \rho_{klij}.$$

Transforming the second set of relations defining $R^0_1 = M_1$ into $\xi_{i,j} + \xi_{j,i} = 0$, we finally obtain the *integrability conditions of the first kind*:

$$\delta^s_i\rho^t_{jkl} - \delta^s_j\rho^t_{ikl} + \delta^s_k\rho^t_{lij} - \delta^t_i\rho^s_{jkl} + \delta^t_j\rho^s_{ikl} - \delta^t_k\rho^s_{lij} - \delta^t_l\rho^s_{kij} = 0.$$

Contracting with respect to s and i and setting

$$\rho_{ij} = \rho_{ji} = \rho^r_{ijr},$$

we find

$$\rho^k_{lij} = \frac{1}{n-1}(\delta^k_j\rho_{li} - \delta^k_i\rho_{lj}).$$

However, we must also have

$$\rho_{rj}\xi^r_i + \rho_{ir}\xi^r_j = 0, \qquad \forall\xi^0_1 \in R^0_1,$$

and so, setting

$$\rho = \omega^{ij}\rho_{ij},$$

we must also have (exercise):

$$\rho_{ij} = \frac{1}{n}\rho\omega_{ij}.$$

But ρ is a scalar, and so we must have a unique *integrability condition of the second kind*:

$$\frac{1}{n(n-1)}\rho = c.$$

Combining the results obtained thus far using an equivariant section, we obtain the *constant curvature condition*, in the form

$$\rho^k_{lij} = c(\delta^k_j\omega_{li} - \delta^k_i\omega_{lj}).$$

Finally, it remains to understand the geometric meaning of the Riemann curvature in the new framework. This will provide an absolutely striking result, explaining in an intrinsic manner all the properties of the Riemann tensor in such a way that (ω, ρ) becomes a section of a natural vector bundle $\mathcal{F}_1$ over the section ω of $\mathcal{F} = S_2 T^*$ with $\det \omega \neq 0$. For this we apply the δ-sequence to the top line of the previous diagram, keeping in mind the top lines of all previous diagrams. We obtain the following commutative and exact diagram:

$$
\begin{array}{ccccccccc}
& & 0 & & & & 0 & & \\
& & \downarrow & & & & \downarrow & & \\
0 & \longrightarrow & S_3 T^* \otimes T & \xrightarrow{\sigma_2(\Phi)} & S_2 T^* \otimes S_2 T^* & \longrightarrow & F_1 & \longrightarrow & 0 \\
& & \downarrow{\scriptstyle\delta} & & \downarrow{\scriptstyle\delta} & & & & \\
0 & \longrightarrow & T^* \otimes S_2 T^* \otimes T & \xrightarrow{\sigma_1(\Phi)} & T^* \otimes T^* \otimes S_2 T^* & \longrightarrow & 0 & & \\
& & \downarrow{\scriptstyle\delta} & & \downarrow{\scriptstyle\delta} & & & & \\
0 \longrightarrow \wedge^2 T^* \otimes M_1 & \longrightarrow & \wedge^2 T^* \otimes T^* \otimes T & \xrightarrow{\sigma(\Phi)} & \wedge^2 T^* \otimes S_2 T^* & \longrightarrow & 0 & & \\
\downarrow{\scriptstyle\delta} & & \downarrow{\scriptstyle\delta} & & \downarrow & & & & \\
0 \longrightarrow \wedge^3 T^* \otimes T & = & \wedge^3 T^* \otimes T & \longrightarrow & 0 & & & & \\
\downarrow & & \downarrow & & & & & & \\
0 & & 0 & & & & & &
\end{array}
$$

Chasing as in the Exercise of Section A, we obtain the short exact sequence

$$ 0 \longrightarrow F_1 \longrightarrow \wedge^2 T^* \otimes M_1 \xrightarrow{\delta} \wedge^3 T^* \otimes T \longrightarrow 0. $$

Hence, from the top line of the diagram or from this sequence we obtain the striking combinatorics:

$$
\begin{aligned}
\dim F_1 &= \frac{n^2(n+1)^2}{4} - \frac{n^2(n+1)(n+2)}{6} \\
&= \frac{n^2(n-1)^2}{4} - \frac{n^2(n-1)(n-2)}{6} = \frac{n^2(n^2-1)}{12},
\end{aligned}
$$

which is known to be the number of independent components of the Riemann tensor.

Meanwhile, in this chase the last diagram identifies F_1 with the kernel:

$$ 0 \longrightarrow F_1 \longrightarrow \wedge^2 T^* \otimes T^* \otimes T^* \xrightarrow{(\delta, \sigma(\Phi))} \wedge^3 T^* \otimes T \times_X \wedge^2 T^* \otimes S_2 T^*, $$

and we recognise at once the two properties of the Riemann tensor recalled above. In particular, we notice that the left part of the diagram is absent in classical differential geometry.

Finally, we have the identification

$$ F_1 = H_1^2(M_1), $$

because $M_2 = 0$ and so $B_1^2(M_1) = 0$. Since the symbol map depends only on ω, the tensor bundle $\mathcal{F}_1$ is well defined as a subbundle of $\wedge^2 T^* \otimes T^* \otimes T^*$ over $\mathcal{F}$.

These results can be extended to the Bianchi identities, see [146], but we shall not need this generalisation.

We now turn to the nonlinear Spencer sequences. Since $M_2 = 0$, we can identify the first and second nonlinear Spencer sequences with the unique sequence

$$ 0 \longrightarrow \Gamma \longrightarrow \mathcal{R}_1 \xrightarrow{\ \overline{D}\ } T^* \otimes \mathrm{R}_1 \xrightarrow{\ \overline{D}'\ } \wedge^2 T^* \otimes \mathrm{R}_1 \,. $$

In the operator $\overline{D}$ the second-order jets must be replaced by γ and first-order jets, while in $\overline{D}'$ the second-order jets must be replaced using the formulas

$$ \chi^k_{ij,s} + \gamma^k_{rj}\chi^r_{i,s} + \gamma^k_{ir}\chi^r_{j,s} - \gamma^r_{ij}\chi^k_{r,s} + \chi^r_{,s}\partial_r \gamma^k_{ij} = 0. $$

These terms are unknown in the literature and in applications, since the Spencer operator D has never been used until now.

We conclude this Example by showing the link between the linear Janet sequence and the linear Spencer sequence, when $n = 3$ and ω is the standard euclidean metric.

Linearising the preceding results we immediately obtain:

$$ \mathcal{D}\xi = (\mathcal{L}(\xi)\omega, \mathcal{L}(\xi)\gamma), $$

$$ D_1 \qquad \begin{cases} \chi^k_{,i} = \partial_i \xi^k - \xi^k_i, \\ \chi^k_{j,i} = \partial_i \xi^k_j, \end{cases} $$

$$ D_2 \qquad \begin{cases} \partial_i \chi^k_{,j} - \partial_j \chi^k_{,i} - \chi^k_{i,j} + \chi^k_{j,i} = 0, \\ \partial_i \chi^k_{l,j} - \partial_j \chi^k_{l,i} = 0, \end{cases} $$

and the corresponding diagram, where

We leave it to the reader to obtain these results from the associated scheme describing the involutivity of R_2, and we notice the three identities:

$$\begin{cases} 6 - 18 + 18 - 6 = 0, \\ 3 - 30 + 60 - 45 + 12 = 0, \\ -3 + 24 - 42 + 27 - 6 = 0. \end{cases}$$

These results were found in the first place by E. and F. Cosserat (already in 1910!), and are quite transparent in the modern reference [177], in which the jet background is lacking.

CHAPTER VI

Differential Galois Theory

The main purpose of this Chapter is to combine

- DIFFERENTIAL ALGEBRA,
- DIFFERENTIAL GEOMETRY, and
- ALGEBRAIC GEOMETRY

in a unique framework, called DIFFERENTIAL ALGEBRAIC GEOMETRY.

The usefulness of the latter in applications will be shown in the next Chapter.

To this end we proceed in two steps. First we present a few differential methods for pure algebra, and show how to incorporate geometry in the standard approach towards classical (nondifferential) Galois theory. Then we will try to add the word 'differential' in front of the various concepts (fields, groups, invariants) found during the first step. As a byproduct, the intrinsic methods already developed will apply at once to differential algebra.

A. Pure Algebra

The purpose of this Section is to present new methods for studying purely algebraic problems. In particular, we will give details on how to apply differential methods in pure algebra, as sketched in the Introduction. We will also give a manner of using a geometric approach towards classical Galois theory, to prove that Galois theory is precisely the theory of principal homogeneous spaces for finite groups. These new methods will be illustrated in Section E by means of explicit examples.

Let k be a field of characteristic zero, i.e. a field containing as a subfield the field $\mathbb{Q}$ of rational numbers.

Keeping in mind the results and notations of Section IV.B, we introduce indeterminates $\chi_1, \ldots, \chi_n$ over k and consider the polynomial ring $k[\chi] = k[\chi_1, \ldots, \chi_n]$. It consists of the polynomials in $\chi_1, \ldots, \chi_n$ with coefficients in k.

For r given polynomials $P_1, \ldots, P_r \in k[\chi]$ we can ask several questions about the *algebraic set* determined by them, i.e. the set of points with coordinates in a given extension of k and annihilated by $P_1, \ldots, P_r$. In particular, we may ask whether or not this set is not empty, and if so, what could be a nice dimension concept for it.

Since it is well known that algebraic sets can be decomposed into finitely many *irreducible subsets*, called *varieties* and determined by prime ideals, the dimension concept must be coherent with this decomposition.

259

In actual practice, the usual definition is as follows, with respect to the ideal $\mathfrak{a} \subset k[\chi]$ generated by $P_1, \ldots, P_r$.

1) Find the *root* $\sqrt{\mathfrak{a}} \supseteq \mathfrak{a}$, consisting of the polynomials having a power in $\mathfrak{a}$. In particular, any ideal such that $\mathfrak{a} = \sqrt{\mathfrak{a}}$ is called *radical*, or *perfect*.
2) Use the fundamental theorem on decomposition, saying that any perfect ideal is a finite intersection of prime ideals, say $\mathfrak{a} = \cap_i \mathfrak{p}_i$.
3) For each prime ideal $\mathfrak{p} \subset k[\chi]$ we can introduce the *integral domain* $k[\chi]/\mathfrak{p}$ and the corresponding field of quotients $K = Q(k[\chi]/\mathfrak{p})$, to obtain the field extension K/k.
4) We can choose a transcendence basis of K/k, i.e. a maximal number of elements of K that are algebraically independent over k, as follows. Let $a_1 \in K$ be such that it does not satisfy any algebraic equation over k. We can look for $a_2 \in K \setminus k(a_1)$ not satisfying any equation over $k(a_1)$, etc., and we end with elements of K that are all algebraic over $k(a_1, \ldots, a_s)$. The integer s is called the transcendence degree of K/k, and we can prove that it does not depend on the construction above; $s = \operatorname{trd} K/k$ is also called the *dimension* of $\mathfrak{p}$.
5) The dimension of $\mathfrak{a}$ is defined as the maximum of the s_i.

It would be nice if we could avoid the search for the root of an ideal and its decomposition into prime components. To this end we adopt the following procedure for transforming a problem concerning algebraic equations into a problem concerning PDE with constant coefficients and one unknown.

1) To any monomial $\chi_\mu = (\chi_1)^{\mu_1} \ldots (\chi_n)^{\mu_n}$ we associate the derivative ∂_μ or, equivalently, the jet coordinate y_μ for one indeterminate y. By k-linearity we can extend this definition to arbitrary polynomials in $k[\chi]$. Hence also to $P_1, \ldots, P_r$, and to the corresponding ideal in $k[\chi]$ will correspond a number of linear PDE in one unknown with constant coefficients and their prolongations.
2) In general, this linear system is not formally integrable. However, by applying the 'up-and-down' process we find an involutive system with the same solutions. The involutive symbol will determine the characters $\alpha_q^1, \ldots, \alpha_q^n$ for q large enough, and we *define* the *dimension* as the number of nonzero characters. It is also the highest power of r in $\dim R_{q+r}$, and it depends only on the involutive symbol M_q of the involutive system R_q. Notice that this system is automatically (sufficiently) regular, because it has constant coefficients.

The link with the purely algebraic problem is easily found by using, if necessary, a linear change of χ in such a way that $\chi_1, \ldots, \chi_s$ becomes a transcendence basis in the case of a single irreducible algebraic set. In that case $\chi_{s+1}, \ldots, \chi_n$ are algebraic over $k(\chi_1, \ldots, \chi_s)$. Accordingly, looking at the corresponding PDE we see that, for all orders q large enough, all jet coordinates of class $\geq s+1$, i.e. *not containing the indices* $1, \ldots, s$ are principal, i.e. have the highest lexicographic ranking in each equation. It follows that $(M_q)^s = \cdots = (M_q)^n = 0$. Alternatively, this result provides a way of computing the number of polynomials of a given degree $q + r$ in the ideal. The actual definition of 'dimension' is *unique*, since it is based on pure linear algebra.

Now we give some applications of these ideas.

As a first striking application, consider a number of polynomials in one indeterminate χ, with coefficients in k. We look for the greatest common divisor (GCD). For this we pass to ODE, and look for the corresponding formally integrable system (involutivity of the symbol is automatically satisfied, since there is only one independent variable). The GCD is given by the ODE of lowest order, since all its prolongations are contained in all other ODE. In particular, if $y = 0$ appears (as a generating lowest order ODE), then 1 belongs to the ideal, which thus has no solution. This result can be extended to polynomials in several independent variables.

The so-called *belonging problem* can also be solved by this technique. Indeed, let $P, P_1, \ldots, P_r \in k[\chi_1, \ldots, \chi_n]$ be given polynomials. We wish to know whether or not P belongs to the ideal generated by $P_1, \ldots, P_r$ in $k[\chi]$. To this end we obtain an involutive system as before, and prolongate it to an order (at most) equal to the degree of P. This leads to a problem of pure linear algebra. It amounts to verifying that the PDE corresponding to P is a linear combination of the PDE determining the involutive system up to that order. Of course, *formal integrability is essential in this verification*.

We give a few comments regarding the effectivity of this method, although we do not wish to study the complexity of the algorithms just described. Nevertheless, in view of the structure of the 'up-and-down' procedure, we see that 2-acyclicity plays a key role although, to the best of our knowledge, it has never been considered in pure algebra. Also, since the use of δ-regular coordinates is unavoidable in the differential approach, we believe that it *is also* unavoidable in the purely algebraic approach, even if it has to be hidden behind certain, so-called 'technical', arguments.

We now turn to a surprising result that will be given a full explanation in the next Section and that will be crucial for the applications in control theory given in the next Chapter.

The study of algebraic sets points out the importance of prime ideals, i.e. ideals such that if a product is in the ideal, one of the factors must be in the ideal too. We shall explain the corresponding property in terms of PDE, and generalize it.

To this end we need a few abstract definitions. When dealing with polynomial rings in one indeterminate, we have seen that any ideal is generated by a single polynomial. More generally we can state

Definition 1. An ideal generated by one element is called a *principal ideal*. A ring A is called a *principal ideal ring* if every ideal in it is principal.

Furthermore, if A is also an *integral domain* (i.e. it contains no divisors of zero, i.e. $ab = 0$ with $a, b \in A$ implies a *or* b is zero), then A is called a *principal ideal domain*.

We now turn to another property possessed by rings of linear differential operators in *several* independent variables and *one* unknown.

Indeed, let $\mathcal{P}, \mathcal{Q}$ be linear partial differential operators acting on the unknown y. We consider the linear inhomogeneous system

$$\mathcal{P}y = u, \qquad \mathcal{Q}y = v,$$

which is similar to the Janet example. Let p be the order of $\mathcal{P}$ and q the order of $\mathcal{Q}$, with $p \leq q$. After r prolongations the number of jet coordinates in y is at least $(p + r + n)!/(p + r)! \, n!$, while the number of jet coordinates in u, v is $2(r + n)!/r! \, n!$.

As polynomials in r, they are both of degree n, with respective coefficients $1/n!$ and $2/n!$. Hence, for r large enough we can use Cramer's rule, to obtain at least one compatibility condition for u and v:

$$\mathcal{A}u = \mathcal{B}v,$$

where $\mathcal{A}$ and $\mathcal{B}$ are convenient linear differential operators (they can be readily determined by means of a computer, as in the Introduction). As a byproduct there is an identity of the form

$$\mathcal{A}P \equiv \mathcal{B}Q.$$

We state:

Definition 2. A ring A is called an *Ore ring* if for $p, q \in A$ there are $a, b \in A$ such that $ap = bq$.

Let A be an arbitrary ring and M an A-module.

Definition 3. An element x of an A-module M is called a *torsion element* if there is $p \in A$ such that $px = 0$. If all elements of M are torsion elements, M is called a *torsion module*. On the other hand, M is called a *torsion free module* if only the zero element is a torsion element.

In the intermediate case we have

Proposition 4. *If A is an Ore ring, the set of torsion elements of an A-module M is a submodule, $\mathrm{tor}(M)$ of M, called the torsion submodule of M.*

PROOF. Let x be a torsion element, with $px = 0$ as in Definition 3, and consider qx for $q \in A$. By Definition 2 there are $a, b \in A$ such that $ap = bq$, and so $bqx = apx = 0$, proving that qx is also a torsion element. Let y be a torsion element with $qy = 0$. Then we similarly have $ap(x + y) = apx + bqy = 0$, and $x + y$ is also a torsion element. $\square$

Definition 5. An A-module M is said to be *finitely generated* if every element is of the form $\sum_{i=1}^{n} a_i x_i$ with $a_i \in A$, $x_i \in M$. A finitely generated A-module M is said to be *free* if it admits a *basis*, i.e. generators x_i as above such that $\sum_{i=1}^{n} a_i x_i = 0$ implies $a_i = 0$.

In the situation considered at the beginning of this Section, if we start with a prime ideal $\mathfrak{p} \subset k[\chi]$, then $ab \in \mathfrak{p}$ implies $a \in \mathfrak{p}$ *or* $b \in \mathfrak{p}$. Passing to linear differential operators in one unknown, we immediately find the Ore property (since there is only one unknown!) and we can define the ring $\mathcal{D} = k[\partial_1, \dots \partial_n]$ of linear differential operators. In our case k consists of 'constants', and the corresponding ring is commutative, although we can consider a more general situation as in the next Section, where k will be a differential field. Accordingly, we may consider the $\mathcal{D}$-module made up by all linear differential operators in one (as in the previous situation) or a certain number of unknowns. A linear system of PDE will determine a submodule of that module, and we will be concerned with such a submodule or with the corresponding finitely generated quotient $\mathcal{D}$-module M. Translating the previous property of prime ideals in terms of operators, we see that, in the commutative case of one unknown considered above, $\mathcal{A}\mathcal{B}y$ belonging to the submodule implies that either $\mathcal{A}y$ or $\mathcal{B}y$ belongs to the submodule. More generally, we say that the submodule is *prime* (*primary*) if, whenever $\mathcal{B}y \notin$ submodule and $\mathcal{A}\mathcal{B}y \in$ submodule, then $\mathcal{A}\times$ total module $\subset$ submodule (respectively, $\mathcal{A}^\nu \times$ total module $\subset$ submodule, for some integer ν). It can be easily seen that this definition depends only on the quotient module M, by rephrasing it as follows: if $\mathcal{A}$ is a zero divisor for M, then $\mathcal{A}$ belongs to the *annihilator* of M (to its root). In fact, for a module over a ring we see that the existence of a *'torsion element'* in the module is equivalent to the existence of a *'zero divisor'* in the ring.

Among the particular situations in which one is sure to be dealing with a prime submodule, the lack of zero divisors is particularly interesting. It occurs when M is torsion free. However, the reader should notice that in the case of one unknown and a strict submodule, the quotient module is automatically a torsion module, even if the submodule is prime.

The study of $\mathcal{D}$-modules is a fashionable subject nowadays. However, it falls outside the scope of this book, since we are concerned only with commutative algebra. Nevertheless, we shall study in more detail the property of being torsion free, as it will become a key step towards an intrinsic definition of controllability, along the lines sketched in the Introduction, that will be extended to nonlinear control systems in the next Chapter, using the results in the next Section. We shall restrict our attention to $\mathcal{D}$-modules.

If M is a torsion module or is torsion free, the situation is clear. We shall now pay attention to the intermediate case, which will be of some importance in applications to control theory in the next Chapter.

Of course, a free module is torsion free. The converse is not evident at all in the case of one unknown, and surely fails to be true for several unknowns. The simplest counterexample is provided by the divergence free condition for a vector field in dimension 2 (exercise). On the other hand, a very simple example of a torsion module is provided by the Cauchy–Riemann system in dimension 2, where each unknown has zero Laplacian (exercise). (See Section V.E for this Example and others.)

We restrict our attention to the case of one unknown and to constant coefficients. In the abstract setting we consider modules over principal ideal domains that are simultaneously commutative rings. In the general situation the following holds [107]:

Proposition 6. *If A is an Ore ring and M is a finitely generated A-module, there is a short exact sequence of A-modules*

$$0 \longrightarrow \mathrm{tor}(M) \longrightarrow M \longrightarrow M/\mathrm{tor}(M) \longrightarrow 0,$$

where $M/\mathrm{tor}(M)$ is finitely generated and torsion free.

PROOF. Taking a finite family (x_i) of generators of M and letting $\bar{x}_i$ be the image of x_i under the canonical projection $M \to M/\mathrm{tor}(M)$, we see that the image of any element $\sum_{i=1}^{n} a_i x_i$ of M can be written $\sum_{i=1}^{n} a_i \bar{x}_i$, and so $M/\mathrm{tor}(M)$ is also finitely generated.

Let $x \in M$ and let $\bar{x}$ be its image. If there is a $b \in A$ such that $b\bar{x} = 0$, then $bx \in \mathrm{tor}(M)$, and there is an $a \in A$ such that $abx = 0$, hence $x \in \mathrm{tor}(M)$, or $\bar{x} = 0$, proving that $M/\mathrm{tor}(M)$ is torsion free. $\square$

In the restricted situation considered right above the previous Proposition we have

Theorem 7. *If A is a commutative principal ideal domain, then a finitely generated torsion free A-module is free.*

PROOF. Let $\{u_1, \ldots, u_r\}$ be a maximal set of linearly independent elements of M taken from a finite set $\{x_1, \ldots, x_n\}$ of generators. Accordingly, for each x_i there is $a_i \in A$ such that $a_i x_i$ belongs to the submodule generated by $u_1, \ldots, u_r$. If $a = a_1 \ldots a_n$, then aM is contained in the latter submodule, and $a \neq 0$, because otherwise one of the a_i should be zero too, contradicting the linear independence of $u_1, \ldots, u_r$. Since M is torsion free, the map $x \to ax$ is an injective homomorphism of M into the finitely generated free module generated by $\{u_1, \ldots, u_r\}$. It remains to prove that any submodule M of a free module F is also free.

In our case, F has a finite basis $\{u_1, \ldots, u_r\}$. Let M_s be the intersection of M with the submodule $(u_1, \ldots, u_s)$ generated by $\{u_1, \ldots, u_s\}$ when $s \leq r$. In particular, $M_1 = M \cap (u_1)$ is of the form $(a_1 u_1)$ for some $a_1 \in A$ because of the principal ideal assumption. Hence, M_1 is either 0 or free of dimension 1. By induction we may assume that M_s is free of dimension $\leq s$. Let $\mathfrak{a}$ be the set of elements $a \in A$ for which there is $x \in M_{s+1}$ that can be written as:

$$x = b_1 u_1 + \cdots + b_s u_s + a u_{s+1},$$

with $b_1, \ldots, b_s \in A$. Then $\mathfrak{a}$ is clearly an ideal of A, so that A is principal and generated by a_{s+1}. If $a_{s+1} = 0$, then $M_{s+1} = M_s$ and by induction the proof is complete. If $a_{s+1} \neq 0$, we let $v \in M_{s+1}$ be such that

$$v = a_1 u_1 + \cdots + a_s u_s + a_{s+1} u_{s+1}.$$

Since $a = c a_{s+1}$ for some $c \in A$, we have $x - cv \in M_s$, and so $M_{s+1} = M_s + (v)$ is a direct sum, because clearly $M \cap (v) = 0$ if $a_{s+1} \neq 0$. By induction we have proved that M is free. $\square$

Corollary 8. *If M is a finitely generated module and N is a submodule, then N is also finitely generated.*

PROOF. As in the proof of Proposition 6 we can always represent M as the quotient module of a free module F. The inverse image of N in F is a finitely generated free module, according to the preceding Theorem. Hence N is also finitely generated. $\square$

In the next Section we shall generalize this result to the nonlinear case.

Corollary 9. *If M is a finitely generated module, then there is a free submodule F of M such that there is a direct sum $M = \mathrm{tor}(M) \oplus F$.*

PROOF. By the previous Theorem, the module $M/\mathrm{tor}(M)$ is finitely generated and free in the short exact sequence of Proposition 6. Let F be the submodule of M generated by arbitrary inverse images (lifts) of generators of $M/\mathrm{tor}(M)$. It follows that F is free and isomorphic to $M/\mathrm{tor}(M)$. Hence its dimension is perfectly determined and is independent of the lifts chosen. Also, clearly $\mathrm{tor}(M) \cap F = 0$. $\square$

In view of the rather restrictive assumptions involved in Theorem 7, the reader should now be convinced of the fact that, in general, we can only give the short exact sequence of Proposition 6, and that it cannot always be split, as in the previous Corollary.

We now turn to another domain of pure algebra, namely classical Galois theory. It is not our intention to give a detailed account of the theory. For this we refer to the many excellent references as [174, 194]. On the contrary, we want to give least possible information via definitions and results to enable the reader to understand that none of the classical arguments can be generalized. We shall sketch a more geometrical approach, first introduced by Vessiot in [192] and giving some insight in the previous arguments, in such a way that the main idea, namely the *use of principal homogeneous spaces*, can be generalized, in the next Section, to more involved situations, in the transition from groups to pseudogroups.

The basic starting point of the theory of algebraic numbers involves computer algebra, as already indicated in the Introduction. Indeed, if we wish to store on a computer the so-called real numbers $\sqrt{2}$ and $\sqrt{3}$, or the imaginary number $i = \sqrt{-1}$, we cannot revert to numerical mathematics but have to use the fact that these are algebraic numbers, being the roots of, respectively, $y^2 - 2$, $y^2 - 3$, and $y^2 + 1$. These polynomials are defined over the field $\mathbb{Q}$ of rational numbers.

In a more mathematical terminology, if K is a field and L is a field containing it, we call L a *field extension* of K. We can speak of the extension L/K by regarding L as a vector space over K, and we call $|L/K| = \dim_K L$ the *degree of L over K*. If y is an indeterminate over K, we let $K[y]$ be the principal ideal ring of polynomials in y with coefficients in K. If $\eta \in L$, we can introduce the *evaluation epimorphism*

$$K[y] \rightarrow K[\eta] : P \rightarrow P(\eta) \in L.$$

If this is also a monomorphism, we say that η is *transcendental* over K. Otherwise we call η *algebraic* over K. The kernel of the evaluation epimorphism is a prime ideal because L is a field. This prime ideal is generated by a single irreducible polynomial $P \in K[y]$, and we say that η is a *root* of P. In actual practice only P is known and the computer has no way of distinguishing 'a priori' between the various roots of P, unless L is given. This is particularly clear for the real root and the two complex roots of $P \equiv y^3 - 2$. As a byproduct we can introduce only the short exact sequence

$$0 \longrightarrow \mathfrak{p} \longrightarrow K[y] \longrightarrow K[\eta] \longrightarrow 0,$$

where $\mathfrak{p} = (P)$ is a prime ideal and η is the image of y under the canonical projection

$$K[y] \rightarrow K[y]/\mathfrak{p}.$$

Since $\mathfrak{p}$ is prime, $K[\eta]$ is an integral domain, and we may introduce its field of quotients:

$$L = Q(K[y]/\mathfrak{p}) = K(\eta).$$

In fact, in this case $K[y]/\mathfrak{p}$ is already a field and a vector space, generated by $1, \eta, \ldots, \eta^{\deg P - 1}$. E.g., $1/(\sqrt{2} - 1) = 1 + \sqrt{2}$ and η is a symbol, like $\sqrt{2}$, $\sqrt{3}$, or i, called the *generic zero* or *generic solution* of $\mathfrak{p}$.

Recapitulating, to construct an algebraic extension L/K we start with a polynomial $P \in K[y]$, check its irreducibility (ensuing that it generates a prime ideal $\mathfrak{p} \in K[y]$), construct the integral domain $K[y]/\mathfrak{p}$ and take its field of quotients. Nowadays various algorithms have been implemented for checking the irreducibility of P.

By elementary linear algebra we deduce

Proposition 10. *If $K \subset L \subset M$ are fields, then*

$$|M/K| = |M/L| \cdot |L/K|.$$

Corollary 11. *If $\eta \neq 0$ and ζ are both algebraic over K, then so are $\eta - \zeta$, $\eta + \zeta$, $\eta\zeta$, and $1/\eta$.*

Now we arrive at a delicate point. Even the average reader is so familiar with elementary algebra that he/she immediately views $\mathbb{Q}(\sqrt{2}, \sqrt{-1})$ as the field of rational functions in square root of 2 and square root of -1. Meanwhile, people forget that this is not trivial, as it is a consequence of the fundamental theorem of algebra, saying that any polynomial equation over $\mathbb{R}$ has its roots in the field $\mathbb{C}$ of complex numbers. In fact, writing $K(\eta, \zeta)$ makes no sense unless η and ζ are elements of an extension L of K and, in that case (taking into account Corollary 11) it is the smallest subfield of L containing K, η, ζ. It is called the field obtained from K by *adjunction* of η, ζ. More generally, if K and L are subfields of M, we let (K, L) be the *composite* of K and L in M, i.e. the smallest subfield of M containing both K and L. A similar evaluation epimorphism

$$K[y, z] \rightarrow K[\eta, \zeta] \subset L$$

can be introduced. In general, however, the prime kernel need not be the ideal generated by the minimal polynomials of η and ζ over K, because the minimal polynomial of ζ over K can be reducible over $K(\eta)$, vice versa. More generally we can state

Definition 12. An extension L/K is called an *algebraic field extension* if every element of L is algebraic over K.

A very useful result in this framework is the following theorem concerning algebraic extensions.

Theorem 13. *Every finitely generated algebraic extension is generated by a single element.*

This element is called a *primitive element*.

PROOF. Let $L = K(\eta^1, \ldots, \eta^m)$, and let $u_1, \ldots, u_m, v$ be indeterminates over K. If R is the minimal polynomial of $\xi = u_1\eta^1 + \cdots + u_m\eta^m$ in $K(u)[v]$, we can find $D \in K[u]$ such that $DR = S \in K[u, v]$, and we have

$$\frac{\partial S(\xi)}{\partial u_k} = \frac{\partial S}{\partial u_k}(\xi) + \eta^k \frac{\partial S}{\partial v}(\xi) = 0$$

with

$$\frac{\partial S}{\partial v}(\xi) = D(u) \cdot \frac{\partial R}{\partial v}(\xi) \neq 0,$$

because R is irreducible in $K(u)[v]$ and therefore does not have a double root. Since K is an infinite field (it contains $\mathbb{Q}$), we can find $c_1, \ldots, c_m \in K$ such that $\partial S / \partial v(\zeta) \neq 0$ with $\zeta = c_1 \eta^1 + \cdots + c_m \eta^m$, and obtain $\eta^k \partial S / \partial v(\zeta) + \partial S / \partial u_k(\zeta) = 0$. This leads to $\eta^k \in K(\zeta)$, and so $L = K(\zeta)$. $\square$

Before sketching Galois theory, we need some abstract definitions.

Definition 14. We denote by $\operatorname{aut}(L/K)$ the group of field automorphisms of L fixing K, i.e. the group of automorphisms $\varphi \colon L \to L$ such that $\varphi(a+b) = \varphi(a) + \varphi(b)$, $\varphi(ab) = \varphi(a)\varphi(b)$, for all $a, b \in L$, and $\varphi(a) = a$ for $a \in K$. Similarly, we denote by $\operatorname{iso}(L/K)$ the *set* of all isomorphisms from L into an extension of L fixing K.

We are now ready for explaining the Galois mechanism. It consists of defining a type of extensions we will deal with and a fundamental theorem describing the properties of such extensions.

Definition 15. A finite algebraic extension is called a *Galois extension* if it satisfies any one of the following, equivalent, conditions:

1) the fixed field of $\operatorname{aut}(L/K)$ is K;
2) $\operatorname{iso}(L/K) = \operatorname{aut}(L/K)$;
3) L is obtained from K by adjunction of all roots of a polynomial over K.

We do not give detailed proofs of the equivalence of these three definitions, since these can be found in any textbook on Galois theory [14, 145, 174, 194]. We shall rather give a few hints and comments, and show that not one of these can be generalized, because of the lack of a geometric interpretation.

The original, constructive, definition of Galois in 1830 is the third one above, with the help of the so-called *Galois resolvent*

$$\prod_{\sigma \in S_n} (v - (u_1 \eta^{\sigma(1)} + \cdots + u_m \eta^{\sigma(m)})) \in K[u, v],$$

where only symmetric functions of the roots $\eta^1, \ldots, \eta^m$ of a given polynomial $P \in K[y]$ are used. As in Theorem 13 we may specialize $u_1, \ldots, u_m$ to values in K, and so obtain a polynomial of at least degree $m!$ over K without multiple roots. The full permutation group on m objects permutes the roots, and so the set of roots can be viewed as a PHS for S_m. However, the Galois resolvent need not be irreducible over K, and one can prove that the irreducible factors again define irreducible PHS for isomorphic subgroups of S_m. Any one of these subgroups is called the *Galois group* Γ of P, and the corresponding irreducible factor is used for defining L/K by residues. This is an explicit way of realizing $\operatorname{aut}(L/K)$ as a subgroup $\Gamma \subseteq S_m$ of permutations. Of course, since $P(\eta) = 0$ for any root, we have

$$\varphi(P(\eta)) = \varphi(P)(\varphi(\eta)) = P(\varphi(\eta)) = 0.$$

Therefore, the third condition implies that any isomorphism of L/K can act on the roots only as a permutation of them and is an automorphism. Simultaneously we have a manner of realizing $\operatorname{aut}(L/K)$ by means of Γ. In examples in Section E we shall see that the Galois extension L/K has nothing to do with the algebraic extension determined by an irreducible P.

Once we know that L/K is a Galois extension, we can establish a correspondence, called *Galois correspondence*, between subgroups of Γ and intermediate fields between K and L, as follows. Any subgroup $\Gamma' \subset \Gamma$ is realized by means of a subgroup of $\text{aut}(L/K)$. This subgroup has fixed field K' between K and L. Conversely, to any intermediate field K' between K and L we can associate the subgroup $\text{aut}(L/K') \subset \text{aut}(L/K)$, which is hence realized by a subgroup $\Gamma' \subset \Gamma$ of permutations. The usefulness of Galois extensions lies in the following Theorem.

Theorem 16 (Fundamental Theorem of Galois Theory). *If L/K is a Galois extension, then the above correspondence gives a bijective order-reversing correspondence between the subgroups of Γ and the intermediate fields between K and L. In particular, L/K' is again a Galois extension for any intermediate field K'. Moreover, an intermediate field K' is a Galois extension of K if and only if Γ' is a normal subgroup of Γ.*

The main use of this Theorem is in replacing the search for the roots of a given polynomial $P \in K[y]$ of degree m by the search for subgroups of the permutation group $\Gamma \subseteq S_m$.

However, regarding the generalization to ODE, or even PDE, we soon discover that none of the equivalent definitions can be generalized! Indeed, it is nonsense to consider all the solutions of a given system of PDE, and, in any case, if we want to deal with pseudogroups, the corresponding transformations have no chance of preserving any algebraic structure at all.

Surprisingly, there is a fourth equivalent definition. It is not standard in the sense that it cannot be found in textbooks, although it involves field-theoretic concepts only.

Definition 17. A finite algebraic extension is called a *Galois extension* if there is a *fundamental isomorphism*:

$$L \otimes_K L \simeq L \oplus \cdots \oplus L \qquad (|L/K| \text{ terms}).$$

The surprising fact is that, although this Definition in terms of tensor product and direct sum of fields seems even more abstract than the three preceding ones, we will be able to interpret it in a simple manner, both geometrically and group theoretically.

To prove the equivalence, let ζ be a primitive element of L/K defined in terms of an irreducible factor $R \in K[v]$ of the Galois resolvent. We let $\mathfrak{p}$ be the prime ideal generated by R in $K[v]$, and let $L\mathfrak{p}$ be its extension in $L[v]$. Accordingly, $L \otimes_K L \simeq L[v]/L\mathfrak{p}$, and we need not take the ring of quotients. However, $L\mathfrak{p}$ is no longer a prime ideal since R, having one root in L, has all its roots in L by Theorem 13 and the third part of Definition 15. Hence R splits over L into factors of degree one having corresponding residue equal to L. Finally, the number of such factors equals the degree of R, which is the number $|\Gamma|$ of elements in Γ or the degree of L over K.

Before giving a geometrical interpretation of this Definition we provide a precise mathematical description of the trick used by Vessiot while introducing a PHS in the construction of Galois extensions. The main problem is to look for *all* the roots $\eta^1, \ldots, \eta^m$ of a polynomial $P \in K[y]$ of degree m. Of course, without loss of generality we may assume that P does not have a double root, i.e. the discriminant δ of P differs from zero. From the theory of symmetric functions we know that $\delta^2 \in K$. We can write:

$$P \equiv (y)^m - \omega^1 (y)^{m-1} + \cdots + (-1)^m \omega^m = 0,$$

where we have used upper indices for ω in order to be in agreement with the notation of Chapter V. Setting $P(y^k) \equiv P_k$, we see that the m equations $P_k = 0$, $k = 1, \ldots, m$, are indeed a linear system for $\omega^1, \ldots, \omega^m$ with as determinant a VanderMonde determinant equal to $\Delta \equiv (y^1 - y^2) \ldots (y^{m-1} - y^m)$, up to sign, and such that $\Delta = \delta$ for the roots of P. Accordingly, we have reduced the problem of finding all the roots of P to that of finding one solution of the automorphic system

$$
\begin{cases}
\Phi^1(y) \equiv y^1 + \cdots + y^m = \omega^1, \\
\ldots\ldots\ldots\ldots \\
\Phi^m(y) \equiv y^1 \ldots y^m = \omega^m.
\end{cases}
$$

Unfortunately, as we shall see in Section E, it may happen that the ideal generated in $K[y^1, \ldots, y^m]$ by the m polynomials $\Phi^k(y) - \omega^k$ is not prime, and therefore cannot be used in determining a Galois extension L of K that would contain all (different) roots $\eta^1, \ldots, \eta^m$ of P. Nevertheless, we can prove that this ideal is perfect if $\delta \neq 0$ (even if P is not irreducible), and thus is the intersection of finitely many prime ideals which are at the same time maximal. From the Introduction we know that the irreducible components of a PHS need not be PHS for subgroups, so it would be pure luck if the preceding PHS splits into irreducible PHS for isomorphic subgroups of S_m. Any one of these subgroups, say $\Gamma \subset S_m$, is called the galois group of P, and we may consider the corresponding prime ideal $\mathfrak{p} \subset K[y^1, \ldots, y^m]$. This construction is in one-to-one correspondence with the splitting of the Galois resolvent into irreducible factors. Since all solutions of $\mathfrak{p}$ are roots of P, it follows that the desired Galois extension is $L = K[y^1, \ldots, y^m]/\mathfrak{p}$ and, certainly, has nothing to do with $K[y]/(P)$ when P itself is irreducible. The subtle argument used by Vessiot to explain this property of a PHS is based on the fact that $\mathfrak{p}$ is prime and maximal, while the action of S_m is rational with $\sigma(y^k) = y^{\sigma(k)}$, for all $\sigma \in S_m$. Indeed, let $\{\eta_1\}$ and $\{\eta_2\}$ be two solutions of $\mathfrak{p}$, i.e. two ordered sets of roots of P. There is exactly one permutation σ in S_m such that $\eta_2 = \sigma(\eta_1)$, and hence $\mathfrak{p}$ and $\sigma(\mathfrak{p})$ have the solution η_1 in common. Since $\mathfrak{p}$ is maximal, we must have $\mathfrak{p} = \sigma(\mathfrak{p})$, and $\mathfrak{p}$ is the defining ideal of an irreducible automorphic system for a subgroup $\Gamma \subset S_m$.

We now relate this point of view to Definition 17 by means of

Definition 18. An *algebraic group* G defined over k is a group G which is also an algebraic set over k.

Now, let X be an irreducible algebraic set, or *variety*, defined over a field K containing k by a prime ideal $\mathfrak{p} \in K[x^1, \ldots, x^n]$. We let the *ring of polynomial functions* on X be the integral domain

$$K[X] = K[x^1, \ldots, x^n]/\mathfrak{p},$$

with field of quotients $L = K(X) = Q(K[X])$.

Definition 19. An *action* of G on X defined over K is a polynomial map $X \times G \to X$ defined over K.

In fact, a polynomial map on the product $X \times G$ is a polynomial in the indeterminates defining X and G modulo the defining ideals of X and G. The distinction between these two sets of indeterminates can be made by using tensor products in such a way that $K[X \times G] = K[X] \otimes_K K[G]$. Also, a *polynomial action* is a polynomial map $y = f(x, a)$ in terms of coordinates x on X and a on G. Finally we notice that the preceding action determines, by duality, a monomorphism of rings:

$$K[X] \to K[X] \otimes_K K[G] = K[X] \otimes_k k[G],$$

using the distributivity of tensor products.

Combining these results, we see that X is a PHS for G if and only if there is a ring isomorphism

$$K[X] \otimes_K K[X] \simeq K[X] \otimes_k k[G].$$

Passing to the ring of quotients, we obtain the *fundamental isomorphism* of rings:

$$Q(L \otimes_K L) \simeq Q(L \otimes_k k[G]),$$

which is nothing else than the dual description of an irreducible PHS for an algebraic group G over K/k.

In the specific situation under consideration in this Section, X is a point set and $K[X] = K(X)$. It follows that the fundamental isomorphism can be written simply as:

$$L \otimes_K L \simeq L \otimes_k k[G],$$

and it remains to study $k[G]$.

However, it is well known that a permutation $\sigma \in S_m$ can be represented by an $(m \times m)$-matrix with in each row and column precisely one entry 1 and all other entries 0. Accordingly, Γ can be identified with an algebraic group G of matrices defined over $\mathbb{Q}$ and completely decomposable over $\mathbb{Q}$, i.e.

$$k[G] \simeq \mathbb{Q} \oplus \cdots \oplus \mathbb{Q} \qquad (|\Gamma| \text{ terms}).$$

Combining the results, the dual description of the automorphic system and the relation $|\Gamma| = |L/K|$ imply the fundamental isomorphism

$$L \otimes_K L \simeq L \otimes_{\mathbb{Q}} \underbrace{(\mathbb{Q} \oplus \cdots \oplus \mathbb{Q})}_{|\Gamma| \text{ terms}},$$

i.e.

$$L \otimes_K L \simeq \underbrace{L \oplus \cdots \oplus L}_{|L/K| \text{ terms}},$$

as in Definition 17.

The fundamental isomorphism proves that Galois theory is nothing else than a theory of PHS.

In [146] we have proved that the Galois correspondence can be extended to a class of algebraic extensions strictly including the Galois extensions. These extensions are called *automorphic extensions*, because the model variety X allowing us to define the extension L/K may now be a PHS for an algebraic group G over K/k, as before, but without assuming that each transformation of G acts as an element of $\text{aut}(L/K)$. E.g., $x^3 - 2 = 0$ transforms to $y^3 - 2 = 0$ under $\{y = x, y = \frac{1}{2}(-1 \pm i\sqrt{3})x\}$, and we have already seen that the corresponding extension is not a Galois extension.

The challenge, therefore, is to recover the Galois theory without ever using $\text{aut}(L/K)$ in the description of the Galois correspondence. Here we only sketch a few arguments in favor of this, since we only wanted to indicate the importance of PHS in classical Galois theory, as the only way of generalizing it. We refer to [146] for more details.

An algebraic subgroup $G' \subset G$ is defined over k by more algebraic relations among the coordinates, hence by a *specialization epimorphism*

$$k[G] \rightarrow k[G']$$

similar to the evaluation epimorphism. In our case, if K' is an intermediate field between K and L, this specialization can be obtained by means of the exact commutative diagram

$$
\begin{array}{ccc}
0 \longrightarrow k[G] & \longrightarrow & L \otimes_K L \\
\downarrow & & \downarrow \\
0 \longrightarrow k[G'] & \longrightarrow & L \otimes_{K'} L \\
\downarrow & & \downarrow \\
0 & & 0
\end{array}
$$

where the kernel of the left vertical epimorphism is the intersection with $k[G]$ of the kernel of the right vertical epimorphism.

Conversely, when G' is an algebraic subgroup of G, by composition we obtain a morphism

$$K[X] \rightarrow K[X] \otimes_k k[G] \rightarrow K[X] \otimes_k k[G'],$$

and hence a monomorphism

$$L \rightarrow Q(L \otimes_k k[G'])$$

which makes it possible to define the intermediate field K' of invariants of G' in L by the formula

$$K' = \{a \in L \mid a \rightarrow a \otimes 1\}.$$

This is coherent with the above diagram of specializations.

The geometric point of view will clear up the use of orbit spaces and the necessity of using PHS to obtain a Galois correspondence. Indeed, let us define an equivalence relation on the PHS X for G by saying that $x \sim y \Leftrightarrow y = ax$, $a \in G' \subset G$. This gives an orbit space X/G'. There is a commutative diagram

$$
\begin{array}{ccc}
X \times G' & \longrightarrow & X \times_{X/G'} X \\
\downarrow & & \downarrow \\
X \times G & \longrightarrow & X \times X
\end{array}
$$

The upper horizontal arrow is an epimorphism, by the definition of X/G', and it is also a monomorphism whenever the lower horizontal arrow is a monomorphism. Hence, *it is essential to begin with a free action*, so that we have an injective graph, *and to use a PHS or the fibered product* $X \times_{X/G} X$, so that we have a surjective graph. This hence gives an isomorphism, leading to the fundamental isomorphism by duality.

B. Differential Algebra

The challenge now is to add the word 'differential' in front of each of the statements of the previous Section. The most surprising result in this Section is the discovery of the meaning and use of the differential algebraic counterpart of the criterion for formal integrability of Chapter III.C.

Definition 1. A *differential ring* is a ring A with n commuting *derivations* $\partial_1, \ldots, \partial_n$ satisfying

$$\partial_i(a + b) = \partial_i a + \partial_i b, \qquad \forall a, b \in A, \quad \forall i = 1, \ldots, n,$$
$$\partial_i(ab) = (\partial_i a)b + a\partial_i b, \qquad \forall a, b \in A, \quad \forall i = 1, \ldots, n,$$
$$\partial_i \partial_j = \partial_j \partial_i, \qquad \forall i, j = 1, \ldots, n.$$

It is called an *ordinary differential ring* if $n = 1$ and a *partial differential ring* if $n > 1$.

The subring $\{a \in A | \partial_i a = 0, \forall i = 1, \ldots, n\}$ is called the *ring of constants* of A. A subring A of a differential ring B is called a *differential subring* if A is stable under each derivation of B. An ideal in a differential ring is called a *differential ideal* if it is stable under each derivation of the ring. A derivation of A can be extended to a derivation of the ring of quotients $Q(A)$ of A by setting

$$\partial_i(a/b) = (b\partial_i a - a\partial_i b)/b^2$$

for all $a, b \in A$, $b \neq 0$.

In actual practice, a very useful situation is as follows. If K is a differential field with derivations $\partial_1, \ldots, \partial_n$ and subfield of constants C, we can introduce indeterminates y^k over K with $k = 1, \ldots, m$ and form the polynomial ring $K\{y\} = K[y_\mu^k | k = 1, \ldots, m, \mu \geq 0]$, setting $y_0^k = y^k$. It can be made a differential ring by introducing the *formal derivations* $d_i y_\mu^k = y_{\mu+1_i}^k$. At this moment we call $y^1, \ldots, y^m$ *differential indeterminates* over K and retain the name *jet coordinates* for the various y_μ^k, as in Chapter II. Again, for simplicity reasons, we shall write

$$K[y_q] = K[y_\mu^k | k = 1, \ldots, m; \ 0 \leq |\mu| \leq q],$$

with $K(y_q) = Q(K[y_q])$, and we set $K\langle y \rangle = Q(K\{y\})$.

Definition 2. If K is a differential subfield of a differential field L, we say that L/K is a *differential extension*. This differential extension is said to be *finitely generated* if there are $\eta^1, \ldots, \eta^m \in L$ such that $L = K\langle \eta^1, \ldots, \eta^m \rangle = K\langle \eta \rangle$.

Of course, we will have to face the same technical problem as in the previous Section, because the solution e^x of $y' - y = 0$ cannot be stored on a computer as a series $1 + \frac{x}{1!} + \frac{x^2}{2!} + \ldots$, but has to be stored as a symbol with the property that upon differentiation of e^x we may replace the result by e^x. Similarly, defining solutions of differential polynomials over K by introducing differential extensions is not very useful and, in any case, we no longer have a huge extension in which we can find the solutions of all possible systems of algebraic equations given by sets of differential polynomials (like the field of complex numbers).

So, as before, the key matter of differential algebra will be the short exact sequence

$$0 \longrightarrow \mathfrak{p} \longrightarrow K\{y\} \longrightarrow K\{\eta\} \longrightarrow 0,$$

which defines a *differential evaluation epimorphism* with $y^k \to \eta^k$ by a residue if we know the prime differential ideal $\mathfrak{p}$ or defines $\mathfrak{p}$ if we know where to find the family η.

As a byproduct, starting with a prime differential ideal $\mathfrak{p} \subset K\{y\}$, we may *define* the differential extension $L = Q(K\{y\}/\mathfrak{p})$ of K, and a generic zero or solution η otherwise, in such a way that $L = K\langle\eta\rangle$.

So, a major problem, already hinted at in the Introduction by way of tricky examples, is to find out how to reduce the study of a number of differential polynomials to the preceding well-defined situation. Equivalently, we would like to have a *criterion for prime differential ideals*, allowing us to recognize whether or not the differential ideal generated in $K\{y\}$ by a certain number of given differential polynomials is prime. If we find the ideal to be prime we stop. Otherwise we have to ensure that the criterion provides the product of two differential polynomials not belonging to the ideal, and in an 'optimal' situation we may look for some sort of successive decomposition into prime differential ideals.

Surprisingly, the answer is given by adapting to differential algebra the criterion of formal integrability.

If $\mathfrak{a} \subset K\{y\}$ is a differential ideal, we can introduce $\mathfrak{a}_q = \mathfrak{a} \cap K[y_q]$, and let $\rho_r(\mathfrak{a}_q)$ be the ideal generated in $K[y_{q+r}]$ by the $d_\nu P$, for all $P \in \mathfrak{a}_q$ and $0 \leq |\nu| \leq r$. We have $\rho_r(\mathfrak{a}_q) \cap K[y_q] = \mathfrak{a}_q$ and $\rho_r(\mathfrak{a}_q) \subseteq \mathfrak{a}_{q+r}$.

The only difference with the differential-geometric situation of Chapter II is that now 'x' does not appear explicitly in the equations, but only via the coefficients of the differential polynomials $P_\tau \in K[y_q]$, although we continue to use the same jet notation. The definition of the symbol M_q can be given as usual; we only have to assume that $\mathcal{R}_q$ is an algebraic variety defined by a prime ideal $\mathfrak{p}_q \subset K[y_q]$, so that M_q and its prolongations are defined over the field $Q(K[y_q]/\mathfrak{p}_q)$. It remains to distinguish the particular situation when M_q or M_{q+1} are vector bundles over $\mathcal{R}_q$. In that case a certain number of determinants, say D_α, cannot all vanish on the solutions of the P_τ. By Hilbert's Nullstellen Theorem we can therefore find polynomials A_α, B_τ in $K[y_q]$ such that the following identity holds:

$$\sum_\alpha A_\alpha D_\alpha + \sum_\tau B_\tau P_\tau = 1.$$

Taking a sufficiently high power of each member, we find that such a relation also holds for arbitrary powers of the D_α.

Primality Criterion for Differential Ideals 3. *Let $\mathfrak{p}_q \subset K[y_q]$ and $\mathfrak{p}_{q+1} \subset K[y_{q+1}]$ be prime ideals such that $\mathfrak{p}_{q+1} = \rho_1(\mathfrak{p}_q)$ and $\mathfrak{p}_{q+1} \cap K[y_q] = \mathfrak{p}_q$. If the symbol M_q of the* *variety $\mathcal{R}_q$ defined by $\mathfrak{p}_q$ is 2-acyclic and if its first prolongation M_{q+1} is a vector bundle over $\mathcal{R}_q$, then $\mathfrak{p} = \rho_\infty(\mathfrak{p}_q)$ is a prime differential ideal with $\mathfrak{p} \cap K[y_{q+r}] = \rho_r(\mathfrak{p}_q)$, $\forall r \geq 0$.*

PROOF. The exact commutative diagram

$$
\begin{array}{ccccccccc}
& & 0 & & 0 & & 0 & & \\
& & \downarrow & & \downarrow & & \downarrow & & \\
0 & \longrightarrow & \mathfrak{p}_q & \longrightarrow & K[y_q] & \longrightarrow & K[y_q]/\mathfrak{p}_q & \longrightarrow & 0 \\
& & \downarrow & & \downarrow & & \downarrow & & \\
0 & \longrightarrow & \mathfrak{p}_{q+1} & \longrightarrow & K[y_{q+1}] & \longrightarrow & K[y_{q+1}]/\mathfrak{p}_{q+1} & \longrightarrow & 0
\end{array}
$$

generically implies, by duality, the surjectivity of $\mathcal{R}_{q+1} \to \mathcal{R}_q \to 0$. Accordingly, the criterion of formal integrability implies the generic surjectivity of $\mathcal{R}_{q+r+1} \to \mathcal{R}_{q+r} \to 0$, and, in particular, the generic surjectivity of $\mathcal{R}_{q+r} \to \mathcal{R}_q \to 0$, for all $r \geq 0$. All eliminations can be done by linear algebra, but projection onto $\mathcal{R}_q$ requires Hilbert's Theorem to prove that any resulting differential polynomial of order q is a linear combination of the various P_r. By duality we obtain $\rho_r(\mathfrak{p}_q) \cap K[y_q] = \mathfrak{p}_q$, for all $r \geq 0$. It remains to prove that $\rho_r(\mathfrak{p}_q)$ is a prime ideal in $K[y_{q+r}]$ for all $r \geq 0$.

Indeed, let $M, N \in K[y_{q+r}]$ with $MN \in \rho_r(\mathfrak{p}_q)$. Selecting a determinant D that does not vanish identically on the zeros of the (nondifferential) polynomials P_r and using the fact that jets of order $\geq q + 1$ appear only in a quasilinear way, we obtain by induction:

$$D^a(\text{principal jet of order } q + r) \in K[y_q][\text{parametric jets of order } q + 1, \ldots, q + r]$$

for a certain power a. Accordingly we have:

$$
\begin{cases}
D^a M + \sum_{\lambda,\rho} A_\rho^\lambda d_\lambda P_\rho = R \in K[y_q][\text{parametric jets of order } q + 1, \ldots, q + r], \\
D^a N + \sum_{\mu,\sigma} A_\sigma^\mu d_\mu P_\sigma = S \in K[y_q][\text{parametric jets of order } q + 1, \ldots, q + r].
\end{cases}
$$

It is known that the extension of $\mathfrak{p}_q$ to the ring $K[y_q][\text{parametric jets of order } q + 1, \ldots, q + r]$ is also a prime ideal. Hence we find that $RS \in$ extension of $\mathfrak{p}_q$, and hence we may assume that $R \in$ extension of $\mathfrak{p}_q$. Accordingly, $D^a M \in \rho_r(\mathfrak{p}_q)$ for each D, and we may use the assumption that M_{q+1} is a vector bundle over $\mathcal{R}_q$ to deduce that $M \in \rho_r(\mathfrak{p}_q)$, because $M = (\sum_\alpha A_\alpha D_\alpha^a + \sum_\tau B_\tau P_\tau)M$. $\square$

We will now adapt this result to the study of perfect differential ideals, which play a very important role in differential algebra.

First of all, in general we have

Proposition 4. *If $\mathfrak{a}$ is a differential ideal in a differential ring A, then $\sqrt{\mathfrak{a}}$ is a differential ideal in A.*

PROOF. We shall prove that $(\partial_i a)^{2r-1} \in A$ whenever $a^r \in A$. Apply ∂_i to a^r, and obtain $ra^{r-1}\partial_i a = \partial_i a^r$, $r(r-1)a^{r-2}(\partial_i a)^r + ra^{r-1}\partial_{ii} a = \partial_{ii}a^r$, etc. For $r = 1$ the first formula gives the required result. For $r = 2$ the second formula, multiplied by $\partial_i a$, gives the required result, and we can proceed by induction to obtain the full proof. $\square$

Proposition 5. *If $\mathfrak{p}$ is a prime differential ideal in $K\{y\}$, then, for q sufficiently large, there is a polynomial $D \in K[y_q]$ such that $D \notin \mathfrak{p}_q$ and*

$$D\mathfrak{p}_{q+r} \subset \sqrt{\rho_r(\mathfrak{p}_q)} \subset \mathfrak{p}_{q+r}, \qquad \forall r \geq 0.$$

PROOF. The symbol M_q of the generic zero $\mathcal{R}_q$ of $\mathfrak{p}_q$ is defined over the field $Q(K[y_q]/\mathfrak{p}_q) \subset Q(K\{y\}/\mathfrak{p}) = L$ because $\mathfrak{p}_q = \mathfrak{p} \cap K[y_q]$. Also, we have $\rho_1(\mathfrak{p}_q) \subset \mathfrak{p}_{q+1}$, and so $M_{q+1} \subset \rho_1(M_q)$. It follows from standard arguments concerning noetherian modules that, for q sufficiently large, $M_{q+r} = \rho_r(M_q)$. Accordingly we find, as in the proof of Criterion 3, a polynomial $D \in K[y_q]$ such that $D \notin \mathfrak{p}_q$. We similarly obtain

$$D^a M \in \rho_r(\mathfrak{p}_q) \subset \mathfrak{p}_{q+r}$$

for any $M \in \mathfrak{p}_{q+r}$, which implies $DM \in \sqrt{\rho_r(\mathfrak{p}_q)}$ by taking the root. $\square$

Theorem 6 (Differential Basis Theorem). *If $\mathfrak{r}$ is a perfect ideal in $K\{y\}$, then $\mathfrak{r} = \sqrt{\rho_\infty(\mathfrak{r}_q)} \overset{def}{=} \{\mathfrak{r}_q\}$ for q sufficiently large.*

PROOF. In the set of perfect differential ideals not possessing this property we select a maximal element $\mathfrak{m}$ by Zorn's Lemma [18, 208].

Suppose we can find $a', a'' \in K\{y\}$ with $a', a'' \notin \mathfrak{m}$ but $a'a'' \in \mathfrak{m}$. Then $\mathfrak{n}' = \{\mathfrak{m}, a'\}$ and $\mathfrak{n}'' = \{\mathfrak{m}, a''\}$ have the property indicated in the Theorem and we can find a q, sufficiently large, such that $\mathfrak{n}' = \{\mathfrak{n}'_q\}$ and $\mathfrak{n}'' = \{\mathfrak{n}''_q\}$. Then $\mathfrak{m} = \{\mathfrak{m}, a'a''\} = \{\mathfrak{m}, a'\} \cap \{\mathfrak{m}, a''\} = \mathfrak{n}' \cap \mathfrak{n}''$ and $\mathfrak{m}_q = \mathfrak{n}'_q \cap \mathfrak{n}''_q$ with $\{\mathfrak{m}_q\} \subset \mathfrak{m}$ because $\mathfrak{m}$ is perfect. However, also $\mathfrak{m} = \{\mathfrak{n}'_q\} \cap \{\mathfrak{n}''_q\} \subset \{\mathfrak{n}'_q \cap \mathfrak{n}''_q\} = \{\mathfrak{m}_q\}$, and so $\mathfrak{m} = \{\mathfrak{m}_q\}$, a contradiction proving that $\mathfrak{m}$ is prime.

By the last Proposition, there is $D \in K[y_q]$, $D \notin \mathfrak{m}_q$ with $D\mathfrak{m} \subset \{\mathfrak{m}_q\} \subset \mathfrak{m}$ for q sufficiently large. By the maximality of $\mathfrak{m}$ we have $\mathfrak{n} = \{\mathfrak{m}, D\} = \{\mathfrak{m}_q, D\}$, and so

$$\mathfrak{m} = \mathfrak{m} \cap \mathfrak{n} = \mathfrak{m} \cap \{\mathfrak{m}_q, D\} = \{\mathfrak{m}_q, D\mathfrak{m}\} \subset \{\mathfrak{m}_q\},$$

which implies the contradiction $\mathfrak{m} = \{\mathfrak{m}_q\}$. $\square$

One should take into consideration the fact that we cannot have $\mathfrak{r} = \rho_\infty(\mathfrak{r}_q)$, but only $\mathfrak{r} = \sqrt{\rho_\infty(\mathfrak{r}_q)}$ in general, as is clear in the examples presented in the Introduction. We can refine this result by copying the proof of the Criterion 3 with M^2 instead of MN. We then obtain:

Perfectness Criterion for Differential Ideals 7. *Let $\mathfrak{r}_q \subset K[y_q]$ and $\mathfrak{r}_{q+1} \subset K[y_{q+1}]$ be perfect ideals such that $\mathfrak{r}_{q+1} = \rho_1(\mathfrak{r}_q)$ and $\mathfrak{r}_{q+1} \cap K[y_q] = \mathfrak{r}_q$. If the symbol M_q of the algebraic set $\mathcal{R}_q$ defined by $\mathcal{R}_q$ is 2-acyclic and its first prolongation M_{q+1} is a vector bundle over $\mathcal{R}_q$, then $\mathfrak{r} = \rho_\infty(\mathfrak{r}_q)$ is a perfect differential ideal with $\mathfrak{r} \cap K[y_{q+r}] = \rho_r(\mathfrak{r}_q)$ for all $r \geq 0$.*

The importance of perfect differential ideals in differential algebra lies in the following Corollary to Theorem 6:

Corollary 8. *Every perfect differential ideal of the differential ring $K\{y\} = K\{y^1, \ldots, y^m\}$ can be expressed as the intersection of finitely many prime differential ideals.*

PROOF. In contrast to the nondifferential case, the differential ring $K\{y\}$ satisfies the ascending chain condition for *perfect* differential ideals *only*, i.e. there is a perfect differential ideal $\mathfrak{m}$ that is maximal with respect to the property that it is not an intersection of finitely many prime ideals [95, 160]. Of course, $\mathfrak{m}$ cannot be prime, and we can find $a, b \in K\{y\}$ such that $a, b \notin \mathfrak{m}$ and $ab \in \mathfrak{m}$. The perfect differential ideals $\{\mathfrak{m}, a\}$ and $\{\mathfrak{m}, b\}$ can be written as intersections of finitely many prime differential ideals. Hence so can $\mathfrak{m} = \{\mathfrak{m}, ab\} = \{\mathfrak{m}, a\} \cap \{\mathfrak{m}, b\}$, giving a contradiction. $\square$

From this Corollary we see that we may confine our attention to prime differential ideals. In particular, the reader should remember that any differential ideal generated by finitely many *linear* differential polynomials is automatically prime. This result immediately proves that any result obtained here can be adapted to the $\mathcal{D}$-modules introduced in the previous Section. In particular, the so-called 'constant coefficient' case, making $\mathcal{D}$ commutative, just means that we consider linear differential operators with coefficients in the subfield of constants C of K.

Definition 9. A family $\eta = (\eta^1, \ldots, \eta^m)$ of elements in a differential extension of a differential field K is said to be *differentially algebraic* (respectively, *differentially transcendental*) over K if the kernel of the differential evaluation epimorphism is a *proper* prime differential ideal (respectively, is the zero ideal).

In a more restrictive situation we have:

Definition 10. A differential extension L of a differential field K is said to be *differentially algebraic* over K if *every* element of L itself is differentially algebraic over K.

Notice that these Definitions come from the nondifferential case by simply adding the word 'differential' in front of the main concepts.

In the nondifferential case, if an extension L/K is finitely generated, then we can select a maximal finite subset S of transcendental elements over K such that L becomes algebraic over $K(S)$. The number of elements in S is called the *transcendence degree* of L/K, and is denoted by $\operatorname{trd} L/K$. This definition cannot be generalized to the differential case because, in general, *a differential extension is not finitely generated as an extension*, unless the corresponding system of PDE is of finite type. Nevertheless, for a differential extension L/K we can always find a maximal subset of elements in L that are differentially transcendental over K and such that L becomes finitely differentially algebraic over $K\langle S \rangle$. The number of elements in S is called the *differential transcendence degree* of L/K, and is denoted by $\operatorname{difftrd} L/K$.

We combine the results of Criterion 3 and Proposition 5 to obtain the *Hilbert polynomials*

$$\begin{cases} \dim \mathrm{M}_{q+r} = \sum_{i=1}^{n} \frac{(r+i-1)!}{r!\,(i-1)!}\alpha_q^i, & \forall r \geq 0, \\[2mm] \dim \mathfrak{p}_{q+r} = \dim \mathfrak{p}_{q-1} + \sum_{i=1}^{n} \frac{(r+i)!}{r!\,i!}\alpha_q^i, & \forall r \geq 0, \end{cases}$$

in terms of the characters $\alpha_q^1, \ldots, \alpha_q^n$ of the involutive symbol M_q for q sufficiently large. If the coordinate system is not δ-regular, we may have to use a linear change of derivations (in differential algebra the independent variables no longer exists) by means of a square, invertible matrix with coefficients in the subfield of constants C of K. From the formulas above we immediately find the relation

$$0 \leq \mathrm{difftrd}\,L/K = \alpha_q^n \leq m.$$

Once more we notice that we cannot compute α_q^n if the system of PDE is not formally integrable. A similar comment can be made for a differential extension $L = Q(K\{y\}/\mathfrak{p})$ of K, since there is no other way to find out that $\mathfrak{p}$ is a prime differential ideal than to use Criterion 3, $\cdots$ which just comes down to the criterion for formal integrability!

The scheme below gives a survey of the links between the differential algebraic and the differential geometric approach to the study of systems of PDE. In this scheme, + indicates that the corresponding case is possible, while 0 indicates that it is impossible. We advise the reader to give examples of each situation described, and to examine well-known systems (Killing equations, Cauchy–Riemann system, divergence-free condition) in this framework.

	symbol	non-	surjective	
			noninjective	injective
	system	over-	under-	determined
difftrd $L/K = 0$ $\begin{cases} \end{cases}$	trd $L/K < \infty$	+	0	+
	trd $L/K = \infty$	+	0	+
difftrd $L/K > 0$		+	+	0

On the other hand, we will now pay attention to (finitely generated) differentially algebraic extensions, in a way similar to that paid in the previous Section to (finitely generated) algebraic extensions. The main result is provided by the following, quite useful but technical, differential counterpart of Proposition A.10.

Proposition 11. *If ζ is differentially algebraic over $K\langle\eta\rangle$ and η is differentially algebraic over K, then ζ is differentially algebraic over K.*

PROOF. Since η is differentially algebraic over K, there is ν_1 such that $d_{\nu_1}\eta \in K(d_\mu\eta|\,\mu < \nu_1)$, and so

$$K(d_\nu\eta|\,|\nu| > |\nu_1|) = K(d_\mu\eta|\,|\mu| \leq |\nu|, \mu \neq \alpha + \nu_1).$$

Similarly, we can find ν_2 such that $d_{\nu_2}\zeta \in K\langle\eta\rangle(d_\mu\zeta \mid \mu < \nu_2)$, which provides a q, sufficiently large, such that

$$d_{\nu_2}\zeta \in K(d_\lambda\eta, d_\mu\zeta \mid |\lambda| \le q, \mu < \nu_2).$$

Therefore,

$$K(d_\nu\zeta \mid |\nu| > |\nu_2|) \subset K(d_\lambda\eta, d_\mu\zeta \mid |\lambda| \le q + |\nu|, |\mu| \le |\nu|, \mu \ne \beta + \nu_2),$$

and if $q + |\nu| > |\nu_1|$ we obtain

$$K(d_\nu\zeta \mid |\nu| > |\nu_2|) \subset K(d_\lambda\eta, d_\mu\zeta \mid |\lambda| \le q + |\nu|, \lambda \ne \alpha + \nu_1, |\mu| \le |\nu|, \mu \ne \beta + \nu_2).$$

If $|\nu|$ is sufficiently large, the number of generators in the left hand field is $(|\nu| + n)!/|\nu|!\,n! = (1/n!)|\nu|^n + \dots$ and exceeds the number of generators in the right hand field, which equals

$$\frac{(q + |\nu| + n)!}{(q + |\nu|)!\,n!} - \frac{(q + |\nu| - |\nu_1| + n)!}{(q + |\nu| - |\nu_1|)!\,n!} + \frac{(|\nu| + n)!}{|\nu|!\,n!} - \frac{(|\nu| - |\nu_2| + n)!}{(|\nu| - |\nu_2|)!\,n!},$$

which is a polynomial in $|\nu|$ of degree *less* than n. The Proposition now follows from the contradiction that would arise if ζ were differentially transcendental over K because, in that case, the $d_\nu\zeta$ should be transcendental over K. □

Corollary 12. *Let L/K be a differential extension and $\xi, \eta \in L$ be differentially algebraic over K (each). Then $\xi + \eta$, $\xi\eta$, ξ/η, $d_i\xi$ are differentially algebraic over K.*

PROOF. Setting $\xi + \eta = \zeta$ we obtain $\xi = \zeta - \eta$. Substituting this into the differential algebraic relations satisfied by ξ over K we find that ζ is differentially algebraic over $K\langle\eta\rangle$, and the result follows from the last Proposition.

A similar proof can be used for $\xi\eta$, ξ/η, and $d_i\xi$. □

We can use this Corollary to introduce two important concepts which can be defined for any differential extension L/K.

Definition 13. The set of elements in L that are algebraic over K is an intermediate differential field, K_0, called the *algebraic closure* of K in L; L/K is called a *regular extension* if $K = K_0$.

In fact, if $b \in K_0$ is a root of the irreducible (minimum) polynomial $P \in K[z]$, we immediately obtain $(d_i b)\partial P/\partial z(b) + (d_i P)(b) = 0$, and so $d_i b \in K(b) \subset K_0$. Notice that K_0 is algebraically closed in L because of transitivity of algebraic dependence.

Definition 14. Similarly, the set of elements in L that are differentially algebraic over K is an intermediate differential field, K', called the *differential algebraic closure* of K in L.

We immediately recognize this Definition as the nonlinear counterpart of Definition A.3. It will play a fundamental role in control theory, in the next Chapter.

Notice that K' is differentially algebraic closed in L because of Proposition 11.

We now give the differential counterpart of Theorem A.13

Definition 15. If K is a differential field with derivations $\partial_1, \ldots, \partial_n$, then we say that $\partial_1, \ldots, \partial_n$ are *algebraically independent* over K if there cannot exist a differential polynomial in $K\{z\}$ vanishing for all elements of K.

One can prove that $\partial_1, \ldots, \partial_n$ are algebraically independent over K if and only if they are *linearly independent* over K, and we will simply say that $\partial_1, \ldots, \partial_n$ are *independent* over K.

The simplest counterexample to independence is, of course, provided by a field of constants.

Proposition 16. *If K is a differential field with independent derivations, then every finitely generated differentially algebraic extension of K is generated by a single element.*

This generating element is called a *primitive element*.

PROOF. Proceeding by induction, we only have to prove that, if η, ζ are differentially algebraic over K (individually), then there is $a \in K$ such that $K\langle \eta, \zeta \rangle = K\langle \eta + a\zeta \rangle$.

Now, if ζ is a differential indeterminate, Corollary 12 implies that $\eta + z\zeta$ is differentially algebraic over $K\langle z \rangle$. Changing notation and introducing a new differential indeterminate u, we can find a differential polynomial $P \in K\{u, v\}$ such that $P(\eta + z\zeta, z) = 0$. If u_ν is the highest jet appearing in P, we may assume that $\partial P / \partial u_\nu (\eta + z\zeta, z) \neq 0$ (use the ordering $(|\nu|, \nu_m, \ldots, \nu_n)$ and obtain

$$\zeta \frac{\partial P}{\partial u_\nu}(\eta + z\zeta, z) + \frac{\partial P}{\partial z_\nu}(\eta + z\zeta, z) = 0$$

by differentiation with respect to z_ν in $K\langle \eta, \zeta \rangle\{z\}$. Since $\partial_1, \ldots, \partial_n$ are independent over K, we can find (but we do not know an effective method for this!) an element $a \in K$ such that $\partial P / \partial u_\nu (\eta + a\zeta, a) \neq 0$ in $K\langle \eta, \zeta \rangle$, and so $\zeta \in K\langle \eta + a\zeta \rangle \Rightarrow K\langle \eta, \zeta \rangle = K\langle \eta + a\zeta \rangle$. □

Remark 17. According to the above-said and the last scheme, apart from the condition on K usually satisfied unless K is a field of constants, we must have difftrd $L/K = 0$ in order to have a primitive element. However, even in simple examples, finding this element may be *extremely* difficult, and this is the reason why applications to control theory have not yet been given [151, 153, 154].

The transition to δ-regular coordinates has already been done in Chapter II, and we have seen that the condition $\alpha_q^n = 0$ amounts to finding $y_{(0,\ldots,q)}^k$ among the principal jets. Introducing the various jets of orders $1, \ldots, q - 1$ with respect to x^n as new differential indeterminates, we immediately obtain:

Proposition 18. *If L/K is a finitely generated differentially algebraic extension with derivations $d_1, \ldots, d_n$, then L/K can be regarded as a finitely generated differential extension with derivations $d'_1, \ldots, d'_{n-1}$ such that $d'_j = c^i_j d_j$ for certain $c^i_j \in C \subset K$.*

Using this Proposition inductively, we can prove the following, technical, Theorem, which may be used to describe any of the intermediate differential fields already considered [95, 145].

Theorem 19. *If L/K is a finitely generated differential extension, then any intermediate field between K and L is finitely generated over K.*

It now remains to introduce the differential counterpart of a *composite differential field* for two 'abstract' differential extensions L/K and M/K. Of course, if L and M are both contained in a differential extension N of K, we can use Corollary 12 to introduce the smallest differential subfield $\langle L, M \rangle$ of N containing both L and M. Hence, *the only difficult problem is to find out what to do when this big differential extension N of K is not given.*

Imagine that N is such a differential composite field, and consider the two chains of inclusions

$$K \subset L \subset N \quad \text{and} \quad K \subset M \subset N.$$

From the universal property of tensor products we can deduce the existence of a morphism

$$L \otimes_K M \to N.$$

This morphism is, moreover, a *differential morphism*, i.e. it preserves the corresponding differential structures if we define $L \otimes_K M$ as a differential ring by setting

$$d(a \otimes b) = (d_L a) \otimes b + a \otimes (d_M b)$$

for any two derivations agreeing on K, i.e. such that $d_L|_K = d_M|_K = \partial$.

It follows that the construction of an abstract differential composite field amounts to looking for prime differential ideals in $L \otimes_K M$.

This is just the inverse Bäcklund problem.

Here we will explain the conceptual approach, since this study will be part of the differential control theory of the next Chapter.

Indeed, let $y = (y^1, \ldots, y^m)$ and $z = (z^1, \ldots, z^r)$ be two families of differential indeterminates over K. If we have a prime differential ideal $\mathfrak{r} \subset K\{y, z\}$, we can introduce the prime differential ideals $\mathfrak{p} = \mathfrak{r} \cap K\{y\}$ and $\mathfrak{q} = \mathfrak{r} \cap K\{z\}$. By the exact commutative diagram

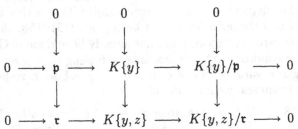

the injective vertical map on the right hand side induces an injection $0 \to L \to N$ that can be identified with the tower of inclusions $K \subset L \subset N$. Similarly, there is a tower of inclusions $K \subset M \subset N$.

This is just the direct Bäcklund problem.

So, we discover that the direct Bäcklund problem, historically the first problem stated in the literature [142], amounts to differential elimination of either y or z.

Conversely, given $\mathfrak{p} \subset K\{y\}$ and $\mathfrak{q} \subset K\{z\}$, i.e. given a nonlinear (algebraic) system of PDE for y and a nonlinear (algebraic) system of PDE for z, we may look for a nonlinear (algebraic) system of PDE for (y, z) such that by respective elimination we recover the original systems in each family of indeterminates, called *resolvent system*. The *mixed system* in (y, z) is then called a *differential correspondence*.

Of course, if the direct Bäcklund problem cannot be solved, then the inverse Bäcklund problem cannot be solved too!

C. Automorphic Systems

In Section A we have studied fields and extensions, in particular Galois extensions. For such extensions we have established a Galois correspondence between intermediate fields and permutation groups. Meanwhile, following Vessiot, we have pointed out the importance of *principal homogeneous spaces* (PHS) by revisiting classical Galois theory in a more geometrical way.

In Section B we have studied differential fields and differential extensions related to systems of algebraic *partial differential equations* (PDE).

In this Section we shall study, from the differential geometrical point of view, PHS for Lie pseudogroups, arising from the spaces of solutions of so-called (by Vessiot) *automorphic systems* of PDE.

We will apply these results in Section D to algebraic automorphic systems, and obtain a differential algebraic interpretation of the *cascade integration* procedure, which takes into account the invariance properties of systems of PDE.

We recall once more that, if X is a manifold and G a Lie group, then we say that X is a PHS for G whenever the graph $X \times G \to X \times X$ of the action of G on X is an isomorphism. Equivalently, we have already pointed out that the *really important reason for using PHS is to have a free action*, because in that case we can consider the isomorphism $X \times G \to X \times_{X/G} X$, provided that X/G can be defined 'in a nice way'.

Of course, if X is a finite set of points and G is a group of permutations represented as a finite group of square, invertible matrices, then the concept of PHS is quite easy to grasp. The same comment can be made when X is a finite-dimensional manifold and G a finite-dimensional Lie group. Hence, the only real difficulty will be when 'X' is the space of solutions of a system of PDE and 'G' a Lie pseudogroup. In fact, the only definition known in this case is as follows: if $y = f(x)$ and $\bar{y} = \bar{f}(x)$ are two solutions of a system of PDE, then there should exist *one, and only one*, transformation $\bar{y} = g(y)$ in the Lie pseudogroup such that $\bar{f} = g \circ f$, at least locally. Of course, this definition cannot be tested by means of symbolic computations, and the purpose of this Section is to modify it so that an effective criterion arises.

We use the notation of Chapter II for nonlinear systems of PDE, and the fibered manifold $\mathcal{E}$ is the trivial product $X \times Y$ of two manifolds. Accordingly, we assume that $\Gamma \subset \mathrm{aut}(Y)$, because in the differential algebraic framework independent variables, and so X, in general no longer exist, since they are replaced by a differential field K.

All difficulty involved in defining PHS now lies in the use of jet bundles and groupoids instead of solutions and pseudogroups.

With obvious notations, the chain rule for derivations provides the following *groupoid action* on the jet level, at any order q:

$$J_q(X,Y) \times_Y \Pi_q(Y,Y) \rightarrow J_q(X,Y).$$

It can be extended to the graph:

$$J_q(X,Y) \times_Y \Pi(Y,Y) \rightarrow J_q(X,Y) \times_X J_q(X,Y),$$

$$\left(\left(x, y, \frac{\partial y}{\partial x} \right), \left(y, \overline{y}, \frac{\partial \overline{y}}{\partial y} \right) \right) \rightarrow \left(\left(x, y, \frac{\partial y}{\partial x} \right), \left(x, \overline{y}, \frac{\partial \overline{y}}{\partial x} \right) \right).$$

Accordingly, we may set

Definition 1. A system $\mathcal{A}_q \subset J_q(X,Y)$ is said to be *invariant* under a Lie groupoid $\mathcal{R}_q \subset \Pi_q(Y,Y)$ if the action of the latter induces a morphism

$$\mathcal{A}_q \times_Y \mathcal{R}_q \rightarrow \mathcal{A}_q.$$

Accordingly, the action is said to be *free* (*transitive*) if the induced graph

$$\mathcal{A}_q \times_Y \mathcal{R}_q \rightarrow \mathcal{A}_q \times_X \mathcal{A}_q$$

is a monomorphism (epimorphism). In particular, if this graph is an isomorphism, we say that $\mathcal{A}_q$ is a PHS for $\mathcal{R}_q$. From now on we suppose that $\mathcal{A}_q$ projects *onto* Y.

By passing to the vertical bundle we obtain a morphism

$$\sharp : \mathcal{A}_q \times_Y \mathrm{R}_q \rightarrow V(\mathcal{A}_q)$$

of vector bundles over $\mathcal{A}_q$. This morphism is a monomorphism if the action is free, but we should not forget that R_q is over the target. In local coordinates we obtain the simple formula

$$v^k_{i_1 \ldots i_q} = \eta^k_{r_1 \ldots r_q} y^{r_1}_{i_1} \cdots y^{r_q}_{i_q}$$

for the restriction of $\sharp$ to the respective symbols.

Passing to the projective limit, we set

Definition 2. The system $\mathcal{A}_q$ is called an *automorphic system* for $\mathcal{R}_q \subset \Pi_q(Y,Y)$ or $\Gamma \subset \mathrm{aut}(Y)$ if $\mathcal{A}_{q+r}$ is a PHS for $\mathcal{R}_{q+r}$ for all $r \geq 0$.

Of course, this definition involves an infinite number of conditions, and the following Theorem is crucial for applications ([145]).

First Criterion for Automorphic Systems 3. *If an involutive system $\mathcal{A}_q \subset J_q(X,Y)$ is a PHS for a Lie groupoid $\mathcal{R}_q \subset \Pi_q(Y,Y)$ and $\mathcal{A}_{q+1} = \rho_1(\mathcal{A}_q)$ is a PHS for $\mathcal{R}_{q+1} = \rho_1(\mathcal{R}_q)$, then $\mathcal{A}_q$ is an automorphic system for $\mathcal{R}_q$, and $\mathcal{R}_q$ is an involutive system of finite Lie equations.*

PROOF. *From now on we assume* $\mathrm{rk}(y_i^k) = \inf(m, n)$. For simplicity of exposition we may assume that the matrix (y_i^k) has the form

$$
m\left\{\underbrace{\begin{pmatrix} 1 & & 0 \\ & \ddots & \\ 0 & & 1 \\ 0 & \cdots & 0 \end{pmatrix}}_{n}, \qquad m\left\{\underbrace{\begin{pmatrix} 1 & & 0 & 0 \\ & \ddots & & \vdots \\ 0 & & 1 & 0 \end{pmatrix}}_{n}\right.\right.
$$

$$\text{case } m > n \qquad\qquad \text{case } n \le m$$

Using the local coordinate expression of ♯, which is now an isomorphism at orders q and $q + 1$, we conclude in both cases that the symbols of $\mathcal{A}_q$ and $\mathcal{R}_q$ have the same nonzero characters. Since these symbols have the same dimension at orders q and $q + 1$, we conclude that the symbol of $\mathcal{R}_q$ is involutive, because the symbol of $\mathcal{A}_q$ is involutive by assumption. It follows that the symbols have the same dimension at any order $q + r$, for all $r \ge 0$.

Finally, chasing in the exact commutative diagram

$$
\begin{array}{ccccccc}
0 & \longrightarrow & \mathcal{A}_{q+1} \times_Y \mathcal{R}_{q+1} & \longrightarrow & \mathcal{A}_{q+1} \times_X \mathcal{A}_{q+1} & \longrightarrow & 0 \\
& & \downarrow & & \downarrow & & \\
0 & \longrightarrow & \mathcal{A}_q \times_Y \mathcal{R}_q & \longrightarrow & \mathcal{A}_q \times_X \mathcal{A}_q & \longrightarrow & 0 \\
& & & & \downarrow & & \\
& & & & 0 & &
\end{array}
$$

we conclude that the left vertical map is an epimorphism, and so $\mathcal{R}_{q+1} \to \mathcal{R}_q$ is also an epimorphism. Accordingly, $\mathcal{R}_q$ is an involutive system of finite Lie equations, and by counting dimensions at order $q + r$, for all $r \ge 0$, we deduce that $\mathcal{A}_q$ is an automorphic system for $\mathcal{R}_q$. The verification of these two steps is very important in practice. □

In examples we shall see that a given system may be automorphic for many groupoids at the same time if these are subgroupoids of the largest groupoid of invariance, in case the action of the latter is not free. Hence, apart from certain rare situations in which the largest groupoid of invariance acts freely and transitively in the sense of Definition 1, it may be difficult to construct automorphic systems. In any case, before looking for automorphic systems, the above comment shows that the least we should do is to look for the largest groupoid of invariance.

In actual practice we write out the defining equations of $\mathcal{A}_q$:

$$
\Phi^\tau\left(x, y, \frac{\partial y}{\partial x}, \dots\right) = 0
$$

and the transformed equations:

$$
\Phi^\tau\left(x, \overline{y}, \frac{\partial \overline{y}}{\partial x}, \dots\right) = 0.
$$

We write down the chain rule for jet composition:

$$\frac{\partial \overline{y}}{\partial x} = \frac{\partial \overline{y}}{\partial y}\frac{\partial y}{\partial x}, \ldots,$$

and substitute these in the transformed equations:

$$\Phi^\tau\left(x, \overline{y}, \frac{\partial \overline{y}}{\partial y}\frac{\partial y}{\partial x}\right) = 0.$$

Using the implicit function theorem (!) we solve for the principal jets as functions of x, y and for the parametric jets, and stipulate (!) that the resulting equations should not depend on x and the parametric jets. This condition provides equations among the $(y, \overline{y}, \partial \overline{y}/\partial y, \ldots)$, which are the defining equations of the largest Lie groupoid of invariance. By way of an example in the introduction we have already seen that the corresponding system of PDE need not be formally integrable at all.

In the differential case *it is almost impossible to apply the above procedure*, and in practice one restricts to the study of systems of algebraic PDE defined over a (partial) differential fields K with subfield of constants C.

In that case, separating $(\overline{y}, \partial \overline{y}/\partial y, \ldots)$ from $(\partial y/\partial x, \ldots)$ amounts to writing out a number of expressions in

$$C\left[\overline{y}, \frac{\partial \overline{y}}{\partial y}, \ldots\right] \otimes_C K\left[\frac{\partial y}{\partial x}, \ldots\right].$$

We can choose a *minimal number* of elements in $K[\partial y/\partial x, \ldots]$ in order to express the transformed system as a system of linear equations over this basis, with coefficients in $C[\overline{y}, \partial \overline{y}/\partial y, \ldots]$. Two such linear systems are equivalent if and only if the ratios of the determining determinants (*Chow coordinates*) are equal to their value at the q-jet of the identity, which is now a rational function of $y, \ldots$ on the condition that $C[y]$ is assumed to be an integral domain contained in $K[\mathcal{A}_q]$ such that $C(y) \subset K(\mathcal{A}_q) \subset L$. By duality we can deduce that we should have an epimorphism $\mathcal{A}_q \to Y$ which becomes the target projection in case $\mathcal{A}_q = \mathcal{R}_q$. The latter is, like the case $X = G$, the simplest example of a PHS.

The procedure above, which was discovered at the end of the 19th century (!) by Jules Drach [50], is also classical in the theory of linear algebraic groups [82, 94], and gives a natural way of exhibiting differential invariants without using the Frobenius Theorem, as we shall now see.

Definition 4. A *differential invariant of order q* is a function on $J_q(X, Y)$ that is invariant under the groupoid action of $\mathcal{R}_q$ defined above.

Working on a finite level can only be done along the way presented above. So it will be of no use for exhibiting differential invariants under the action, since it only gives differential invariants of the groupoid itself. Relations between the study at order q and that at order $q + 1$ are quite technical and have been completed in [145] by means of the Spencer operator.

On the infinitesimal level we may copy the work done in Chapter V.C, but now with $m \neq n$. Accordingly, we can still use Frobenius' Theorem for the vertical distribution

$\sharp(R_q(Y))$ and the prolongations $\sharp(R_{q+r}(Y))$ for any $r \geq 0$ in a rather generic way, i.e. for the orbits of maximal dimensions. Since the work of Lie it is known that a fundamental set of differential invariants of order $q + 1$ must contain the formal derivatives of the differential invariants of a fundamental set of order q. This implies the *existence* of a generating fundamental set of sufficiently high order. However, this order has never been explicitly determined. Of course, it is understood that the 'best' orbits are obtained under a free action, as each of them becomes a PHS under the groupoid action.

Accordingly, the following result is essential for determining classifying spaces for groupoid actions as the spaces of solutions of certain systems of PDE.

Second Criterion for Automorphic Systems 5. *If the involutive groupoid $\mathcal{R}_q \subset \Pi_q(Y, Y)$ acts freely on $J_q(X, Y)$, then the prolongations $\mathcal{R}_{q+r}$ act freely on $J_{q+r}(X, Y)$, for all $r \geq 0$, and the order of a generating fundamental set of differential invariants is $q + 1$.*

PROOF. If $M_q(Y) \subset S_q T^*(Y) \otimes T(Y)$ is the symbol of $R_q(Y) \subset J_q(T(Y))$, we known that there is a morphism over $J_q(X, Y)$:

$$\sharp: J_q(X, Y) \times_Y M_q(Y) \to V(J_q(X, Y)) \cap S_q T^* \otimes T(Y)$$

and an isomorphism over $\mathcal{R}_q$:

$$M_q(Y) \times_Y \mathcal{R}_q \to V(\mathcal{R}_q) \cap S_q T^*(Y) \otimes T(Y).$$

Accordingly, we have the following commutative diagram of affine bundles:

$$
\begin{array}{ccc}
J_q(X, Y) \times_Y M_{q+1}(Y) & \longrightarrow & J_q(X, Y) \times_X S_{q+1} T^* \otimes T(Y) \\
\downarrow & & \downarrow \\
J_{q+1}(X, Y) \times_Y \mathcal{R}_{q+1} & \longrightarrow & J_{q+1}(X, Y) \times_X J_{q+1}(X, Y) \\
\downarrow & & \downarrow \\
J_q(X, Y) \times_Y \mathcal{R}_q & \longrightarrow & J_q(X, Y) \times_X J_q(X, Y)
\end{array}
$$

A chase reveals that the middle horizontal arrow is a monomorphism if and only if the upper horizontal arrow is a monomorphism, equivalently: if and only if there is a monomorphism

$$M_{q+1}(Y) \to S_{q+1} T^* \otimes T(Y)$$

over $J_q(X, Y) \times_Y \mathcal{R}_q$, and in fact over $J_1(X, Y) \times_Y \mathcal{R}_1$ since only first-order jets are involved. In local coordinates we have

$$v^k_{i_1 \ldots i_q} = \eta^k_{r_1 \ldots r_q} y^{r_1}_{i_1} \cdots y^{r_q}_{i_q}.$$

As in the proof of the First Criterion, we discover that the monomorphism is *automatic* if $n \geq m$. If $n < m$ it was deduced from the monomorphism $T^*(Y) \to T^*(X) = T^*$

expressed in local coordinates by the formula $u_i = y_i^k v_k$ and from a chase in the exact commutative diagram

$$
\begin{array}{ccc}
0 & & 0 \\
\downarrow & & \downarrow \\
\mathrm{M}_{q+1}(Y) & \longrightarrow & S_{q+1}T^* \otimes T(Y) \\
\downarrow{\scriptstyle \delta} & & \downarrow{\scriptstyle \delta} \\
0 \longrightarrow T^*(X) \otimes \mathrm{M}_q(Y) & \longrightarrow & T^* \otimes S_q T^* \otimes T(Y)
\end{array}
$$

where, for simplicity, the pullbacks are not indicated. In particular, we see that we *must* have $\alpha_q^{n+1} = 0, \ldots, \alpha_q^m = 0$.

Finally, the vertical bundle of the orbits at order $q+r$ is a pullback of $\sharp(\mathrm{R}_{q+r}(Y))$ for $r \geq 0$. Accordingly, the corresponding symbol is a pullback of $\sharp(\mathrm{M}_{q+r}(Y))$, *but only for* $r \geq 1$, and we shall give examples of this strange fact in Section E. It follows that the symbol of the orbits is involutive at order $q + 1$ only since then it is isomorphic to $\mathrm{M}_{q+1}(Y)$, which is involutive because $\mathrm{M}_q(Y)$ is involutive by assumption. The remainder of the proof now follows from the First Criterion. $\square$

As in Chapter IV we can construct a natural bundle $\mathcal{F}$ and a natural epimorphism $\Phi \colon J_q(X, Y) \to \mathcal{F}$, but *the reader should not forget that we have to use $q + 1$ instead of q whenever the action is free at order q.* If the section ω of $\mathcal{F}$ is arbitrary, but satisfying compatibility conditions that could be stated as in Chapter IV, then the system

$$
\mathcal{A}_q \qquad \Phi^\tau(y_q) = \omega^\tau(x) \qquad \text{(Lie form)}
$$

is said to be a *general automorphic system.*

Conversely, if $\mathcal{A}_q$ is an automorphic system, then ω is well determined, and we say that $\mathcal{A}_q$ is a *special automorphic system.* The restriction of Φ to a given invariant system of order q may impose more or less restrictions on the section. As a byproduct we see that *if a groupoid of invariance of a system acts freely,* then the classifying space for the generic orbits is the space of solutions of a certain system of PDE for ω on $\mathcal{F}$.

We have already met automorphic systems for finite groups in classical Galois theory, and for linear algebraic or Lie groups in the Picard–Vessiot theory presented in the Introduction. We would like to continue this Section with an example of an automorphic system for a Lie pseudogroup. This example is sufficiently generic to be of interest. More specific situations of this type are given in Section E.

Consider the first-order system of PDE

$$
w + H(t, x, z, p) = 0
$$

for the single unknown function z of two independent variables x (space) and t (time), with $p = \partial z / \partial x$ and $w = \partial z / \partial t$. Among the simplest situations known to exist in analytical dynamics, we first find the case

$$\frac{\partial H}{\partial z} = 0,$$

corresponding to the Hamilton–Jacobi equation, then the case

$$\frac{\partial H}{\partial z} = 0, \qquad \frac{\partial}{\partial t}\left(\frac{\partial H}{\partial x} \Big/ \frac{\partial H}{\partial p}\right) = 0,$$

corresponding to the possibility of separation of the variables x, t during integration (this was discovered by Levi-Civita), and then the case

$$\frac{\partial H}{\partial z} = 0, \qquad \frac{\partial H}{\partial x} = 0$$

of a cyclic variable or the case

$$\frac{\partial H}{\partial z} = 0, \qquad \frac{\partial H}{\partial t} = 0$$

of conservation of energy. In [145, 146] we have proved that this nested chain of cases corresponds to a decreasing chain of Lie pseudogroups. Here we will only explain the main idea, and, to illustrate the First Criterion for automorphic systems, we will treat in some detail the case corresponding to the Hamilton–Jacobi equation.

Definition 6. A *complete integral* $z = f(t, x; a, b)$ is a family of solutions depending on two parameters (a, b) in such a way that

$$\frac{\partial(z, p)}{\partial(a, b)} \neq 0 \qquad \text{(Hessian condition)}$$

whenever $p = \partial f / \partial x$.

Examples of complete integrals are given in Section E.

Theorem 7. *The search for a complete integral of the PDE $w + H(t, x, z, p) = 0$ is equivalent to the search for a single solution of the automorphic system obtained by eliminating $\rho(t, x, z, p)$ in the Pfaffian system*

$$dz - p\,dx + H(t, x, z, p)\,dt = \rho(dZ - P\,dX).$$

The corresponding pseudogroup is the pseudogroup of contact transformations in (X, Z, P) that reproduces the 1-form $dZ - P\,dX$ up to a function factor.

PROOF. If $z = f(t, x; a, b)$ is a complete integral, then

$$dz - p\,dx + H(t, x, z, p)\,dt = \frac{\partial f}{\partial a}\,da + \frac{\partial f}{\partial b}\,db.$$

Using the implicit function theorem and the Hessian condition, we may set

$$a = X(t, x, z, p), \qquad b = Z(t, x, z, p),$$

and obtain the Pfaffian system with $\rho = \partial f / \partial b$, $P = -(\partial f / \partial a)/(\partial f / \partial b)$.

Conversely, under the assumption

$$\frac{\partial(Z, X)}{\partial(z, p)} \neq 0,$$

which is equivalent to the Hessian assumption, we can use the implicit function theorem and obtain

$$z = f(t, x; a, b), \qquad p = g(t, x; a, b).$$

For $a = $ const, $b = $ const we have $dX = 0$, $dZ = 0$, and so $dz - p\, dx + H(t, x, z, p)\, dt = 0$, i.e. $g = \partial f / \partial x$, in particular.

For another solution (denoted by a $\leftrightarrow$) we have

$$dz - p\, dx + H(t, x, z, p)\, dt = \overline{p}(d\overline{Z} - \overline{P}\, d\overline{X}),$$

and so

$$d\overline{Z} - \overline{P}\, d\overline{X} = \frac{\rho}{\overline{\rho}}(dZ - P\, dX).$$

Closing this system, we immediately obtain

$$d\overline{X} \wedge d\overline{Z} \wedge d\overline{P} = \left(\frac{\rho}{\overline{\rho}}\right)^2 dX \wedge dZ \wedge dP$$

and, closing again, we deduce that $\rho/\overline{\rho}$, which initially is a function of (t, x, z, p), is in fact a function of (X, Z, P). This is precisely the definition of a contact transformation permuting each pair of solutions of the Pfaffian system. $\square$

For the Hamilton–Jacobi equation it follows that, since z is absent, if $z = f(t, x)$ is a solution, then so is $z = f(t, x) + b$ for any constant b.

Accordingly (see Section E for details),

Corollary 8. *The search for a complete integral $z = f(t, x; a) + b$ of the Hamilton–Jacobi equation is equivalent to the search for a single solution of the Pfaffian system*

$$dz - p\, dx + H(t, x, p)\, dt = dZ - P\, dX.$$

This is an automorphic system for the Lie pseudogroup of contact transformations preserving volume.

PROOF. We first prove that the Pfaffian system

$$dz - p\, dx + H(t, x, z, p)\, dt = dZ - P\, dX$$

is compatible if and only if $\partial H / \partial z = 0$. Indeed, by closing the system we obtain

$$dx \wedge dp + dH \wedge dt = dX \wedge dP.$$

After termwise exterior multiplication we obtain

$$dz \wedge dx \wedge dp + dz \wedge dH \wedge dt - p\, dx \wedge dH \wedge dt + H\, dt \wedge dx \wedge dp = dZ \wedge dX \wedge dP.$$

Closing again, we obtain

$$-dp \wedge dx \wedge dH \wedge dt + dH \wedge dt \wedge dx \wedge dp = 2\frac{\partial H}{\partial z} dz \wedge dt \wedge dx \wedge dp = 0,$$

and the desired condition selecting the Hamilton–Jacobi equation.

In this situation we obtain $\rho = \partial f/\partial b = 1$ and the desired Pfaffian system which is an automorphic system for the Lie pseudogroup preserving the 1-form $dZ - P\,dX$, i.e. such that

$$dZ - \overline{P}\,d\overline{X} = dZ - P\,dX.$$

By closing this system we obtain

$$d\overline{X} \wedge d\overline{P} = dX \wedge dP,$$

and so, by exterior multiplication,

$$d\overline{X} \wedge d\overline{Z} \wedge d\overline{P} = dX \wedge dZ \wedge dP,$$

i.e. conservation of the volume 3-form. □

In actual practice, the following Remark may be useful.

Remark 9. Computing the dimensions, we have

$$\begin{cases} \dim \mathcal{A}_q \times_Y \mathcal{R}_q = \dim \mathcal{A}_q + \dim \mathcal{R}_q - \dim Y, \\ \dim \mathcal{A}_q \times_X \mathcal{A}_q = \dim \mathcal{A}_q + \dim \mathcal{A}_q - \dim X. \end{cases}$$

Hence, whenever we have a PHS, the equality of dimensions becomes

$$\dim \mathcal{R}_q - \dim Y = \dim \mathcal{A}_q - \dim X,$$

in which only the dimensions of the fibers appear.

D. Cascade Integration

It is not our purpose to give in this Section even a rough sketch of differential Galois theory, since it has been treated elsewhere and involves quite a number of delicate arguments [145]. Instead, our aim is simply to explain by simple situations how to establish a correspondence between groups and fields, and ultimately between pseudogroups and differential fields. In this way, a new link is made between differential geometry and differential algebra, which could be useful in applications. The basic motivation for establishing this link is provided by the following explicit example.

Problem. Study the integration of the third-order system $\mathcal{A}_3$ defined by the linear ordinary differential equation

$$P \equiv y_{xxx} - a(x)y_x = 0,$$

where $a(x)$ is an arbitrary function of x.

We give three ways of solving this problem, and the main conceptual result lies in discovering how these depend on each other. This point of view is subsequently generalized to arbitrary systems of PDE, and we will consider the special case of algebraic PDE.

Solution 1. Graduate courses provide many tricks for integrating ODE. Among them, a few standard 'changes of variables' are well known, and any student will be inclined to set

$$\frac{y_{xx}}{y_x} = u$$

in order to find that u is a solution of the Riccati equation

$$u_x + u^2 - a(x) = 0.$$

Then, a second change of variables

$$\frac{y_x}{y} = v$$

proves that v is a solution of the Riccati equation

$$v_x + v^2 - uv = 0,$$

provided u is known. As a byproduct we have 'splitted' the integration of the initial third-order system into that of three first-order systems, called '*simpler systems*', to wit: the two Riccati equations above and the first-order ODE

$$y_x - vy = 0.$$

Of course, we see that these three systems depend on each other, since we first have to integrate the Riccati equation for u, substitute u into the Riccati equation for v, solve this last Riccati equation, and finally substitute v into the above first-order ODE and solve it for y. Nothing more can be said, apart from the fact that we could have set

$$y_x = w$$

to obtain the linear (special Riccati) ODE

$$w_x - uw = 0,$$

with the same comment as above.

Solution 2. In this example we use the group invariance technique described in the previous Section. Of course, the given system is linear, and we hence find a Lie group action. So we use this example only for the sake of clarity, and postpone to Section E nonlinear examples involving pseudogroups.

We make the change of unknowns $\overline{y} = y(y)$, and substitute jets for derivatives in order to find, via the chain rule,

$$\overline{y}_x = \frac{\partial \overline{y}}{\partial y} y_x,$$

$$\overline{y}_{xx} = \frac{\partial^2 \overline{y}}{\partial y \partial y} (y_x)^2 + \frac{\partial \overline{y}}{\partial y} y_{xx},$$

$$\overline{y}_{xxx} = \frac{\partial^3 \overline{y}}{\partial y \partial y \partial y} (y_x)^3 + 3 \frac{\partial^2 \overline{y}}{\partial y \partial y} y_x y_{xx} + \frac{\partial \overline{y}}{\partial y} y_{xxx}.$$

The so-called *translated* or *transformed ODE* can be represented by the zero linear differential polynomial

$$\overline{P} \equiv \overline{y}_{xxx} - a(x)\overline{y}_x = 0,$$

and becomes

$$\frac{\partial^3 \overline{y}}{\partial y \partial y \partial y}(y_x)^3 + 3\frac{\partial^2 \overline{y}}{\partial y \partial y}y_x y_{xx} + \frac{\partial \overline{y}}{\partial y}y_{xxx} - a(x)\frac{\partial \overline{y}}{\partial y}y_x = 0.$$

The shortest basis for the decomposition of $\overline{P}$ is

$$(y_x)^3, \quad y_x y_{xx}, \quad y_{xxx} - a(x)y_x.$$

Accordingly, the quotients

$$\Sigma \equiv \frac{\frac{\partial^3 \overline{y}}{\partial y \partial y \partial y}}{\frac{\partial \overline{y}}{\partial y}}, \qquad \Omega \equiv \frac{\frac{\partial^2 \overline{y}}{\partial y \partial y}}{\frac{\partial \overline{y}}{\partial y}}$$

are the differential invariants of the largest groupoid of invariance. In Section E we shall see that it need not be involutive, or even formally integrable. Now we discover that

$$d_y\Omega + \Omega^2 = \Sigma,$$

and so the only generating differential invariant is Ω. This leads to the Lie equation

$$\mathcal{R}_2 \quad \frac{\partial^2 \overline{y}}{\partial y \partial y} = 0$$

and to the Lie group of transformations

$$\Gamma \quad \overline{y} = ay + b, \qquad a, b = \text{const},$$

with underlying Lie group G of dimension 2 and parameters (a, b). We obtain the prolongations

$$\overline{y}_x = ay_x,$$
$$\overline{y}_{xx} = ay_{xx}.$$

Accordingly, the action on second-order jets is free, and we are in a position to apply the second Criterion for being an automorphic system by using $\Phi \equiv y_{xx}/y_x$ and $d_x\Phi$, although in *this* specific situation the use of Φ only would suffice. We find the monomorphism

$$0 \to \mathcal{A}_3 \times_Y \mathcal{R}_3 \to \mathcal{A}_3 \times_X \mathcal{A}_3,$$

and $\dim \mathcal{R}_3 - \dim Y = 2 < 3 = \dim \mathcal{A}_3 - \dim X$. Accordingly, $\mathcal{A}_3$ is *not* a PHS for $\mathcal{R}_3$ and so surely not an automorphic system. Hence, the general automorphic system is

$$\Phi \equiv \frac{y_{xx}}{y_x} = u,$$

where u is partially constrained by the first Riccati equation found in Solution 1. So, this Riccati equation is just a classifying space for the space of orbits as a kind of 'transversal' manifold, as in the following picture:

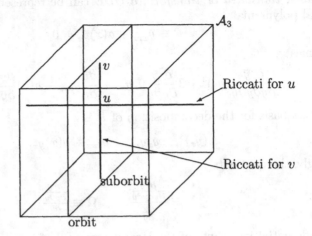

Of course, each orbit space is a PSH for the group, since the action is known to be free.

We may now select the subgroup

$$\Gamma' \quad \overline{y} = ay,$$

with generating differential invariant $\Psi = y_x/y$. In the orbit determined by u, the suborbit space is classified by introducing the automorphic system

$$\Psi \equiv \frac{y_x}{y} = v,$$

where now v is partially constrained by the second Riccati equation in Solution 1. So, we see that the so-called 'changes of variables' amount to the successive construction of (partially) constrained automorphic systems. A similar comment applies to the other subgroup,

$$\Gamma'' \quad \overline{y} = y + b,$$

and we see that the cascade decomposition is not unique, but is determined by group inclusion. The main novelty in this approach lies in the identification of classifying spaces for orbits with solution spaces of (systems of) PDE or ODE. We finally notice that the orbit spaces for $\Gamma, \Gamma', \Gamma''$ are automorphic systems contained in $\mathcal{A}_3$, given their respective constraints.

Solution 3. Let us introduce an ordinary differential 'ground' field k, and suppose that $a \in k$ (we cannot write $a(x) \in k$ unless $k = \mathbb{Q}(x)$, the ordinary differential field of rational functions in x). We may now regard P as a linear ordinary differential polynomial (in fact, this is the reason for the notation) generating a prime differential ideal $\mathfrak{p}$ in $k\{y\}$. We can introduce the differential extension L/k with $L = Q(k\{y\}/\mathfrak{p}) \simeq k(y, y_x, y_{xx})$ and $d_x y = y_x, d_x y_x = y_{xx}, d_x y_{xx} = a y_x$. Of course, $u = y_{xx}/y_x \in L$ is differentially algebraic over k because y is differentially algebraic over k and $k\langle u \rangle = K$ is an intermediate differential field. Here we have introduced a slight abuse of notation, as we may have written $k(u) \subset k(y, y_x, y_{xx})$. Indeed, if we refer to L, then we should use a residue η, and define $K = k\langle \sigma \rangle$ with $\sigma = d_{xx}\eta/d_x\eta$ in L. Continuing with the previous notation, we may introduce the intermediate differential extension $K' = k\langle v \rangle$, and thus obtain the chain of inclusions

$$k \subset K \subset K' \subset L$$

because $u = (d_x v + v^2)/v$ by Solution 1. A similar comment applies with $K'' = k\langle w \rangle$ and $w = y_x$, since we should have $u = d_x w / w$. Here we have used the cascade decomposition to split L/k into the three differential extensions L/K', K'/K, K/k, all having transcendence degree 1. At the same time we notice that there is a bijective order-reversing correspondence between the subgroups of Γ and the fields intermediate between K and L *only*. The transition from groups to fields has been used in Solution 2, and we will now prove the converse. In fact, if

$$\frac{\overline{y}_{xx}}{\overline{y}_x} = \frac{y_{xx}}{y_x},$$

then $\left(\frac{\partial^2 \overline{y}}{\partial y \partial y} / \frac{\partial \overline{y}}{\partial y} \right) y_x = 0$. Thus, L/K has a *model variety* (i.e. a variety defined by the same prime ideal over K as the prime ideal used to define L/K), which is an automorphic system for Γ, although *this property does not hold for L/k*. Similarly, if

$$\frac{\overline{y}_x}{\overline{y}} = \frac{y_x}{y},$$

then $\frac{1}{\overline{y}} \frac{\partial \overline{y}}{\partial y} = \frac{1}{y}$ and L/K' has a model variety which is an automorphic system for Γ'. Again, similarly, if

$$\overline{y}_x = y_x,$$

then $\frac{\partial \overline{y}}{\partial y} = 1$ and L/K'' has a model variety which is an automorphic system for Γ''. Accordingly,

Definition 4. A differential extension L/K is called a *differential automorphic extension* if it has a model variety which is an automorphic system for some Lie pseudogroup Γ.

Notice that this Definition can be effectively tested by combining the criterion for being a prime differential ideal and the first criterion for being an automorphic system. In fact, if the involutive groupoid $\mathcal{R}_q$ is transitive, then M_{q+1} is a vector bundle defined over L (we should not forget that there is no notion of 'independent variable' in differential algebra). A necessary and sufficient criterion for this is that the involutive automorphic system itself is transitive.

It follows from these considerations that *the only possible candidates for a differential Galois theory are the differential automorphic extensions*.

We now turn to the consideration of the nature of the Lie pseudogroups involved in such extensions. We notice that the chain rule for jet composition is purely polynomial, given that the system projects onto Y, i.e. if it is transitive, the variables y and $\overline{y}$ are independent, and the groupoid of invariance is defined by algebraic equations among the $(y, \overline{y}, \partial \overline{y}/\partial y, \dots)$ that involve only the field of constants k of the differential field K over which the system is defined. In Solution 3 the differential field k

has been used to distinguish the differential extension L/k determined by the given system from the differential automorphic extension L/K, in order to obtain the same Galois correspondence $K \leftrightarrow \Gamma$, $K' \leftrightarrow \Gamma'$, $K'' \leftrightarrow \Gamma''$ as in the classical Galois theory of Section A.

Definition 5. An *algebraic Lie pseudogroup* is a Lie pseudogroup defined by a system of algebraic equations in (all) the jet coordinates.

The study of algebraic Lie pseudogroups is a vast open area of research, and preparatory material is provided in [145]. Let us give a few basic facts, adapting the notation of Chapter IV to the algebraic case by considering a fixed field of constants k.

An algebraic Lie pseudogroup is defined by a certain number of equations in (x, y_q) over k, where, of course, $\det(y_i^k) \neq 0$. In fact, it suffices to require the equations to be algebraic over $k(x)$, because we may use the translated equations to find differential invariants that are rational functions in $k(y_q)$. Then evaluation of $\Phi(y_q) = \omega(x)$ at the identity proves that $\omega(x) \in k(x)$, and we may multiply out the denominators so that we can work in $k[x, y_q]$. Because of this, and some other reasons to be given later, it seems preferable to work in the ring

$$k(x) \otimes_k k(y)[y_\mu^k \mid 1 \leq |\mu| \leq q].$$

Notice that as $q \to \infty$, the projective limit of these is a differential ring, because

$$d_i a(y) = \frac{\partial a(y)}{\partial y^k} y_i^k.$$

These conventions respect the 'symmetry' between interchange of source and target. (We should have introduced a localization, because of the Jacobian condition, but did not do so because of simplicity reasons.) Taking the quotient of this differential ring with respect to the differential ideal generated by the defining equations of the algebraic pseudogroup Γ and all their prolongations leads to a differential ring, $k[\Gamma]$, into which we can inject both $k(x)$ *and* $k(y)$.

Remark 6. In general, the ring of quotients $k(\Gamma)$ is a only a direct sum of integral domains and becomes a field only in case Γ has precisely one connected component of the identity transformation. The groupoid

$$\mathcal{R}_1 \quad \{y_2^1 = 0, y_1^2 = 0\} \cup \{y_1^1 = 0, y_2^2 = 0\}$$

is a good example of a situation with two components (these are disjoint because of the jacobian condition); the first component is the connected component of the identity.

Turning to the particular situation of Lie groups of transformations and Lie group actions, we have seen that for q sufficiently large there is an isomorphism

$$X \times G \simeq \mathcal{R}_q.$$

If G and the action are defined over k, by working in the ring $k[x, y_q]$ we should find an isomorphism

$$k[x] \otimes_k k[G] \simeq k[\mathcal{R}_q].$$

Passing to the projective limit, given our convention regarding the corresponding differential ring, we only have the isomorphism

$$k(x)(G) \simeq k(\Gamma).$$

We now apply the previous results to the study of irreducible automorphic systems defined over the differential field K, with field of constants k, and write $(y, \overline{y})$ instead of (x, y). The reader should not forget that, in contrast to [96], *the situation useful in applications is concerned with two differential rings*, the first one, $K\{y\}$, allowing one to define the automorphic system and having a differential structure with respect to 'x' (even if it does not turn up unless $K = k(x)$), the second, $k(y) \otimes_k k(\overline{y})[\partial \overline{y}/\partial y, \ldots]$, allowing one to define the groupoid of invariance and having a differential structure with respect to 'y'. Taking into account the definition of automorphic systems of PDE by means of projective limits, we immediately obtain the fundamental isomorphism

$$Q(L \otimes_K L) \simeq Q(L \otimes_{k(y)} k[\Gamma]),$$

which characterizes a differential automorphic extension. We will make some comments on this isomorphism.

First we notice that the right hand term is defined only when there is an injection $0 \to k(y) \to L$, i.e. when the involutive system $\mathcal{A}_q$ projects onto Y, as already said. In this case the right hand term depends on $k(\Gamma)$ only.

Secondly, the left hand term is a differential ring under the law

$$d_i(a \otimes b) = (d_i a) \otimes b + a \otimes (d_i b),$$

while the right hand term is a differential ring under the law

$$d_i(a \otimes \gamma) = (d_i a) \otimes \gamma + a y_i^k \otimes \frac{d\gamma}{dy^k}.$$

Here we have to take into account the specific differential structure of $k[\Gamma]$. Again in contrast to [96], in Section E we shall give an example showing that in general $k[\Gamma]$ does not have a differential structure with respect to 'x'.

In the case of Lie groups of transformations, distributivity of tensor products gives

$$L \otimes_{k(y)} k(\Gamma) = L \otimes_k k(G),$$

and the fundamental isomorphism becomes

$$Q(L \otimes_K L) \simeq Q(L \otimes_k k[G]),$$

in agreement with the excellent reference [24].

The above approach has been adopted so as to use (differential) fields only. Accordingly, the various existing theories can be classified as follows:

[Classical Galois Theory] $\operatorname{trd} L/K = 0, \quad |L/K| < \infty;$

[Picard–Vessiot Theory] $\operatorname{trd} L/K < \infty;$

[Differential Galois Theory] $\operatorname{trd} L/K = \infty.$

Therefore, the use of the fundamental isomorphism is *necessary* for the any unification, while the use of the criteria presented in this Chapter is *necessary* for any effective study.

The epimorphism

$$\mathcal{A}_q \times_X \mathcal{A}_q \to \mathcal{R}_q \to 0$$

gives rise to a monomorphism

$$o \to k[\Gamma] \to L \otimes_K L$$

which can be used for specialization, as in section A. Hence, *the major problem in differential Galois theory is to give a geometric interpretation of the preceding inclusion*. The answer to this question will reveal that the understanding by all authors [89, 95] that Picard–Vessiot theory is the only geometric framework that can provide such an answer, is a misunderstanding. To grasp this point, we briefly return to our standard notation for PHS. Indeed, if X is a PHS for some Lie group G, then the isomorphism $X \times G \simeq X \times X$ is defined by $y = f(x, a)$ in such a way that we may obtain $a = a(x, y)$ from the *identity* $y \equiv f(x, a(x, y))$, and a jacobian condition

$$\frac{\partial(x, y)}{\partial(x, a)} = \det \begin{pmatrix} 1 & \dfrac{\partial f}{\partial x} \\ 0 & \dfrac{\partial f}{\partial a} \end{pmatrix} = \det \left(\frac{\partial f}{\partial a} \right) \neq 0$$

is defined because $\dim X = \dim G$. Let us now consider a vector field ξ and the vector field $\eta = j_1(f)(\xi)$ transformed according to the formula

$$\eta^k(f(x)) = \xi^i(x) \partial_i f^k(x)$$

for any given a. Then ξ commutes with the infinitesimal generators of the action whenever it is unchanged under the action, i.e.

$$\xi^k(f(x)) \equiv \xi^i(x) \partial_i f^k(x)$$

for any a. Let us introduce the vector field

$$\delta = \xi^i(x) \frac{\partial}{\partial x^i} + \xi^k(y) \frac{\partial}{\partial y^k}$$

over $X \times X$, and apply it to the previous identity. We obtain

$$\xi^k(y) = \xi^i(x) \frac{\partial f^k}{\partial x^i}(x, a) + (\delta \cdot a^\sigma) \frac{\partial f^k}{\partial a^\sigma}(x, a),$$

i.e. $\delta \cdot a^\sigma = 0$ (using the jacobian condition). So, we see that *when dealing with a PHS (and only then)*, the parameters can be viewed as functions on $X \times X$ that are annihilated by the extension δ to $X \times X$ of any vector field on X commuting with any infinitesimal generator of the action.

This remark was more or less discovered by Vessiot, and has been adapted to differential Galois theory in later papers [192]. As a byproduct we will see in Section E that it is quite a misleading coincidence that the parameters of the general linear group can be regarded as differential constants in the Picard–Vessiot theory. However, the existence of a distribution Δ_q on $\mathcal{A}_q$ commuting with the distribution $\Theta_q = \sharp(\mathcal{R}_q)$, even in the case of an ordinary PHS for a Lie group, is a quite delicate result involving Galois cohomology [95]. Therefore we do not present it here. Passing to the projective limit we obtain distributions Δ and Θ such that $[\Theta, \Delta] = 0$. This makes sense because any intermediate differential field K' between K and L is finitely generated, and we may use a geometric picture with q sufficiently large. We have $\Theta(K) = 0$, i.e. $\theta \cdot a = 0$ for all $\theta \in \Theta$ and all $a \in K$. It follows that

$$\Theta(\Delta(K)) = \Theta(\Delta(K)) - (\Delta(\Theta(K))) = [\Theta, \Delta](K) = 0.$$

A similar result holds for the distribution $\Theta' \subset \Theta$ determined by a Lie subpseudogroup $\Gamma' \subset \Gamma$. Also, we know that the formal derivative of a differential invariant is a differential invariant itself. Accordingly, we only need consider intermediate differential fields that are stable by Δ; we use the notation 'cst' to refer to Δ. So, we have

$$k(y) = \mathrm{cst}(L), \qquad k[\Gamma] \subset \mathrm{cst}(L \otimes_K L),$$

and the fundamental isomorphism even provides the equality:

$$k(\Gamma) = \mathrm{cst}(Q(L \otimes_K L)),$$

because of the following technical Lemma

Lemma 7. *If L/K is a differential extension with $C = \mathrm{cst}(K)$ and $D = \mathrm{cst}(L)$, then K and D are linearly disjoint over C in L. Moreover, if $L = Q(K \otimes_C D)$, then there is a bijection between intermediate differential fields and intermediate fields of constants, given by*

$$C' = \mathrm{cst}(K') \longleftrightarrow K' = Q(K \otimes_C C').$$

PROOF. We prove this Lemma for ordinary differential fields, but the proof can easily be adapted to any field with derivations ([24]).

Let $d_1, \ldots, d_r$ be a minimal set of elements in D that are linearly independent over C but linearly dependent over K. Then there are $k_1, \ldots, k_r \in K$ such that

$$k_1 d_1 + \cdots + k_r d_r = 0$$

in L, and we may assume $k_r = 1$. Applying d we immediately obtain

$$(dk_1)d_1 + \cdots + (dk_{r-1})d_{r-1} = 0,$$

contradicting the minimality condition, and $k_1, \ldots, k_r \in C$.

Now, let C' be between C and D. Any element k' of $K' = (K, C') \subset L$ can be written as

$$k' = \sum_\alpha a_\alpha k_\alpha \bigg/ \sum_\alpha b_\alpha k_\alpha,$$

with $a_\alpha, b_\alpha \in C' \subset D$, by using a convenient basis $\{k_\alpha\}$ of K over C. Hence, if $k' \in \mathrm{cst}(K') \subset D$, then

$$\sum_\alpha (a_\alpha - k' b_\alpha) k_\alpha = 0$$

in $(K, D) \subseteq L$, and so $a_\alpha - k' b_\alpha = 0$. Since at least one of the b_α must be nonzero, say b_1, we obtain $k' = a_1/b_1 \in C'$, and thus $C' = \mathrm{cst}(K')$.

Conversely, suppose $C' = \mathrm{cst}(K')$, and consider $(K, C') \subseteq K'$. Any $k' \in K' \subset L = (K, D)$ can be written, as before, as

$$k' = \sum_\alpha a_\alpha k_\alpha \Big/ \sum_\alpha b_\alpha k_\alpha ,$$

with $a_\alpha, b_\alpha \in D$, $k_\alpha \in K$. It follows that $\sum_\alpha a_\alpha k_\alpha - \sum_\alpha b_\alpha(k' k_\alpha) = 0$, and so (a_α, b_α) are linearly dependent over K', and so over C' by the first part of the proof. Therefore we can find $a'_\alpha, b'_\alpha \in C'$ such that $\sum_\alpha a'_\alpha k_\alpha - \sum_\alpha b'_\alpha(k' k_\alpha) = 0$ with $\sum_\alpha b'_\alpha k_\alpha \neq 0$ because the k_α are linearly independent over C, and thus over C' by the first part of the proof. Hence we find $k' = \sum_\alpha a'_\alpha k_\alpha / \sum_\alpha b'_\alpha k_\alpha \in (K, C')$. Thus, $K' = (K, C')$. □

We know that $\sharp(\mathrm{R}_q)$ is a Lie algebra over the functions on Y. Therefore, with a slight abuse of language, Θ is a Lie algebra, over $k(y)$, of derivations of L annihilating K. Similarly, one can prove that Δ is a Lie algebra, over K, of derivations of L stabilizing K.

Definition 8. Δ is called the *reciprocal Lie algebra* of Θ.

Collecting the results, we are now ready for sketching a proof of the following key Theorem (see [145] for more details).

Theorem 9 (Fundamental Theorem of Differential Galois Theory). *If L/K is a differential automorphic extension for an algebraic pseudogroup Γ with reciprocal Lie algebra Δ stabilizing K, then there is a bijective order-reversing correspondence between the stable intermediate differential fields and the algebraic subpseudogroups of Γ.*

PROOF. Since Γ acts freely on L/K, the second criterion for being an automorphic system and a transition to the projective limit prove that any subpseudogroup $\Gamma' \subset \Gamma$ also acts freely on L/K. Accordingly, it acts simply transitively on a generic orbit, which is a model for L/K' as an automorphic system for Γ' if we construct K' as in the beginning of this Section. More precisely, we may set $L = K\langle \eta \rangle$ and dualize the action by a monomorphism

$$0 \to K\{y\} \to L \otimes_{k(y)} k[\Gamma]$$

that can be extended to a monomorphism

$$0 \to L \xrightarrow{\tilde{\tau}} Q(L \otimes_{k(y)} k[\Gamma]),$$

or composed with the specialization

$$k[\Gamma] \to k[\Gamma'] \to 0$$

to similarly obtain a monomorphism

$$0 \to L \xrightarrow{\tilde{\tau}'} Q(L \otimes_{k(y)} k[\Gamma']).$$

By definition, we have

$$K = \{a \in L \mid \tilde{\tau}(a) = a \otimes 1\},$$

and therefore we may set

$$K' = \{a \in L \mid \tilde{\tau}'(a) = a \otimes 1\}.$$

Conversely, let K' be a stable intermediate differential field. The kernel of the specialization epimorphism

$$L \otimes_K L \to L \otimes_{K'} L \to 0$$

is thus a stable (differential) ideal in $L \otimes_K L$ that can be intersected with $k[\Gamma]$ to give the specialization epimorphism

$$k[\Gamma] \to k[\Gamma'] \to 0,$$

as in Section A. It follows that

$$k[\Gamma'] \subset \text{cst}(L \otimes_{K'} L),$$

and we obtain a monomorphism

$$0 \to Q(L \otimes_{k(y)} k[\Gamma']) \to Q(L \otimes_{K'} L),$$

by Lemma 7. Composing $\tilde{\tau}'$ with this monomorphism gives

$$\tilde{\tau}'(a) = a \otimes 1 \Rightarrow a \otimes 1 = 1 \otimes a \text{ in } L \otimes_{K'} L \Rightarrow a \in K',$$

and the differential field of invariants of Γ' in L is just K'. This proves that L/K' is an automorphic extension for Γ'. Hence we obtain

$$Q(L \otimes_{K'} L) \simeq Q(L \otimes_{k(y)} k[\Gamma']),$$

and so

$$k(\Gamma') = \text{cst}(Q(L \otimes_{K'} L)),$$

by Lemma 7. $\square$

We end this Section by making a few remarks concerning 'normal' subpseudogroups.

Definition 10. The *normalizer* $N(\Gamma)$ of a Lie pseudogroup Γ in $\text{aut}(X)$ is the largest Lie pseudogroup such that

$$\Gamma \lhd N(\Gamma) \subset \text{aut}(X).$$

Setting $\tilde{\Gamma} = N(\Gamma)$, we thus obtain in $\text{aut}(X)$:

$$f \circ g \circ f^{-1} \in \Gamma, \qquad \forall f \in \tilde{\Gamma}, \quad \forall g \in \Gamma.$$

Contrary to claims in the literature [117], we do not believe that it is possible to define normality on the groupoid level. The simplest example is provided by the normalizer of the pseudogroup of volume-preserving transformations, i.e. the pseudogroup preserving the volume up to a constant factor. This example proves that the definition of normality at the groupoid level, given in [117], is not compatible with that on the pseudogroup level, and must therefore be considered with great care.

In fact, if Γ is defined by a groupoid $\mathcal{R}_q \subset \Pi_q(X, X)$, we may define $\tilde{\Gamma}$ by the groupoid

$$\tilde{\mathcal{R}}_{q+1} = \{f_{q+1} \in \Pi_{q+1}(X, X) \mid f_{q+1}(R_{q+1}) = R_{q+1}\}.$$

On the infinitesimal level we have

$$\tilde{R}_{q+1} = \{\xi_{q+1} \in J_{q+1}(T) \mid L(\xi_{q+1}) R_q \subset R_q\}.$$

As a byproduct, if $\pi_q^{q+1}: R_{q+1} \to R_q$ is surjective, then we obtain the purely algebraic definition

$$\tilde{\mathcal{R}}_{q+1} = \{\xi_{q+1} \in J_{q+1}(T) \mid \{\xi_{q+1}, \eta_{q+1}\} \in R_q, \; \forall \eta_{q+1} \in R_{q+1}\},$$

and the following technical Proposition, proved in [144], holds.

Proposition 11. *If* $\pi_q^{q+1}: R_{q+1} \to R_q$ *is surjective and the symbol* M_q *of* R_q *is* 2-*acyclic, then* $\tilde{R}_{q+1}$ *is formally integrable with symbol* $\tilde{M}_{q+1} = M_{q+1}$.

It follows that $\tilde{\Gamma}$ is an algebraic pseudogroup whenever Γ is such.

E. Examples

Example 1. Consider the parametric curve

$$y = x^2, \qquad z = x^3$$

in 3-dimensional space. The corresponding system is defined by the two PDE

$$y_{111} - y_3 = 0, \qquad y_{11} - y_2 = 0,$$

where we have adopted the correspondence

$$x \to x_1, \qquad y \to x_2, \qquad z \to x_3.$$

The system of coordinates is surely not δ-regular with respect to the lexicographic order and we interchange 1 and 3 to obtain the system

$$R_3 \quad \begin{cases} y_{333} - y_1 = 0, \\ y_{33} - y_2 = 0. \end{cases}$$

We leave it to the reader to verify that the system

$$R_3^{(2)} \quad \begin{cases} y_{333} - y_1 = 0 & \quad 1 \;\; 2 \;\; 3 \\ y_{233} - y_{22} = 0 & \quad 1 \;\; 2 \;\; \bullet \\ y_{223} - y_{12} = 0 & \quad 1 \;\; 2 \;\; \bullet \\ y_{222} - y_{11} = 0 & \quad 1 \;\; 2 \;\; \bullet \\ y_{133} - y_{12} = 0 & \quad 1 \;\; \bullet \;\; \bullet \\ y_{123} - y_{11} = 0 & \quad 1 \;\; \bullet \;\; \bullet \\ y_{122} - y_{113} = 0 & \quad 1 \;\; \bullet \;\; \bullet \\ y_{33} - y_2 = 0 & \quad \bullet \;\; \bullet \;\; \bullet \\ y_{23} - y_1 = 0 & \quad \bullet \;\; \bullet \;\; \bullet \\ y_{22} - y_{13} = 0 & \quad \bullet \;\; \bullet \;\; \bullet \end{cases}$$

is involutive, with symbol $M_3^{(2)}$ and having characters $(3, 0, 0)$, in agreement with the fact that the variety is 1-dimensional.

Example 2. Over the field k we consider the ideal generated by the following 8 polynomials:

$$(\chi_4)^2, \quad \chi_3\chi_4, \quad (\chi_3)^2, \quad \chi_2\chi_4 - \chi_1\chi_3, \chi_2\chi_3 + \chi_1\chi_4,$$
$$(\chi_2)^2 - \alpha\chi_1\chi_3 - \beta\chi_1\chi_4, \chi_1\chi_2, (\chi_1)^2,$$

where $\alpha, \beta \in k$. The root of this ideal is generated by $\chi_1, \chi_2, \chi_3, \chi_4$, and is thus a prime ideal of dimension zero (it has only the solution $\chi = 0$).

Since only homogeneous polynomials of degree two are involved, the symbol M_2 and the system R_2 are defined by the same equations:

$$
\begin{cases}
v_{44} & = 0 \\
v_{34} & = 0 \\
v_{33} & = 0 \\
v_{24} - v_{13} & = 0 \\
v_{23} + v_{14} & = 0 \\
v_{22} - \alpha v_{13} - \beta v_{14} & = 0 \\
v_{12} & = 0 \\
v_{11} & = 0
\end{cases}
\qquad
\begin{array}{cccc}
1 & 2 & 3 & 4 \\
1 & 2 & 3 & \bullet \\
1 & 2 & 3 & \bullet \\
1 & 2 & \bullet & \bullet \\
1 & 2 & \bullet & \bullet \\
1 & 2 & \bullet & \bullet \\
1 & \bullet & \bullet & \bullet \\
1 & \bullet & \bullet & \bullet
\end{array}
$$

The prolongations with respect to the dots (*nonmultiplicative variables*) are not consequences of prolongations with respect to the other (*multiplicative*) variables. So, M_2 is *not* involutive, because $20 - 18 = 2$ and by direct checking we see that $M_3 = 0$ and $\dim M_2 = 10 - 8 = 2$. Hence, to prove that M_2 is 2-acyclic we only have to prove the injectivity of the first map δ in the sequence

$$0 \longrightarrow \wedge^2 T^* \otimes M_2 \longrightarrow \wedge^3 T^* \otimes T^* \longrightarrow \wedge^4 T^* \longrightarrow 0$$
$$\boxed{12} \qquad\qquad \boxed{16} \qquad\qquad \boxed{1}$$

i.e. to prove that the linear equations

$$v_{ri,jk} + v_{rj,ki} + v_{rk,ij} = 0$$

have only the zero solution.

Initially, we know that there can only be $\dim \wedge^3 T^* \otimes T^* - \dim \wedge^4 T^* = 16 - 1 = 15$ linearly independent such equations. Among the resulting ones we have $v_{13,34} = 0$, $v_{14,34} = 0$, and an easy computation shows that the only other two equations involving α, β are:

$$\alpha v_{13,13} + \beta v_{14,13} = 0, \qquad \alpha v_{13,14} + \beta v_{14,14} = 0.$$

But there are also the two equations

$$v_{14,13} - v_{13,14} = 0, \qquad v_{14,14} + v_{13,13} = 0,$$

and, finally, the two equations

$$\alpha v_{13,14} - \beta v_{13,13} = 0, \qquad \beta v_{13,14} + \alpha v_{13,13} = 0,$$

which are linearly independent if and only if $\alpha^2 + \beta^2 \neq 0$. We may choose $\alpha = \beta = 1$ (so that this condition is satisfied), but M_2 is 2-acyclic in general. As a byproduct, the short exact sequences

$$0 \longrightarrow M_2 \longrightarrow S_2 T^* \longrightarrow F_0 \longrightarrow 0$$

$$0 \longrightarrow S_3 T^* \longrightarrow T^* \otimes F_0 \longrightarrow F_1 \longrightarrow 0$$

imply that there are $\dim F_1 = 4 \times 8 - 20 = 12$ first-order compatibility conditions.

Example 3. Consider the two polynomials $\chi_1 + \chi_2$, $\chi_1\chi_2 - 1$ over $\mathbb{Q}$ and the associated linear system R_2:

$$\begin{cases} y_{12} - y = 0 \\ y_2 + y_1 = 0 \end{cases} \quad \boxed{\begin{array}{cc} 1 & 2 \\ \bullet & \bullet \end{array}}$$

this system is not formally integrable, and the solved form for $R_2^{(1)}$ is

$$\begin{cases} y_{22} + y = 0 \\ y_{12} - y = 0 \\ y_{11} + y = 0 \\ y_2 + y_1 = 0 \end{cases} \quad \boxed{\begin{array}{cc} 1 & 2 \\ 1 & \bullet \\ 1 & \bullet \\ \bullet & \bullet \end{array}}$$

The symbol $M_2^{(1)} = 0$ is trivially involutive. We leave it to the reader to verify that $R_2^{(1)}$ is involutive.

Example 4. Similarly, consider the three polynomials

$$\chi_1 + \chi_2 + \chi_3, \quad \chi_1\chi_2 + \chi_2\chi_3 + \chi_3\chi_1, \quad \chi_1\chi_2\chi_3 - 1$$

and the associated linear system R_3:

$$\begin{cases} y_1 + y_2 + y_3 = 0, \\ y_{12} + y_{23} + y_{13} = 0, \\ y_{123} - y = 0. \end{cases}$$

The symbol of the corresponding formally integrable system is of finite type because its characteristic variety is zero since it is the solution of the homogeneous system

$$\chi_1 + \chi_2 + \chi_3 = 0, \quad \chi_1\chi_2 + \chi_2\chi_3 + \chi_3\chi_1 = 0, \quad \chi_1\chi_2\chi_3 = 0$$

with only root $\chi_1 = \chi_2 = \chi_3 = 0$.

We leave it to the reader to prove that $R_3^{(2)}$ is formally integrable (use the substitution $y_3 = -y_2 - y_1$) with $\dim M_3^{(2)} = 1$, although $M_3^{(2)}$ is not 2-acyclic and $M_4^{(2)} = 0$. In fact, notice that M_3 is involutive, although not in δ-regular coordinates since it is defined by the single equation $v_{123} = 0$ instead of by $v_{333} = 0$ as it should have been in δ-regular coordinates. The symbol of the corresponding formally integrable system is zero because the characteristic variety reduces to $\chi = 0$.

Example 5 (S. Arnborn). Consider the four polynomials

$$\chi_1 + \chi_2 + \chi_3 + \chi_4, \quad \chi_1\chi_2 + \chi_2\chi_3 + \chi_3\chi_4 + \chi_4\chi_1,$$
$$\chi_1\chi_2\chi_3 + \chi_2\chi_3\chi_4 + \chi_3\chi_4\chi_1 + \chi_4\chi_1\chi_2, \quad \chi_1\chi_2\chi_3\chi_4 - 1.$$

The associated linear system R_4 is defined by the following four PDE:

$$\begin{cases} y_1 + y_2 + y_3 + y_4 = 0, \\ y_{12} + y_{23} + y_{34} + y_{14} = 0, \\ y_{123} + y_{234} + y_{134} + y_{124} = 0, \\ y_{1234} - y = 0, \end{cases}$$

but now the *symbol of the corresponding formally integrable system is not of finite type*, because its characteristic variety is not reduced to $\chi = 0$. Indeed, at least one of the χ must be zero, say $\chi_1 = 0$. Then we obtain:

$$\chi_2 + \chi_3 + \chi_4 = 0, \quad \chi_3(\chi_2 + \chi_4) = 0, \quad \chi_2\chi_3\chi_4 = 0,$$

i.e.

$$\chi_1 = 0, \quad \chi_3 = 0, \quad \chi_2 + \chi_4 = 0,$$

and the 2 components

$$\{\chi_1 = 0, \chi_3 = 0, \chi_2 + \chi_4 = 0\}, \qquad \{\chi_2 = 0, \chi_4 = 0, \chi_1 + \chi_3 = 0\}$$

of the characteristic variety both have dimension 1. Instead of studying the preceding system with 4 independent variables, we first set $y_4 = -y_1 - y_2 - y_3$, and consider the system R_4 below, with 3 independent variables:

$$\begin{cases} y_{33} + 2y_{13} + y_{11} = 0, \\ y_{233} + y_{223} + y_{133} + 2y_{123} + y_{122} + y_{113} + y_{112} = 0, \\ y_{1233} + y_{1223} + y_{1123} + y = 0. \end{cases}$$

Using two prolongations of the first-order PDE and one prolongation of the second-order PDE, we arrive at the following fourth-order PDE for $R_4^{(2)}$:

$$\begin{cases}
y_{3333} + 2y_{1333} + y_{1133} & = 0 \\
y_{2333} + 2y_{1233} + y_{1123} & = 0 \\
y_{2233} + 2y_{1223} + y_{1122} & = 0 \\
y_{2223} + y_{1222} - y_{1123} - y_{1112} & = 0 \\
y_{1333} + 2y_{1133} + y_{1113} & = 0 \\
y_{1233} + 2y_{1123} + y_{1112} & = 0 \\
y_{1223} - 2y_{1123} - y_{1112} + y & = 0 \\
y_{1133} + 2y_{1113} + y_{1111} & = 0 \\
y_{1123} + y_{1122} - y_{1113} + y_{1112} - y_{1111} - y = 0
\end{cases}$$

1	2	3
1	2	•
1	2	•
1	2	•
1	•	•
1	•	•
1	•	•
1	•	•
1	•	•

Although we shall see that $R_4^{(2)}$ is formally integrable, we cannot use the criterion to test this property because the symbol $M_4^{(2)}$ is neither involutive nor 2-acyclic. To see this, notice that prolongation with respect to the multiplicative variables must be completed by adding the two fifth-order PDE

$$\begin{cases} y_{11222} + y_{11122} - y_1 - y_2 = 0, \\ y_{11113} + y_{11111} - y_1 - y_3 = 0, \end{cases}$$

and a similar result may be obtained at the next order, by adding the sixth-order PDE

$$y_{111112} - 2y_{11} - y_{13} + y_{12} + y_{23} = 0.$$

These results can also be obtained directly by a symbolic computation.

Accordingly, we have

$$\dim M_4^{(2)} = 6, \quad \dim M_5^{(2)} = 5, \quad \dim M_6^{(2)} = 4,$$

and it follows that the δ-sequence

$$0 \longrightarrow M_6^{(2)} \overset{\delta}{\longrightarrow} T^* \otimes M_5^{(2)} \overset{\delta}{\longrightarrow} \wedge^2 T^* \otimes M_4^{(2)} \overset{\delta}{\longrightarrow} \wedge^3 T^* \otimes M_3^{(1)}$$

cannot be exact. In this sequence we have introduced the symbol $M_3^{(1)}$ of the third-order system obtained by prolongation of the first-order PDE of R_4 and adding the third-order PDE in such a way that $M_3^{(1)}$ is defined by the equations

$$\begin{cases} v_{333} - 3v_{113} - 2v_{111} & = 0 \\ v_{233} + 2v_{123} + v_{112} & = 0 \\ v_{223} + v_{122} - v_{113} - v_{111} & = 0 \\ v_{133} + 2v_{113} + v_{111} & = 0 \end{cases} \quad \begin{array}{ccc} 1 & 2 & 3 \\ 1 & 2 & \bullet \\ 1 & 2 & \bullet \\ 1 & \bullet & \bullet \end{array}$$

and we find $\dim M_3^{(1)} = 10 - 4 = 6$. Incidentally we see that $M_3^{(1)}$ is involutive, with characters $(5, 1, 0)$, and we can verify that the corresponding characteristic variety is defined by the four polynomial equations

$$\begin{cases} (\chi_3 + \chi_1)^2(\chi_3 - 2\chi_1) = 0, \\ \chi_2(\chi_1 + \chi_3)^2 = 0, \\ (\chi_1 + \chi_3)(\chi_2 - \chi_1)(\chi_2 + \chi_1) = 0, \\ \chi_1(\chi_1 + \chi_3)^2 = 0. \end{cases}$$

The root is easily seen to contain

$$\chi_1(\chi_1 + \chi_3) = 0, \qquad \chi_3(\chi_1 + \chi_3) = 0,$$

and is thus generated by $\chi_1 + \chi_3 = 0$ only. The corresponding variety has dimension 2, which is equal to the number of nonzero characters.

Counting dimensions in the preceding sequence, we see that it cannot be exact, because the space of coboundaries in $\wedge^2 T^* \otimes M_4^{(2)}$, has dimension $3 \times 5 - 4 = 11$ while the space of cocycles has dimension at least $3 \times 6 - 6 = 12$.

The delicate point is now to notice that the coordinate system is not δ-regular, even not for $M_4^{(2)}$. Indeed, by the linear change of coordinates

$$\bar{x}^1 = x^1, \quad \bar{x}^2 = x^2 + \alpha x^1, \quad \bar{x}^3 = x^3,$$

two of the nine defining equations can be written as

$$\begin{cases} \alpha^2(1-\alpha)v_{\overline{2223}} + \alpha^2(1+\alpha-\alpha^2)v_{\overline{2222}} + \text{class } \overline{\mathrm{I}} = 0, \\ \alpha(1-\alpha)v_{\overline{2223}} - \alpha^3 v_{\overline{2222}} + \text{class } \overline{\mathrm{I}} = 0. \end{cases}$$

Hence, if $\alpha \neq -1, 0, 1$, we obtain

$$v_{\overline{2222}} + \text{class } \overline{\mathrm{I}} = 0,$$

proving that the *second and third characters must be equal to zero*, in agreement with the above quoted result concerning the solution space.

In this way one can prove that $M_6^{(2)}$ is involutive, and we finally need $2 + 2 = 4$ effective prolongations in order to find formal solutions. Accordingly, compatibility conditions for $R_4^{(2)}$ are described by third-order PDEs, a fact that is a priori not evident at all.

Example 6. In general, the Galois extension that can be associated with an irreducible polynomial $P \in K[y]$ of degree m by adding to K the m (different) roots of P has nothing to do with the extension of K determined by the prime ideal generated by P in $K[y]$. In fact, the Galois extension can be a vector space over K of dimension $m!$ (for a general polynomial) while the extension determined by P is a vector space over K of dimension m. Nevertheless, both extensions sometimes coincide, although their group-theoretical interpretations may, of course be distinct. The present Example is devoted to the study of the case $m = 3$. Consider the polynomial

$$P \equiv y^3 - py + q \in K[y],$$

and suppose it is irreducible over K. In that case we may set, as usual, $L = K[y]/P = K[\eta] = K(\eta)$, and we shall try to avoid using the roots α, β, γ of P directly in a convenient (!) extension of K. Accordingly, instead of saying that $L = K(\alpha, \beta, \gamma)$ and identifying α with η, we shall merely say that P splits into three factors over $L = K(\eta)$. We first prove that these factors are distinct, i.e. an irreducible P has no multiple roots. Indeed, if α would be a multiple root, we would have

$$\alpha^3 - p\alpha + q = 0, \qquad 3\alpha^2 - p = 0,$$

i.e. $2p\alpha - 3q = 0$. If $p = 0$, then $\alpha = 0$ and $q = 0$ would lead to $P \equiv y^3$, which is reducible over $\mathbb{Q} \subset K$. On the other hand, if $p \neq 0$, then $\alpha = 3q/2p \in K$, and P would be reducible over K. Thus, if we assume that P is irreducible over K, then it cannot have a multiple root. We introduce

$$\delta = (\alpha - \beta)(\beta - \gamma)(\gamma - \alpha).$$

Notice that $\delta^2 = 4p^3 - 27q^2 \neq 0$. It follows that L/K is a Galois extension if and only if $\delta \in K$. Indeed, if $L = K(\alpha, \beta, \gamma) = K(\eta)$, with identification $\alpha = \eta$, then the Galois group of P, i.e. the Galois group of L/K, is a subgroup of the group of permutations S_3 on $\{\alpha, \beta, \gamma\}$ consisting of 3 elements, since $|L/K| = m = 3$. Hence it can only be the normal subgroup $U_3 \lhd S_3$ with elements $\{e, (123), (132)\}$. It follows that $\delta \in K$. Conversely, if $\delta \in K$, then the Galois group must be a subgroup of U_3 with 3 elements, hence must be U_3 itself.

We have just proved that L/K is a Galois extension if and only if $\delta \in K$, i.e. $4p^3 - 27q^2$ has a square root in K. Now we will study this algebraic condition.

In $K(\eta)[y] = L[y]$ the polynomial P has a factorization

$$P \equiv (y - \eta)(y^2 + \eta y + \eta^2 - p),$$

and the factor of degree two splits over K if and only if we may find $u, v, w \in K$ such that

$$4p - 3\eta^2 = (u + v\eta + w\eta^2)^2,$$

because L is generated by $\{1, \eta, \eta^2\}$ as a vector space over K. Taking into account that $\eta^3 = p\eta - q$ and developing, we obtain, over the same basis,

$$\begin{cases} v^2 + 2uw + pw^2 + 3 = 0, \\ 2uv - qw^2 + 2pvw = 0, \\ u^2 - 2qvw - 4p = 0, \end{cases}$$

and finally

$$\begin{cases} p = -\dfrac{v^2 + 2uw + 3}{w^2}, \\ q = -2v\dfrac{v^2 + uw + 3}{w^3}, \\ (2v^2 + uw + 2)(2v^2 + uw + 6) = 0. \end{cases}$$

If u, v, w are chosen such that $2v^2 + uw + 2 = 0$, we find

$$w^3 P(y) \equiv (wy + 2v)(wy - v + 1)(wy - v - 1),$$

and P cannot be irreducible. Hence we *must* have $2v^2 + uw + 6 = 0$. Setting $r = v/w$, $s = 1/w$ we find

$$p = 3(r^2 + 3s^2), \qquad q = 2r(r^2 + 3s^2),$$
$$\delta = \pm 18s(r^2 + 3s^2),$$

and we have $\delta \neq 0$, only if $s \neq 0$ and the equation $z^2 + 3 = 0$ has no root in K.

These computations have been done under the assumption that P is irreducible over K. However, the above choice of (p, q) does not ensure that P is irreducible. Indeed, choosing $r = 9ab^2 - a^3$, $s = 3b(a^2 - b^2)$ with $a, b \in K$ we easily obtain the root $2a(a^2 + 3b^2)$ in K.

When $K = \mathbb{Q}$, choosing $r = s = 1/2$ implies $p = 3$, $q = 1$. We obtain $P \equiv y^3 - 3y + 1$, which is irreducible over $\mathbb{Q}$. Indeed, if $c/d \in \mathbb{Q}$ is a root with $c, d \in \mathbb{Z}$ relatively prime, then we should have $c^3 - 3cd^2 + d^3 = 0$, whence c would divide d^3, contrary to the assumption that c, d are relatively prime. We may choose $u = -4$, $v = 1$, $w = 2$ and observe that

$$y^2 + \eta y + \eta^2 - 3 = \left(y + \frac{\eta}{2}\right)^2 + \frac{3}{4}\eta^2 + 3$$
$$= \left(y + \frac{\eta}{2}\right)^2 - \frac{1}{4}(-4 + \eta + 2\eta^2)^2$$
$$= (y + 2 - \eta^2)(y - 2 + \eta + \eta^2).$$

Hence, if η is a root, the other two roots are $\eta^2 - 2$ and $2 - \eta - \eta^2$, respectively. These cannot be equal, because $\delta^2 = 81 \neq 0$, and we can choose $\delta = 9 \in \mathbb{Q}$.

The reader is encouraged to treat the case $K = \mathbb{Q}$, $L = \mathbb{Q}(\sqrt{2}, i) = \mathbb{Q}(\sqrt{2} + i)$ with $|L/K| = 4$ relative to the Klein 4-group $V_4 = \{e, (12)(34), (13)(24), (14)(23)\}$.

Example 7. This is the best Example known to us concerning a nontrivial application of the second criterion for automorphic systems.

Consider the involutive groupoid

$$\mathcal{R}_2 \qquad \frac{\partial(\bar{y}^1, \bar{y}^2)}{\partial(y^1, y^2)} = 1, \qquad \frac{\partial^2 \bar{y}^k}{\partial y^r \partial y^s} = 0$$

which determines the Lie pseudogroup

$$\Gamma \qquad \bar{y}^l = A_k^l y^k + B^k \qquad \text{with } \det A = 1.$$

We prove that $\mathcal{R}_2$ acts freely on $J_2(X, Y)$ with $\dim X = n = 1$, $\dim Y = m = 2$, using

$$R_2 \qquad \eta_r^r = 0, \qquad \eta_{rs}^k = 0.$$

Applying $\sharp$, we see that $\sharp(R_2)$ has the following basis of 5 generators:

$$\left\{ \frac{\partial}{\partial y^1}, \frac{\partial}{\partial y^2}, y_x^1 \frac{\partial}{\partial y_x^1} + y_{xx}^1 \frac{\partial}{\partial y_{xx}^1} - y_x^2 \frac{\partial}{\partial y_x^2} - y_{xx}^2 \frac{\partial}{\partial y_{xx}^2}, \right.$$

$$\left. y_x^1 \frac{\partial}{\partial y_x^2} + y_{xx}^1 \frac{\partial}{\partial y_{xx}^2}, y_x^2 \frac{\partial}{\partial y_x^1} + y_{xx}^2 \frac{\partial}{\partial y_{xx}^1} \right\}.$$

The rank depends on the matrix

$$\begin{pmatrix} y_x^1 & -y_x^2 & y_{xx}^1 & -y_{xx}^2 \\ 0 & y_x^1 & 0 & y_{xx}^1 \\ y_x^2 & 0 & y_{xx}^2 & 0 \end{pmatrix},$$

which is easily seen to have generic rank 3. Hence $\sharp(R_2)$ has rank $3 + 2 = 5$, and the action is free.

At order 2 we may exhibit the differential invariant Φ, and consider the system $\mathcal{A}_2$:

$$\mathcal{A}_2 \qquad \Phi(y_2) \equiv y_x^2 y_{xx}^1 - y_x^1 y_{xx}^2 = \omega(x),$$

which is seen to be a PHS for $\mathcal{R}_2$. In fact,

$$\begin{cases} \dim \mathcal{A}_2 - \dim X = 6 - 1 = 5, \\ \dim \mathcal{R}_2 - \dim Y = 6 - 1 = 5. \end{cases}$$

The action of $\mathcal{R}_3 = \rho_1(\mathcal{R}_2)$ on $\rho_1(\mathcal{A}_2)$ is free because we have the same number of independent generators, which are only prolongated. However,

$$\begin{cases} \dim \rho_1(\mathcal{A}_2) - \dim X = 6 + 2 - (1 + 1) = 6, \\ \dim \rho_1(\mathcal{R}_2) - \dim Y = \dim \mathcal{R}_2 - \dim Y = 5. \end{cases}$$

Hence there must be a new differential invariant, say Ψ, of order 3 distinct from $d_x\Phi$. Consider:

$$\mathcal{A}_3 \quad \begin{cases} \Phi \equiv y_x^2 y_{xx}^1 - y_x^1 y_{xx}^2 = \omega(x), \\ d_x\Phi \equiv y_x^2 y_{xxx}^1 - y_x^1 y_{xxx}^2 = \rho(x), \\ \Psi \equiv y_{xx}^2 y_{xxx}^1 - y_{xx}^1 y_{xxx}^2 = \sigma(x). \end{cases}$$

We have:

$$\begin{cases} \dim \mathcal{A}_3 - \dim X = 6 + 2 - 3 = 5, \\ \dim \mathcal{R}_3 - \dim Y = 2 + 4 - 1 = 5. \end{cases}$$

Hence $\mathcal{A}_3$ is a general automorphic system for $\mathcal{R}_3$, provided that the compatibility condition

$$\partial_x\omega(x) - \rho(x) = 0$$

holds. All characters of $\mathcal{A}_3$ and $\mathcal{R}_3$ are zero, because both systems are of finite type.

We see that the rank of the preceding equation drops by one and becomes equal to 2 if $y_x^1 y_{xx}^2 - y_x^2 y_{xx}^1 = 0$. Hence we have to assume that $\omega(x) \neq 0$ for any special automorphic system. In fact, looking at the action of Γ on first and second jets, we see that the above Φ plays the role of Wronskian determinant in the Picard–Vessiot theory, and should therefore be nonzero. This would also ensure that the symbol of $\mathcal{A}_3$ has rank 2, hence vanishes.

Similar results would be obtained with the larger Lie pseudogroup of affine transformations, by dropping the unimodularity condition and Φ while taking into account the following technical formula for the matrix with as entries the components of the six generators:

$$\det \begin{pmatrix} y_x^1 & 0 & y_{xx}^1 & 0 \\ 0 & y_x^1 & 0 & y_{xx}^2 \\ y_x^2 & 0 & y_{xx}^2 & 0 \\ 0 & y_x^2 & 0 & y_{xx}^2 \end{pmatrix} = (y_x^1 y_{xx}^2 - y_x^2 y_{xx}^1)^2.$$

The details are left to the reader as an exercise.

Example 8. In this Example we show how to establish the Picard–Vessiot theory for the ODE $y_{xx} = 0$. This ODE cannot be treated by Kolchin's method. The two confusions made in the literature concerning the Picard–Vessiot theory, despite the existence of [24], are clearly exhibited in this Example.

First of all, as already said in the introduction, the key idea is to replace the search for two linearly independent solutions by the search for a single solution of the finite-type involutive automorphic system

$$\mathcal{A}_2 \quad y_{xx}^1 = 0, \quad y_{xx}^2 = 0$$

for the action of GL(2) defined by the involutive groupoid

$$\mathcal{R}_2 \quad \bar{y}^l = \frac{\partial \bar{y}^l}{\partial y^k} y^k, \quad \frac{\partial^2 \bar{y}^k}{\partial y^r \partial y^s} = 0$$

with zero symbol. Of course,

$$\begin{cases} \dim \mathcal{A}_2 - \dim X = 4, \\ \dim \mathcal{R}_2 - \dim Y = 4, \end{cases}$$

and we have to take into account the Wronskian condition

$$\det \begin{vmatrix} y^1 & y_x^1 \\ y_2 & y_x^2 \end{vmatrix} = y_1 y_x^2 - y^2 y_x^1 \neq 0$$

for the solutions.

Prolongation of the action of GL(2) to first-order jets gives

$$\begin{pmatrix} \overline{y}^1 & \overline{y}_x^1 \\ \overline{y}_2 & \overline{y}_x^2 \end{pmatrix} = \begin{pmatrix} a & b \\ c & d \end{pmatrix} \begin{pmatrix} y^1 & y_x^1 \\ y_2 & y_x^2 \end{pmatrix},$$

which we symbolically write as

$$\overline{Y}_1 = A Y_1 \Rightarrow A = \overline{Y}_1 Y_1^{-1}.$$

Since our automorphic system is defined by a linear differential ideal in $\mathbb{Q}\{y_1, y_2\}$, we find a differential automorphic extension L/K with $L \simeq K(y^1, y^2, y_x^1, y_x^2)$ and derivations

$$d_x y^k = y_x^k, \qquad d_x y_x = 0.$$

In fact, we may choose $K = \mathbb{Q}(x)$ and $k = \mathbb{Q}$.

Since we have a Lie group, rather than a Lie pseudogroup, of transformations, we can introduce GL(2) as an abstract irreducible variety G defined over k, without any need to consider a, b, c, d as 'constants' $\cdots$ anywhere! As a byproduct we have the fundamental isomorphism

$$Q(L \otimes_K L) \simeq Q(L \otimes_k k[G]),$$

and we shall deal directly with this.

The infinitesimal generators of the prolonged action of GL(2) to first-order jets are the four vertical vector fields:

$$\left\{ y^1 \frac{\partial}{\partial y^1} + y_x^1 \frac{\partial}{\partial y_x^1}, y^1 \frac{\partial}{\partial y^2} + y_x^1 \frac{\partial}{\partial y_x^2}, y^2 \frac{\partial}{\partial y^1} + y_x^2 \frac{\partial}{\partial y_x^1}, y^2 \frac{\partial}{\partial y^2} + y_x^2 \frac{\partial}{\partial y_x^2} \right\}.$$

The existence of the reciprocal Lie algebra is provided by Galois cohomology, and we can choose the following basis of four vertical generators (however, their *determination does not seem to be effective*):

$$\left\{ y^1 \frac{\partial}{\partial y^1} + y^2 \frac{\partial}{\partial y^2}, y^1 \frac{\partial}{\partial y_x^1} + y^2 \frac{\partial}{\partial y_x^2}, y_x^1 \frac{\partial}{\partial y^1} + y_x^2 \frac{\partial}{\partial y^2}, y_x^1 \frac{\partial}{\partial y_x^1} + y_x^2 \frac{\partial}{\partial y_x^2} \right\},$$

combined with ∂_x. It is purely accidental, as can be seen from other examples, that ∂_x can be added to the *third* generator to yield d_x, since actually ∂_x should be combined with the *fourth* generator because the lift to first-order jets of an infinitesimal change of source 'x' is:

$$\xi(x) \frac{\partial}{\partial x} - \frac{\partial \xi(x)}{\partial x} \left(y_x^1 \frac{\partial}{\partial y_x^1} + y_x^2 \frac{\partial}{\partial y_x^1} \right)$$

and we may choose $\xi(x)$ and $\partial \xi(x)/\partial x$ in K, which is the ground field of the reciprocal Lie algebra. In fact, we may choose two different $\xi(x)$ or, equivalently, replace the derivative by a section $\xi_x(x)$ to obtain the previous decomposition of generators for the reciprocal Lie algebra.

It follows that $\operatorname{cst}(K) = \operatorname{cst}(L) = k$ *even though we are not dealing with differential constants!* E.g., we find

$$a = \frac{y_x^2 \overline{y}^1 - y^2 \overline{y}_x^1}{y^1 y_x^2 - y^2 y_x^1},$$

which can be written as

$$a = \frac{y_x^2}{y^1 y_x^2 - y^2 y_x^1} \otimes y^1 - \frac{y^2}{y^1 y_x^2 - y^2 y_x^1} \otimes y_x^1,$$

and we can verify that this term is annihilated by the extension of each generator of the reciprocal Lie algebra, such as

$$y^1 \frac{\partial}{\partial y_x^1} + y^2 \frac{\partial}{\partial y_x^2} + \overline{y}^1 \frac{\partial}{\partial \overline{y}_x^1} + \overline{y}^2 \frac{\partial}{\partial \overline{y}_x^2}.$$

Of course, the reader discovers that this point of view is quite far from the one of Kolchin in [95, 96]; however, it closely follows the view of Bialynicki–Birula [24]. By the above computation, we have indeed

$$k[G] \subset \operatorname{cst}(L \otimes_K L),$$

and all assumptions of the general theory are satisfied.

Introducing the intermediate differential field $K' = K(y_x^1)$, the equality $\overline{y}_x^1 = y_x^1$ for the action of $\operatorname{GL}(2)$ implies $a = 1$, $b = 0$ and determines the subgroup G' consisting of the invertible (2×2)-matrices

$$\begin{pmatrix} 1 & 0 \\ c & d \end{pmatrix}.$$

It is obvious that the action of G' preserves y^1 and that the differential field of invariants of G' in L is $K'' = K(y^1, y_x^1)$, which strictly contains K'. Hence the Galois correspondence does not work for any intermediate differential field.

Finally, we notice that K' is *not* stable under the reciprocal Lie algebra, while K'' is.

In this example we can directly verify the properties possessed by $k[G]$. Indeed,

$$d_x \overline{Y}_1 = d_x(AY_1) = (d_x A) Y_1 + A d_x Y_1 = A d_x Y_1,$$
$$(d_x A) Y_1 = 0, \quad \det(Y_1) \neq 0 \Rightarrow d_x A = 0,$$

i.e. *in this particular case* $k[G]$ *has a trivial differential structure with respect to* d_x. In view of other examples this is merely a coincidence!

We can also directly determine G' by the specialization $k[G] \to k[G']$. As an illustration we will do this for 'a'. Since $K' = K(y_x^1)$, we may identify y_x^1 with $\overline{y}_x^1$ in $L \otimes_{K'} L$, i.e. $y_x^1 \otimes 1$ and $1 \otimes y_x^1$, while this is not possible in $L \otimes_K L$. Hence,

$$a = \frac{y_x^2 \overline{y}^1 - y^2 y_x^1}{y^1 y_x^2 - y^2 y_x^1},$$

and we do not obtain $a = 1$. This is the reason why, in order to also have

$$k[G'] \subset \text{cst}(L \otimes_{K'} L),$$

we *must* use a stable intermediate field (which is then *automatically* a differential field, as noted earlier). It follows that stabilizing K' amounts to introducing $K'' = K(y^1, y_x^1)$, and *now* we can also identify y^1 and $\bar{y}^1$, i.e. $y^1 \otimes 1$ and $1 \otimes y^1$, to obtain

$$a = \frac{y_x^2 y^1 - y^2 y_x^1}{y^1 y_x^2 - y^2 y_x^1} = 1,$$

as required. Once again, this philosophy is very new $\cdots$ but it is the only one fitting in with the existence of differential Galois theory for PDE!.

Finally we will make some comments regarding the difference between '*general*' and '*special*' second-order ODE. Indeed, considering the general situation, we should have

$$K = \mathbb{Q}\langle \Phi^1, \Phi^2 \rangle, \qquad L = \mathbb{Q}\langle y^1, y^2 \rangle,$$

with, as usual,

$$\Phi^1 = \frac{\begin{vmatrix} y^1 & y^2 \\ y_{xx}^1 & y_{xx}^2 \end{vmatrix}}{\begin{vmatrix} y^1 & y^2 \\ y_x^1 & y_x^2 \end{vmatrix}}, \qquad \Phi^2 = \frac{\begin{vmatrix} y_x^1 & y_x^2 \\ y_{xx}^1 & y_{xx}^2 \end{vmatrix}}{\begin{vmatrix} y^1 & y^2 \\ y_x^1 & y_x^2 \end{vmatrix}},$$

and we should obtain the *identity*

$$y_{xx}^1 - \Phi^1 y_x^1 + \Phi^2 y^1 \equiv 0.$$

Hence, setting $K' = K\langle \Phi^1, \Phi^2, y_x^1 \rangle$, we have indeed $K' = K\langle \Phi^1, \Phi^2, y^1 \rangle = K''$, in contrast to the special case. In fact, in *that* case we can prove that any intermediate differential field is stable. Other examples, however, do not confirm this result in general.

Example 9. Here we consider the automorphic extension L/K for the pseudogroup

$$\Gamma \quad \bar{y}^1 = g(y^1), \qquad \bar{y}^2 = \frac{y^2}{(\partial g(y^1)/\partial y^1)}.$$

First we will treat this Example in the differential-geometrical setting. To this end we consider, with $n = 1$, $m = 2$, the involutive system

$$\mathcal{A}_1 \quad y^2 y_x^1 = \omega(x)$$

with character 1.

We have already proved that $\mathcal{A}_1$ is invariant under a groupoid which is not involutive, and we consider the corresponding involutive groupoid

$$\mathcal{R}_1 \quad \bar{y}^2 \frac{\partial \bar{y}^1}{\partial y^1} = y^2, \qquad \frac{\partial \bar{y}^1}{\partial y^2} = 0, \qquad \frac{\partial(\bar{y}^1, \bar{y}^2)}{\partial(y^1, y^2)} = 1,$$

which acts freely. We have

$$R_1 \quad y^2\frac{\partial \eta^1}{\partial y^1} + \eta^2 = 0, \quad \frac{\partial \eta^1}{\partial y^2} = 0, \quad \frac{\partial \eta^1}{\partial y^1} + \frac{\partial \eta^2}{\partial y^2} = 0,$$

$$M_1 \quad v_1^1 = 0, \quad v_2^1 = 0, \quad v_1^1 + v_2^2 = 0,$$

with characters $(1, 0)$.

We see that

$$\begin{cases} \dim \mathcal{A}_1 - \dim X = 4 - 1 = 3, \\ \dim \mathcal{R}_1 - \dim Y = 6 - 3 = 3, \\ \dim \mathcal{A}_2 - \dim X = 3 + 1 = 4, \\ \dim \mathcal{R}_2 - \dim Y = 3 + 1 = 4. \end{cases}$$

The first criterion for automorphic systems implies that $\mathcal{A}_1$ is an automorphic system for Γ. In fact, the solution spaces of $\mathcal{A}_1$ and $\mathcal{R}_1$ both depend on one arbitrary function of one variable.

We may now assume that $\omega \in K$ for some differential ground field K, and we notice that Γ is an algebraic pseudogroup. We shall consider the successive differential fields

$$k = \mathbb{Q}, \qquad K = \mathbb{Q}\langle y^2 y_x^1\rangle, \qquad L = \mathbb{Q}\langle y^1, y^2\rangle,$$

where L/K is a general differential automorphic extension.

On the one hand $\sharp(R_1)$ is generated by the three vertical vector fields

$$\left\{ \frac{\partial}{\partial y^1}, y^2\frac{\partial}{\partial y^2} - y_x^1\frac{\partial}{\partial y_x^1} + y_x^2\frac{\partial}{\partial y_x^2}, y_x^1\frac{\partial}{\partial y_x^2} \right\}.$$

The reciprocal distribution is generated by the two vertical vector fields

$$\left\{ \delta_1 = y_x^1\frac{\partial}{\partial y_x^1} - y_x^2\frac{\partial}{\partial y_x^2}, \delta_2 = y^2\frac{\partial}{\partial y_x^2} \right\}$$

to which we may add ∂_x as if we were considering the prolongation of changes of 'x'.

We now study $k[\Gamma]$ at order 1. We have

$$\frac{\partial \bar{y}^1}{\partial y^1} = \frac{y^2}{\bar{y}^2}, \qquad \frac{\partial \bar{y}^1}{\partial y^2} = 0, \qquad \frac{\partial \bar{y}^2}{\partial y^2} = \frac{\bar{y}^2}{y^2}.$$

Now

$$\bar{y}_x^2 = \frac{\partial \bar{y}^2}{\partial y^1}y_x^1 + \frac{\partial \bar{y}^2}{\partial y^2}y_x^2$$

implies

$$\frac{\partial \bar{y}^2}{\partial y^1} = \frac{y^2\bar{y}_x^2 - \bar{y}^2 y_x^2}{y^2 y_x^1}.$$

Notice that all these elements of $k[G]$ are annihilated by the extensions of δ_1 and δ_2 to $\mathbb{Q}(y^1, y^2, y_x^1, y_x^2) \otimes_{\mathbb{Q}} \mathbb{Q}(\bar{y}^1, \bar{y}^2, \bar{y}_x^1, \bar{y}_x^2)$.

We will now prove that $k[\Gamma]$ does not have a differential structure with respect to 'x'. Indeed,

$$d_x\left(\frac{\bar{y}^2}{y^2}\right) = \frac{\bar{y}_x^2}{y^2} - \frac{\bar{y}^2 y_x^2}{(y^2)^2},$$

and this element is *not* annihilated by the extension of δ_1.

Finally we prove that only stable intermediate differential fields provide a Galois correspondence. Indeed, let us choose $K' = \mathbb{Q}\langle y^2 y_x^1, y_x^2 \rangle$. We find

$$\bar{y}_x^2 = y_x^2 \Rightarrow \frac{\partial \bar{y}^2}{\partial y^2} = 1, \quad \frac{\partial \bar{y}^2}{\partial y^1} = 0,$$

and so

$$\Gamma' \quad \bar{y}^1 = y^1 + a, \quad \bar{y}^2 = y^2.$$

It follows that the field of invariants of Γ' is $K'' = \mathbb{Q}\langle y^2 y_x^1, y^2 \rangle$, which strictly contains K'. Notice that K' is stable under δ_1 but *not* under δ_2, while K'' is stable under both δ_2 and δ_1.

We point out that the reciprocal distribution works for any intermediate differential field, but must be known *a priori*.

Example 10. Here we give a nice illustration of Theorem C.7, and provide all details. We take the variables x, t, z, $p = \partial z/\partial x$, $w = \partial z/\partial t$ and consider the following first-order algebraic PDE for z:

$$pw - txz = 0 \Rightarrow w - \frac{txz}{p} = 0.$$

$\boxed{1}$ We have the complete integral

$$z = \left(\frac{a}{4}t^2 + \frac{1}{4a}x^2 + b\right)^2 \geq 0,$$

which leads to

$$p = \frac{x}{a}\left(\frac{a}{4}t^2 + \frac{1}{4a}x^2 + b\right)$$

and to the Hessian condition

$$\frac{\partial(z,p)}{\partial(a,b)} = p\left(\frac{t^2}{2} + \frac{p^2}{2z}\right) \neq 0.$$

It is purely accidental that, in this case, we can give the explicit corresponding solution of the automorphic system:

$$\begin{cases} X = \pm\sqrt{z}\dfrac{x}{p}, \\[2mm] Z = \pm\sqrt{z}\left(1 - \dfrac{xt^2}{4p} - \dfrac{xp}{4z}\right), \\[2mm] P = \dfrac{p^2}{4z} - \dfrac{t^2}{4}, \end{cases}$$

with $\rho = \pm 2\sqrt{z}$. We can immediately verify that

$$dz - p\,dx - \frac{txz}{p}\,dt = \pm 2\sqrt{z}(dZ - P\,dX).$$

$\boxed{2}$ The other complete integral is

$$z = \left(\frac{1}{2}t^2 + \bar{a}\right)\left(\frac{1}{2}x^2 + \bar{b}\right),$$

and it leads to

$$p = x\left(\frac{1}{2}t^2 + \bar{a}\right)$$

and to the Hessian condition

$$\frac{\partial(z, p)}{\partial(\bar{a}, \bar{b})} = -p \neq 0.$$

Again it is purely accidental that we can give the explicit corresponding solution of the automorphic system:

$$\begin{cases} \overline{X} = \dfrac{p}{x} - \dfrac{1}{2}t^2, \\[2mm] \overline{Z} = \dfrac{xz}{p} - \dfrac{1}{2}x^2, \\[2mm] \overline{P} = -\dfrac{x^2 z}{p^2}, \end{cases}$$

with $\bar{p} = p/x$.
We can immediately verify that

$$dz - p\,dx - \frac{txz}{p}\,dt = \frac{p}{x}(d\overline{Z} - \overline{P}\,d\overline{X}).$$

The exterior condition

$$d\overline{Z} - \overline{P}\,d\overline{X} = 2X(dZ - P\,dX)$$

follows from $\boxed{1}$ and $\boxed{2}$ and we have the two relations

$$\frac{Z}{X} = \frac{p}{x} - \frac{t^2}{4} - \frac{x^2}{4a^2} = \frac{p}{x} - \frac{t^2}{4} - \frac{p^2}{4z} = \left(\frac{p}{x} - \frac{t^2}{2}\right) - \left(\frac{p^2}{4z} - \frac{t^2}{4}\right),$$

$$XZ = \frac{x^2 z}{p^2}\left(\frac{p}{x} - \frac{t^2}{4}\right) - \frac{x^2}{4} = \left(\frac{xz}{p} - \frac{1}{2}x^2\right) + \frac{x^2 z}{p^2}\left(\frac{p^2}{4z} - \frac{t^2}{4}\right),$$

leading to the contact transformation

$$\begin{cases} \overline{X} = \dfrac{Z}{X} + P, \\[2mm] \overline{Z} = XZ - X^2 P, \\[2mm] \overline{P} = -X^2, \end{cases}$$

along the results of Section C.

Example 11. In Section C we have seen an example of an automorphic system for the Lie pseudogroup of contact transformations in 3 variables preserving the volume. It was defined by the Pfaffian system

$$dz - p\,dx + H(t,x,p)\,dt = dZ - P\,dX,$$

and *in a purely functional way* we have proved that it is an automorphic system. Now we shall prove *in a purely formal way* that it is an automorphic system, using directly the first criterion for automorphic systems. In fact, we have successively,

$$\mathcal{A}_1 \begin{cases} \dfrac{\partial Z}{\partial t} - P\dfrac{\partial X}{\partial t} = H(t,x,p), \\[2mm] \dfrac{\partial Z}{\partial x} - P\dfrac{\partial X}{\partial x} = -p, \\[2mm] \dfrac{\partial Z}{\partial z} - P\dfrac{\partial X}{\partial z} = 1, \\[2mm] \dfrac{\partial Z}{\partial p} - P\dfrac{\partial X}{\partial p} = 0, \end{cases}$$

$$\mathcal{R}_1 \begin{cases} \dfrac{\partial \overline{Z}}{\partial X} - \overline{P}\dfrac{\partial \overline{X}}{\partial X} = -P, \\[2mm] \dfrac{\partial \overline{Z}}{\partial Z} - \overline{P}\dfrac{\partial \overline{X}}{\partial Z} = 1, \\[2mm] \dfrac{\partial \overline{Z}}{\partial P} - \overline{P}\dfrac{\partial \overline{X}}{\partial P} = 0. \end{cases}$$

Accordingly we have (as manifolds):

$$\begin{cases} \dim \mathcal{A}_1 = (4 + 3 + 12) - 4 = 15, \\ \dim \mathcal{R}_1 = (3 + 3 + 9) - 3 = 12, \end{cases}$$

$$\begin{cases} \dim \mathcal{A}_1 \times_Y \mathcal{R}_1 = \dim \mathcal{A}_1 + \dim \mathcal{R}_1 - \dim Y = 15 + 12 - 3 = 24, \\ \dim \mathcal{A}_1 \times_X \mathcal{A}_1 = \dim \mathcal{A}_1 + \dim \mathcal{A}_1 - \dim X = 15 + 15 - 4 = 26, \end{cases}$$

and *the criterion is not satisfied.*

In fact, neither $\mathcal{A}_1$ nor $\mathcal{R}_1$ are formally integrable. We have

$$z = f(t,x;a) + b \Rightarrow p = g(t,x;a),$$

and so

$$a = X(t,x,p), \qquad -\frac{\partial f}{\partial a} = P(t,x,p).$$

Accordingly we obtain

$$\mathcal{A}_1^{(1)} \begin{cases} \dfrac{\partial Z}{\partial t} - P\dfrac{\partial X}{\partial t} = H(t, x, p), \\[2mm] \dfrac{\partial Z}{\partial x} - P\dfrac{\partial X}{\partial x} = -p, \\[2mm] \dfrac{\partial Z}{\partial z} = 1, \quad \dfrac{\partial X}{\partial z} = 0, \quad \dfrac{\partial P}{\partial z} = 0, \\[2mm] \dfrac{\partial Z}{\partial p} - P\dfrac{\partial X}{\partial p} = 0, \\[2mm] \dfrac{\partial P}{\partial t}\dfrac{\partial X}{\partial p} - \dfrac{\partial P}{\partial p}\dfrac{\partial X}{\partial t} = \dfrac{\partial H}{\partial p}, \\[2mm] \dfrac{\partial P}{\partial x}\dfrac{\partial X}{\partial t} - \dfrac{\partial P}{\partial t}\dfrac{\partial X}{\partial x} = -\dfrac{\partial H}{\partial x}, \\[2mm] \dfrac{\partial P}{\partial p}\dfrac{\partial X}{\partial x} - \dfrac{\partial P}{\partial x}\dfrac{\partial X}{\partial p} = 1, \end{cases}$$

and so

$$\dim \mathcal{A}_1^{(1)} = (4 + 3 + 12) - 9 = 10.$$

Similarly,

$$\mathcal{R}_1^{(1)} \begin{cases} \dfrac{\partial \overline{Z}}{\partial X} - \overline{P}\dfrac{\partial \overline{X}}{\partial X} = -P, \\[2mm] \dfrac{\partial \overline{Z}}{\partial Z} = 1, \quad \dfrac{\partial \overline{P}}{\partial Z} = 0, \quad \dfrac{\partial \overline{X}}{\partial Z} = 0, \\[2mm] \dfrac{\partial \overline{Z}}{\partial P} - \overline{P}\dfrac{\partial \overline{X}}{\partial P} = 0, \\[2mm] \dfrac{\partial \overline{P}}{\partial P}\dfrac{\partial \overline{X}}{\partial X} - \dfrac{\partial \overline{P}}{\partial X}\dfrac{\partial \overline{X}}{\partial P} = 1, \end{cases}$$

and so

$$\dim \mathcal{R}_1^{(1)} = (3 + 3 + 9) - 6 = 9.$$

Accordingly,

$$\begin{cases} \dim \mathcal{A}_1^{(1)} \times_Y \mathcal{R}_1^{(1)} = 10 + 9 - 3 = 16, \\[2mm] \dim \mathcal{A}_1^{(1)} \times_X \mathcal{A}_1^{(1)} = 10 + 10 - 4 = 16, \end{cases}$$

and so, by Remark C.9,

$$\dim \mathcal{R}_1^{(1)} - \dim Y = \dim \mathcal{A}_1^{(1)} - \dim X = 6.$$

Now $\mathcal{A}_1^{(1)}$ projects onto $X \times Y$ and $\mathcal{R}_1^{(1)}$ projects onto $Y \times Y$. Hence $\mathcal{A}_1^{(1)} \times_Y \mathcal{R}_1^{(1)}$ and $\mathcal{A}_1^{(1)} \times_X \mathcal{A}_1^{(1)}$ both project onto $X \times Y \times Y$. Accordingly, the injectivity of the graph of the action follows from the injectivity of $\sharp$ restricted to the symbol of $\mathcal{R}_1^{(1)}$ because $n = 4 > 3 = m$. Finally, an injective morphism between spaces of the same dimension is an isomorphism, and it remains to study the first prolongation. This is the hard step. However, because of the second criterion with $n > m$ we know that the action is free. Hence we only have to prove that we have spaces of the same dimension in the prolongated graph of the action. By Remark C.9, we only have to prove that

$$\dim \rho_1(\mathcal{R}_1^{(1)}) - \dim Y = \dim \rho_1(\mathcal{A}_1^{(1)}) - \dim X,$$

i.e.

$$\dim \rho_1(\mathcal{R}_1^{(1)}) - \dim \mathcal{R}_1^{(1)} = \dim \rho_1(\mathcal{A}_1^{(1)}) - \dim \mathcal{A}_1^{(1)}.$$

We shall prove in fact that $\mathcal{R}_1^{(1)}$ and $\mathcal{A}_1^{(1)}$ are involutive systems and deduce the preceding equality from the equality of the dimensions of the respective prolongated symbols.

First of all, this is a nice application of the prolongation Theorem in Chapter III.C. Indeed,

$$M_1 \quad \begin{cases} v_X^{\overline{Z}} - \overline{P} v_X^{\overline{X}} = 0 \\ v_Z^{\overline{Z}} - \overline{P} v_Z^{\overline{X}} = 0 \\ v_P^{\overline{Z}} - \overline{P} v_P^{\overline{X}} = 0 \end{cases} \quad \begin{array}{|ccc|} \hline P & Z & X \\ P & Z & \bullet \\ P & \bullet & \bullet \\ \hline \end{array}$$

and M_1 is involutive. Accordingly, $\rho_1(\mathcal{R}_1^{(1)}) = \mathcal{R}_2^{(1)}$, and similarly $\rho_1(\mathcal{A}_1^{(1)}) = \mathcal{A}_2^{(1)}$. We now study $\mathcal{R}_1^{(1)}$ by means of the linearized system $R_1^{(1)}$. We find

$$R_1^{(1)} \quad \begin{cases} \eta_Z^{\overline{Z}} & = 0 \\ \eta_Z^{\overline{X}} & = 0 \\ \eta_Z^{\overline{P}} & = 0 \\ \eta_X^{\overline{Z}} - P\eta_X^{\overline{X}} - \eta^{\overline{P}} = 0 \\ \eta_X^{\overline{X}} + \eta_P^{\overline{P}} & = 0 \\ \eta_P^{\overline{Z}} - P\eta_P^{\overline{X}} & = 0 \end{cases} \quad \begin{array}{|ccc|} \hline P & Z & X \\ P & Z & X \\ P & Z & X \\ P & X & \bullet \\ P & X & \bullet \\ P & \bullet & \bullet \\ \hline \end{array}$$

We can immediately verify that $R_1^{(1)}$, and thus $\mathcal{R}_1^{(1)}$, is involutive with characters $(2, 1, 0)$. Accordingly, the dimension of the prolongated symbol is $18 - 14 = 4$, and thus we have

$$\dim \rho_1(\mathcal{R}_1^{(1)}) - \dim \mathcal{R}_1^{(1)} = 4.$$

A similar calculation can be performed for $\mathcal{A}_1^{(1)}$ in the nonlinear framework with principal jets

$$Z_t, X_t, P_t, Z_z, X_z, P_z, Z_x, X_x, Z_p,$$

and we find the characters $(2, 1, 0, 0)$ for the involutive system $\mathcal{A}_1^{(1)}$, leading hence to

$$\dim \rho_1(\mathcal{A}_1^{(1)}) - \dim \mathcal{A}_1^{(1)} = 4.$$

So, we have *exactly* the conditions of the first criterion, and we may conclude that $\mathcal{A}_1^{(1)}$ is an involutive automorphic system for the involutive groupoid $\mathcal{R}_1^{(1)}$.

Example 12. Looking for the solutions of the Pfaffian system

$$dz - p\,dx + H(t, x, p)\,dt = dZ - P\,dX$$

studied in the previous Example, we now want to obtain complete integrals of the form

$$z = u(t; a) + v(x; a) + b,$$

where we have separated the variables x and t. Accordingly, we should have

$$p = g(x; a)$$

and so

$$a = X(x, p) \Rightarrow \frac{\partial X}{\partial z} = 0, \quad \frac{\partial X}{\partial t} = 0.$$

Accordingly, the new system of PDE to introduce is:

$$\mathcal{A}_1 \begin{cases} \dfrac{\partial Z}{\partial t} = H(t, x, p), \\[2mm] \dfrac{\partial X}{\partial t} = 0, \\[2mm] \dfrac{\partial Z}{\partial z} = 1, \\[2mm] \dfrac{\partial X}{\partial z} = 0, \\[2mm] \dfrac{\partial Z}{\partial x} - P\dfrac{\partial X}{\partial x} = -p, \\[2mm] \dfrac{\partial Z}{\partial p} - P\dfrac{\partial X}{\partial p} = 0. \end{cases}$$

This system has an involutive symbol but is not formally integrable, since we can prolongate the first PDE with respect to x, the fifth PDE with respect to t and subtract, or prolongate the first PDE with respect to p, prolongate the sixth PDE with respect to t and subtract, to obtain the two additional PDE

$$\frac{\partial P}{\partial t}\frac{\partial X}{\partial x} = \frac{\partial H}{\partial x}, \quad \frac{\partial P}{\partial t}\frac{\partial X}{\partial p} = \frac{\partial H}{\partial p}.$$

This implies

$$\frac{\partial X}{\partial x} \bigg/ \frac{\partial X}{\partial p} = \frac{\partial H}{\partial x} \bigg/ \frac{\partial H}{\partial p},$$

and therefore the condition for separating variables:

$$\frac{\partial}{\partial t}\left(\frac{\partial H}{\partial x} \bigg/ \frac{\partial H}{\partial p}\right) = 0,$$

which was first discovered by Levi-Civita, as already mentioned. We notice that the use of group theory allows us to recover such a differential condition for the Hamiltonian function as an integrability/compatibility condition for the latter automorphic system.

CHAPTER VII

Control Theory

In the first Section of this Chapter we recall, using standard notations, the main notions of *ordinary differential control theory*, also called *classical control theory*, which deals with ODE. We will not give proofs of the results, since these can be found in numerous textbooks, but do give references. The purpose of the second Section is to show that all notions mentioned in the first Section do indeed come from the formal theory of PDE. As a byproduct, not only do we prove that all notions of (classical) ordinary differential control theory can be generalized to partial differential control theory, but also that these generalizations are instances of notions in the partial differential framework, of course independently of any functional analysis or dynamical systems approach. The link with differential algebra will also be exhibited. Finally, in the third Section we will give many examples illustrating the two previous Sections.

A. Ordinary Differential Control Theory

Ordinary differential control theory, or simply OD control theory, studies input/output relations defined by ODE. A standard picture, to which we will return in the next Section, relates the input to the output by a black box along representative arrows, as follows:

The simplest way of viewing such a box is to imagine the following situation, taken from an electrical circuit:

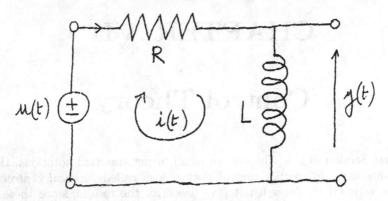

with resistance R and coil L. In this example the input $u(t)$ is the difference of the potential *applied* at the left to the circuit while the output $y(t)$ is the difference of the potential *measured* at the right.

In fact, from basic electricity theory we get

$$\begin{cases} u(t) = Ri(t) + Li'(t), \\ y(t) = Li'(t), \\ u(t) = Ri(t) + y(t). \end{cases}$$

Eliminating the intensity $i(t)$ of current, we finally obtain the control system

$$y' + \frac{R}{L}y = u',$$

which will have a sufficient number of features to be useful for later purposes.

First of all, we notice that this control system does not depend on R *and* L, but only on their quotient R/L.

Secondly, the control system is first order in u *and* y.

Thirdly, because of linearity we have isolated the output on the left and the input on the right.

Let us introduce the Heaviside function $\Upsilon(t)$:

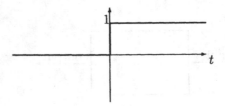

and set $u(t) = e^{-2t}\Upsilon(t)$. We then find

$$y(t) = \left(a \exp\left(-\frac{R}{L}t\right) + \frac{1}{1 - \frac{R}{2L}} e^{-2t} \right) \Upsilon(t),$$

where the constant a depends on $y(0)$ only.

It follows from this simple example that knowledge of the output depends on

- the control system;
- the input;
- the initial data.

In practice it may be useful to assume that the input is of the form $u(t) = A\cos(\omega t + \varphi)$, where the amplitude A and phase φ are arbitrary but the pulsation ω is given. hence we may introduce the additional ODE

$$u'' + \omega^2 u = 0,$$

and regard the control system as being made up by the *two* given ODE.

One could also consider the movement $y(t)$ of a mass m acted upon by a force $u(t)$ while suspended by a spring with elasticity constant k and combined with a dashpot of constant f. The resulting second-order control system is

$$my'' + fy' + ky = u,$$

and the corresponding picture is as follows:

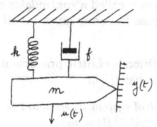

In most cases one can transform a high-order ODE into a system of first-order ODE and eventually modify the output in order to obtain a control system of the form

$$\dot{y} = Ay + Bu,$$

where A is a square $(m \times m)$-matrix and B is a rectangular $(m \times r)$-matrix. To be in agreement with previous notations, we call the independent variable x or t, in dependence on the physical meaning (space or time), and use the equivalent jet notations:

$$\dot{y} = y' = y_x,$$

although in Section B we will make minor changes in order to obtain coherent notations.

In the first system we can make the change of output

$$y \to y + u$$

to obtain the control system

$$\dot{y} = -\frac{R}{L}(y + u),$$

while in the second system we may set

$$y = y^1, \qquad \dot{y} = y^2$$

to obtain the control system

$$\begin{cases} \dot{y}^1 = y^2, \\ \dot{y}^2 = -\dfrac{f}{m}y^2 - \dfrac{k}{m}y^1 + \dfrac{1}{m}u. \end{cases}$$

This control system, in its standard form, is said to define an *input→state→output relation*, because for it $y(0)$ is the only initial data. From the formal point of view, however, *state is only an artifact for transforming a high-order system into a first-order one*. However, the existence of such a procedure led people to believe that certain notions in control theory should be related to the study of trajectories of the vector field $Ay + Bu$ in m-dimensional output space. Indeed, when u varies, this vector field also varies, and one may therefore state, even for nonlinear control systems of the form

$$\dot{y} = F(y, u),$$

the following crucial definition of a first basic notion.

Definition 1. A control system is called a *controllable control system* if there is an input that makes it possible to pass from any given *initial* state to any *final* state, in finite time.

Of course, this functorial definition is of little practical use unless one is able to test it. For linear control systems one has [110]:

Proposition 2. *A linear control system in standard form $\dot{y} = Ay + Bu$ is controllable if and only if* $\operatorname{rk}(B, AB, \ldots, A^{m-1}B) = m$.

The standard proof uses the explicit integration formula

$$y(t) = \exp(t - t_0)Ay(t) + \int_{t_0}^{t} \exp(t - \tau)Bu(\tau)\, d\tau$$

and the Cayley–Hamilton theorem for $(m \times m)$-matrices, stating the existence of numbers $\alpha_0, \ldots, \alpha_m$ such that

$$\alpha_m A^m + \cdots + \alpha_1 A + \alpha_0 I = 0.$$

Therefore it is not only impossible to extend this Definition/Proposition to PDE, but it also seems strange that a Definition given in terms of trajectories (integration) can be described by a formal test (rank of a matrix) which does not appeal to initial data.

The situation becomes even more complicated when people tried to extend this Proposition to more general control systems, as follows [78, 84, 128]:

Definition 3. A nonlinear control system of the form $\dot{y} = a(y) + b(y)u$ is said to be a *weakly-controllable control system* if the so-called *weak distribution* $a, b, [a, b], \ldots$ generated by a, b and all the respective brackets has (maximal) rank m. It is called a *strongly-controllable control system* if the so-called *strong distribution*, obtained by taking out a from the weak distribution, has (maximal) rank m.

A first comment is that the weak distribution is a Lie algebra of vector fields, while the strong one is not. However, if one deals with $a(y) = Ay$, $b(y) = B$, then only the use of the strong distribution is coherent with Proposition 2. In any case, once again the test is formal and does not appeal to explicit integration. Hence we should not only question the proper definition of controllability, but also the way to test it. It also seems strange to distinguish between a strong and a weak concept while the systems $\dot{y} = 1$ and $\dot{y} = 0$ are surely both uncontrollable!

Finally we notice that the previous Definition/test of controllability applies to control systems in standard form, and cannot be extended to the more general input/output situations encountered in practice and in our two examples. However, it is our belief that controllability is a 'built-in' property of a control system, i.e. a formal property depending only on the equations given and not on their possible integration. Hence we may ask:

Question 4. Does there exist a formal definition of controllability that

- does not depend on a state representation;
- can be extended from ODE to PDE?

Of course, this definition should be tested in such a way that the criterion obtained for the situations of Proposition 2 and Definition 3 is recovered or justified.

To answer the above Question, we need a definition that takes into account the way one measures in physical systems. Indeed, a measure is a scalar value of a certain experimental quantity obtained by means of a convenient corresponding apparatus (a thermometer measures temperature, a manometer measures pressure, etc.). Accordingly we give

Definition 5. An *observable* is any (scalar) function of inputs, outputs and their derivatives up to a certain order.

Once we have chosen an observable, there are only two possible situations:

- the observable is *free*, i.e. does not satisfy any ODE itself;
- the observable is *constrained* by at least one ODE.

Of course, if one can find a constrained observable, one cannot 'act' on it, since the input no longer participates in the ODE it must satisfy. Hence, for a control system to be controllable it is necessary that all observables be free. As we will prove below, it is highly nontrivial that this is also sufficient, and we state (compare with Definition 1):

Definition 6. A control system is called a *totally controllable control system* if all observables are free. If all observables are constrained it is called a *totally uncontrollable control system*. In all other cases it may be called a *partially controllable control system* or *partially uncontrollable control system*, depending on the property under consideration.

Of course, this Definition does not depend on state representations, and in section B we shall see that it comes, in a straightforward manner, from differential algebra. For linear control systems we have (cf. Proposition 2):

Proposition 7. *A linear control system of the (standard) form $\dot{y} = Ay + Bu$ is totally controllable if and only if*

$$\mathrm{rk}(B, AB, \ldots, A^{m-1}B) = m.$$

PROOF. By differentiation of the control system any observable can be brought to the form $z = f(y, u, \dot{u}, \ddot{u}, \ldots)$. If z is constrained by a relation of the form $\dot{z} = \Phi(z)$, we must have $z = f(y)$, for otherwise u cannot disappear in the identity. Since we are dealing with the linear situation, we may introduce a constant m-vector a and its adjoint $\tilde{a}$ and set $z = \tilde{a}y$.

Taking into account the Cayley–Hamilton theorem for A, i.e.

$$\alpha_m A^m + \cdots + \alpha_1 A + \alpha_0 I = 0,$$

and differentiating the control system, we obtain

$$\alpha_1 \quad \dot{y} = Ay + Bu,$$
$$\alpha_2 \quad \ddot{y} = A^2 y + ABu + B\dot{u},$$
$$\ldots\ldots\ldots$$
$$\alpha_m \quad y^{(m)} = A^m y + A^{m-1}Bu + \cdots + Bu^{(m-1)}.$$

For $u, \dot{u}, \ldots$ to disappear, we should have, in succession, $\tilde{a}B = 0$, $\tilde{a}AB = 0 \ldots$, $\tilde{a}A^{m-1}B = 0$. This immediately gives

$$\alpha_m z^{(m)} + \cdots + \alpha_1 \dot{z} + \alpha_0 z = 0.$$

Hence, if the rank of the controllability matrix is not equal to m, there is at least one observable which is constrained by an ODE and the system is not totally controllable. $\square$

We similarly have [78, 154]:

Proposition 8. *A nonlinear control system of the form $\dot{y} = a(y) + b(y)u$ is totally controllable if and only if the strong controllability distribution has maximal rank m.*

PROOF. As in the proof of the previous Proposition, a constrained observable must be of the form $z = f(y)$. Introducing the standard Lie derivative, we successively obtain

$$\dot{z} = L(a)f + L(b)fu \Rightarrow L(b)f = 0,$$
$$\ddot{z} = L(a)L(a)f + L(b)L(a)fu$$
$$= L(a)L(a)f + (L(a)L(b) - L([a,b]))fu$$
$$= L(a)L(a)f - L([a,b])fu \Rightarrow L([a,b])f = 0,$$

etc. Hence f is annihilated by the distribution $b, [a, b], \ldots$, and we obtain

$$z = f(y) = \varphi_0(y),$$
$$\dot{z} = L(a)f(y) = \varphi_1(y),$$
$$\ddot{z} = L(a)L(a)f(y) = \varphi_2(y),$$

$$\cdots\cdots\cdots$$

It follows that $z, \dot{z}, \ddot{z}, \ldots$ are also annihilated by this distribution since, e.g., we have

$$L(b)L(a)f = L(a)L(b)f + L([b, a])f = 0.$$

Hence the maximum order of the ODE constraining z is equal to the corank of the strong controllability distribution, which plays a role only in the study of controllability. In the sequel we will explain the role of the so-called weak controllability distribution and the resulting confusion. $\square$

For linear control systems we can introduce the $\mathcal{D}$-module M, the quotient of the $\mathcal{D}$-module of linear differential operators in the input and output by the $\mathcal{D}$-submodule generated by the given control system and all prolongations. We see that Definition 6 comes down to the following abstract definition [65]:

Definition 9. A linear control system is said to be a *controllable control system* if M is torsion free.

In the next Section this Definition will appear as a particular instance of a more general Definition which is also valid for nonlinear time-dependent control systems. For the moment we restrict ourselves to time-independent control systems. We apply to control theory the few algebraic results found at the beginning of Section VI.A. In that case M is a module over a commutative ring which is also a principal ideal domain, and we deduce from Theorem VI.A.7 that the finitely generated module M is free if and only if it is torsion free. Translating this abstract result into the language of control theory, we obtain

Proposition 10. *A control system is parametrizable if and only if it is totally controllable.*

PROOF. By a parametrization we mean a way of expressing input *and* output by derivatives of a certain number of *arbitrary* functions, in such a way that the control system becomes the compatibility condition for this linear inhomogeneous system of derivatives. In fact, this is a practical rephrasing of the property that M be free, i.e. we can find a certain number of 'free' (basis) linear combinations ξ of input/output and their derivatives such that each input/output can be expressed by a linear combination in the ξ and derivatives in such a way that the control system becomes the set of compatibility conditions for these expressions.

We can express this construction by means of a differential sequence

$$E \xrightarrow{\mathcal{D}} F_0 \xrightarrow{\mathcal{D}_1} F_1$$
$$\xi \longrightarrow \eta = (u, y) \longrightarrow \zeta.$$

Of course, counting the differential transcendence degrees, in case the u are free and the y are differentially algebraic over the u, we have

$$\begin{cases} \dim E \geq \#(\text{inputs}), \\ \dim F_1 \geq \#(\text{outputs}), \end{cases}$$

because neither $\mathcal{D}$ nor $\mathcal{D}_1$ is necessarily formally integrable, as we will see in (examples in) Section C.

In a more classical setting, introducing the control system in operator form:

$$\mathcal{A}y = \mathcal{B}u$$

and the parametrization in the operator form

$$y = \mathcal{P}\xi, \qquad u = \mathcal{Q}\xi,$$

we obtain the *operator identity*

$$\mathcal{A}\mathcal{P} \equiv \mathcal{B}\mathcal{Q}.$$

In the case under consideration we may introduce the corresponding χ-polynomial matrices, to obtain the polynomial identity

$$A(\chi)P(\chi) \equiv B(\chi)Q(\chi).$$

If we do not consider formal integrability of the control system but only surjectivity of the operator system $\mathcal{D}_1$, then $A(\chi)$ and $Q(\chi)$ are both square matrices, and we obtain a left and right (coprime) factorization of the so-called *transfer matrix*:

$$H(\chi) = A^{-1}(\chi)B(\chi) = P(\chi)Q^{-1}(\chi),$$

obtained by Laplace transformation from the original control system:

$$\mathcal{A}y = \mathcal{B}u \xrightarrow{\text{Laplace}} A(\chi)Y = B(\chi)U$$
$$\downarrow$$
$$Y = H(\chi)U$$

However, it is essential to notice that the previous approach does not distinguish between input and output and reduces the results to purely module-theoretical concepts. $\square$

To clarify the various notions introduced above, we give a tentative picture, illustrated by a few Examples that will be worked out in Section C.

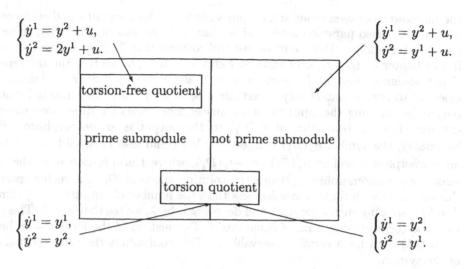

$$\begin{cases} \dot{y}^1 = y^2 + u, \\ \dot{y}^2 = 2y^1 + u. \end{cases} \qquad\qquad \begin{cases} \dot{y}^1 = y^2 + u, \\ \dot{y}^2 = y^1 + u. \end{cases}$$

torsion-free quotient

prime submodule   not prime submodule

torsion quotient

$$\begin{cases} \dot{y}^1 = y^1, \\ \dot{y}^2 = y^2. \end{cases} \qquad\qquad \begin{cases} \dot{y}^1 = y^2, \\ \dot{y}^2 = y^1. \end{cases}$$

The full module consists of all linear operators in the input and output. The submodule is generated by the linear control system and its prolongations. Here we pay attention to the corresponding quotient module when studying the torsion. In particular, we notice that for any torsion system of the form $\dot{y} = Ay$ we have $\alpha_m y^{(m)} + \cdots + \alpha_1 \dot{y} + \alpha_0 y = 0$ for each component, because of the Cayley–Hamilton theorem. Therefore $\dot{y} - Ay$ generates a prime submodule if and only if all the eigenvalues of A coincide.

It remains to explain how it is possible to *effectively* construct the parametrization. For this we will use the sequence dual to the one introduced in the proof of the last Proposition. More precisely, we will take $\mathcal{D}_1$ formally surjective, *even if it is not formally integrable*, so that we have a differential sequence

$$0 \longrightarrow E \xrightarrow{\ \mathcal{D}\ } F_0 \xrightarrow{\ \mathcal{D}_1\ } F_1 \longrightarrow 0$$
$$\xi \longrightarrow \eta \longrightarrow \zeta$$

with $\dim E = \#(\text{inputs})$, $\dim F_1 = \#(\text{outputs})$.

The duality is obtained by introducing the dual bundles E^*, F_0^*, F_1^* and dual operators $\tilde{\mathcal{D}}, \tilde{\mathcal{D}}_1$ obtained from $\mathcal{D}, \mathcal{D}_1$ via integration by parts in such a way that there is a dual sequence

$$0 \longleftarrow E^* \xleftarrow{\ \tilde{\mathcal{D}}\ } F_0^* \xleftarrow{\ \tilde{\mathcal{D}}_1\ } F_1^* \longleftarrow 0$$
$$\nu \longleftarrow \mu \longleftarrow \lambda$$

with relations

$$\langle \tilde{\mathcal{D}}\mu, \xi \rangle = \langle \mu, \mathcal{D}\xi \rangle, \qquad \langle \tilde{\mathcal{D}}_1\lambda, \eta \rangle = \langle \lambda, \mathcal{D}_1\eta \rangle$$

for sections with compact support, using Stokes' theorem. In any case, according to the general theory of Chapter IV and since we have only one independent variable, the symbol is involutive and so $\mathcal{D}_1$ can be of order ≥ 1 if and only if $\mathcal{D}$ is *not* formally integrable. This fact will be clearly illustrated in Section C. Hence, if we know that the control system is totally controllable (a property not depending on separation of the unknowns between input and output variables, which are *all together* denoted by η), then it is also parametrizable and we can, without loss of generality, bring $\mathcal{D}$ to formal integrability. From now on we will assume that $\mathcal{D}$ is involutive. It follows from Chapter IV that $\mathcal{D}_1$ is of first order and involutive, hence that the *corresponding Janet sequence stops at* F_1 because $n = 1$ and $\dim F_0 = \dim E + \dim F_1$. It is essential to notice that $\mathcal{D}$ may nevertheless *be of high order*. Separating input from output by choosing the input to be a transcendence basis for sheer convenience, we discover that the restriction of $\sigma_\chi(\mathcal{D}_1)$ to the output is an isomorphism. Hence, by duality, the symbol $\sigma_\chi(\tilde{\mathcal{D}}_1)$ is injective. We could also have said that $\sigma_\chi(\mathcal{D}_1)$ is an epimorphism, and so $\sigma_\chi(\tilde{\mathcal{D}}_1) = -\sigma_\chi(\tilde{\mathcal{D}}_1)$, where transposition is in the matrix sense, is a monomorphism. Hence the solution space of $\tilde{\mathcal{D}}_1$ is a vector space over the constants with finite dimension less than the number of outputs, i.e. $\leq \dim F_1 = \dim F_1^*$. Isolating such a constant and denoting it by λ, we see that $\dot{\lambda} = 0$. This proves that the corresponding sum of equations of $\mathcal{D}_1$ must be of the form $\dot{z} = 0$, because $\dot{z}\lambda = (z\dot{\lambda}) - z\dot{\lambda}$ for a certain observable z. This contradicts the total controllability of the system.

A simple direct proof runs as follows. Suppose the linear first-order control system is of the form $\dot{y} = Ay + Bu$, and multiply each equation by λ, regarded as a row vector. We have:

$$\lambda(\dot{y} - Ay - Bu) = (\dot{\lambda}y) - (\dot{\lambda} + \lambda A)y - \lambda Bu,$$

and the resulting dual operator $\tilde{\mathcal{D}}_1$ is defined by:

$$\dot{\lambda} + \lambda A = 0, \qquad \lambda B = 0.$$

In general, this system is not formally integrable, and we may add $\dot{\lambda}B = -\lambda AB = 0$, $\ddot{\lambda}B = -\dot{\lambda}AB = \lambda A^2 B, \ldots$ hence we see that the *solution space of this system is zero if and only if the control system is totally controllable*. This new parametrization may allow us to define the new (totally controllable) control system $\tilde{\mathcal{D}}$, which may be of high order because we have just proven that, in general, $\tilde{\mathcal{D}}_1$ *is not at all formally integrable*. This is the reason why $\mathcal{D}$ may be of high order! Recapitulating, to construct the parametrization $\mathcal{D}$ of a totally controllable control system $\mathcal{D}_1$, we can first construct $\tilde{\mathcal{D}}_1$, find the compatibility conditions $\tilde{\mathcal{D}}$, and subsequently recover $\mathcal{D}$ by duality. Of course, this construction is quite specific for ordinary control theory, but gives new insight on duality and its applications. Many Examples illustrating this technique will be given in Section C.

We now turn to the second important concept in control theory, namely *observability*. Using again the example of the electrical circuit (see above), we may choose to measure $i(t)$ instead of $y(t)$, given an input $u(t)$. In that case $y = u - Ri$, and conversely $Ri = u - y$, a result proving that knowing (u, i) is equivalent to knowing (u, y) in a purely deterministic way. However, when we turn to the second example, reading

$$\ddot{y} + \dot{y} + y = u$$

in standard dot notation and with $m = f = k = 1$, we see that the equivalent system

$$\begin{cases} \dot{y}^1 = y^2, \\ \dot{y}^2 = -y^1 - y^2 + u, \end{cases}$$

can be observed by means of $y = y^1$ since, by requirement, $y^2 = \dot{y}$. It can also be observed by means of $y^2 = \dot{y}$, since in this case $\dot{y}^2 + y^2 + y = u$.

Also, the first control system can be observed by $z = \ddot{y}$, since for $R = L$ we have in succession:

$$\dot{y} + y + u = 0, \qquad z + \dot{y} + \dot{u} = 0,$$

and so

$$y = z + \dot{u} - u.$$

Similarly, the second control system can also be observed by $z = \ddot{y}$ since then

$$z + \dot{y} + y = u, \qquad \dot{z} + z + \dot{y} = \dot{u},$$

and so

$$y = \dot{z} + u - \dot{u}.$$

However, for $m = f = 1$, $k = 0$ we see that the corresponding control system

$$\ddot{y} + \dot{y} = u$$

can be observed neither by $z = \dot{y}$ nor by $z = \ddot{y}$.

Collecting these results, we give the following

Definition 11. A control system with inputs u and outputs y can be observed by a number of new outputs z (which are supposed to be functions of u, y and their derivatives) if the y can be recovered as functions of u, z and their derivatives.

In contrast to the definition of controllability, the definition of observability is easily seen to admit immediate generalization to PDE.

In the linear case the following criterion is well known.

Proposition 12. *The linear control system $\dot{y} = Ay + Bu$ can be observed by means of the new outputs $z = Cy + Du$ if and only if*

$$\operatorname{rk}(\tilde{C}, \tilde{A}\tilde{C}, \dots, \tilde{A}^{m-1}\tilde{C}) = m$$

for the transposed matrices.

PROOF. We obtain, in succession,

$$z = Cy + \dots,$$
$$\dot{z} = CAy + \dots,$$
$$z^{(m-1)} = CA^{m-1}y + \dots,$$

and the Proposition follows from linear algebra. The Cayley–Hamilton theorem shows that $\tilde{A}^m\tilde{C}$ cannot increase the rank of the observability matrix. $\square$

We will now study the close relation between total controllability of a control system and observability of its dual, using the test above.

Setting $\mu = (z, v)$, the dual control system $\tilde{\mathcal{D}}$ is parametrized by the relations

$$\begin{cases} v = \dot{\lambda} + \lambda A, \\ z = \lambda B, \end{cases} \Leftrightarrow \begin{cases} \dot{\tilde{\lambda}} = -\tilde{A}\tilde{\lambda} + v, \\ \tilde{z} = \tilde{B}\tilde{\lambda}. \end{cases}$$

Notice that the *observability condition* for this system is precisely the *controllability condition* for the initial control system, since $\tilde{A} = A$, $\tilde{B} = B$. Hence we recover classical results in a new setting [166].

The question is now: How many outputs do we need to observe a control system? It appears that the following result was not known before its announcement in [151] as an elementary consequence of the theorem on the primitive element in differential algebra (Section VI.B). Indeed, if we assume that the control system is defined by algebraic ODE and that each output is differentially algebraic over the inputs, then we are *precisely* in the situation in which a primitive element may exist, and we obtain:

Proposition 13. *An algebraic control system of the type above can be observed by a single output.*

Remark 14. Of course, another existence condition is provided by a specific property of the ground differential field, a property not always met in practice. E.g., the control system $\dot{y}^1 = y^1$, $\dot{y}^2 = 2y^2$ can be observed by $z = y^1 + y^2$ because

$$z = y^1 + y^2, \qquad \dot{z} = y^1 + 2y^2,$$

and the jacobian of $(z, \dot{z})$ with respect to (y^1, y^2) is 1. However, the control system $\dot{y}^1 = y^1$, $\dot{y}^2 = y^2$, which is also defined over $\mathbb{Q}$, is only observable over $\mathbb{Q}(t)$ by a single output, say $z = y^1 + ty^2$, since

$$z = y^1 + ty^2, \qquad \dot{z} = y^1 + (1 + t)y^2,$$

and the jacobian of $(z, \dot{z})$ with respect to (y^1, y^2) is 1 again. In actual practice, especially when dealing with nonlinear algebraic control systems, it is almost impossible to find by hand such a primitive element, and for this reason the result above had not been known. Any effective computation needs a random choice of elements in the ground differential field, according to the constructive proof already quoted. In any case, we notice that the condition that each output be differentially algebraic over the inputs is *very* restrictive and cannot be supposed to hold in general. In classical contexts, it is assumed to hold by 'tradition', since it fits in with many engineering concepts. However, from a conceptual point of view the specific property of a control system allowing observability by a single output must be carefully distinguished.

In the linear time-independent situation we have seen that Laplace transformation reduced the situation to the linear relation $Y = H(\chi)U$, where $H(\chi)$ is the transfer matrix. In that case one is led to a study of the injectivity/surjectivity of $H(\chi)$. Accordingly, we may state

Definition 15. A control system is said to be a *left-invertible control system* if one can recover the input from the output in a deterministic way (injectivity). It is called a *right-invertible control system* if any output can be obtained, i.e. if there is no ODE among the outputs which contains the inputs (surjectivity). The maximum number of outputs that can be given independently of the inputs is called the *differential output rank* of the control system.

We will study these definitions so as to prepare the results of the next Section. First of all, in view of the above definitions, we have seen that, most of the time, input and output are mixed while playing a perfectly symmetric role otherwise. The lack of reciprocity is forbidden in the Definition above because of ... tradition! Indeed, if one may have the possibility of knowing the input from the output, the converse cannot be possible because it is assumed ... not to be, in the sense that each output is assumed to be *effectively* differential over the inputs. We will see counterexamples of this situation in Section C. Then, if we may eliminate the input by looking for an ODE between the outputs only, we could do the same by eliminating the outputs to find an ODE for the inputs only; however, once again this is forbidden because it is assumed ... that the inputs are free!

The situation concerning left invertibility is not so clear in the literature, since certain authors restrict the definition to the case in which each input is differential over the outputs. These distinct situations must be carefully distinguished via a detailed system analysis.

It remains to study the way to test the Definitions above. Once more, the situation which depends only on the input/output relation can be covered by using a state bringing the system to its standard first-order representation, even in the linear case. This is particularly clear within the various definitions in the literature of so-called *structure at infinity*.

Definition 16. The *structure at infinity* of a control system is a sequence of increasing integers culminating at the differential output rank.

We will see that this is closely related to the inversion problem. The concept has been introduced after 1970 for the case of a single input u/single output y control system given in standard form $u \to x \to y$ by the equations

$$\dot{x} = Ax + bu, \qquad y = cx,$$

where b is a column vector and c is a row vector. The main problem is to look for u as a function of $x, y, \dot{y}, \ldots$. One proceeds as follows. Differentiate and substitute to obtain

$$\dot{y} = cAx + cbu, \qquad \ddot{y} = cA^2x + cAbu + cb\dot{u}, \ldots$$

So, the recovery of u from the state x and output derivatives $\dot{y}, \ddot{y}, \ldots$ depends only on the lowest k for which the *number* $cA^k b$ is nonzero.

Equivalently using the Laplace transform (with our usual parameter χ), we have

$$\chi X = AX + bU, \quad Y = cX \Rightarrow Y = c(\chi I - A)^{-1}bU = H(\chi)U,$$

and k is the unique integer such that $\chi^{k+1} H(\chi)$ has a finite nonzero value at infinity. This explains also the name of the concept. Indeed, one only has to put

$$(\chi I - A)^{-1} = \frac{1}{\chi}\left(I - \frac{1}{\chi}A\right)^{-1} = \frac{1}{\chi}\left(I + \frac{1}{\chi}A + \cdots + \frac{1}{\chi^k}A^k + \ldots\right)$$

and use Taylor expansion with respect to $1/\chi \to 0$ as $\chi \to \infty$.

During the last decade, the above concept has been generalized to the particular class of nonlinear control systems that are affine with respect to the input [63, 70, 91, 124]. See the recent book [128] for a short historical survey with corresponding references. This progress was made possible by the introduction of differential-geometric methods in control theory. At the same time, a computational algorithm has been given for *inverting* an affine nonlinear system, i.e. to obtain the input from knowledge of the state and output derivatives, in a way extending the algorithm known for linear systems [22, 23, 44, 63, 91, 123, 170]. We will sketch the result, and illustrate it by various Examples in Section C. We will also show how it is possible to revise this algorithm along the lines of Section B.

Consider an affine control system of the form

$$\dot{x} = a(x) + b(x)u, \qquad y = c(x).$$

Since a, b are vector fields over the state space 'x', we may introduce the Lie derivative and obtain

$$\dot{y} = L(a)c(x) + L(b)c(x)u.$$

If the matrix $L(b)c$ has (generic) rank σ_1, we may exhibit σ_1 such equations that are linearly independent in u, say

$$\dot{y}_1' = a_1(x) + b_1(x)u$$

with matrix $b_1(x)$ of (generic) rank σ_1, and eliminate u among the other equations, to obtain equations of the form

$$\ddot{y}_1'' = c_1(x, \dot{y}_1')$$

not involving u anymore. Again differentiating these equations, we obtain $\sigma_2 \geq \sigma_1$ equations that are linearly independent in u, of the form

$$\begin{cases} \dot{y}_1' = a_1(x) + b_1(x)u, \\ \ddot{y}_2'' = a_2(x, \dot{y}_1', \ddot{y}_1') + b_2(x, \dot{y}_1')u, \end{cases}$$

and other equations not containing u anymore. Continuing in this way we will obtain $0 \leq \sigma_1 \leq \cdots \leq \sigma_k \leq \cdots \leq \#(\text{inputs})$ equations that are linearly independent in u. This procedure must stop because the number of inputs is finite. Now, the number of initial data $x(0)$ is equal to the number n of state variables x. Therefore, *each* output y satisfies an ODE of order n at most over the inputs, and the corresponding sequence of integers stabilizes at σ_n. Finally, the $\dot{y}_1', \ddot{y}_2', \ldots, y_k^{(k)}$ are differentially (algebraically) independent because they are at least linearly independent in u. Hence the limit value σ_n is equal to the differential output rank, i.e. the number of outputs that can be given freely (independently of the inputs). Also, counting the differential transcendence degrees in the differential algebraic case (a, b, c rational functions of x over a differential field K) gives

$$\text{difftrd } K\langle u, y\rangle / K = \text{difftrd } K\langle u, y\rangle / K\langle u\rangle + \#(\text{inputs})$$
$$= \text{difftrd } K\langle u, y\rangle / K\langle y\rangle + \text{differential output rk},$$

where, for simplicity, (u, y) denotes the generic couple (input, output)-solution of the system. Since $\text{difftrd } K\langle u, y\rangle / K\langle u\rangle = 0$ by assumption (each state, and hence each output, is differentially algebraic over the inputs), we find

$$\sigma_n \leq \inf(\#(\text{inputs}), \#(\text{outputs})).$$

However, one can also say that σ_n among the inputs are functions of x and output derivatives. Since x is differentially algebraic over the inputs, Section VI.B implies that these σ_n inputs are differentially algebraic over the outputs and the remaining ($\#(\text{inputs}) - \sigma_n$) inputs. Accordingly,

$$\text{difftrd } K\langle u, y\rangle / K\langle y\rangle = \#(\text{inputs}) - \sigma_n$$

and so

$$\text{difftrd } K\langle y\rangle / K = \#(\text{inputs}) - (\#(\text{inputs}) - \sigma_n) = \sigma_n,$$

as before.

Rephrasing the previous eliminations in terms of the jacobian matrices

$$J_k = \frac{\partial(\dot{y}, \ddot{y}, \ldots, y^{(k)})}{\partial(u, \dot{u}, \ldots, u^{(k-1)})}$$

we immediately obtain [22, 23]

$$\text{generic rk } J_k = \sigma_1 + \cdots + \sigma_k,$$

and this procedure can be applied to more general situations too.

Examining the above approach, which has led to the definition and computation of the structure at infinity, we see that, on the one hand, we have tried to eliminate the state by sufficiently often differentiating the outputs, as if we were looking for a high-order input/output control system [165]. However, we allow ourselves to explicitly retain the state. On the other hand, we perform exactly as if we were trying to find the inputs from the outputs in a deterministic way, again explicitly retaining the states.

Hence we arrive at the following conceptual question:

Question 17. Is it possible to define another finite increasing sequence of integers depending only on the input/output description and ending with a number describing an intrinsic property of the control system?

We now study the various possible presentations of a control system under changes of input or output.

The simplest operation is the change inputs and outputs among themselves, as in the previous Chapter. Once again, this enforces the simultaneous treatment of input *and* output, without any separation between them. As a byproduct we can study the invariance of the control system, or the equivalence with various models, e.g. linear ones.

We study these two situations separately, using simple examples, in order to prove that the main delicate problem in the study of invariance or equivalence is ... the absence of formal integrability of the system of PDE, since in general there are many inputs/outputs.

Suppose we have a control system

$$\dot{y}^k = a^k(y) + b_r^k(y)u^r.$$

The transformation $\overline{y} = g(y)$ preserves the control system if and only if

$$\frac{\partial \overline{y}^l}{\partial y^k}(a^k(y) + b_r^k(y)u^r) = a^l(\overline{y}) + b_r^l(\overline{y})u^r$$

for any input, i.e.

$$\frac{\partial \overline{y}^l}{\partial y^k}a^k(y) = a^l(\overline{y}), \qquad \frac{\partial \overline{y}^l}{\partial y^k}b_r^k(y) = b_r^l(\overline{y}).$$

Hence g must preserve the vectors a and b. So, it must also preserve $[a, b], \ldots$, i.e. g must preserve the Lie algebra generated by a and b, i.e. g must preserve the so-called weak controllability distribution. Accordingly, the infinitesimal transformations preserving the control system are elements of a Lie algebra $\Theta = \Theta_{\text{out}}$ which is infinite dimensional, in general. For this reason the weak controllability distribution has often been mixed up with the *centralizer* $C(\Theta)$ of Θ, which consists of all vector fields commuting with any vector field in Θ and such that $[\Theta, C(\Theta)] = 0$.

Definition 18. $C(\Theta)$ is called the *control Lie algebra*.

Notice that, *in this case*, the main reason for introducing $a, b, [a, b]$ and the various other brackets lies in bringing the system of PDE defining the invariance to a formally integrable form by prolongation/projection, as in Chapter III.

In the case of invariance, the system of PDE defining the finite transformations has at least the identity solution, and is thus always compatible. In the case of equivalence the situation may be more involved, as is shown by the following example involving 1 independent variable x, 1 input u and 2 outputs y^1, y^2 or z^1, z^2. We want to know whether the control system $y^2y_x^1 = u$ can be transformed into the linear control system $z_x^1 = u$. To this end we need

$$\frac{\partial z^1}{\partial y^1}y_x^1 + \frac{\partial z^1}{\partial y^2}y_x^2 = u \Leftrightarrow y^2y_x^1 = u.$$

Hence we should have

$$\frac{\partial z^1}{\partial y^1} = y^2, \qquad \frac{\partial z^1}{\partial y^2} = 0.$$

However, differentiating the first PDE with respect to y^2, the second with respect to y^1, and subtracting, we obtain $1 = 0$ as new zero-order equation. This proves the impossibility of solving this equivalence problem.

Another situation in which the corresponding system of ODE is incompatible is found in the study of constraints on the output. E.g., consider again the control system

$$\dot{y}_x^1 = y^2 + u, \qquad \dot{y}_x^2 = y^1 + u.$$

We want to impose differential conditions on the output as follows:

$$\begin{cases} y^1 + y^2 = 1 & \Rightarrow u = 1/2, \\ y^1 - y^2 = 1 & \Rightarrow 1 = 0 \text{ impossible}, \\ y^1 - y^2 = 0 & \Rightarrow \text{always possible}. \end{cases}$$

The underlying conceptual idea is that any solution of the output space admits a pullback restriction in the input space. In the linear case, the whole set of ODE can be written out explicitly by writing the output on the left and the input on the right in the corresponding differential operators. The point then is to find the compatibility conditions for the right hand side. For this, one has to bring the left hand side to formal integrability (the hard step) without differentiation, in order to have a 2-acyclic or involutive symbol since this condition is automatically fulfilled for $n = 1$. As soon as a nonlinearity enters the picture, things may be extremely difficult. The following Example, given by Vessiot in [190] is quite instructive.

Example 19. Let y^1, y^2, y^3 be three linearly independent solutions of the same linear third-order ODE

$$y_{xxx}^k - p(x)y_{xx}^k + q(x)y_x^k - r(x)y^k = 0 \qquad (k = 1, 2, 3).$$

These three ODE can be regarded as an ODE control system with inputs (p, q, r) and outputs (y^1, y^2, y^3). A priori, the inputs are arbitrary, but now we will look for differential constraints on (p, q, r) by allowing the initial data to be such that

$$(y^3)^2 - y^1 y^2 = 0$$

for the outputs. The resulting system of ODE is not formally integrable (it contains zero- and third-order ODE). We leave it to the reader to differentiate the constraint, using a computer algebra system, and find the only differential condition

$$9\partial_{xx}p - 18p\partial_x p - 27\partial_x q + 4(p)^3 - 18pq + 54r = 0.$$

In Section C we will also find this condition by group-theoretical methods.

We conclude this Section with a study of '*causality*' in classical control theory. The following Definition is used to select a class of ordinary differential control systems.

Definition 20. A control system is called a *causal control system* if knowledge of the input up to a certain time and convenient initial conditions suffice to determine the output up to that time.

A careful study of the Laplace transformation shows that we can *test this functorial Definition in a purely formal way*. E.g., with the usual notations of jet theory, if x denotes the independent variable (time), u the input and y the output, then a simple comparison of the respective orders of the derivatives involved shows that the systems $y_x + u = 0$ and $y_x + u_x = 0$ are causal, while the system $y_x + u_{xx} = 0$ is not causal. However, in all three cases the input may be regarded as a generator for a differential transcendence basis for the three differential extensions generated by the three corresponding differential polynomials appearing on the left hand sides. This comment cannot be made if certain conditions are imposed upon the input signal. E.g.,

$$u_{xx} + u = 0, \quad y_x + u_{xx} = 0 \Rightarrow y_x - u = 0.$$

Difficulties also arise in the nonlinear situation.

Among the recent tentatives towards a formal definition of causality, we will examine the one proposed by J.C. Willems for the linear case in [200].

First of all , inputs and outputs are regarded as unknowns of a mixed system of ODE. Therefore we want to distinguish input from output by means of the following formal criterion, making it possible to write a causal system in the particular, solved, form

$$\mathcal{A}y = \mathcal{B}u$$

by assuming

1) there does not exist an ODE between the u alone;

2) $A(\chi)$ is a square matrix with $\det A(\chi) \neq 0$ and $\lim_{\chi \to \infty} A(\chi)^{-1}B(\chi) < \infty$.

The second assumption shows that the number of initial conditions is finite (*only involving output*) and generalizes the comparison of orders (both input and output together are involved).

B. Partial Differential Control Theory

The purpose of this Section is not to extend to PDE the definitions and results of Section A and concerning ODE, but rather to exhibit independently definitions and results for PDE and to prove that these coincide with the ones already given. To this end, and for obtaining a more striking presentation, this Section will exactly follow the lines of the previous Section.

Ordinary differential control theory as presented in Section A dealt with the study of input/output relations defined by systems of ODE. Similarly, *partial differential control theory* deals with the study of input/output relations defined by systems of PDE. Roughly speaking, we will be concerned with *any* number of inputs and outputs, functions of *any* number of independent variables, and relations given by linear or nonlinear systems of PDE of *any* order. To agree with the notation used in this book, we slightly change the notations of Section A. We introduce an *input fibered manifold* $\mathcal{E}$ over the *base space* X with local coordinates (x^i, y^k), where $i = 1, \ldots, n$ and $k = 1, \ldots, m$. Similarly we introduce an *output fibered manifold* $\mathcal{F}$ over X with local coordinates (x^i, z^r), where $i = 1, \ldots, n$ and $r = 1, \ldots, p$.

Definition 1. A *partial differential control system*, or *PD control system*, is a mixed system $\mathcal{R}_q \subset J_q(\mathcal{E} \times_X \mathcal{F})$.

The importance of this Definition in applications lies in the fact that *any* problem in engineering is of the type above. To underline the importance of this remark, examples taken from elasticity, electromagnetism, fluid mechanics, ... will be given in Section C.

According to the results of Chapter III, we may assume that $\mathcal{R}_q$ is formally integrable, and even involutive, *as a whole*, i.e. with respect to *all variables* (y, z) *together*. The canonical projections of $\mathcal{R}_\infty$ onto $J_\infty(\mathcal{E})$ and $J_\infty(\mathcal{F})$ are (*in general*) the prolongations of the involutive *input resolvent system* $\mathcal{P}_{q+r} \subset J_{q+r}(\mathcal{E})$ and *output resolvent system* $\mathcal{Q}_{q+r} \subset J_{q+r}(\mathcal{F})$.

Definition 2. *Input* is any solution $y = f(x)$ of the input resolvent system, and *output* is any solution $z = g(x)$ of the output resolvent system.

Remark 3. This Definition is consistent with prolongations, since indeed

$$\mathcal{R}_{q+r} \subseteq \mathcal{P}_{q+r} \times_X \mathcal{Q}_{q+r} \subset J_{q+r}(\mathcal{E} \times_X \mathcal{F}),$$

while $\mathcal{R}_{q+r}$ projects onto $\mathcal{P}_{q+r}$ and $\mathcal{Q}_{q+r}$ *separately*. For this reason a differential control system is sometimes called a *differential correspondence*, by analogy with the following standard Definition.

Definition 4. If X and Y are manifolds (varieties), then a submanifold (subvariety) $Z \subseteq X \times Y$ is called a *correspondence between X and Y* if Z projects onto X and Y *separately*.

In this Definition, X and Y play a perfectly symmetric role, a fact explaining why input and output *must* play a perfectly symmetric role in control theory.

Two problems may arise in this setting. The first problem is, at least from a purely historical point of view, called the *direct Bäcklund problem*. It deals with the algorithmic determination of resolvent systems, given the differential control system. In Section C we will see that this problem originated in the study of curves and surfaces in euclidean 3-space. It is indeed a problem of differential elimination, and can be treated along the lines sketched in Chapter III by many examples. In general, to solve it there is no alternative to assuming the control system to be sufficiently regular with respect to input, output, or input and output together. In the linear case the solution in uniquely determined by a linear differential operator, possibly of high order, as in the Janet Example. In the nonlinear case the search for the input resolvent system can be done by regarding the control system as a system for the output with given input. The study of the corresponding formal integrability gives, *in general*, rise to a 'tree' of possibilities. However, no general rule can be given, unless the system is linear with respect to the output and has coefficients depending only on x and the input. The search for the output resolvent system is done in a similar way, by exchanging the roles of input and output. One should notice that the systems for input, for output, and for input/output may be individually and independently formally integrable or not, as can be seen from Examples in Section C.

The second problem, viewed also from the historical point of view, is called the *inverse Bäcklund problem* and is much more difficult. It concerns the determination of a differential correspondence when the corresponding resolvent systems are given. Of course, testing a particular tentative differential correspondence, with the aim to adapt certain parameters or arbitrary functions involved, is part of the direct Bäcklund problem. The merit of the correspondence is to reduce the 'size' of the output space by knowledge of a compatible input. Indeed, if an input (a solution of the input resolvent system, see Definition 2) is given and substituted into the control system, we obtain a subsystem of the output resolvent system. Examples of this procedure are given in Section C.

We can use local coordinates (x, y_q) for $J_q(\mathcal{E})$, (x, z_q) for $J_q(\mathcal{F})$, and (x, y_q, z_q) for $J_q(\mathcal{E} \times_X \mathcal{F})$. The control system is defined by a system of nonlinear PDE of the form

$$\mathcal{R}_q \qquad \Phi^\tau(x, y_q, z_q) = 0.$$

However, in most applications $\mathcal{E} = X \times Y$ and $\mathcal{F} = X \times Z$ are trivial fibered manifolds, with $\mathcal{E} \times_X \mathcal{F} = X \times Y \times Z$.

As a first question we may wonder about the definition of 'state' in this new setting. To explain our new point of view, we refer to the results concerning duality in Section A and consider a linear control system.

From Chapter IV we know that in the beginning of a Janet sequence

$$E \xrightarrow{\;\mathcal{D}\;} F_0 \xrightarrow{\;\mathcal{D}_1\;} F_1$$

formal exactness at F_0 means that $\mathcal{D}_1$ is in some sense well parametrized by $\mathcal{D}$, i.e. the kernel of $\mathcal{D}_1$ is formally equal to the image of $\mathcal{D}$. For an involutive $\mathcal{D}$ we know that $\mathcal{D}_1$ is of order one, involutive, and such that for $\mathcal{D}_1 = \Psi_1 \circ j_1$, $B_1 = \ker \Psi_1$, we have a surjective map $B_1 \to F_0 \to 0$ (of course, $B_1 \subset J_1(F_0)$).

In Section A we have seen that, on the one hand, controllability of the control system is equivalent to existence of a parametrization. On the other hand, it was much easier to handle duality by having the control system in its first-order 'standard' form. Surprisingly, this form does not only have the above property of surjectivity, but also a property which is quite specific for operators like $\mathcal{D}_1$. Indeed, looking back to the property of the symbol sequence of $\mathcal{D}$ studied in Section IV.B, we know that $\sigma_\chi(\mathcal{D}_1)$ is never injective ($\forall \chi \in T^*$), in contrast to $\sigma_\chi(\mathcal{D})$, and we have just discovered that this property is possessed by *any* control system (whether controllable or not), since it is a consequence of causality. Indeed, taking into account the input and the output together in the fiber of F_0, the characteristic matrix becomes

$$\left. \begin{pmatrix} \chi & & 0\;0 & \cdots & 0 \\ & \ddots & & \vdots & \vdots \\ 0 & & \chi\;0 & \cdots & 0 \end{pmatrix} \right\} \#(\text{outputs})$$
$$\underbrace{}_{\#(\text{outputs})} \underbrace{}_{\#(\text{inputs})}$$

Taking into account the previous results, we discover that the generalization of the so-called standard form to PDE control systems is nothing but the reduction to a first-order system of an involutive system $\mathcal{R}_q$ via the embedding $\mathcal{R}_{q+1} \subset J_1(\mathcal{R}_q)$ with $\mathcal{R}_{q+1} \to \mathcal{R}_q \to 0$ exact. Notice that both input *and* output together must be taken into account.

Since this procedure only involves a change of dependent variables, the notion of observable remains the same.

Definition 5. An *observable* is any (scalar) function of inputs, outputs and their derivatives up to a certain order.

When the input and output are solutions of the control system, there are two possibilities (like in Section A):

- The observable is *free*, i.e. does not satisfy any PDE itself.
- The observable is *constrained* by at least one PDE.

Of course, if we can find a constrained observable, we cannot 'act' on it, since the input no longer appears in the PDE it has to satisfy. Hence, for a control system to be controllable, it is necessary that all observables be free. So, we give the following Definition.

Definition 6. A control system is called a *totally controllable control system* if all observables are free. It is called a *totally uncontrollable control system* if all observables are constrained. In all other cases it is called a *partially controllable control system*, or an *partially uncontrollable control system*, depending on the property that needs to be stressed.

Since reduction to first order amounts to changing dependent variables, the Definition above does not depend on the state representation given.

It remains to establish a link between controllability and parametrization in the linear setting.

First of all, a *control system that is parametrizable is also controllable*. Indeed, let ξ be the 'parameters' and η a scalar observable. By linearly changing the independent variables, if necessary, we see that η contains at least one of the derivatives $\partial_{n...n}\xi$ of maximal order, and so η is free (a scalar operator has no compatibility condition!).

Conversely, consider a totally controllable control system. At least one of the unknowns is free, and we can use the formula

$$\alpha_q^n = \min_\chi \dim \ker \sigma_\chi(\mathcal{D})$$

of Section IV.B to conclude that $\sigma_\chi(\mathcal{D}_1)$ is surely *not injective*. Accordingly, $\sigma_\chi(\tilde{\mathcal{D}}_1)$ is surely not injective, and so $\tilde{\mathcal{D}}_1$ must be overdetermined, with a certain number of compatibility conditions expressed by an operator $\tilde{\mathcal{D}}$ that can be neither formally integrable nor involutive, since $\tilde{\mathcal{D}}_1$ can be neither integrable nor involutive. Again dualizing $\tilde{\mathcal{D}}$, we obtain an operator $\mathcal{D}$ which has *at least* the compatibility conditions expressed by $\mathcal{D}_1$ *but* may have others, their full set being expressed by an operator $\mathcal{D}_1'$.

We will now comment upon the relation between the operator $\mathcal{D}$ and its dual $\tilde{\mathcal{D}}$. In the nondifferential case, if we have a map $\varphi \colon E \to F$ and its dual $\tilde{\varphi} \colon F^* \to E^*$, then there are relations

$$\dim \ker \tilde{\varphi} = \dim \operatorname{coker} \varphi = \dim \ker \varphi + \dim F - \dim E.$$

As a byproduct, setting

$$\operatorname{rk} \varphi = \dim \operatorname{im} \varphi = \dim E - \dim \ker \varphi,$$
$$\operatorname{rk} \tilde{\varphi} = \dim \operatorname{im} \tilde{\varphi} = \dim F - \dim \ker \tilde{\varphi},$$

we immediately obtain the relation

$$\operatorname{rk} \varphi = \operatorname{rk} \tilde{\varphi}.$$

We will generalize this equation to differential operators by introducing concepts from differential algebra.

First of all, a linear differential operator $\mathcal{D}$ acting on a certain number of unknowns determines a linear differential ideal, which is therefore prime and makes it possible to introduce a differential extension. With a slight abuse of language, we will denote the differential transcendence degree of such a differential extension by $\operatorname{difftrd} \mathcal{D}$. Of course, if we have $\mathcal{D}\xi = \eta$, then the right hand side η is a solution of the compatibility conditions $\mathcal{D}_1 \eta = 0$ for a certain linear differential operator $\mathcal{D}_1$. If $\mathcal{D}$, and so $\mathcal{D}_1$, has coefficients in a differential field K, then there is a tower of differential extensions:

$$K \subset K\langle \eta \rangle \subset K\langle \xi \rangle.$$

Counting the differential transcendence degrees, we obtain the important formula

$$\operatorname{difftrd} \mathcal{D} + \operatorname{difftrd} \mathcal{D}_1 = \#(\text{unknowns}).$$

Introducing the differential sequence

$$E \xrightarrow{\mathcal{D}} F_0 \xrightarrow{\mathcal{D}_1} F_1,$$

which need not be a Janet sequence (although we must take care of the fact that $\mathcal{D}_1$ has to generate *all* possible compatibility conditions), we may now successively define:

$$\begin{cases} \operatorname{diff} \operatorname{rk} \mathcal{D} = \dim E - \operatorname{difftrd} \mathcal{D}, \\ \operatorname{diff} \operatorname{rk} \mathcal{D}_1 = \dim F_0 - \operatorname{difftrd} \mathcal{D}_1, \end{cases}$$

and obtain the important formula

$$\operatorname{diff} \operatorname{rk} \mathcal{D} + \operatorname{diff} \operatorname{rk} \mathcal{D}_1 = \dim F_0.$$

Of course, *if $\mathcal{D}$ is involutive* we have

$$\begin{cases} \operatorname{difftrd} \mathcal{D} = \min_\chi \dim \ker \sigma_\chi(\mathcal{D}), \\ \operatorname{diff} \operatorname{rk} \mathcal{D} = \max_\chi \dim \operatorname{im} \sigma_\chi(\mathcal{D}), \end{cases}$$

but the previous formula is completely general, given the single assumption regarding $\mathcal{D}_1$ made above.

Now, the main difficulty in comparing $\mathcal{D}$ and $\tilde{\mathcal{D}}$ is that $\tilde{\mathcal{D}}$ need not be formally integrable or involutive at all, even when $\mathcal{D}$ is involutive. The many Examples in Section C should convince the reader. Hence we will study the effect of *prolongation* and *projection* on the computation of the differential rank of an operator.

We will first deal with prolongation. Introduce the corresponding equations and dualizing variables, as follows:

$$\begin{cases} \Phi^\tau = 0 & \lambda_\tau \\ d_i \Phi^\tau = 0 & \lambda_\tau^i \end{cases}$$

Since we have the relations

$$\lambda_\tau^i d_i \Phi^\tau = d_i(\lambda_\tau^i \Phi^\tau) - (\partial_i \lambda_\tau^i)\Phi^\tau,$$

the dual operator for the first prolongation is obtained from the initial operator by the substitution

$$\lambda_\tau \rightarrow \lambda_\tau - \partial_i \lambda_\tau^i.$$

Accordingly, the differential transcendence degree is increased by the number of λ_τ^i, because these are free. However, the number of unknowns on which the adjoint operator acts increases by the same number, and so the differential rank is unchanged under prolongation.

Let us similarly deal with projection, and assume that, starting with the PDE $\Phi^\tau = 0$ of order q, there is one linear combination, say $A_\tau^i(x)d_i\Phi^\tau$, which is still of order q. Introducing the dualizing variable λ, we have

$$\lambda A_\tau^i(x)d_i\Phi^\tau = d_i(\lambda A_\tau^i \Phi^\tau) - (\partial_i(A_\tau^i(x)\lambda))\Phi^\tau.$$

Hence, the dual operator for projection is obtained from the initial operator by the substitution

$$\lambda_\tau \rightarrow \lambda_\tau - \partial_i(A_\tau^i(x)\lambda),$$

and the same comment as before proves that the differential rank is unchanged.

Now, if $\mathcal{D}$ is involutive, its differential rank is equal to the maximum number of equations that can be solved for the derivatives of the unknowns with respect to the independent variable x^n. By duality, the same property will be possessed by the equations in λ, and so

$$\text{diff rk } \mathcal{D} \leq \text{diff rk } \tilde{\mathcal{D}}.$$

Bringing $\tilde{\mathcal{D}}$ to formal integrability and involution does not change the differential rank, as already noticed, and the same inequality holds for $\mathcal{D}$ and $\tilde{\mathcal{D}}$ interchanged. Accordingly, we have

$$\text{diff rk } \mathcal{D} = \text{diff rk } \tilde{\mathcal{D}}.$$

We will use this in proving the following Theorem.

Theorem 7. *A linear partial differential operator is parametrizable if and only if it is controllable.*

PROOF. Consider the differential sequences

$$E \xrightarrow{\mathcal{D}} F_0 \xrightarrow{\mathcal{D}_1} F_1$$
$$E^* \xleftarrow{\tilde{\mathcal{D}}} F_0^* \xleftarrow{\tilde{\mathcal{D}}_1} F_1^*$$

obtained from $\mathcal{D}_1$ by first constructing $\tilde{\mathcal{D}}_1$, then $\tilde{\mathcal{D}}$, generating *all* compatibility conditions of $\tilde{\mathcal{D}}_1$, and finally $\mathcal{D}$ by dualizing $\tilde{\mathcal{D}}$.

From the above formula we see that

$$\text{diff rk } \mathcal{D} = \text{diff rk } \tilde{\mathcal{D}}, \qquad \text{diff rk } \mathcal{D}_1 = \text{diff rk } \tilde{\mathcal{D}}_1,$$

and the relation

$$\text{diff rk } \tilde{\mathcal{D}} + \text{diff rk } \tilde{\mathcal{D}}_1 = \dim F_0^*.$$

Accordingly, we have
$$\text{diff rk}\,\mathcal{D} + \text{diff rk}\,\tilde{\mathcal{D}} = \dim F_0.$$
If we introduce the differential operator $\mathcal{D}_1'$ generating *all* compatibility conditions of $\mathcal{D}$, then
$$\text{diff rk}\,\mathcal{D} + \text{diff rk}\,\mathcal{D}_1' = \dim F_0,$$
and so
$$\text{diff rk}\,\mathcal{D}_1 = \text{diff rk}\,\mathcal{D}_1',$$
which leads to
$$\text{difftrd}\,\mathcal{D}_1 = \text{difftrd}\,\mathcal{D}_1'.$$
Setting $\mathcal{D}_1\eta = \zeta$, $\mathcal{D}_1'\eta = \zeta'$, we have the following inclusions of differential fields:
$$K \subset K\langle\zeta\rangle \subset K\langle\zeta'\rangle \subset K\langle\eta\rangle,$$
and so
$$\text{difftrd}(K\langle\zeta'\rangle/K\langle\zeta\rangle) = 0.$$
It follows that *each* new compatibility condition of $\mathcal{D}_1'$ is differentially algebraic over the ones of $\mathcal{D}_1$ and may provide a constrained observable. Hence, if $\mathcal{D}_1$ is controllable, i.e. if the corresponding control system is controllable, then this situation *cannot* take place and $\mathcal{D}_1'$ *must* coincide with $\mathcal{D}_1$, which is thus parametrized by $\mathcal{D}$. $\quad\square$

This situation is particularly well illustrated by all the Examples at the end of Section C.

We now study controllability in the nonlinear setting, and, in particular, in the differential algebraic setting. For simplicity we assume (using the Criterion in Section VI.B) that the control system is defined over a *ground differential field* K with n derivations $\partial_1, \ldots, \partial_n$ by a prime differential ideal $\mathfrak{r} \subset K\{y, z\}$. In this setting we may introduce the *input differential ideal* $\mathfrak{p} = \mathfrak{r} \cap K\{y\}$ and the *output differential ideal* $\mathfrak{q} = \mathfrak{r} \cap K\{z\}$. Notice that $\mathfrak{p}$ and $\mathfrak{q}$ are prime if $\mathfrak{r}$ is prime but, of course, the difficulty in practice will be to find a finite basis for $\mathfrak{p}$ and $\mathfrak{q}$ whenever a finite basis for $\mathfrak{r}$ is known. Following the end of Section VI.B, we introduce the *input differential field* $L = Q(K\{y\}/\mathfrak{p})$ and the *output differential field* $M = Q(K\{z\}/\mathfrak{q})$, as well as the *input/output differential field* $N = Q(K\{y, z\}/\mathfrak{r})$. We have the following key picture of control theory:

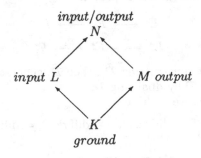

As before, *two problems may arise in this setting*. The first one is to determine $\mathfrak{p}$ and $\mathfrak{q}$ if $\mathfrak{r}$ is given. This is the *direct Bäcklund problem* for differential algebra. The technique has been sketched in Chapter III for the differential case. In this case, in which we know that $\mathfrak{r}$ is prime, we know that $\mathfrak{p}$ and $\mathfrak{q}$ are also prime. Hence we can always follow the *most generic* branches of the tree appearing in the study of differential eliminations. However, presently we are far from having computer algebra packages for this purpose, unless, as already said, the control system is linear in the input and output *separately*, even if it is not linear in the input and output *at the same time*.

The second problem is to find N if L and M are given. It amounts to finding a prime differential ideal in $L \otimes_K M$, since there is a universal map

$$L \otimes_K M \to N.$$

This is the *inverse Bäcklund problem* for differential algebra. There is no general algorithm that can be used.

Now we will set up controllability in the differential algebraic framework.

In the linear case, the $\mathcal{D}$-module approach of Section A can remain untouched, since we are dealing with Ore rings, as already noticed. As a byproduct we recall that the PD control system (input and output mixed) determines, by residuation, a $\mathcal{D}$-module M, and the PD control system is said to be controllable if M is torsion free.

In the algebraic case, it follows from the above comments that *controllability should only involve the differential extension N/K*. Restating the search for constrained/free observables in this setting, we give the following Definition.

Definition 8. A differential algebraic control system is said to be *algebraically controllable* if K is differentially algebraically closed in N. It is said to be *algebraically parametrizable* if N is contained in a purely differentially transcendental extension of K. As a byproduct, a parametrizable control system is automatically controllable.

Remark 9. If K' is the differential algebraic closure of K in N, then K' is differentially algebraically closed in N (see Section VI.B). Hence N/K' is the 'minimal' controllable extension that can be obtained from N/K. This explains the name '*minimal realization*' in classical control theory. The idea behind it is to break the differential extension N/K into a totally uncontrollable extension K'/K and a totally controllable extension N/K', as will be explained in the Examples of Section C.

Remark 10. If the control system is written in classical notation with one input u and one output y, related by $u\ddot{y} - \dot{u} = 0$, then the observable $z = \dot{y} - \log u$ satisfies $\dot{z} = 0$ and the control system is not controllable, although it is algebraically controllable, because $\log u$ does not belong to N. Hence one must be careful in dealing with algebraic analogies. Differential algebra should not be a target but just a manner for creating/understanding new concepts.

More generally, taking into account Definition 6 and Example C.2, we notice that a nonlinear control system is controllable if and only if the corresponding linearised control system is controllable and therefore parametrizable according to Theorem 7, which can also be used to test such a property.

In the remainder of this Section we shall prove that (at least in the differential algebraic case) *knowledge of a system only depends on the four differential fields* K, L, M, N and their relative positions or properties, such as inclusion, intersection, differential transcendence degrees, etc.

We now turn to the second important concept in control theory, namely *observability*. Once again, differential algebra will be a guide towards new concepts. In our case N is a finitely generated differential extension of K. Accordingly, it has been defined by generators, but many other possible choices can be made for generating N over K. Among these, we mention the choice of a transcendence basis of N/K mentioned in Section VI.B and the subsequent generation of N over $K\langle$differential transcendence basis$\rangle$ by a primitive element. In particular, if the inputs are free and difftrd $N/L = 0$, then we can observe N/L by a single conveniently chosen output. In any case, we always have

$$\text{difftrd}\,N/K = \text{difftrd}\,N/L + \text{difftrd}\,L/K$$
$$= \text{difftrd}\,N/M + \text{difftrd}\,M/K.$$

Definition 11. difftrd L/K is called the *differential input rank*, and difftrd M/K is called the *differential output rank*, of the control system.

Of course, for sufficiently regular control systems the same definition applies in the differential case, by computing the corresponding n-character of the input and output resolvent systems; the preceding sum formulas are still valid. However, we want to stress that there is no need for assuming that the inputs constitute a differential transcendence basis of N/K, in the same way that we do not require that the outputs constitute a differential transcendence basis of N/K (see the many Examples in Section C). Of course, we recall that the *various differential transcendence degrees cannot be computed if we have not brought the corresponding systems to formal integrability or involutivity*.

This comment will allow us to revisit the so-called 'structure at infinity' studied in Section A. Indeed, consider the control system as a system for *input* when *output* is given (take care!), i.e. in the algebraic case, let us look at the differential extension N/M. Recall that the classical approach is to regard the control system as a system for *output* when *input* is given, i.e. to look at the differential extenson N/L. However, the roles of input and output are symmetric and if one study is useful, then so must be the other.

Let us imagine thus that the output z is given and satisfies the output resolvent system, i.e. in the algebraic case, let us assume that the differential extension M/K is known and that the system for the input y is therefore defined over M. Accordingly, the input y will satisfy a number of PDE of orders $0, \ldots, q$, assuming that the initial system is involutive at order q. Making, if necessary, a linear change of coordinates to obtain a δ-regular coordinate system, we may look for the maximum number of equations at each order, say $\tau_0, \ldots, \tau_q$, that can be solved with respect to derivatives (or jets) of y with indices equal to n only, in a manner similar to the computation of the nth character. Since we have to keep adding new equations to the prolongations of the preceding ones, we have

$$0 \leq \tau_0 \leq \cdots \leq \tau_q.$$

In the algebraic case we also have

$$\tau_q = \#(\text{input}) - \text{difftrd}\, N/M.$$

If the inputs constitute a differential transcendence basis of N/K, then

$$\text{differential output rank} = \text{difftrd}\, M/K$$
$$= \text{difftrd}\, N/K - \text{difftrd}\, N/M$$
$$= \text{difftrd}\, L/K - \text{difftrd}\, N/M$$
$$= \#(\text{input}) - \text{difftrd}\, N/M$$
$$= \tau_q.$$

We discover the intrinsic meaning of the structure at infinity, and at the same time we have a tentative inversion algorithm.

So, the *direct approach* in control theory follows the *left tower* $K \subset L \subset N$ of inclusions of differential fields, while the *inverse approach* in control theory follows the *right tower* $K \subset M \subset N$ of inclusions of differential fields.

Using the simple example of the control system with one input u and one output z considered in Section A, we will indicate the confusion that might arise.

Let us recall that we deal with (take care to the change of notation in the parametrization)

$$\dot{y} = Ay + bu, \qquad z = cy,$$

where b is a column vector and c is a row vector. We have

$$\alpha_0 \quad z = cy,$$
$$\alpha_1 \quad \dot{z} = cAy + cbu,$$
$$\cdots\cdots\cdots$$
$$\alpha_m \quad z^{(m)} = cA^m y + cA^{m-1}bu + \cdots + cA^k bu^{(m-k+-1)},$$

where k is the smallest integer such that the *number* $cA^k b$ is not equal to zero. Hence, using the Cayley–Hamilton formula,

$$\alpha_m A^m + \cdots + \alpha_1 A + \alpha_0 I = 0,$$

we see that y can be eliminated. We immediately find the control system in (u, z). It is a system of order m in z if we look at N/L, *but* a system of order $m - k - 1$ in u if we look at N/M. As a byproduct we have

$$\tau_0 = 0, \ldots, \tau_{m-k-2} = 0, \tau_{m-k-1} = 1,$$

and the present approach does not depend upon parametrization (nor state). The intrinsic explanation of this is that if R_q is an involutive system of order q, then $R_{q+1} \subset J_1(R_q)$ is also an involutive system of order 1, with the *same symbol* as R_{q+1} and so the *same characters*, because the corresponding Hilbert polynomials are identical. We finally notice that the use of the transfer matrix is not compatible or coherent with formal integrability, and must be dealt with with great care in the intrinsic approach.

We now come to the problem of *invertibility*. We have to prove that our approach makes it possible to understand the various definitions in the literature, at least for the algebraic case.

Definition 12. A control system is called a *left-invertible control system* if the input can be recovered in a deterministic way from the output, i.e. if $L \subseteq M = N$ in the differential algebraic case.

Of course, we have to carefully distinguish this case from the one in which *each* input individually is differentially algebraic over the outputs, a concept that also has a meaning in the purely differential case. In the differential algebraic case the latter can be simply expressed by difftrd $N/M = 0$. We notice that this result, *which is not at all necessary*, is the counterpart of the condition difftrd $N/L = 0$, which is hence also *not at all necessary*.

Similarly, we may assume that the inputs are free, and so difftrd $L/K = \#$(inputs). hence the counterpart is to assume that difftrd $M/K = \#$(outputs), and we discover that, in view of our picture of fields, the term 'right-invertible', adopted in the standard literature, is quite misleading and must be changed in the future in order to agree with a general scheme. Indeed, if left invertible is concerned with the inclusion $L \subset M$, then right invertible should be concerned with $M \subset L$. Also, in the same way as the condition difftrd $M/K = \#$(outputs) (i.e. the outputs are free) is *not at all necessary*, conversely the condition difftrd $L/K = \#$(inputs) is *not at all necessary* (i.e. the inputs need not be free). This comment takes a particularly clear physical meaning in Section C, where we will study the Maxwell equations. It is also backed up when we match two control systems between (y, z) and (z, w), since we need have the same resolvent system in z, or at least the space of solutions of the second system must include the space of solutions of the first. The resulting differential correspondence between (y, w) is obtained by eliminating z, but no procedure is perfectly effective today, unless we are dealing with linear control systems. The basic idea is to join the two systems, consider the resulting system as a system for z with coefficients depending on (y, w), and finally study the formal integrability of this system to find a differential correspondence for (y, w). Of course, in the differential algebraic case we have an ideal in $K\langle y, z \rangle$ which is extended to $K\langle y, z, w \rangle$, and an ideal in $K\langle z, w \rangle$ which is also extended to $K\langle y, z, w \rangle$. These two ideals are then added to obtain a yet larger ideal in $K\langle y, z, w \rangle$, and we have to take the intersection with $K\langle y, w \rangle$. The last step is the hardest.

We finally notice that the study of invariance or equivalence of control systems leads to systems of PDE exactly as in the differential Galois theory of Chapter III, and we will not repeat it. We point out, however, that the possibility to define systems of PDE on differential fields in the differential algebraic case is an important feature allowing us to change input, output, or both.

We conclude this Section with a tentative definition of *causality*. The basic intuitive engineering idea is that outputs must appear in equations with order exceeding or equal to the order of equations for inputs. To give a precise mathematical definition, we look at the characteristic matrix (in the nonlinear case we use linearization or the vertical machinery).

Definition 13. A formally integrable control system is called a *causal control system* if the characteristic matrix and its restriction to the output only have the same rank.

Again, for the computation of the output or input rank the assumption of formal integrability in this Definition is crucial.

C. Examples

Example 1. Consider, with one independent variable x, one input u and two outputs y^1, y^2 the linear OD control system

$$y_x^1 = y^2 + u, \qquad y_x^2 = \alpha y^1 + u,$$

where α is a constant parameter. In the standard notation of Section A, we have

$$A = \begin{pmatrix} 0 & 1 \\ \alpha & 0 \end{pmatrix}, \quad B = \begin{pmatrix} 1 \\ 1 \end{pmatrix} \Rightarrow (B, AB) = \begin{pmatrix} 1 & 1 \\ 1 & \alpha \end{pmatrix}.$$

The determinant of the last matrix is $\alpha - 1$, and so the control system is controllable if and only if $\alpha \neq 1$.

Any observable is of the form $z = f(y^1, y^2, u, u_x, \dots)$. However, if z is constrained by one ODE at least, then $z = f(y^1, y^2)$. In the module-theoretical approach, to look for torsion elements we may choose $z = y^2 + \lambda y^1$, and we have an ODE of the form $z_x + \mu z = 0$. The input u is free. By subtracting the two ODE, we obtain for the outputs

$$y_x^2 - y_x^1 + y^2 - \alpha y^1 = 0.$$

Hence the differential output rank is 1 for all α. By substitution we obtain

$$y_x^2 + \lambda y_x^1 + \mu y^2 + \lambda \mu y^1 = 0.$$

Identification of the last two ODE gives

$$\lambda = -1, \quad \mu = 1 \Rightarrow \alpha = -\lambda \mu = 1.$$

Indeed, $z = y^2 - y^1$ satisfies $z_x + z = 0$ for $\alpha = 1$, and the minimum realization breaks the total differential extension N/K, where $K = \mathbb{Q}$, $N = K\langle \text{input}, \text{output} \rangle$ into the differential extension K'/K defined by $z_x + z = 0$ and the differential extension N/K' defined by

$$y_x^1 = y^1 + z + u,$$

where K' is the differential algebraic closure of K in N. A similar decomposition can be obtained from the module-theoretical point of view.

We now turn to the construction of the dual control system, and use the new variables $\eta = (\eta^1 = y^1, \eta^2 = y^2, \eta^3 = u)$, combined with $\lambda = (\lambda_1, \lambda_2)$ since we have two ODE. By duality, we obtain from the system

$$\begin{cases} \dot{\eta}^1 - \eta^2 - \eta^3 = 0 \,\big|\, -\lambda_1 \\ \dot{\eta}^2 - \alpha \eta^1 - \eta^3 = 0 \,\big|\, -\lambda_2 \end{cases}$$

the system (up to sign)

$$\begin{cases} \eta^1 & \Rightarrow \dot{\lambda}_1 + \alpha\lambda_2 = \mu_1, \\ \eta^2 & \Rightarrow \dot{\lambda}_2 + \lambda_1 = \mu_2, \\ \eta^3 & \Rightarrow \lambda_1 + \lambda_2 = \mu_3. \end{cases}$$

To find the image, i.e. the corresponding conditions for $\mu = (\mu_1, \mu_2, \mu_3)$, we have to study the formal integrability (so, the involutivity) of the corresponding homogeneous system

$$\begin{cases} \dot{\lambda}_1 + \alpha\lambda_2 = 0, \\ \dot{\lambda}_2 + \lambda_1 = 0, \\ \lambda_1 + \lambda_2 = 0. \end{cases}$$

Differentiating the zero-order equation and substituting, we obtain the additional zero-order equation

$$\alpha\lambda_2 + \lambda_1 = 0.$$

hence, the operator $\lambda \to \mu$ is injective if and only if $\alpha = 1$ (according to the equations $(\alpha - 1)y^1 = 0$, $(\alpha - 1)y^2 = 0$), in agreement with the general theory.

When $\alpha = 1$ the system is involutive and so the only compatibility condition is

$$\dot{\mu}_3 + \mu_3 - \mu_1 - \mu_2 = 0.$$

Using multiplication by ξ and integration by parts, the dual operator $\xi \to \eta$ is easily seen to be (up to sign)

$$\begin{cases} \xi = \eta^1, \\ \xi = \eta^2, \\ \dot{\xi} - \xi = \eta^3, \end{cases}$$

and the image η is a solution of

$$\dot{\eta}^1 - \eta^2 - \eta^3 = 0, \qquad \eta^2 - \eta^1 = 0,$$

i.e. we do not recover our initial control system (no wonder, it is not parametrizable since it is not controllable ($\alpha = 1!$)).

On the other hand, when $\alpha \neq 1$ the system is *not* formally integrable, and so the compatibility conditions are obtained as follows. First we obtain

$$\begin{cases} (\alpha - 1)\lambda_1 = \dot{\mu}_3 + \alpha\mu_3 - \mu_1 - \mu_2, \\ (\alpha - 1)\lambda_2 = -\dot{\mu}_3 - \mu_3 + \mu_1 + \mu_2, \end{cases}$$

and so we obtain the *single* compatibility condition (although it seems as though there should be *two*):

$$\ddot{\mu}_3 - \dot{\mu}_1 - \dot{\mu}_2 - \alpha\mu_3 + \mu_1 + \alpha\mu_2 = 0.$$

Accordingly, multiplying again by ξ and integrating by parts, the dual operator $\xi \to \eta$ is easily seen to be (with correct sign):

$$\begin{cases} \dot{\xi} + \xi = \eta^1, \\ \dot{\xi} + \alpha\xi = \eta^2, \\ \ddot{\xi} - \alpha\xi = \eta^3, \end{cases}$$

and the image η is a solution of our initial control system, which is now effectively parametrized by a second-order operator that is injective if and only if $\alpha \neq 1$, in agreement with the general theory of Section A.

We will now prove that these results do not depend on the manner of observing the system. Indeed, choosing $y = y^1$, we obtain the second-order control system

$$\ddot{y} - \alpha y - \dot{u} - u = 0$$

in differential operator form, and we could have separated the input from the output. We obtain the polynomial 'matrices'

$$A(\chi) = \chi^2 - \alpha, \qquad B(\chi) = \chi + 1,$$

so that the transfer matrix becomes

$$H(\chi) = \frac{\chi + 1}{\chi^2 - \alpha}.$$

The causality condition is satisfied and $H(\chi)$ reduces to $1/(\chi - 1)$ when $\alpha = 1$. There is observability because $y^1 = y$, $y^2 = \dot{y} - u$.

We will now study total controllability. For this we notice that any observable is of the form

$$z = f(y, \dot{y}, u, \dot{u}, \dots).$$

If z is constrained by an ODE, examination of the derivatives of u shows that $z = f(y, \dot{y}, u)$. So we may set $z = \dot{y} + \lambda y + \mu z$, and substitute this into $\dot{z} + \nu z = 0$ to obtain

$$\ddot{y} + (\lambda + \nu)\dot{y} + \lambda\nu y + \mu\dot{u} + \mu\nu u = 0.$$

After identification with the control system we obtain

$$\lambda + \nu = 0, \quad \lambda\nu = -\alpha, \mu = -1, \mu\nu = -1,$$

i.e.

$$\lambda = -1, \quad \mu = -1, \quad \nu = 1 \Rightarrow \alpha = 1.$$

We can immediately verify that the observable $z = \dot{y} - y - u$ satisfies $\dot{z} + z = 0$, and we recover the result of the study above, using an equivalent first-order system.

We will now study the dual point of view, by introducing $\eta = (\eta^1 = y, \eta^2 = u)$. Multiplying the control system by λ and integrating by parts, we obtain

$$\begin{cases} \ddot{\lambda} - \alpha\lambda = \mu_1, \\ \dot{\lambda} - \lambda = \mu_2. \end{cases}$$

For $\alpha = 1$ the corresponding homogeneous system is involutive and the only *first-order* compatibility condition is

$$\dot{\mu}_2 + \mu_2 - \mu_1 = 0,$$

giving rise to the dual operator $\mu \to \nu$:

$$\dot{\mu}_2 - \mu_2 - \mu_1 = \nu.$$

Multiplying by ξ and integrating by parts gives (up to sign):

$$\begin{cases} \xi = \eta^1, \\ \dot{\xi} - \xi = \eta^2, \end{cases}$$

and therefore the image

$$\dot{\eta}^1 - \eta^2 - \eta^1 = 0$$

instead of the original control system

$$\ddot{\eta}^1 - \eta^1 - \dot{\eta}^2 - \eta^2 = 0.$$

Notice that this image is only a subsystem of the initial control system.

On the other hand, when $\alpha \neq 1$ the corresponding homogeneous system is *not* involutive, and the only *second-order* compatibility condition is

$$\ddot{\mu}_2 - \dot{\mu}_1 + \mu_1 - \alpha\mu_2 = 0.$$

It gives rise to the dual operator $\mu \to \nu$:

$$\ddot{\mu}_2 - \dot{\mu}_1 + \mu_1 - \alpha\mu_2 = \nu.$$

Multiplying by ξ and integrating by parts gives (with correct sign):

$$\begin{cases} \dot{\xi} + \xi = \eta^1, \\ \ddot{\xi} - \alpha\xi = \eta^2, \end{cases}$$

and so the image must be obtained by taking into account that the corresponding homogeneous system is *not* involutive when $\alpha \neq 1$. By differentiation we immediately obtain

$$(\alpha + 1)\xi = \dot{\eta}^1 - \eta^2 - \eta^1,$$

and so

$$\ddot{\eta}^1 - \alpha\eta^1 - \dot{\eta}^2 - \eta^2 = 0,$$

which is the original control system.

So, once again we have obtained an effective parametrization by duality.

Using this Example, we can verify that a torsion-free module is free indeed, by choosing ξ as independent generator.

We will now study the invariance of the given control system. It will provide the two specific values $0, 1$ of α, while controllability only gave the value 1. So, invariance theory gives a more refined way of studying the control system.

According to the general theory, the defining system of finite Lie equations is

$$
\begin{cases}
y^2 \dfrac{\partial \bar{y}^1}{\partial y^1} + \alpha y^1 \dfrac{\partial \bar{y}^1}{\partial y^2} = \bar{y}^2, \\[2mm]
y^2 \dfrac{\partial \bar{y}^2}{\partial y^1} + \alpha y^1 \dfrac{\partial \bar{y}^2}{\partial y^2} = \alpha \bar{y}^1, \\[2mm]
\dfrac{\partial \bar{y}^1}{\partial y^1} + \dfrac{\partial \bar{y}^1}{\partial y^2} = 1, \\[2mm]
\dfrac{\partial \bar{y}^2}{\partial y^1} + \dfrac{\partial \bar{y}^2}{\partial y^2} = 1.
\end{cases}
$$

The corresponding linearized system is

$$
\begin{cases}
y^2 \dfrac{\partial \eta^1}{\partial y^1} + \alpha y^1 \dfrac{\partial \eta^1}{\partial y^2} - \eta^2 = 0, \\[2mm]
y^2 \dfrac{\partial \eta^2}{\partial y^1} + \alpha y^1 \dfrac{\partial \eta^2}{\partial y^2} - \alpha \eta^1 = 0, \\[2mm]
\dfrac{\partial \eta^1}{\partial y^1} + \dfrac{\partial \eta^1}{\partial y^2} = 0, \\[2mm]
\dfrac{\partial \eta^2}{\partial y^1} + \dfrac{\partial \eta^2}{\partial y^2} = 0.
\end{cases}
$$

Both systems are *not* formally integrable. In terms of vector fields we have

$$
a = y^2 \frac{\partial}{\partial y^1} + \alpha y^1 \frac{\partial}{\partial y^2}, \quad b = \frac{\partial}{\partial y^1} + \frac{\partial}{\partial y^2} \Rightarrow [b, a] = \frac{\partial}{\partial y^1} + \alpha \frac{\partial}{\partial y^2}.
$$

Since $[\eta, a] = 0, [\eta, b] = 0 \Rightarrow [\eta[a, b]] = 0$, we obtain $[\eta, [b, a] - b] = 0$, and therefore

$$
(\alpha - 1)\frac{\partial \eta^1}{\partial y^2} = 0, \qquad (\alpha - 1)\frac{\partial \eta^2}{\partial y^2} = 0.
$$

For $\alpha = 1$ the system $[\eta, a] = 0, [\eta, b] = 0$ is formally integrable, together with the system of finite transformations. Let us set $\bar{y}^1 = y^1 + g(y^1, y^2), \bar{y}^2 = y^2 + h(y^1, y^2)$. We find

$$
\begin{cases}
\dfrac{\partial g}{\partial y^1} + \dfrac{\partial g}{\partial y^2} = 0, \\[2mm]
\dfrac{\partial h}{\partial y^1} + \dfrac{\partial h}{\partial y^2} = 0, \\[2mm]
y^2 \dfrac{\partial g}{\partial y^1} + y^1 \dfrac{\partial g}{\partial y^2} = h, \\[2mm]
y^2 \dfrac{\partial h}{\partial y^1} + y^1 \dfrac{\partial h}{\partial y^2} = g.
\end{cases}
$$

It follows that

$$
g = g(y^1 - y^2), \qquad h = h(y^1 - y^2),
$$
$$
(y^2 - y^1)g' = h, \qquad (y^2 - y^1)h' = g,
$$

which leads to

$$g = (y^1 - y^2)g' + (y^1 - y^2)g''$$

i.e.

$$g = \lambda(y^1 - y^2) + \frac{\mu}{y^1 - y^2} \Rightarrow h = -\lambda(y^1 - y^2) + \frac{\mu}{y^1 - y^2},$$

and finally to the Lie group of transformations:

$$\begin{cases} \overline{y}^1 = y^1 + \lambda(y^1 - y^2) + \dfrac{\mu}{y^1 - y^2}, \\ \overline{y}^2 = y^2 - \lambda(y^1 - y^2) + \dfrac{\mu}{y^1 - y^2}. \end{cases}$$

In case $\alpha \neq 1$ we obtain

$$\frac{\partial \eta^1}{\partial y^2} = 0, \quad \frac{\partial \eta^2}{\partial y^2} = 0 \Rightarrow \frac{\partial \eta^1}{\partial y^1} = 0, \quad \frac{\partial \eta^1}{\partial y^2} = 0, \quad \eta^2 = 0, \quad \alpha \eta^1 = 0.$$

Hence, for $\alpha = 0$ we obtain the group

$$\overline{y}^1 = y^1 + \nu, \qquad \overline{y}^2 = y^2,$$

while for $\alpha \neq 0, 1$, we obtain the identity transformation

$$\overline{y}^1 = y^1, \qquad \overline{y}^2 = y^2.$$

More generally, we will prove that a control system of the form $\dot{y}^l = A^l_k y^k + B^l_r u^r$ admits only the identity transformation if and only if it is controllable with $\det A \neq 0$.

Indeed, calling $\theta^l_\alpha \partial/\partial y^l$ the column vectors of the controllability matrix, we can deduce from the corresponding infinitesimal Lie equations that

$$\theta^l_\alpha \frac{\partial \eta^k}{\partial y^l} = 0 \Rightarrow \frac{\partial \eta^k}{\partial y^l} = 0,$$

because $\mathrm{rk}(\theta^l_\alpha) = m$. The only equations left over from $[\eta, a] = 0$ are therefore

$$A^l_k \eta^k = 0,$$

and lead to $\eta = 0$ whenever $\det A \neq 0$.

In the situation above we have

$$\det A = \det \begin{pmatrix} 0 & 1 \\ \alpha & 0 \end{pmatrix} = -\alpha,$$

and this result justifies the previous computations.

Example 2. The purpose of this Example is to show the relations between controllability and formal integrability.

Consider the nonlinear algebraic control system

$$u\ddot{y} - \dot{u} = \alpha \in \mathbb{Q},$$

with one input u and one output y. Since $\dot{u}$ appears in the ODE, if an observable $z = f(y, \dot{y}, u, \dot{u}, \dots)$ would satisfy $\dot{z} = \Phi(z)$, then necessarily $z = f(u, y, \dot{y})$, using a well-known argument that we do not repeat here. Differentiating z and taking into account the control system, we get

$$\dot{u}\frac{\partial f}{\partial u} + y\frac{\partial f}{\partial y} + \frac{\alpha + \dot{u}}{u}\frac{\partial f}{\partial \dot{y}} \equiv \Phi(f(u, y, \dot{y})).$$

Looking at the term $\dot{u}$, we *must* have

$$u\frac{\partial f}{\partial u} + \frac{\partial f}{\partial \dot{y}} = 0,$$

and so we are left with

$$\dot{y}\frac{\partial f}{\partial y} - \alpha\frac{\partial f}{\partial u} \equiv \Phi(f(u, y, \dot{y})).$$

So, all (2×2)-minors of the matrix

$$\begin{pmatrix} \dot{y}\dfrac{\partial^2 f}{\partial u \partial y} - \alpha\dfrac{\partial^2 f}{\partial u \partial u} & \dot{y}\dfrac{\partial^2 f}{\partial y \partial y} - \alpha\dfrac{\partial^2 f}{\partial u \partial y} & \dot{y}\dfrac{\partial^2 f}{\partial y \partial \dot{y}} + \dfrac{\partial f}{\partial y} - \alpha\dfrac{\partial^2 f}{\partial u \partial \dot{y}} \\[3mm] \dfrac{\partial f}{\partial u} & \dfrac{\partial f}{\partial y} & \dfrac{\partial f}{\partial \dot{y}} \end{pmatrix}$$

must vanish, and we finally obtain

$$u\dot{y}\frac{\partial^2 f}{\partial u \partial y} - \alpha u\frac{\partial^2 f}{\partial u \partial u} + \dot{y}\frac{\partial^2 f}{\partial y \partial \dot{y}} + \frac{\partial f}{\partial y} - \alpha\frac{\partial^2 f}{\partial u \partial \dot{y}} = 0,$$

taking into account the PDE for f given above. Differentiating this very PDE with respect to u, respectively y, and subtracting the results, we arrive at the following system of two PDE for f:

$$\begin{cases} \dfrac{\partial f}{\partial \dot{y}} + u\dfrac{\partial f}{\partial u} = 0, \\[3mm] \dfrac{\partial f}{\partial y} + \alpha\dfrac{\partial f}{\partial u} = 0. \end{cases}$$

This system need not be formally integrable. Indeed, differentiating the first PDE with respect to y, the second with respect to $\dot{y}$, and subtracting, we find

$$\alpha\frac{\partial^2 f}{\partial u \partial \dot{y}} - u\frac{\partial^2 f}{\partial u \partial y} = 0.$$

Differentiating the first PDE with respect to u we find

$$\frac{\partial^2 f}{\partial u \partial \dot{y}} + u\frac{\partial^2 f}{\partial u \partial u} + \frac{\partial f}{\partial u} = 0,$$

while differentiation of the second PDE with respect to u gives

$$\frac{\partial^2 f}{\partial u \partial y} + \alpha\frac{\partial^2 f}{\partial u \partial u} = 0.$$

By substitution into the first PDE we finally obtain

$$\alpha\frac{\partial f}{\partial u} = 0.$$

For $\alpha \neq 0$ this system is *not* formally integrable, and the corresponding formally integrable (involutive) system with the same solution is

$$\frac{\partial f}{\partial u} = 0, \qquad \frac{\partial f}{\partial y} = 0, \qquad \frac{\partial f}{\partial \dot{y}} = 0.$$

The general solution is $f = $ cst, and so we cannot find any nontrivial constrained observable. In such a case the initial control system is controllable.

If $\alpha = 0$, then the above system becomes

$$\frac{\partial f}{\partial y} = 0, \qquad \frac{\partial f}{\partial \dot{y}} + u\frac{\partial f}{\partial u} = 0.$$

It is formally integrable (involutive), with general solution

$$z = f(u, \dot{y}) = h(\dot{y} - \log u).$$

The observable $z = \dot{y} - \log u$ satisfies $\dot{z} = 0$ and the control system is not controllable, without any reference to a state representation.

A similar technique applies (with difficulty!) to more general situations in the differential algebraic case.

Notice that for $\alpha = 0$ the control system is '*geometrically*' uncontrollable, but '*algebraically*' controllable, because $\dot{y} - \log u \notin \mathbb{Q}\langle u, y \rangle$. Indeed, in this case $K = \mathbb{Q}$, $N \simeq \mathbb{Q}\langle u \rangle(y, \dot{y})$, and K is differentially algebraically closed in N although the control system is surely not controllable. This shows the limitations of differential algebraic methods.

We will now study the relation between this nonlinear control system and its linearization

$$u\ddot{Y} + \ddot{y}U - \dot{U} = 0,$$

obtained by using uppercase letters. This linearization must be regarded as a linear system of ODE defined over the differential field N, *independently of the value of* α. Our study of its controllability must take into account the fact that, *now*, the coefficients effectively depend on the independent variable x (or t, if we adhere to classical notations).

Let us look for an observable of the form

$$Z = \dot{Y} + \lambda(y, \dot{y}, u, \dots)Y + \mu(y, \dot{y}, u, \dots)U$$

satisfying an ODE of the form

$$\dot{Z} + \nu(y, \dot{y}, u, \dots)Z = 0.$$

Developing and substituting gives

$$\ddot{Y} + (\lambda + \nu)\dot{Y} + \left(\frac{d\lambda}{dx} + \lambda\nu\right)Y + \mu\dot{U} + \left(\frac{d\mu}{dx} + \mu\nu\right)U = 0.$$

After identification we obtain the conditions

$$\lambda + \nu = 0, \qquad \frac{d\lambda}{dx} + \lambda\nu = 0, \qquad \mu = -\frac{1}{u}, \qquad \frac{d\mu}{dx} + \mu\nu = \frac{\ddot{y}}{u}.$$

It follows that

$$\lambda = -\nu = -\frac{\dot{u}}{u} + \ddot{y} = \frac{\alpha}{u},$$

and we are left with the only condition

$$\frac{\alpha \dot{u}}{u^2} + \frac{\alpha^2}{u^2} = 0,$$

which amounts to

$$\alpha(\dot{u} + \alpha) = 0.$$

For $\alpha \neq 0$ this *identity* in N cannot take place, and the linearized system is controllable.

For $\alpha = 0$ the linearized system is *not* controllable, since it admits the constrained linear observable

$$Z = \dot{Y} - \frac{U}{u},$$

which satisfies

$$\dot{Z} = 0,$$

although K is differentially algebraically closed in N. This result contradicts an announcement made in [65]. We finally notice that $\dot{Y} - (U/u)$ is nothing else but the linearized version of $\dot{y} - \log u$, and it is defined over N, although $\log u$ is not algebraic.

When $\alpha = 1$, we leave it to the reader to verify that the parametrization is:

$$\begin{cases} U = u^2 \ddot{\xi} + \dfrac{u^2 \ddot{u}}{\dot{u}+1} \dot{\xi} + u^2 \left(\overbrace{\dfrac{\ddot{u}}{\dot{u}+1}}^{\bullet} \right) \xi, \\ Y = u\dot{\xi} + \left[\dfrac{u\ddot{u}}{\dot{u}+1} - (\dot{u}+1) \right] \xi. \end{cases}$$

Example 3. With one input u and three outputs y^1, y^2, y^3 we consider the following nonlinear control system

$$\begin{cases} \dot{y}^1 = y^2 + y^3 u, \\ \dot{y}^2 = -y^1, \\ \dot{y}^3 = -y^1 u. \end{cases}$$

The two corresponding vector fields are

$$a = y^2 \frac{\partial}{\partial y^1} - y^1 \frac{\partial}{\partial y^2}, \qquad b = y^3 \frac{\partial}{\partial y^1} - y^1 \frac{\partial}{\partial y^3},$$

and we recognize two of the infinitesimal generators of the group of rotations in 3-space. Accordingly, we obtain

$$[a, b] = y^3 \frac{\partial}{\partial y^2} - y^2 \frac{\partial}{\partial y^3},$$

and we cannot construct new brackets. E.g., we have

$$[a, [a, b]] = -b, \qquad [b, [a, b]] = a.$$

If we look at controllability, Section A implies that we have to look at the rank of
the distribution generated by $b, [a, b], a$, i.e. the rank of the matrix

$$
\begin{pmatrix}
y^3 & 0 & y^2 \\
0 & y^3 & -y^1 \\
-y^1 & -y^2 & 0
\end{pmatrix}.
$$

Clearly, its generic rank is $2 < 3$. Accordingly, the control system is not controllable
and we have the constrained observable

$$
z = (y^1)^2 + (y^2)^2 + (y^3)^2,
$$

satisfying

$$
\dot{z} = 0.
$$

Now, if we look at invariance of this control system, we see that the control Lie
algebra is generated by $\{a, b, [a, b]\}$, i.e. by the 3 infinitesimal generators of the rotation
group. It thus remains to recover Θ from $C(\Theta)$. Recall that $[\Theta, C(\Theta)] = 0$. Hence,
setting $C_r^k(y)\partial/\partial y^k$ for the infinitesimal generators (with $r = 1, 2, 3$), we have

$$
C_r^k \frac{\partial \eta^l}{\partial y^k} - \eta^k \frac{\partial C_r^l(y)}{\partial y^k} = 0.
$$

Using the identity $y^r C_r^k(y) = 0$ already found, we obtain

$$
\eta^k y_r \frac{\partial C_r^l}{\partial y^r} = \eta^k \frac{\partial(y^r C_r^l)}{\partial y^k} - \eta^k C_k^l = 0 \Rightarrow \eta^k C_k^l(y) = 0.
$$

So we find that $\eta^k(y) = \lambda(y)y^k$ with:

$$
C_r^k(y)y^l \frac{\partial \lambda}{\partial y^k} = 0 \Rightarrow \lambda = \text{cst},
$$

and so Θ is generated by

$$
y^1 \frac{\partial}{\partial y^1} + y^2 \frac{\partial}{\partial y^2} + y^3 \frac{\partial}{\partial y^3},
$$

which is easily seen to be convenient.

Example 4 (Vessiot). We study the following OD control system with constraint:

$$
\begin{cases}
y_{xxx}^1 - p(x)y_{xx}^1 + q(x)y_x^1 - r(x)y^1 = 0, \\
y_{xxx}^2 - p(x)y_{xx}^2 + q(x)y_x^2 - r(x)y^2 = 0, \\
y_{xxx}^3 - p(x)y_{xx}^3 + q(x)y_x^3 - r(x)y^3 = 0, \\
(y^3)^2 - y^1 y^2 = 0,
\end{cases}
$$

where we consider linearly independent solutions only. In that case the first 3 ODE
prove that, for fixed p, q, r, one passes from one solution (y^1, y^2, y^3) to another $(\overline{y}^1, \overline{y}^2, \overline{y}^3)$
by the standard action of GL(3), as follows:

$$
\begin{pmatrix}
\overline{y}^1 \\
\overline{y}^2 \\
\overline{y}^3
\end{pmatrix}
=
\begin{pmatrix}
\alpha & \beta & \gamma \\
\alpha' & \beta' & \gamma' \\
\alpha'' & \beta'' & \gamma''
\end{pmatrix}
\begin{pmatrix}
y^1 \\
y^2 \\
y^3
\end{pmatrix},
$$

where $\alpha, \beta, \ldots, \gamma''$ are constants. Of course, adding the quadratic constraint results in shrinking down the group of invariance. In fact, setting

$$y^1 = u^2, \qquad y^2 = v^2, \qquad y^3 = uv,$$

we have found a way for giving a parametrization of the surface. We now compute the subgroup of GL(3) preserving the full control system with constraint. Although the computation is quite tedious, we do give the details.

First of all, from invariance we can deduce that

$$(\alpha'' y^1 + \beta'' y^2 + \gamma'' y^3)^2 - (\alpha y^1 + \beta y^2 + \gamma y^3)(\alpha' y^1 + \beta' y^2 + \gamma' y^3) = 0.$$

Separating the terms and using $(y^3)^2 - y^1 y^2 = 0$, we immediately obtain

$$(y^1)^2 \rightarrow (\alpha'')^2 - \alpha\alpha' = 0,$$
$$(y^2)^2 \rightarrow (\beta'')^2 - \beta\beta' = 0,$$
$$y^1 y^2 \rightarrow (\gamma'')^2 - \gamma\gamma' + 2\alpha''\beta'' - \alpha\beta' - \beta\alpha' = 0,$$
$$y^1 y^3 \rightarrow 2\alpha''\gamma'' - \alpha\gamma' - \alpha'\gamma = 0,$$
$$y^2 y^3 \rightarrow 2\beta''\gamma'' - \beta\gamma' - \beta'\gamma = 0.$$

Let us set

$$\alpha = a^2, \quad \beta = b^2, \quad \alpha' = (a')^2, \quad \beta' = (b')^2 \Rightarrow \alpha'' = aa', \quad \beta'' = bb'.$$

We are left with the three equations

$$\begin{cases} (\gamma'')^2 - \gamma\gamma' - (ab' - ba')^2 = 0, \\ 2aa'\gamma'' - a^2\gamma' - (a')^2\gamma = 0, \\ 2bb'\gamma'' - b^2\gamma' - (b')^2\gamma = 0. \end{cases}$$

Computing γ and γ' from the last two equations, we obtain

$$\gamma = \frac{2ab}{ab' + ba'}\gamma'', \qquad \gamma' = \frac{2a'b'}{ab' + ba'}\gamma''.$$

Substituting this into the first equation gives

$$\gamma = 2ab, \qquad \gamma' = 2a'b', \qquad \gamma'' = ab' + ba',$$

i.e.

$$\bar{u}^2 = \bar{y}^1 = \alpha y^1 + \beta y^2 + \gamma y^3 = a^2 u^2 + b^2 u^2 + 2abuv,$$

and finally,

$$\begin{cases} \bar{u} = au + bv, \\ \bar{v} = a'u + b'v. \end{cases}$$

It follows that the Lie group of invariance reduces from $GL(3)$ to $GL(2)$ when the constraint is taken into account. The orbits can be described using the well-known basis of differential invariants

$$\frac{\begin{vmatrix} u & u_{xx} \\ v & v_{xx} \end{vmatrix}}{\begin{vmatrix} u & u_x \\ v & v_x \end{vmatrix}} = M(x), \qquad \frac{\begin{vmatrix} u_x & u_{xx} \\ v_x & v_{xx} \end{vmatrix}}{\begin{vmatrix} u & u_x \\ v & v_x \end{vmatrix}} = N(x).$$

Equivalently, u and v must be solutions of the *same* second-order linear ODE, i.e.

$$\begin{cases} u_{xx} - M(x)u_x + N(x)u = 0, \\ v_{xx} - M(x)v_x + N(x)v = 0. \end{cases}$$

Similarly, solving the first 3 ODE in the control system, which are linear in p, q, r, while taking into account the nonzero Wronskian condition for y^1, y^2, y^3, we obtain, e.g.,

$$\frac{\begin{vmatrix} y^1 & y^1_x & y^1_{xxx} \\ y^2 & y^2_x & y^2_{xxx} \\ y^3 & y^3_x & y^3_{xxx} \end{vmatrix}}{\begin{vmatrix} y^1 & y^1_x & y^1_{xx} \\ y^2 & y^2_x & y^2_{xx} \\ y^3 & y^3_x & y^3_{xx} \end{vmatrix}} = p(x).$$

The quotients of the determinants is a differential invariant for the action of $GL(3)$. Hence, replacing y^1, y^2, y^3 by their expressions in u, v, the resulting rational function of (u, v) must be a differential invariant for the action of $GL(2)$. Hence it can be expressed as a function of M, N and their derivatives. Using

$$y^1 = u^2, \qquad y^1_x = 2uu_x, \qquad y^1_{xx} = 2uu_{xx} + 2(u_x)^2, \ldots$$

and substituting

$$u_{xx} = M(x)u_x - N(x)u,$$

we finally get

$$\begin{cases} p = 3M, \\ q = -\partial_x M + 2M^2 + 4N, \\ r = -2\partial_x N + 4MN. \end{cases}$$

This computation is compatible with the Wronskian assumption, as indeed

$$\begin{vmatrix} y^1 & y^1_x & y^1_{xx} \\ y^2 & y^2_x & y^2_{xx} \\ y^3 & y^3_x & y^3_{xx} \end{vmatrix} = -2 \left(\begin{vmatrix} u & u_x \\ v & v_x \end{vmatrix} \right)^3.$$

So, we have obtained the following commutative diagram, in which the order of the operators is boxed:

$$(u, v) \xrightarrow{\boxed{2}} (M, N)$$
$$\boxed{0} \downarrow \qquad\qquad \downarrow \boxed{1}$$
$$(y^1, y^2, y^3) \xrightarrow{\boxed{3}} (p, q, r)$$

The compatibility conditions for the right vertical operator are obtained as follows: The first equation implies $M = p/3$. Substituting this into the second equation, we obtain

$$4N = q + \frac{1}{3}\partial_x p - \frac{2}{9}p^2.$$

Substituting this into the last equation gives

$$9\partial_{xx}p - 18p\partial_x p - 27\partial_x q + 4(p)^3 - 18pq + 54r = 0,$$

i.e. the desired constraint on the input.

Example 5. The following linear PD control system, provided by M. Janet is quite striking. Let us consider two independent variables (x^1, x^2), two inputs (y^1, y^2), and two outputs (z^1, z^2), linked by the mixed system

$$R_2 \quad \begin{cases} z^1_{11} + z^2_{12} - z^2 - y^1 = 0, \\ z^1_{12} + z^2_{22} + z^1 - y^2 = 0. \end{cases}$$

Regarded as a system in y, we could give z arbitrary, since the first equation gives y^1 and the second y^2. Hence the system is formally integrable of order zero.

The situation is rather different when we regard the system as one in z. Although the characteristic determinant (in z) is

$$\det \begin{pmatrix} (\chi_1)^2 & \chi_1\chi_2 \\ \chi_1\chi_2 & (\chi_2)^2 \end{pmatrix} = 0,$$

the system is nevertheless not formally integrable. In Chapters II and IV we have seen that we can invert the system as follows:

$$\begin{cases} y^1_{11} + y^2_{12} - y^2 - z^1 = 0, \\ y^1_{12} + y^2_{22} + y^1 - z^2 = 0. \end{cases}$$

Using vector bundles E and F instead of the fibered manifold $\mathcal{E}$ and $\mathcal{F}$, we have, for $q \geq 2$:

$$\mathcal{P}_q = J_q(E), \qquad \mathcal{Q}_q = J_q(F), \qquad \mathcal{R}_q \subset J_q(E \times_X F),$$

and so there is no resolvent system for the input or output.

Since the system has constant coefficients, we can use the trivial differential field $K = \mathbb{Q}$. Over it,

$$L = K\langle y \rangle, \qquad M = K\langle z \rangle,$$

although there is a double isomorphism

$$L \simeq M \simeq N$$

which makes sense only if L and M are put into N via the control system.

The control system is both left and right invertible in our sense *and* in the classical sense, although it is nontrivial. This situation can appear only in the setting of PD control systems.

Example 6 (Bertrand curves). Imagine two curves, γ and γ' in 3-space, with corresponding points M and M', such that if $\vec{t}, \vec{t}'$ are the tangent unit vectors and $\vec{n}, \vec{n}'$ are the normal unit vectors to γ and γ' at M and M', respectively, then

$$|\vec{MM'}| = a = \text{cst}, \qquad \vec{t} \cdot \vec{t}' = 0, \qquad \vec{MM'} \parallel \vec{n} \parallel \vec{n}'.$$

Analytical geometry with Cartesian coordinates (x, y, z) for M and (x', y', z') for M' gives functions of a common parameter through the curvilinear abcisses s and s'. The difficult problem now is to eliminate (x, y, z) or (x', y', z') in order to find the '*resolvent system*' for each curve. Quite fortunately, this difficult problem in differential algebraic elimination theory has a very simple geometric solution, which we now give.

Using the Frénet formulas, we have

$$\vec{MM'} = a\vec{n}, \qquad \frac{d\vec{t}}{ds} \parallel \vec{n}, \qquad \frac{d\vec{n}}{ds} = -\kappa\vec{t} + \tau\vec{b},$$

where $(\vec{t}, \vec{n}, \vec{b})$ is a direct orthonormal frame at M and κ is the *curvature* of γ at M. Combining the assumptions, we thus have

$$\frac{d\vec{M'}}{ds'} \cdot \frac{ds'}{ds} = \frac{d\vec{M}}{ds} + a\frac{d\vec{n}}{ds} \Rightarrow \frac{ds'}{ds}\vec{t}' = \vec{t} + a(-\kappa\vec{t} + \tau\vec{b}),$$

$$\vec{t} \cdot \vec{t}' = 0, \quad \vec{t} \cdot \vec{b} = 0 \Rightarrow a\kappa = 1, \quad a\kappa' = 1.$$

So, both curves must have curvature $1/a$. The simplest example is that of two helices on a cylinder of diameter a; namely, $z = \frac{a}{2}\theta$ and $z' = \frac{a}{2}(\theta + \pi)$ in cylindrical coordinates.

Example 7 (Lie–Bianchi surfaces). Let us have two surfaces, Σ and Σ', in 3-space, with points M and M' having Cartesian coordinates (x, y, z) and (x', y', z'), respectively, such that when

$$dz = p\, dx + q\, dy, \qquad dz' = p'\, dx' + q'\, dy',$$

then we have

$$\begin{cases} (x - x')^2 + (y - y')^2 + (z - z')^2 = a^2, \\ p(x - x') + q(y - y') - (z - z') = 0, \\ p'(x - x') + q'(y - y') - (z - z') = 0, \\ pp' + qq' + 1 = 0. \end{cases}$$

There is a common parametrization by coordinates (u, v), leading to a very delicate problem in differential elimination. Its delicate computation is given in [145] and will not be repeated. Using the standard notation r, s, t for the second-order derivatives of z with respect to x and y, we must have

$$a(rt - s^2) + 1 + p^2 + q^2 = 0$$

as an intrinsic constant curvature condition for Σ, and similarly for Σ' since the result obtained by the elimination procedure is independent of the parametrization.

Example 8 (Korteweg–de Vries equation). This Example is presented in such a way that scientists working in the field of solitons are warned for misleading facts and ideas.

Consider two independent variables (x^1, x^2) with $x^1 = x$ =space, $x^2 = t$ =time, one input y and one output z, related as follows via the nonlinear PD control system consisting of two differential polynomials:

$$\mathcal{R}_2 \quad \begin{cases} P_1 \equiv 2y + z_1 + (z)^2 = 0, \\ P_2 \equiv y_{11} + 2y(z)^2 + 4(y)^2 - 2zy_1 - \frac{1}{2}z_2 = 0. \end{cases}$$

The input and output may be solitary waves, or *solitons*, constrained by resolvent systems that we are about to determine.

The first PDE may be solved with respect to z_1, while the second may be solved with respect to z_2. After crossed prolongation we find equations that can only contain z. Luckily, z disappears, and we are left with the input resolvent system

$$\mathcal{P}_3 \quad y_{111} + 12yy_1 + y_2 = 0,$$

which is also called the *Korteweg–de Vries equation*. The elimination of y is much easier, since the first differential polynomial allows us to express y as a function of z and z_1. Substituting this into the second differential polynomial gives the following output resolvent system, also called the *modified Korteweg–de Vries equation*:

$$\mathcal{Q}_3 \quad z_{111} - 6(z)^2 z_1 + z_2 = 0.$$

Notice that $\mathcal{R}_2$ is *not* formally integrable, but that $\mathcal{R}_2^{(1)}$ is involutive. Now, if $\mathcal{R}_2$ is regarded as a system for the output with input given and satisfying $\mathcal{P}_3$, then it is involutive and even of finite type. This is *purely accidental*, and we should not search for a general idea underlying the existence of any *connection*; namely, a map from the zero jet of z to the first jets of z that gives the identity on the zero jet of z when followed by projection [142]. So, it is also *purely accidental* that the original *direct Bäcklund problem*, i.e. the search for individual properties of two surfaces with first-order constant element related by requirement (like Bertrand curves or the Lie–Bianchi Example considered above), involves 2 independent variables, 2 unknowns, and 4 equations (two of these identify the independent variables, given a common parametrization of the two surfaces). The original Bäcklund problem is concerned with the study of a differential correspondence, and *the number of independent or dependent variables may be arbitrary*: this is a typical example of a PD control system, *and nothing else*. Now, the system for the input with output given and satisfying $\mathcal{Q}_3$ is *not* formally integrable, but, once again, *its left invertibility is purely accidental*.

In our usual picture of fields we have, *in this case*, $K \subset L \subset M = N$. As a byproduct, the so-called notion of *pseudopotential* or *covering* [97] is nothing else but an attempt to bring the elementary concept of *differential extension* in differential algebra ... back to differential geometry, but *without quoting its evident origin*. Let us see how these simple facts relate to the above Example.

Indeed, L/K is defined by $\mathcal{P}_3$. We may then look for a system of the type

$$\begin{cases} z_1 = F(z; y, \dots), \\ z_2 = G(z; y, \dots), \end{cases}$$

where, possibly, jets of y are introduced into F and G. This system must be involutive if y is a solution of $\mathcal{P}_3$. The reader who has matured Chapter V needs no comment in order to understand the *actual* method just described, and had better call it the *inverse Bäcklund problem*. Of course, *there is no way of prescribing the form of the equations defining M/L*. The above approach, using a connection, is just one among thousands. The only possibility would be to fix a type of system defining the extension (it should contain arbitrary functions) and to use the formal integrability criterion to adapt these functions to the desired target. Of course, when both L/K and M/K are given, there are infinitely many possibilities for N/K such that N contains both M and L, *this is truly the inverse Bäcklund problem*, unless one restricts the type of PDE appearing in the differential correspondence. In any case, because of the need for a direct verification, the *inverse Bäcklund problem cannot be solved if we do not know how to solve the direct Bäcklund problem*.

Another reason for refusing the need of connections is the perfectly symmetrical role that must be played by input and output in any OD/PD control system.

However, a differential correspondence may be used as usual. Namely, if we pick a solution of the input resolvent system and substitute it into the differential correspondence, then we find a *subsystem* of the output resolvent system, and this may help in determining solutions of the output resolvent system.

In our Example, it is obvious that $y = 0$ is a solution of $\mathcal{P}_3$. Substitution of it into $\mathcal{R}_2$ gives the subsystem

$$z_1 + (z)^2 = 0, \qquad z_2 = 0$$

of $\mathcal{Q}_3$, for indeed

$$z_{11} = -2zz_1 = 2(z)^3 \Rightarrow z_{111} = -6(z)^4 = 6(z)^2 z_1.$$

We finally notice that the two resolvent systems may be identical. In that case we will speak of a *differential autocorrespondence*.

The best known example of this is given by the sine-Gordon first-order differential correspondence

$$\mathcal{R}_1 \quad \begin{cases} y_1 + z_1 - \sin(y - z) = 0, \\ y_2 - z_2 - \sin(y + z) = 0. \end{cases}$$

It gives rise to the two second-order resolvent systems

$$\mathcal{P}_2 \quad y_{12} - \tfrac{1}{2}\sin 2y = 0,$$
$$\mathcal{Q}_2 \quad z_{12} - \tfrac{1}{2}\sin 2z = 0,$$

which are identical. However, in that case differential algebraic methods can provide an intuitive understanding of a few concepts only, since we do not deal with differential polynomials.

We end this Example by pointing out that the phrase *Bäcklund transformation* are quite misleading, ... since it has the exact meaning of standard *arrow* in ordinary differential control theory!

Example 9 (Burgers equation). Let $(x, t) =$ (position, time) be two independent variables with one input y and one output z satisfying the Cole–Hopf correspondence

$$2z_x + y = 0, \qquad 4z_t + 2y_x - y^2 = 0.$$

The resolvent systems are

$$y_t + yy_x - y_{xx} = 0$$

(i.e. the Navier–Stokes equation for a 1-dimensional flow, also called the Burgers equation) and

$$z_t - (z_x)^2 - z_{xx} = 0,$$

which becomes the heat equation

$$\theta_t - \theta_{xx} = 0$$

if we set $z = \log \theta$. The control system is involutive in z if y satisfies the input resolvent system; it is *not* involutive in y, even if z satisfies the output resolvent system, since we have a zero-order equation and a first-order equation. However, the control system is involutive in (y, z), since it can be solved with respect to the two jets (y_x, z_x) at the same time, and so it is not possible to perform any crossed derivative prolongation. Of course, since both y and z are differentially algebraic over $\mathbb{Q}$, the control system is totally uncontrollable. This property can be verified directly by looking at the differential transcendence degree, which is zero because the characteristic matrix

$$\begin{pmatrix} 0 & 2\chi_x \\ 2\chi_x & 4\chi_t \end{pmatrix}$$

has generic rank 2. The restriction of the characteristic matrix to the output has generic rank one, which equals the number of outputs. So the control system is also causal. We can choose z as primitive element determined by its resolvent system, and the system is left invertible because $y = -2z_x$. The input is defined over the output by one zero-order and two first-order PDE, obtained by prolongation, when the output satisfies the output resolvent system.

Example 10 (Maxwell equations). Here we will regard the second set of Maxwell equations as a PD control system. Let $(x^1, x^2, x^3, x^4 = t)$ be space-time variables, $(\vec{j}, \rho)$ be the source input, with $\vec{j}$ the density of electric current and ρ the density of electric charge, and let $(\vec{H}, \vec{D})$ be the induction output, where $\vec{H}$ is magnetic induction and $\vec{D}$ is electric induction.

The well-known corresponding PD control system is

$$\begin{cases} \vec{\nabla} \wedge \vec{H} - \partial_t \vec{D} = \vec{j}, \\ \vec{\nabla} \cdot \vec{D} = \rho. \end{cases}$$

As a matter of fact, taking the space-time divergence allows us to eliminate $(\vec{H}, \vec{D})$ and obtain the PDE describing conservation of current, namely:

$$\frac{\partial \rho}{\partial t} + \vec{\nabla} \cdot \vec{j} = 0.$$

No differential condition can relate $\vec{H}$ and $\vec{D}$ themselves, because, given any couple $(\vec{H}, \vec{D})$, the two PDE of the control system make it possible to determine $(\rho, \vec{j})$ individually.

It follows that any observable, as a function of $(\rho, \vec{j}, \vec{H}, \vec{D})$ and their derivatives, can indeed be regarded as a function of $(\vec{H}, \vec{D})$, which are free. Hence any observable is free and the system is controllable.

With $K = \mathbb{Q}$, the input/output differential field is indeed $N \simeq K\langle \vec{H}, \vec{D}\rangle$, hence the differential transcendence degree of N/K is 6. We will compute this number directly for the control system brought to formal integrability *in all unknowns* by adding the condition of conservation of current, which is also a first-order PDE. The characteristic matrix along the order $\rho < \vec{j} < \vec{H} < \vec{D}$ is:

$$\begin{pmatrix} \chi_1 & \chi_2 & \chi_3 & \chi_4 & 0 & 0 & 0 & 0 & 0 & 0 \\ 0 & 0 & 0 & 0 & 0 & -\chi_3 & \chi_2 & -\chi_4 & 0 & 0 \\ 0 & 0 & 0 & 0 & \chi_3 & 0 & -\chi_1 & 0 & -\chi_4 & 0 \\ 0 & 0 & 0 & 0 & -\chi_2 & \chi_1 & 0 & 0 & 0 & -\chi_4 \\ 0 & 0 & 0 & 0 & 0 & 0 & 0 & \chi_1 & \chi_2 & \chi_3 \end{pmatrix}$$

Multiplying the columns by $(0, \chi_1, \chi_2, \chi_3, \chi_4)$ and summing, we obtain a zero column. Hence, the generic rank is 4, which proves that the differential transcendence degree is $10 - 4 = 6$, as required.

The restriction of the above matrix to the output is given by the 6 columns on the right, and has rank 3. Hence the control system is not causal. For this reason, to obtain causality we must add the first set of Maxwell equations *and* the constitutive relations, in order to use only the potential, and even in that case we must also add the Lorentz divergence conditions on the 4-potential.

Example 11 (Euler equations for an incompressible fluid). Consider the following PD control system on space-time:

$$\partial_t \vec{v} + (\vec{v} \cdot \vec{\nabla})\vec{v} + \vec{\nabla}p = 0, \qquad \vec{\nabla} \cdot \vec{v} = 0,$$

where p denotes pressure and $\vec{v}$ the velocity of the fluid. For simplicity we have taken the mass per unit volume equal to 1.

For a 1-dimensional space we have

$$\partial_t v + v \partial_x v + \partial_x p = 0, \qquad \partial_x v = 0.$$

Hence, differentiation with respect to x gives

$$\partial_x v = 0, \qquad \partial_{xx} p = 0.$$

It follows that p and v are both differentially algebraic, and the system is totally uncontrollable.

For a 2-dimensional space the situation is more complicated. Let us write out the control system as follows:

$$\begin{cases} \partial_t v^1 + v^1 \partial_1 v^1 + v^2 \partial_2 v^1 + \partial_1 p = 0, \\ \partial_t v^2 + v^1 \partial_1 v^2 + v^2 \partial_2 v^2 + \partial_2 p = 0, \\ \partial_1 v^1 + \partial_2 v^2 = 0. \end{cases}$$

Taking the derivative with respect to x^1 of the second PDE, with respect to x^2 of the first, and subtracting, we find

$$\partial_t (\partial_1 v^2 - \partial_2 v^1) + v^1 \partial_{11} v^2 + v^2 \partial_{12} v^2 - v^1 \partial_{12} v^1 - v^2 \partial_{22} v^1 = 0,$$

since we have eliminated pressure. Differentiating again with respect to x^2 and using $\partial_2 v^2 = -\partial_1 v^1$, we find

$$\partial_t \Delta v^1 + v^1 \Delta \partial_1 v + v^2 \Delta \partial_2 v^1 - \partial_1 v^1 \Delta v^1 - \partial_2 v^1 \Delta v^2 = 0,$$

i.e. a PDE of the form

$$(\partial_2 v^1) \Delta v^2 - (\Delta \partial_2 v^1) v^2 = A(v^1),$$

with A a nonlinear operator acting on v^1. Differentiating this PDE with respect to x^2 we find

$$(\partial_{22} v^1) \Delta v^2 - (\Delta \partial_{22} v^1) v^2 = B(v^1),$$

with B a nonlinear operator acting on v^1. Solving this linear system in $(v^2, \Delta v^2)$ gives

$$v^2 = C(v^1), \qquad \Delta v^2 = D(v^1),$$

with, again, C and D nonlinear operators acting on v^1. Hence, v^1 must satisfy the 6th order PDE

$$\Delta C - D = 0.$$

A similar computation can be done with v^2. The computation for p is left to the reader as a (tricky) exercise, and also leads to a 6th order PDE.

The result for 3-space is still open.

Notice that *elimination does not commute with linearization.* Indeed, after linearization, where we may drop the quadratic term in velocity, we obtain

$$\partial_t \vec{v} + \vec{\nabla} p = 0, \qquad \vec{\nabla} \cdot \vec{v} = 0,$$

i.e. by successively taking the divergence and the Laplacian of the first PDE:

$$\Delta p = 0, \qquad \partial_t \Delta v^1 = 0, \qquad \partial_t \Delta v^2 = 0,$$

i.e. 2nd and 3rd order PDE instead of the 6th order ones above.

We will now study the characteristic matrix in the case of 3-space. The method can, however, be extended to arbitrary n-space. We have $n+1$ equations for $n+1$ unknowns, and the matrix becomes

$$\begin{pmatrix} \chi_1 & \chi_2 & \chi_3 & 0 \\ \chi_4 + \vec{\chi} \cdot \vec{v} & 0 & 0 & \chi_1 \\ 0 & \chi_4 + \vec{\chi} \cdot \vec{v} & 0 & \chi_2 \\ 0 & 0 & \chi_4 + \vec{\chi} \cdot \vec{v} & \chi_3 \end{pmatrix},$$

where we have set $\chi^4 = t$, $\vec{\chi} \cdot \vec{v} = \chi_1 v^1 + \chi_2 v^2 + \chi_3 v^3$. The determinant of this matrix is

$$(\chi_4 + \vec{\chi} \cdot \vec{v})^2 (\chi_1^2 + \chi_2^2 + \chi_3^2).$$

Hence, its generic rank is 3, showing that the system is determined and thus formally integrable. Accordingly, $\alpha_1^4 = 0$ and the differential transcendence degree is zero, although the coordinate system is not δ-regular because $\chi_1 = \chi_2 = \chi_3 = 0$ implies that the characteristic determinant vanishes. Hence $t = 0$ does not fit the initial data, in contrast to what could be believed from this 'looking-like' evolution equation. It also follows that any observable is constrained in arbitrary dimension, and so the control system is totally uncontrollable. This result seems to be neither known nor used up till now. However, we do believe that it will become a key tool for understanding turbulence [112].

Retaining the mass per unit volume ρ and adding the equation

$$\frac{\partial \rho}{\partial t} + (\vec{v} \cdot \vec{\nabla})\rho = 0$$

leads to the necessity of multiplying the characteristic determinant by $\chi_4 + \vec{\chi} \cdot \vec{v}$, and does not lead to a change in the nature of the above results.

Example 12 (Bénard cells). This Example, perhaps one of the most fascinating from hydrodynamics, will be treated with great detail, since it is one of the best examples for illustrating the usefulness of applying the new techniques of differential algebraic geometry in engineering physics. In fact, we will follow the classical computations, to be found in [36, 104], but the reader will notice that the new interpretation provided below can be adapted to other situations.

Bénard discovered in 1900 the following basic phenomenon: there appear stationary convective cells of hexagonal section in a thin layer of fluid between two horizontal plates at different temperatures and heated from below, whenever a certain critical constant temperature gradient is applied and the Rayleigh number (a dimensionless number to be defined below) is sufficiently large.

Assume that the lower plate is at temperature T_0, the upper plate is at temperature T_1, their distance being L, $\Theta = AgL = T_0 - T_1 > 0$, where g is gravity, and $x^3 = z$ is the vertical axis of cartesian coordinates. We will use the following notations:

ρ: mass per unit volume
$v = 1/\rho$:   specific volume
p: pressure
T: absolute temperature
u: specific internal energy
s: specific entropy
$c_p = c_p(T, p)$:   heating coefficient at constant pressure
$\vec{g}$:   gravity
η: viscosity $(\Rightarrow \nu = \eta/a)$
κ: thermal conductivity (constant) $(\Rightarrow \lambda = \kappa/ac_p)$
α: thermal dilatation coefficient (constant)
$\vec{v}$:   velocity

In this study we neglect the isothermic compressibility of the liquid, and we set

$$\rho = a(1 - \alpha(T - T_0)) \qquad \text{with } a = \rho(T_0).$$

First of all, if the fluid is in mechanical equilibrium, the hydrostatic law

$$-\vec{\nabla}p + \rho\vec{g} = 0$$

leads to $p = p(z)$ and $\rho = (1/g)\partial p/\partial z$, which implies $T = T(p, \rho) = T(z)$ in any state equation relating (p, ρ, T).

Recall the local version of the first principle of thermodynamics,

$$du = -p\,dv + c_p\,dT + h\,dp$$

as well as the second principle,

$$ds = \frac{c_p}{T}\,dT + \frac{h}{T}\,dp.$$

Looking at the crossed derivatives, we have

$$\begin{cases} \dfrac{\partial h}{\partial T} + \dfrac{\partial v}{\partial T} = \dfrac{\partial c_p}{\partial p}, \\[2mm] \dfrac{\partial h}{\partial T} - \dfrac{h}{T} = \dfrac{\partial c_p}{\partial p}, \end{cases}$$

and hence follows the Clapeyron formula

$$h = -T\left(\frac{\partial v}{\partial T}\right)_p.$$

Using the above principle, we have

$$v = v(T(p, s), p) = v(p, s).$$

In a spontaneous ascending adiabatic movement, a unit mass element passing from z to $z + dz$ must be heavier than the element that occupied this place before in order to attain equilibrium (Archimedes' effect). So,

$$v(p + dp, s + ds) - v(p + dp, s) > 0 \Rightarrow \left(\frac{\partial v}{\partial s}\right)_p \frac{ds}{dz} > 0 \qquad \text{because } dz > 0.$$

The implicit function theorem now implies

$$\left(\frac{\partial v}{\partial s}\right)_p = \left(\frac{\partial v}{\partial T}\right)_p \left(\frac{\partial T}{\partial s}\right)_p = \frac{T}{c_p}\left(\frac{\partial v}{\partial T}\right)_p,$$

with $c_p > 0$, $v > 0$, $T > 0$, $\left(\frac{\partial v}{\partial T}\right)_p > 0$, and finally

$$\frac{ds}{dz} > 0,$$

i.e.

$$\frac{ds}{dz} = \left(\frac{\partial s}{\partial T}\right)_p \frac{dT}{dz} + \left(\frac{\partial s}{\partial p}\right)_T \frac{dp}{dz}.$$

From the thermodynamic principles we obtain

$$\left(\frac{\partial s}{\partial T}\right)_p = \frac{c_p}{T}, \qquad \left(\frac{\partial s}{\partial p}\right)_T = \frac{h}{T} = -\left(\frac{\partial v}{\partial T}\right)_p,$$

and so we find the equilibrium condition

$$\frac{dT}{dz} \geq -\frac{gT}{vc_p}\left(\frac{\partial v}{\partial T}\right)_p.$$

We may now introduce the rate of deformation:

$$v_{ij} = \frac{1}{2}(\partial_i v_j + \partial_j v_i),$$

and recall that viscous irreversibility gives rise to the following rate of change of entropy under the Stokes assumption for viscosity:

$$\rho T\dot{s} = \kappa\Delta T + 2\eta v^{ij}v_{ij} - \frac{2}{3}\eta(\vec{\nabla}\cdot\vec{v})^2.$$

This equation must be combined with

$$T\dot{s} = c_p\dot{T} - T\left(\frac{\partial v}{\partial T}\right)_p \dot{p}.$$

Finally we have to take into account the conservation of mass

$$\frac{\partial\rho}{\partial t} + (\vec{v}\cdot\vec{\nabla})\rho + \rho(\vec{\nabla}\cdot\vec{v}) = 0$$

and the Navier–Stokes equation for a Newtonian viscous fluid:

$$-\vec{\nabla}p + \eta\Delta\vec{v} + \frac{\eta}{3}\vec{\nabla}(\vec{\nabla}\cdot\vec{v}) + \rho\vec{g} = \rho\left(\frac{\partial\vec{v}}{\partial T} + (\vec{v}\cdot\vec{\nabla})\vec{v}\right).$$

We linearize these equations around a steady mechanical and thermal equilibrium state, denoted by a 'bar' over the corresponding letters. We have:

$$-\vec{\nabla}\bar{p} + \bar{\rho}\vec{g} = 0, \qquad \Delta\bar{T} = 0,$$

and so (see the equilibrium condition above)

$$\overline{T} = -Az + T_0 \qquad \text{with } A > 0.$$

This leads to

$$\overline{p}(z) = a(1 + \alpha Az), \quad \overline{p}(z), \quad \overline{T}(z).$$

We may set

$$T - T_0 = (T - \overline{T}) + (\overline{T} - T_0) = \theta - Az$$

and obtain

$$\rho = \overline{\rho} - \alpha a\theta = a(1 - \alpha(\theta - Az)).$$

With zero reference velocity we can introduce the perturbations (variations) $p = \overline{p} + \pi$, $T = \overline{T} + \theta$, and we finally obtain

$$\alpha \left(Av_3 - \frac{d\theta}{dt} \right) + (1 - \alpha(\theta - Az))\vec{\nabla} \cdot \vec{v} = 0.$$

The Clapeyron formula now gives

$$\frac{\partial c_p}{\partial p} = -T \left(\frac{\partial^2 v}{\partial T \partial T} \right)_p = \frac{2\alpha^2 T}{a(1 - \alpha(\theta - Az))^3},$$

and so

$$c_p = c(T) + \frac{2\alpha^2 Tp}{a(1 - \alpha(T - T_0))^3}.$$

Substituting this into the Navier–Stokes equation and in the heat equation, we obtain

$$\begin{cases} a(1 - \alpha(T - T_0))c_p\dot{T} - \dfrac{\alpha T}{1 - \alpha(T - T_0)}\dot{p} = \kappa\Delta T + \eta v^{ij}v_{ij} - \dfrac{2}{3}\eta(\vec{\nabla}\cdot\vec{v})^2, \\ -\vec{\nabla}p + \eta\Delta\vec{v} + \dfrac{\eta}{3}\vec{\nabla}(\vec{\nabla}\cdot\vec{v}) + a(1 - \alpha(T - T_0))\vec{g} = a(1 - \alpha(T - T_0))\dot{\vec{v}}, \end{cases}$$

with, respectively,

$$\begin{cases} \vec{\nabla}p = \vec{\nabla}\overline{p} + \vec{\nabla}\pi = a(1 + \alpha Az)\vec{g} + \vec{\nabla}\pi, \\ \dot{p} = \dot{\overline{p}} + \dot{\pi} = (\vec{v}\cdot\vec{\nabla})\overline{p} + \dot{\pi} = a(1 + \alpha Az)\vec{v}\cdot\vec{g} + \dot{\pi}, \\ -\vec{\nabla}p + a(1 - \alpha(T - T_0))\vec{g} = -\alpha a\theta\vec{g} - \vec{\nabla}\pi. \end{cases}$$

Introducing the distance L between the plates and setting $c = c(T_0)$, we may introduce the dimensionless quantities 'with accent'

$$\pi' = \frac{\pi}{\Pi}, \quad \theta' = \frac{\theta}{\Theta}, \quad x' = \frac{x}{L}, \quad v'^i = \frac{v^i}{V}, \quad t' = \frac{t}{\tau},$$

where

$$V = \frac{\kappa}{a^2 c^2 L}, \quad \Pi = aV^2, \quad \tau = \frac{L}{V}.$$

We also introduce the following dimensionless quantities:

$$\begin{cases} \epsilon_1 = \alpha\Theta, & \epsilon_2 = \dfrac{\kappa^2}{a^2 c^3 L^2 \Theta}, \\[2mm] \text{Prandtl number} \quad P = \dfrac{\eta c}{\kappa}, \\[2mm] \text{Rayleigh number} \quad R = \dfrac{g\alpha A L^4 a^2 c}{\eta\kappa}. \end{cases}$$

Experiment shows that

$$\epsilon_1 \approx 10^{-4}, \quad \epsilon_2 \approx 10^{-11}, \quad P \approx 1, \quad R \approx 10^3.$$

Accordingly, we may look for Taylor expansions in ϵ_1 and ϵ_2 that depend on P and R.

Using the variables introduced above, in cartesian coordinates $(\vec{\rho}_1, \vec{\rho}_2, \vec{\rho}_3)$ the control system becomes

$$-\epsilon_1\left(\frac{d\theta'}{dt'} - v'^3\right) + (1 - \epsilon_1(\theta' - x'^3))(\vec{\nabla}' \cdot \vec{v}\,') = 0,$$

$$(1 - \epsilon_1(\theta' - x'^3))\frac{d\vec{v}\,'}{dt'} = -\vec{\nabla}' \cdot \pi' + P R \theta' \vec{e}_3 + P(\Delta'\vec{v}\,' + \vec{\nabla}'(\vec{\nabla} \cdot \vec{v}\,')),$$

$$(1 - \epsilon_1(\theta' - x'^3))c_p'\left(\frac{d\theta'}{dt'} - v'^3\right) = \Delta'\theta' + \epsilon_2 P\left(v'^{ij}v'_{ij} - \frac{2}{3}(\vec{\nabla}' \cdot \vec{v}\,')^2\right)$$

$$+ \frac{T}{(1 - \epsilon_1(\theta' - x'^3))}\left(\epsilon_1\frac{d\pi'}{dt'} - P R(1 + \epsilon_1 x'^3)v'^3\right).$$

We will now look for a solution $(\pi', \theta', \vec{v}\,')$ of the form

$$\begin{cases} \pi' = \pi'_{0,0} + \pi'_{1,0}\epsilon_1 + \pi'_{0,1}\epsilon_2 + \dots, \\[1mm] \theta' = \theta'_{0,0} + \theta'_{1,0}\epsilon_1 + \theta'_{0,1}\epsilon_2 + \dots, \\[1mm] \vec{v}\,' = \vec{v}\,'_{0,0} + \vec{v}\,'_{1,0}\epsilon_1 + \vec{v}\,'_{0,1}\epsilon_2 + \dots. \end{cases}$$

After substitution, the PDE at order zero in (ϵ_1, ϵ_2) becomes

$$\begin{cases} \vec{\nabla}' \cdot \vec{v}\,'_{0,0} = 0, \\[2mm] \dfrac{d\vec{v}\,'_{0,0}}{dt'} = -\nabla'\pi'_{0,0} + P R \theta'_{0,0}\vec{e}_3 + P\Delta'\vec{v}\,'_{0,0}, \\[2mm] c^*\left(\dfrac{d\theta'_{0,0}}{dt'} - v'^3\right) = \Delta'\theta'_{0,0}, \end{cases}$$

where we have set $c^* = c(T_0 + \theta_{0,0})/c$.

Returning to the initial notations, we discover that *in a first approximation with respect to (ϵ_1, ϵ_2), the study of free convection can be performed using the Boussinesq* *system*

$$\begin{cases} \vec{\nabla} \cdot \vec{v} = 0, \\ \dfrac{d\vec{v}}{dt} = \nu \Delta \vec{v} - \alpha \theta \vec{g} - \dfrac{1}{a} \vec{\nabla} \pi, \\ \dfrac{d\theta}{dt} = \lambda \Delta \theta - \dfrac{A}{g} \vec{g} \cdot \vec{v}. \end{cases}$$

If $\Theta = T_1 - T_0$ is not too large, then by the equilibrium condition found above the fluid is at rest and there is a pure conduction. Otherwise convection starts at a certain critical value of Θ, or rather of R (in fact, R > 1710).

Neglecting the terms $(\vec{v} \cdot \vec{\nabla})\vec{v}$ and $(\vec{v} \cdot \vec{\nabla})\theta$, we obtain the stationary linear PDE

$$\begin{cases} \vec{\nabla} \cdot \vec{v} = 0, \\ \nu \Delta \vec{v} - \dfrac{1}{a} \vec{\nabla} \pi - \alpha \theta \vec{g} = 0, \\ \lambda \Delta \theta - \dfrac{A}{g} \vec{g} \cdot \vec{v} = 0. \end{cases}$$

The boundary conditions on the plates are

$$\theta = 0, \qquad v^3 = 0, \qquad \frac{\partial v^3}{\partial x^3} = 0,$$

the last equation coming from $\vec{\nabla} \cdot \vec{v} = 0$ and the fact that the viscous fluid is 'attached' to the plates with $v^1 = v^2 = 0$ for any x^1, x^2 on each plate. We obtain, in succession,

$$\vec{\nabla} \wedge (\nu \Delta \vec{v}) = \nu \Delta (\vec{\nabla} \wedge \vec{v}) = \vec{\nabla} \wedge (\alpha \theta \vec{g}),$$

$$\nu \Delta (\vec{\nabla} \wedge \vec{\nabla} \wedge \vec{v}) = \nu \Delta (\vec{\nabla}(\vec{\nabla} \cdot \vec{v}) - \Delta \vec{v}) = -\nu \Delta^2 \vec{v}$$

$$= \vec{\nabla} \wedge \vec{\nabla} \wedge (\alpha \theta \vec{g}) = \vec{\nabla} \cdot (\vec{\nabla} \cdot (\alpha \theta \vec{g})) - \Delta(\alpha \theta \vec{g})$$

$$\Rightarrow \nu \Delta \Delta v^3 = \alpha g \partial_{33} \theta - \alpha g \Delta \theta = -\alpha g (\partial_{11} + \partial_{22}) \theta$$

$$\Rightarrow \Delta \Delta \Delta \theta - \frac{R}{L^4}(\partial_{11}\theta + \partial_2 \theta) = 0.$$

In view of the heat equation, at first sight it seems surprising that the same equation is obtained for decoupling v^3, although θ and v^3 have completely different physical meanings.

More generally, we will prove that any observable is constrained by at least one PDE, even in the time-varying situation. Setting $x^4 = t$, we will study the linearized Boussinesq system. It is *not* formally integrable, and the difficulty of studying it lies in this very fact. Taking the divergence of the Navier-Stokes equation while taking into account the conservation of mass, we obtain the new second-order PDE

$$\alpha(\vec{g} \cdot \vec{\nabla})\theta + \Delta \pi = 0.$$

Hence, setting $\chi^2 = (\chi_1)^2 + (\chi_2)^2 + (\chi_3)^2$, the characteristic matrix becomes (we only consider second-order terms):

$$\begin{pmatrix} -\nu\chi^2 & 0 & 0 & 0 & 0 \\ 0 & -\nu\chi^2 & 0 & 0 & 0 \\ 0 & 0 & -\nu\chi^2 & 0 & 0 \\ 0 & 0 & 0 & -\lambda\chi^2 & 0 \\ 0 & 0 & 0 & 0 & -\chi^2 \end{pmatrix}.$$

It has full generic rank 5, even in the stationary case, and the differential transcendence degree of the input/output extenson is zero. This proves that any observable is differentially algebraic over $\mathbb{Q}$ or $\mathbb{R}$. Notice that the linearized Boussinesq control system has constant coefficients. Hence, resolvent systems can be obtained for *each* unknown just by computing the full determinant of the matrix obtained by replacing each derivative by the corresponding polynomial in χ. After straightforward transformations we obtain the matrix

$$\begin{pmatrix} \chi_1 & \chi_2 & \chi_3 & 0 & 0 \\ \chi_4 - \nu\chi^2 & 0 & 0 & 0 & \dfrac{1}{a}\chi_1 \\ 0 & \chi_4 - \nu\chi^2 & 0 & 0 & \dfrac{1}{a}\chi_2 \\ 0 & 0 & \chi_4 - \nu\chi^2 & -\alpha g & \dfrac{1}{a}\chi_3 \\ 0 & 0 & -A & \chi_4 - \nu\chi^2 & 0 \end{pmatrix}$$

with determinant

$$-\frac{1}{a}(\chi_4 - \nu\chi^2)[(\chi_4 - \nu\chi^2)(\chi_4 - \lambda\chi^2)\chi^2 - \alpha Ag(\chi_1^2 + \chi_2^2)].$$

Hence, if an observable is constrained by a PDE, this PDE must factor through the differential operators

$$\partial_t - \nu\Delta \quad \text{or} \quad (\partial_t - \nu\Delta)(\partial_t - \lambda\Delta)\Delta - \alpha Ag(\partial_{11} + \partial_{22}).$$

Indeed, if we consider the observable

$$\zeta = \partial_1 v^2 - \partial_2 v^1,$$

which is the vertical component of $\vec{\nabla} \wedge \vec{v}$, we obtain

$$\partial_t \zeta - \nu\Delta\zeta = 0$$

as second-order constraint for ζ. A similar comment holds in the stationary case, where factorization is through

$$\Delta \quad \text{or} \quad \nu\lambda\Delta\Delta\Delta - \alpha Ag(\partial_{11} + \partial_{22}),$$

and we recover the previous result, because

$$\frac{\alpha Ag}{\nu\lambda}\frac{\alpha a Agc}{\nu\kappa} = \frac{\alpha a^2 gAc}{\eta\kappa} = \frac{\text{R}}{L^4}.$$

Now we will show how to use these results to study the instability of the temperature. The same method could be used for any other observable.

To this end we notice that the temperature, or rather its perturbation, is constrained by

$$\partial_{tt}\Delta\theta - (\lambda + \nu)\partial_t\Delta\Delta\theta + \lambda\nu\Delta\Delta\Delta\theta - \alpha A g(\partial_{11} + \partial_{22})\theta = 0.$$

Substituting a solution of the form

$$F(x^3)\exp i(k_1 x^1 + k_2 x^2)\exp pt,$$

we find, following the straightforward but tedious computations in [36], that p is real for any positive Rayleigh number. It follows that the transition from stability to instability must occur via a stationary state, i.e. via a solution of the 6th-order equation

$$\Delta\Delta\Delta\theta - \frac{R}{L^4}(\partial_{11} + \partial_{22})\theta = 0.$$

In any case, after using the new scale of length L, the above equation depends on R only. The 6 boundary conditions (3 on each plate) can be transferred to θ, because their knowledge is sufficient to solve the full system. This implies that we must solve a linear system of 6 equations with determinant depending on R only. This is the crucial fact in explaining why instability depends on the Rayleigh number only, in agreement with experiments.

More precisely, since v^3 is proportional to $\Delta\theta$ in a stationary state (because of the heat equation), we may transfer the boundary conditions to

$$\theta = 0, \qquad \partial_{33}\theta = 0, \qquad \partial_{333}\theta = 0$$

on each plate. For F we obtain the equation

$$(d_{33} - k^2)F + \frac{R k^2}{L^4}F = 0.$$

The general solution is a linear combination of $\exp(\mu x^3/L)$ and $\exp(-\mu x^3/L)$, where

$$\mu^2 = k^2 L^2 - R^{1/3}(kL)^{2/3}\sqrt[3]{1},$$

with 3 different values for the cubic root of unity. Hence there is a basis of 6 solutions, and the 6 boundary conditions will provide a linear homogeneous system of 6 equations in 6 unknowns, the coefficients of the decomposition in the basis. Accordingly, the determinant of the system must be zero, and this gives a functional relation $(R, kL) = 0$. The resulting function $R = R(kL)$ has a minimum at a certain value of kL. This minimum gives a criterion for instability, while the value of k gives the periodicity (though not the symmetry!) of the movement in the (x^1, x^2)-plane.

Example 13 (Thermal instability in a rotating incompressible fluid). Again we follow the lines of the excellent reference [36]. We add to the Boussinesq control system a term describing the Coriolis force under a rotation $\vec{\Omega}$ of the experimental apparatus. The linearized control system becomes

$$
\begin{cases}
\vec{\nabla} \cdot \vec{v} = 0, \\
\partial_t \vec{v} - \nu \Delta \vec{v} + \alpha \theta \vec{g} + \frac{1}{a} \vec{\nabla} \pi + 2\vec{\Omega} \wedge \vec{v} = 0, \\
\partial_t \theta - \kappa \Delta \theta + A\vec{g} \cdot \vec{v} = 0,
\end{cases}
$$

and the elimination procedure is much more delicate than in the case of pure instability, although the procedure is similar. To simplify the discussion, we assume that $\vec{\Omega}$ is parallel to $\vec{g}$. In this case we can prove that θ is constrained by an 8th-order PDE!

Indeed, using the same procedure as in the Bénard problem, taking into account the new term, we obtain

$$
\begin{cases}
(\partial_t - \nu\Delta)\Delta v^3 - \alpha g(\partial_{11} + \partial_{22})\theta + 2\Omega \partial_3 \zeta = 0, \\
(\partial_t - \nu\Delta)\zeta - 2\Omega \partial_3 v^3 = 0, \\
(\partial_t - \kappa\Delta)\theta - Agv^3 = 0,
\end{cases}
$$

where $\zeta = \partial_1 v^2 - \partial_2 v^1$ is the vertical component of $\vec{\nabla} \wedge \vec{v}$.

Applying the operator $\partial_t - \nu\Delta$ to the first PDE, we obtain the additional term

$$
4\Omega^2 \partial_{33} v^3 = 4\frac{\Omega^2}{Ag}(\partial_t - \kappa\Delta)\partial_{33}\theta,
$$

and the leading term becomes

$$
\frac{\nu^2 \kappa}{Ag}\Delta\Delta\Delta\Delta\theta,
$$

as desired.

This Example proves that if we regard $\Omega = \text{cst}$ as a parameter on which the given PDE depends, then *specialization and elimination do not commute*. Moreover, specialization followed by elimination always gives a smaller system (less solutions!) than does elimination followed by specialization.

Example 14 (stability of Couette flow). We will now consider small oscillations of a cylindrical column of rotating fluid under the assumption of axisymmetry. According to the detailed reference [36], the 4 linearized PDE for the 3 unknown components of velocity (u_r, u_θ, u_z) and the perturbation π of pressure are as follows in cylindrical coordinates:

$$
\begin{cases}
\partial_t u_r - 2\Omega u_\theta + \partial_r \pi = 0, \\
\partial_t u_\theta + 2\Omega u_r = 0, \\
\partial_t u_z + \partial_z \pi = 0, \\
\partial_r u_r + \frac{1}{r}u_r + \partial_z u_z = 0.
\end{cases}
$$

Under the assumption that the angular velocity Ω is constant, we obtain by differential elimination the following 4th order PDE for u_θ:

$$
\partial_{ttzz} u_\theta - 4\Omega^2 \partial_{zz} u_\theta - \partial_{ttrr} u_\theta - \frac{1}{r}\partial_{ttr} u_\theta + \frac{1}{r^2}\partial_{tt} u_\theta = 0.
$$

More generally, the characteristic matrix is

$$\begin{pmatrix} \chi_t & 0 & 0 & \chi_r \\ 0 & \chi_t & 0 & 0 \\ 0 & 0 & \chi_t & \chi_z \\ \chi_r & 0 & \chi_z & 0 \end{pmatrix}$$

Surprisingly, it is symmetric, with determinant equal to $\chi_t^2(\chi_r^2+\chi_z^2)$. Hence its generic rank is 4, and the control system is totally uncontrollable. Again, the coordinate system is *not* δ-regular.

Example 15 (Free vibration of an inelastic cable). In this engineering Example, although the PDE are very simple, the explicit computation is very delicate. We believe that this approach must nevertheless be used in the study of the *galop* phenomenon that we will now describe.

In very cold areas that are also flat, like Canada or Siberia, long-distance electric transportation often fails to work because of a very specific instability caused by a combination of constant wind and the deposit on one side of the electric wires of a thin shell of ice. Under these circumstances the cable starts to vibrate with various modes, including torsion vibration, and breaks after a while, leading to damages that are hardly repairable during the winter season. In view of the following treatment of 'pure' vibration, we believe that the study of instability, like for the Bérnard problem, must be done by decoupling the observables involved. This computation can surely not be achieved without using computer algebra, as we will see.

Let $x^1 = s$ be the curvilinear abcissa along the cable, and $x^2 = t$ the time, with $\omega = (\omega_{ij})$ the standard euclidean metric on $\mathbb{R}^n$. The basic control system is

$$\partial_{tt}y^k = \partial_s(u\partial_s y^k), \qquad \omega_{kl}\partial_s y^k \partial_s y^l = 1,$$

with $k = 1,\ldots,m$. The m first PDE describe the dynamics of the cable, while the second PDE describes the inelasticity condition. In these equations u is the ordinary tension of the cable, considered as the input, while the position $y = (y^1,\ldots,y^m)$ is the output.

For simplicity we use jet notation $\partial_{st}y^k = y^k_{st},\ldots$. Convenient prolongation of the various PDE leads successively to

$$\begin{cases} \omega_{kl}(y_s^k y_{stt}^l + y_{st}^k y_{st}^l) = 0, \\ \omega_{kl}(y_s^k y_{sss}^l + y_{ss}^k y_{ss}^l) = 0. \end{cases}$$

From the initial PDE of the dynamics,

$$y_{tt}^k = uy_{ss}^k + u_s y_s^k,$$

we deduce

$$y_{stt}^k = uy_{sss}^k + 2u_s y_{ss}^k + u_{ss}y_s^k,$$

and finally we obtain the new second-order PDE

$$u_{ss} + \omega_{kl}(y_{st}^k y_{st}^l - uy_{ss}^k y_{ss}^l) = 0.$$

In computing the last character, and thus the differential transcendence degree, we find that the main problem is that the initial system is not formally integrable (we have just given a new second-order PDE!). Setting

$$\chi = \chi_s \, ds + \chi_t \, dt$$

and taking into account the two second-order PDE found above, the characteristic matrix becomes

$$
\begin{pmatrix}
\chi_t^2 - u\chi_s^2 & & & 0 \\
0 & \ddots & 0 & \vdots \\
& & \chi_t^2 - u\chi_s^2 & 0 \\
\star & \cdots & \star & \chi_s^2
\end{pmatrix},
$$

where $\star$ indicates various second-order polynomials. The determinant of this matrix is

$$(\chi_t^2 - u\chi_s^2)^m \chi_s^2,$$

and the generic rank is equal to $m + 1$. Accordingly, any observable is constrained by at least one PDE, a fact highly nonevident at first sight, as we will illustrate for $m = 1, 2$.

For $m = 1$ we have indeed

$$\partial_{tt} y = \partial_s(u\partial_s y), \qquad \partial_s y = 1,$$

and so

$$\partial_{ss} u = 0.$$

The case $m = 2$ already is extremely delicate. Setting $y_s^1 = \cos\theta$, $y_s^2 = \sin\theta$ and substituting, we obtain the following new control system in (u, θ):

$$
\begin{cases}
\theta_{tt} - u\theta_{ss} - 2u_s\theta_s = 0, \\
u_{ss} + (\theta_t)^2 - u(\theta_s)^2 = 0.
\end{cases}
$$

This system is linear and inhomogeneous in u, and involves only derivatives with respect to s, so we can readily eliminate u. Indeed, setting

$$\alpha = 2(\theta_s)^2 \left[\frac{\theta_{sss}}{\theta_s} - \frac{3}{2}\left(\frac{\theta_{ss}}{\theta_s}\right)^2 + 2(\theta_s)^2 \right],$$

and then taking the derivative of the first PDE with respect to s and using the second PDE to eliminate u_{ss}, we obtain

$$
\begin{cases}
\alpha u = 2\theta_s\theta_{stt} - 3\theta_{ss}\theta_{tt} + 4(\theta_s\theta_t)^2, \\
\alpha u_s = \theta_{tt}\theta_{sss} - \theta_{ss}\theta_{stt} + 2(\theta_s)^3\theta_{tt} - 2\theta_s(\theta_t)^2\theta_{ss}.
\end{cases}
$$

The easy elimination of u then leads to a single, highly nonlinear differential polynomial of order 4 for θ which is invariant under $\theta \to -\theta$, and thus determines a single PDE of order 5 for y^1 and y^2.

The elimination of θ is even more delicate. Taking the derivatives of the second PDE with respect to s and t, we obtain three PDE that can be solved for $\theta_{ss}, \theta_{tt}, \theta_{st}$. Among the two crossed derivatives that can be effected in order to check the compatibility of these PDE, one leads to an identity to zero, while the other leads to a second PDE between θ_s and θ_t. A tedious computation of the resultant finally leads to a single 5th-order PDE in u. From the mechanical point of view this fact is highly nonevident at first sight.

The computational difficulty in the case $m = 3$ seems to explode, and still is a challenge for computer algebra and its applications.

The computation of the differential transcendence degree, or of the last character, does not depend on the generators since with coordinates (u, θ) the characteristic determinant is $(\chi_t - u\chi_s^2)\chi_s^2$. This leads to the same conclusion as before.

Example 16. The purpose of this Example is to clarify the relation between duality and formal integrability in classical control theory.

Consider the following control system, with one input u and two outputs (y^1, y^2),

$$\begin{cases} \ddot{y}^2 + y^2 + u = 0, \\ \dot{y}^1 + u = 0. \end{cases}$$

In the classical approach using the transfer matrix, we separate input and output and introduce the matrices

$$A(\chi) = \begin{pmatrix} 0 & \chi^2 + 1 \\ \chi & 0 \end{pmatrix}, \quad B(\chi) = \begin{pmatrix} 1 \\ 1 \end{pmatrix} \Rightarrow H(\chi) = \begin{pmatrix} 1/\chi \\ 1/(\chi^2 + 1) \end{pmatrix}.$$

The differential output rank is 1 because, by subtraction, there is the following ODE between (y^1, y^2):

$$\ddot{y}^2 - \dot{y}^1 + y^2 = 0.$$

The control system is controllable. Indeed, any constrained observable is of the form $z = f(y^1, y^2, \dot{y}^2)$, and in this linear case we can consider $z = \dot{y}^2 + \lambda y^2 + \mu y^1$. Substituting this into $\dot{z} + \nu z = 0$ leads to

$$\ddot{y}^2 + (\lambda + \nu)\dot{y}^2 + \lambda \nu y^2 + \mu \dot{y}^1 + \mu \nu y^1 = 0.$$

By identification we get

$$\lambda + \nu = 0, \quad \lambda \nu = 1, \quad \mu = -1, \quad \mu \nu = 0,$$

i.e. $\nu = 0$, so that $\lambda = 0$ and we find $\lambda \nu = 0 = 1$, which is impossible! Hence we may look for a parametrization by using the duality method. We proceed as follows. Multiplying the two equations by λ_1, λ_2 and integrating by parts, we obtain

$$\begin{cases} \ddot{\lambda}_1 + \lambda_1 = 0, \\ -\dot{\lambda}_2 = 0, \\ \lambda_1 + \lambda_2 = 0. \end{cases}$$

This system is not formally integrable. From $\dot{\lambda}_2 = 0$ and $\dot{\lambda}_1 + \dot{\lambda}_2 = 0$ we deduce that $\dot{\lambda}_1 = 0$, $\ddot{\lambda}_1 = 0$, and so $\lambda_1 = \lambda_2 = 0$. Hence the differential operator $\lambda \rightarrow \mu$ is given by

$$\begin{cases} \ddot{\lambda}_1 + \lambda_1 = \mu_1 & \text{dual of } \eta^1 = y^2, \\ -\dot{\lambda}_2 = \mu_2 & \text{dual of } \eta^2 = y^1, \\ \lambda_1 + \lambda_2 = \mu_3 & \text{dual of } \eta^3 = u. \end{cases}$$

Therefore the image is the 3rd-order system

$$\dddot{\mu}_3 + \ddot{\mu}_2 + \dot{\mu}_3 - \dot{\mu}_1 + \mu_2 = 0.$$

Multiplying by ξ and integrating by parts immediately gives

$$\begin{cases} -\ddot{\xi} - \xi = u = \eta^3, \\ \dot{\xi} + \xi = y^1 = \eta^2, \\ \xi = y^2 = \eta^1. \end{cases}$$

The homogeneous system implies $\dot{\xi} = 0 \Rightarrow \ddot{\xi} = 0 \Rightarrow \xi = 0$, and the operator is injective, with image

$$\begin{cases} \ddot{y}^2 + y^2 + u = 0, \\ \dot{y}^1 + u = 0, \end{cases}$$

and we have obtained the required parametrization. Let us now reduce the initial control system to the following formally integrable control system:

$$\begin{cases} \ddot{y}^2 + y^2 + u = 0, \\ \ddot{y}^1 + \dot{u} = 0, \\ \dot{y}^1 + u = 0. \end{cases}$$

Multiplying these three equations by $\lambda_1, \lambda_2, \lambda_3$ and integrating by parts, we obtain

$$\begin{cases} \ddot{\lambda}_1 + \lambda_1 = \mu^1 & \text{dual of } \eta^1 = y^2, \\ \ddot{\lambda}_2 - \dot{\lambda}_3 = \mu^2 & \text{dual of } \eta^2 = y^1, \\ \lambda_1 - \dot{\lambda}_2 + \lambda_3 = \mu^3 & \text{dual of } \eta^3 = u. \end{cases}$$

The corresponding homogeneous system is *not* formally integrable. The general solution satisfies $\lambda_1 - \dot{\lambda}_2 + \lambda_3 = 0$, and so $\dot{\lambda}_1 = 0$, i.e. $\ddot{\lambda}_1 = 0$, and finally $\lambda_1 = 0$, with $\dot{\lambda}_2 - \lambda_3 = 0$. Therefore the operator $\lambda \rightarrow \mu$ is *not* injective. Hence injectivity of the dual operator is not a byproduct of controllability, as one could have believed from the general theory for standard linear systems. In fact, if $\Phi = 0$ is one equation, then *any* combination $\lambda \dot{\Phi} + \dot{\lambda} \Phi$ is the derivative of $\lambda \Phi$. The image of the operator above is

$$\dddot{\mu}_3 + \ddot{\mu}_2 + \dot{\mu}_3 - \dot{\mu}_1 + \mu_2 = 0,$$

and we may proceed as before, to find the same parametrization.

In fact, considering the inhomogeneous system

$$\begin{cases} \ddot{\eta}^1 + \eta^1 + \eta^3 = \zeta^1, \\ \ddot{\eta}^2 + \dot{\eta}^3 = \zeta^2, \\ \dot{\eta}^2 + \eta^3 = \zeta^3, \end{cases}$$

we obtain the compatibility condition

$$\dot{\zeta}^3 - \zeta^2 = 0,$$

i.e. the operator $\eta \to \zeta$ is *not* surjective. Multiplying by κ and integrating by parts gives

$$\begin{cases} 0 = \lambda_1, \\ -\kappa = \lambda_2, \\ -\dot{\kappa} = \lambda_3, \end{cases}$$

with image $\lambda_1 = 0$, $\dot{\lambda}_2 - \lambda_3 = 0$, as before. Hence we discover that the original difficulty lies in the fact that $\mathcal{D}_1 \colon \eta \to \zeta$ is not surjective, not because it is not formally integrable (it is!) but because $\mathcal{D} \colon \xi \to \eta$ is not formally integrable. As a byproduct, if we bring $\mathcal{D}$ to formal integrability, and thus involutivity, then $\mathcal{D}_1$ will be automatically formally integrable, involutive, *and* surjective (because $n = 1$).

Let us introduce j_3:

$$\begin{cases} \dddot{\xi} = \theta^4 = -\eta^1 - \eta^3, \\ \ddot{\xi} = \theta^3 = \dot{\eta}^1, \\ \dot{\xi} = \theta^2 = \eta^1, \\ \xi = \theta^1 = \eta^2 - \dot{\eta}^1. \end{cases}$$

We find the inversion formula

$$\eta^1 = \theta^2, \quad \eta^2 = \theta^1 + \dot{\theta}^2, \quad \eta^3 = -\theta^4 - \theta^2.$$

It follows that our initial control system can be observed by θ instead of η. However, now we have the first-order control system

$$\begin{cases} \dot{\theta}^1 - \theta^2 = 0, \\ \dot{\theta}^2 - \theta^3 = 0, \\ \dot{\theta}^3 - \theta^4 = 0. \end{cases}$$

It is equivalent to the initial one and can, of course, be written in standard form by setting $z^1 = \theta^1$, $z^2 = \theta^2$, $z^3 = \theta^3$, $v = \theta^4$:

$$\begin{cases} \dot{z}^1 = z^2, \\ \dot{z}^2 = z^3, \qquad \dot{z} = Cz + Dv. \\ \dot{z}^3 = v, \end{cases}$$

This system has been obtained from a parametrization and therefore it is controllable, with controllability matrix of rank

$$\mathrm{rk}(D, CD, C^2 D) = \mathrm{rk} \begin{pmatrix} 0 & 0 & 1 \\ 0 & 1 & 0 \\ 1 & 0 & 0 \end{pmatrix} = 3.$$

The new operator $\mathcal{D}_1$ is surjective, and the corresponding $\tilde{\mathcal{D}}_1$ is injective.

This Example clearly shows that $\tilde{\mathcal{D}}_1$ *is injective only when* $\mathcal{D}_1$ *is surjective and controllable at the same time.*

Example 17. This Example will illustrate duality theory in the partial differential situation. First of all we will prove that we can dualize a differential sequence. To this end, let

$$\begin{cases} (\mathcal{D}\xi)^\tau = A_k^{\tau i}(x)\partial_i \xi^k + a_k^\tau(x)\xi^k, \\ (\mathcal{D}_1 \eta)^\alpha = B_\tau^{\alpha j}(x)\partial_j \eta^\tau + b_\tau^\alpha(x)\eta^\tau. \end{cases}$$

The condition $\mathcal{D}_1 \circ \mathcal{D} \equiv 0$ is written as

$$\begin{cases} \partial_{ij}\xi^k \rightarrow B_\tau^{\alpha j} A_k^{\tau i} = 0, \\ \partial_i \xi^k \rightarrow B_\tau^{\alpha r} \partial_r A_k^{\tau i} + B_\tau^{\alpha i} a_k^\tau + b_\tau^\alpha A_k^{\tau i} = 0, \\ \xi^k \rightarrow b_\tau^\alpha a_k^\tau + B_\tau^{\alpha r} \partial_r a_k^\tau = 0. \end{cases}$$

By duality (integration by parts), and with the notations of Sections A and B, we have

$$\begin{cases} (\tilde{\mathcal{D}}\mu)_k = -A_k^{\tau i}\partial_i \mu_\tau + (a_k^\tau - \partial_i A_k^{\tau i})\mu_\tau, \\ (\tilde{\mathcal{D}}_1 \lambda)_\tau = -B_\tau^{\alpha j}\partial_j \lambda_\alpha + (b_\tau^\alpha - \partial_j B_\tau^{\alpha j})\lambda_\alpha. \end{cases}$$

Substituting this into the composition $\tilde{\mathcal{D}} \circ \tilde{\mathcal{D}}_1$ and expanding the result, we obtain the following terms:

order 2: $\partial_{ij}\lambda_\alpha \rightarrow A_k^{\tau i} B_\tau^{\lambda j} = 0.$

order 1:

$$\begin{aligned} \partial_j \lambda_\alpha \rightarrow & A_k^{\tau i}\partial_i B_\tau^{\alpha j} + A_k^{\tau j}\partial_i B_\tau^{\alpha i} + (\partial_i A_k^{\tau i})B_\tau^{\alpha j} - a_k^\tau B_\tau^{\alpha j} - A_k^{\tau j}b_\tau^\alpha \\ = & \partial_i(A_k^{\tau i} B_\tau^{\alpha j}) - (B_\tau^{\alpha i}\partial_i A_k^{\tau j} + a_k^\tau B_\tau^{\alpha j} + A_k^{\tau j}b_\tau^\alpha) \\ = & 0. \end{aligned}$$

order 0:

$$\begin{aligned} \lambda_\alpha \rightarrow & A_k^{\tau i}\partial_{ij}B_\tau^{\alpha j} - A_k^{\tau i}\partial_i b_\tau^\alpha - (\partial_i A_k^{\tau i})b_\tau^\alpha - a_k^\tau \partial_j B_\tau^{\alpha j} + (\partial_i A_k^{\tau i})(\partial_j (B_\tau^{\alpha j}) + a_k^\tau b_\tau^\alpha \\ = & \partial_i(A_k^{\tau i}\partial_j B_\tau^{\alpha j}) - \partial_i(A_k^{\tau i}b_\tau^\alpha) - \partial_i(B_\tau^{\alpha j}a_k^\tau) \\ = & -\partial_i(B_\tau^{\alpha j}\partial_j A_k^{\tau i} + A_k^{\tau i}b_\tau^\alpha + B_\tau^{\alpha i}a_k^\tau) \\ = & 0. \end{aligned}$$

This proves that

$$\mathcal{D}_1 \circ \mathcal{D} \equiv 0 \Leftrightarrow \tilde{\mathcal{D}} \circ \tilde{\mathcal{D}}_1 \equiv 0,$$

and so

$$\sigma_\chi(\mathcal{D}_1) \circ \sigma_\chi(\mathcal{D}) \equiv 0 \Leftrightarrow \sigma_\chi(\tilde{\mathcal{D}}) \circ \sigma_\chi(\tilde{\mathcal{D}}_1) \equiv 0.$$

Introducing the various dual spaces and maps (transposed), we see that whenever we have a ker-coker sequence

$$0 \longrightarrow \ker \Phi \longrightarrow A \overset{\Phi}{\longrightarrow} B \longrightarrow \operatorname{coker} \Phi \longrightarrow 0,$$

as in Chapter I, then we also have

$$0 \longleftarrow (\ker \Phi)^* \longleftarrow A^* \overset{\tilde{\Phi}}{\longleftarrow} B^* \longleftarrow (\operatorname{coker} \Phi)^* \longleftarrow 0.$$

Therefore, for any exact sequence

$$A \overset{\Phi}{\longrightarrow} B \overset{\Psi}{\longrightarrow} C$$

we obtain a dual sequence

$$A^* \overset{\tilde{\Phi}}{\longleftarrow} B^* \overset{\tilde{\Psi}}{\longleftarrow} C^*.$$

Therefore,

$$\begin{aligned}
\dim \ker \tilde{\Phi} - \dim \operatorname{im} \tilde{\Psi} &= \dim \operatorname{coker} \Phi - \dim \operatorname{im} \tilde{\Psi} \\
&= \dim \operatorname{coker} \Phi - (\dim B^* - \dim \operatorname{coker} \tilde{\Psi}) \\
&= \dim \operatorname{coker} \Phi - \dim B^* + \dim \operatorname{coker} \tilde{\Psi} \\
&= \dim \operatorname{coker} \Phi - \dim B + \dim \ker \Psi \\
&= \dim \ker \Psi - \dim \operatorname{im} \Phi \\
&= 0.
\end{aligned}$$

Hence duality preserves exactness of sequences.

We will apply these results to PD control theory. Indeed, consider as operator $\mathcal{D}_1$ the following operator, which has already been studied in Chapter IV for other reasons:

$$\begin{cases}
\partial_2 \eta^2 + \partial_2 \eta^3 - \partial_1 \eta^3 - \dot{\partial}_1 \eta^2 = \zeta^1 \\
\partial_2 \eta^1 - \partial_2 \eta^3 - \partial_1 \eta^3 - \partial_1 \eta^2 = \zeta^2 \\
\partial_1 \eta^1 - 2\partial_1 \eta^3 - \partial_1 \eta^2 = \zeta^3
\end{cases}
\begin{array}{c} \lambda_1 \\ \lambda_2 \\ \lambda_3 \end{array}$$

We may ask whether such a control system is controllable or not, *independently of the way of selecting input and output.*

First we notice that the characteristic matrix, which also has already been studied in Chapter IV and is recalled here for the sake of clarity,

$$\begin{pmatrix}
0 & \chi_2 - \chi_1 & \chi_2 - \chi_1 \\
\chi_2 & -\chi_1 & -(\chi_1 + \chi_2) \\
\chi_1 & -\chi_1 & -2\chi_1
\end{pmatrix}$$

has rank two and is therefore not injective. It follows that the dual matrix is not surjective, and so the symbol $\sigma_\chi(\tilde{\mathcal{D}}_1)$ is not surjective, since it is the transposed matrix of $\sigma_\chi(\mathcal{D}_1)$. For $\tilde{\mathcal{D}}_1$ we obtain the following operator, after multiplication of the previous equations by $(\lambda_1, \lambda_2, \lambda_3)$ and integration by parts as usual,

$$
\begin{cases}
\eta^1 \to -\partial_1\lambda_2 - \partial_1\lambda_3 = \mu_1, \\
\eta^2 \to -\partial_2\lambda_1 + \partial_1\lambda_1 + \partial_1\lambda_2 + \partial_1\lambda_3 = \mu_2, \\
\eta^3 \to -\partial_2\lambda_1 + \partial_1\lambda_1 + \partial_2\lambda_2 + \partial_1\lambda_2 + 2\partial_1\lambda_3 = \mu_3.
\end{cases}
$$

The image of this operator is

$$
\mu_3 - \mu_2 + \mu_1 = 0 \mid \xi.
$$

By duality we should get

$$
\eta^1 = \xi, \quad \eta^2 = -\xi, \quad \eta^3 = \xi,
$$

and this formula does not parametrize $\mathcal{D}_1$, which is therefore surely not controllable. Indeed, let us set $\eta^1 + \eta^2 = A$, $\eta^2 + \eta^3 = B$. We immediately obtain

$$
\partial_2 B - \partial_1 B = 0, \quad \partial_2 A - \partial_1 A = 0, \quad \partial_1 A - 2\partial_1 B = 0.
$$

Hence the 2 observables A and B are constrained and the system is only partially controllable.

Example 18. In this Example we apply the duality method for the Spencer operator.

For $n = 2$ we consider the operator $\mathcal{D} = \text{div}$ acting on T. We recall that the Spencer operator D_1 is given by

$$
j_1 \begin{cases}
\xi^1 = \eta^1 \\
\xi^2 = \eta^2 \\
\partial_1\xi^1 = \eta^3 \\
\partial_2\xi^1 = \eta^4 \\
\partial_1\xi^2 = \eta^5 \\
\partial_2\xi^2 = -\eta^3
\end{cases}
\Rightarrow D_1
\quad D \begin{cases}
\partial_1\eta^1 - \eta^3 = 0 & \lambda_1 \\
\partial_1\eta^2 - \eta^5 = 0 & \lambda_2 \\
\partial_2\eta^1 - \eta^4 = 0 & \lambda_3 \\
\partial_2\eta^2 + \eta^3 = 0 & \lambda_4 \\
\partial_1\eta^4 - \partial_2\eta^3 = 0 & \lambda_5 \\
-\partial_1\eta^3 - \partial_2\eta^5 = 0 & \lambda_6
\end{cases}
$$

We have the differential sequence

$$
0 \longrightarrow \Theta \xrightarrow{\ j_1\ } C_0 \xrightarrow{\ D_1\ } C_1
$$
$$
\boxed{5} \qquad\quad \boxed{6}
$$

where D_1 is involutive because $\mathcal{D}$ is involutive. In contrast to the differential sequence

$$
0 \longrightarrow T \xrightarrow{\ j_1\ } C_0(T) \xrightarrow{\ D_1\ } C_1(T)
$$
$$
\boxed{2} \qquad\quad \boxed{6} \qquad\qquad \boxed{6}
$$

which shows that D_1 is 'parametrized' by j_1, in the preceding situation there is nothing similar, and we may question the existence of a parametrization for D_1.

Multiplying the 6 equations by $(\lambda_1, \ldots, \lambda_6)$ and integrating by parts as usual, we find

$$
\begin{cases}
\eta^1 \to -\partial_1 \lambda_1 - \partial_2 \lambda_3 = -\mu_1, \\
\eta^2 \to -\partial_1 \lambda_2 - \partial_2 \lambda_4 = \mu_2, \\
\eta^3 \to -\lambda_1 + \lambda_4 + \partial_2 \lambda_5 + \partial_1 \lambda_6 = \mu_3, \\
\eta^4 \to -\lambda_3 - \partial_1 \lambda_5 = \mu_4, \\
\eta^5 \to -\lambda_2 + \partial_2 \lambda_6 = \mu_5.
\end{cases}
$$

For $D_1 \colon C_0(T) \to C_1(T)$ we notice that $\tilde{D}_1$ is involutive. However, in the restricted case of $D_1 \colon C_0 \to C_1$, we notice that $\tilde{D}_1$ is *not* involutive and that we have to prolongate once in order to obtain a trivially involutive (zero) symbol. Accordingly, *there is one compatibility condition, of order 2*, namely (exercise):

$$
\partial_{12}\mu_3 + \partial_{22}\mu_4 - \partial_{11}\mu_5 + \partial_1 \mu_2 - \partial_2 \mu_1 = 0 \,|\xi.
$$

Multiplication by ξ and integration by parts gives the second-order parametrization

$$
\begin{cases}
\mu_1 \to \partial_2 \xi = \eta^1, \\
\mu_2 \to -\partial_1 \xi = \eta^2, \\
\mu_3 \to \partial_{12} \xi = \eta^3, \\
\mu_4 \to \partial_{22} \xi = \eta^4, \\
\mu_5 \to -\partial_{11} \xi = \eta^5,
\end{cases}
$$

which is quite unexpected in this framework. The existence of such a parametrization immediately proves that D_1 is torsion free (a result not evident a priori!). It could have been obtained directly by using the parametrization $\xi^1 = \partial_2 \xi$, $\xi^2 = -\partial_1 \xi$ of ker div.

Example 19. In this Example we apply the duality method for recovering the Killing equations from the Riemann tensor. For simplicity of exposition we do this in dimension 2, and so we consider the linear second-order PDE

$$
\partial_{11}\epsilon_{22} + \partial_{22}\epsilon_{11} - 2\partial_{12}\epsilon_{12} = 0 \,|\lambda.
$$

Multiplication by λ and integration by parts gives

$$
\begin{cases}
\epsilon_{11} \to \partial_{22}\lambda = \mu_1, \\
\epsilon_{12} \to -2\partial_{12}\lambda = \mu_2, \\
\epsilon_{22} \to \partial_{11}\lambda = \mu_3.
\end{cases}
$$

The image is

$$
\begin{cases}
\partial_1 \mu_1 + \frac{1}{2}\partial_2 \mu_2 = 0 \\
\partial_2 \mu_3 + \frac{1}{2}\partial_1 \mu_2 = 0
\end{cases}
\begin{array}{l}
\xi^1 \\
\xi^2
\end{array}
$$

Multiplication by (ξ^1, ξ^2) and integration by parts gives

$$
\begin{cases}
\mu_1 \to \partial_1 \xi^1 = \epsilon_{11}, \\
\mu_2 \to \frac{1}{2}(\partial_1 \xi^2 + \partial_2 \xi^1) = \epsilon_{12}, \\
\mu_3 \to \partial_2 \xi^2 = \epsilon_{22},
\end{cases}
$$

i.e. the standard 'parametrization'.

On this Example we can verify that the curvature condition determines a torsion-free module (it is parametrizable) which is nevertheless not free.

Example 20. We give an Example of the application of duality when the coefficients explicitly depend on the independent variables. To this end we consider contact transformations in 3-space, and we will recover the existence of a generating function for the infinitesimal transformations, even though the corresponding Lie operator is used in the first place in the Janet sequence, considered as a resolution of the sheaf of infinitesimal contact transformations.

The finite transformations are defined by the Pfaffian system

$$(dy^1 - y^3 \, dy^2) = \rho(x)(dx^1 - x^3 \, dx^2).$$

The corresponding (not formally integrable) system is

$$\begin{cases} \dfrac{\partial y^1}{\partial x^2} - y^3 \dfrac{\partial y^2}{\partial x^2} + x^3 \left(\dfrac{\partial y^1}{\partial x^1} - y^3 \dfrac{\partial y^2}{\partial x^1} \right) = 0, \\[2mm] \dfrac{\partial y^1}{\partial x^3} - y^3 \dfrac{\partial y^2}{\partial x^3} = 0. \end{cases}$$

By linearization we obtain the following (again not formally integrable) linear system of PDE:

$$\begin{cases} \dfrac{\partial \eta^1}{\partial x^2} - x^3 \dfrac{\partial \eta^2}{\partial x^2} + x^3 \dfrac{\partial \eta^1}{\partial x^1} - (x^3)^2 \dfrac{\partial \eta^2}{\partial x^1} - \eta^3 = 0 & \bigg| \; \lambda_1 \\[2mm] \dfrac{\partial \eta^1}{\partial x^3} - x^3 \dfrac{\partial \eta^2}{\partial x^3} = 0 & \bigg| \; \lambda_2 \end{cases}$$

Multiplying these equations by λ_1 and λ_2, respectively, and then integrating by parts as usual (take care!), we arrive at

$$\begin{cases} \eta^1 \to -(\partial_2 \lambda_1 + x^3 \partial_1 \lambda_1 + \partial_3 \lambda_2) = \mu_1, \\ \eta^2 \to x^3(\partial_2 \lambda_1 + x^3 \partial_1 \lambda_1 + \partial_3 \lambda_2) + \lambda_2 = \mu_2, \\ \eta^3 \to -\lambda_1 = \mu_3. \end{cases}$$

So we have

$$\lambda_1 = -\mu_3, \qquad \lambda_2 = \mu_2 + x^3 \mu_1.$$

Substituting this into the first equation of the above system, we see that the image is defined by a unique first-order PDE, as follows:

$$\partial_3 \mu_2 + x^3 \partial_3 \mu_1 - \partial_2 \mu_3 - x^3 \partial_1 \mu_3 + 2\mu_1 = 0 \,|\xi.$$

Multiplying by ξ and integrating by parts (take care!), we arrive at

$$\begin{cases} \mu_1 \to -x^3 \partial_3 \xi + \xi = \eta^1, \\ \mu_2 \to -\partial_3 \xi = \eta^2, \\ \mu_3 \to \partial_2 \xi + x^3 \partial_1 \xi = \eta^3. \end{cases}$$

We recover the standard well-known parametrization of the infinitesimal contact transformations η by means of the generating function ξ. We can immediately verify that the corresponding differential module is torsion free *and* free, since it is generated by the free observable $\xi = \eta^1 - x^3\eta^2$.

Example 21. We provide an Example of parametrization in which, nevertheless, the symbol sequences are not exact. Consider the system

$$\mathcal{D}_1 \quad \begin{cases} \ddot{\eta}^1 - \eta^3 = 0 \,\big|\, \lambda_1 \\ \dot{\eta}^2 + \eta^1 = 0 \,\big|\, \lambda_2 \end{cases}$$

Multiplying by (λ_1, λ_2) and integrating by parts as usual, we obtain

$$\tilde{\mathcal{D}}_1 \quad \begin{cases} \eta^1 \to \ddot{\lambda}_1 + \lambda_2 = \mu_1, \\ \eta^2 \to -\dot{\lambda}_2 = \mu_2, \\ \eta^3 \to -\lambda_1 = \mu_3, \end{cases}$$

with image

$$\tilde{\mathcal{D}} \quad \ddot{\mu}_3 + \dot{\mu}_1 + \mu_2 = 0.$$

Multiplying by ξ and integrating by parts gives the parametrization

$$\mathcal{D} \quad \begin{cases} \mu^1 \to -\dot{\xi} = \eta^1, \\ \mu^2 \to \xi = \eta^2, \\ \mu^3 \to -\ddot{\xi} = \eta^2. \end{cases}$$

Nevertheless, we have, in a generic sense,

$$1 = \dim \operatorname{im} \sigma_\chi(\mathcal{D}) < \dim \ker \sigma_\chi(\mathcal{D}_1) = 2.$$

However, completing $\mathcal{D}_1$ to the formally integrable

$$\mathcal{D}'_1 \quad \begin{cases} \ddot{\eta}^1 - \eta^3 = 0 \,\big|\, \lambda_1 \\ \ddot{\eta}^2 + \eta^1 = 0 \,\big|\, \lambda_3 \\ \dot{\eta}^2 + \eta^1 = 0 \,\big|\, \lambda_2 \end{cases}$$

Multiplying now by $(\lambda_1, \lambda_2, \lambda_3)$, an integration by parts gives

$$\tilde{\mathcal{D}}'_1 \quad \begin{cases} \eta^1 \to \ddot{\lambda}_1 + \lambda_2 - \dot{\lambda} - 3 = \mu_1, \\ \eta^2 \to \ddot{\lambda} - 3 - \dot{\lambda}_2 = \mu_2, \\ \eta^3 \to -\lambda_1 = \mu_3, \end{cases}$$

with image

$$\tilde{\mathcal{D}}' = \tilde{\mathcal{D}} \qquad \ddot{\mu}_3 + \dot{\mu}_1 + \mu_2 = 0$$

obtained by the substitution $\lambda - 2 \to \lambda_2 - \dot{\lambda}_3$ and therefore the same parametrization $\mathcal{D} = \mathcal{D}'$. In that case we have, in a generic sense

$$\dim \operatorname{im} \sigma_\chi(\mathcal{D}') = \dim \ker \sigma_\chi(\mathcal{D}'_1) = 1,$$

and so the symbol sequence is exact.

Example 22. We give an Example in which the dual sequence merely interchanges the succession of operators. Indeed, using the notations of this Chapter, we consider the equation

$$\mathcal{D} \quad \begin{cases} \partial_{11}\xi + \partial_{22}\xi = \eta^1, \\ x^1\partial_1\xi + x^2\partial_2\xi = \eta^2. \end{cases}$$

This operator is not formally integrable, but we know from Chapter IV that the image is defined by the single PDE

$$\mathcal{D}_1 \qquad \partial_{11}\eta^2 + \partial_{22}\eta^2 - x^1\partial_1\eta^1 - x^2\partial_2\eta^1 - 2\eta^1 = 0.$$

Multiplying by λ and integrating by parts (be careful) as usual, we obtain

$$\tilde{\mathcal{D}}_1 \quad \begin{cases} \eta^1 \to x^1\partial_1\lambda + x^2\partial_2\lambda = \mu_1, \\ \eta^2 \to \partial_{11}\lambda + \partial_{22}\lambda = \mu_2, \end{cases}$$

with image again defined by a single PDE:

$$\tilde{\mathcal{D}} \qquad \partial_{11}\mu_1 + \partial_{22}\mu_1 - x^1\partial_1\mu_2 - x^2\partial_2\mu_2 - 2\mu_2 = 0.$$

Multiplying by ξ and integrating by parts (with care) as usual, we obtain the initial operator.

Example 23. The following tricky Example proves that a given parametrization need not be recovered by using the algorithm of Theorem B.7.

Let us start with the Lie operator $\mathcal{D}\xi = \eta$:

$$x^2\partial_1\xi^1 + \xi^2 = \eta^1, \qquad \partial_2\xi^1 = \eta^2, \qquad \partial_1\xi^1 + \partial_2\xi^2 = \eta^3,$$

with compatibility condition $\mathcal{D}_1\eta = 0$:

$$\partial_2\eta^1 - x^2\partial_1\eta^2 - \eta^3 = 0. \quad |\lambda$$

By duality we obtain

$$\tilde{\mathcal{D}}_1 \quad \begin{cases} \eta^1 \to -\partial_2\lambda = \mu_1, \\ \eta^2 \to x^2\partial_1\lambda = \mu_2, \\ \eta^3 \to -\lambda = \mu_3, \end{cases}$$

with image defined by

$$\tilde{\mathcal{D}} \quad \begin{cases} \partial_2\mu_3 - \mu_1 = 0 \,\big|\, \xi^2 \\ x^2\partial_1\mu_3 + \mu_2 = 0 \,\big|\, \xi^1 \end{cases}$$

The dual of $\tilde{\mathcal{D}}$ is

$$\begin{cases} \mu_1 \to -\xi^2 = \eta^1, \\ \mu_2 \to \xi^1 = \eta^2, \\ \mu_3 \to -\partial_2\xi^2 - x^2\partial_1\xi^1 = \eta^3, \end{cases}$$

but does not coincide with $\mathcal{D}$. However, dualizing $\mathcal{D}$ we obtain the equation

$$\begin{cases} \xi^1 \to -x^2\partial_1\mu_1 - \partial_2\mu_2 - \partial_1\mu_3 = -\partial_2(x^2\partial_1\mu_3 + \mu_2) = 0, \\ \xi^2 \to \mu_1 - \partial_2\mu_3 = 0, \end{cases}$$

which are just consequences of the ones defining $\tilde{\mathcal{D}}$.

Example 24. Consider the involutive first-order system

$$\mathcal{D}_1' \quad \begin{cases} \partial_2\eta^3 - \partial_3\eta^2 = 0 \,\big|\, \lambda_1 \\ \partial_3\eta^1 - \partial_1\eta^3 = 0 \,\big|\, \lambda_2 \\ \partial_1\eta^2 - \partial_2\eta^1 = 0 \,\big|\, \lambda_3 \end{cases}$$

Multiplying the equations by $(\lambda_1, \lambda_2, \lambda_3)$ and integrating by parts as usual, we find

$$\tilde{\mathcal{D}}_1' \quad \begin{cases} \eta^1 \to \partial_2\lambda_3 - \partial_3\lambda_2 = \mu_1, \\ \eta^2 \to \partial_3\lambda_1 - \partial_1\lambda_3 = \mu_2, \\ \eta^3 \to \partial_1\lambda_2 - \partial_2\lambda_1 = \mu_3, \end{cases}$$

with image defined by the equation

$$\tilde{\mathcal{D}} \qquad \partial_1\mu_1 + \partial_2\mu_2 + \partial_3\mu_3 = 0 \quad |\xi.$$

Multiplying by ξ and integrating by parts, we find

$$\mathcal{D} \quad \begin{cases} \mu_1 \to -\partial_1\xi = \eta^1, \\ \mu_2 \to -\partial_2\xi = \eta^2, \\ \mu_3 \to -\partial_3\xi = \eta^3, \end{cases}$$

i.e. the correct 'gradient' parametrization of the initial 'rotational' system, which is hence controllable.

Let us now *drop* the third equation in the original system and proceed as before, starting afresh with

$$\mathcal{D}_1 \quad \begin{cases} \partial_2\eta^3 - \partial_3\eta^2 = 0 \,\big|\, \lambda_1 \\ \partial_3\eta^1 - \partial_1\eta^3 = 0 \,\big|\, \lambda_2 \end{cases}$$

In the first situation, the characteristic matrix

$$\begin{pmatrix} 0 & -\chi_3 & \chi_2 \\ \chi_3 & 0 & -\chi_1 \\ -\chi_2 & \chi - 1 & 0 \end{pmatrix}$$

has rank 2, like the characteristic matrix made up by the first two lines. Hence, the differential transcendence degrees of the two differential extensions defined by the linear systems are 1 in both cases.

Multiplying by (λ_1, λ_2) and integrating by parts as usual, we find

$$\tilde{\mathcal{D}}_1 \quad \begin{cases} \eta^1 \to -\partial_3\lambda_2 = \mu_1, \\ \eta^2 \to \partial_3\lambda_1 = \mu_2, \\ \eta^3 \to \partial_1\lambda_2 - \partial_2\lambda_1 = \mu_3, \end{cases}$$

with image again defined by the same divergence-type equation $\tilde{\mathcal{D}}$ as before, leading to the same parametrization $\mathcal{D}$, with compatibility conditions $\mathcal{D}_1' \neq \mathcal{D}_1$. However, setting $\mathcal{D}_1\eta = \zeta$, we have

$$\partial_1\zeta^1 + \partial_2\zeta^2 + \partial_3\zeta^3 \equiv 0,$$

and so the constraint $\partial_3 \zeta^3 = 0$ whenever $\zeta^1 = \zeta^2 = 0$. This proves that $\mathcal{D}_1$ is *not* controllable.

Example 25. As a nontrivial example of duality theory, we will study the Janet example. Using the notation of this Chapter, we know that

$$\mathcal{D} \quad \begin{cases} \partial_{33}\xi - x^2 \partial_{11}\xi = \eta^1, \\ \partial_{22}\xi = \eta^2, \end{cases}$$

admits the third-order compatibility condition

$$\mathcal{D}_1 \qquad A \equiv \partial_{233}\eta^2 - x^2 \partial_{112}\eta^2 - 3\partial_{11}\eta^2 - \partial_{222}\eta^1 = 0,$$

and another 6th-order one, $B = 0$, which is *not* a differential consequence of the quoted one and satisfies

$$\partial_{3333}A - 2x^2 \partial_{1133}A + (x^2)^2 \partial_{1111}A - 2\partial_2 B = 0.$$

Forgetting about its origin, we may look for the PDE $A = 0$ and wonder whether or not it is controllable.

Multiplying A by λ and integrating by parts (be careful) as usual, we find

$$\begin{cases} \eta^1 \to \partial_{222}\lambda = \mu_1, \\ \eta^2 \to -\partial_{233}\lambda + x^2 \partial_{112}\lambda - 2\partial_{11}\lambda = \mu_2, \end{cases}$$

with (be careful) image defined by the single PDE

$$\partial_{33}\mu_1 + \partial_{22}\mu_2 - x^2 \partial_{11}\mu_1 = 0.$$

Multiplying by ξ and integrating by parts as usual, we find

$$\begin{cases} \mu_1 \to \partial_{33}\xi - x^2 \partial_{11}\xi = \eta^1, \\ \mu_2 \to \partial_{22}\xi = \eta^2, \end{cases}$$

which is exactly the Janet system. Accordingly we see that $A = 0$ is *not* controllable, because whenever $A = 0$ we obtain $\partial_2 B = 0$ and the observable B is thus constrained.

Since the Janet system is not involutive, we notice that the chosen coordinate system is only partly δ-regular for it (one equation is solved with respect to ∂_{33}) but not at all even partly regular for the compatibility conditions (A is not solved with respect to ∂_{333}). Nevertheless, in a generic sense we have

$$\dim \operatorname{im} \sigma_\chi(\mathcal{D}) = \dim \ker \sigma_\chi(\mathcal{D}_1) = 1,$$

in agreement with the general theory.

This Example clearly shows that sometimes the only way to find out the controllability of a system is to look for a parametrization of it.

Example 26 (Free motion of a pendulum). Let us consider the movement of a pendulum of mass m and length L under the gravity $(0, g)$ in a coordinate system (x, y). Denoting by T the torsion of the pendulum, the equations of motion become:

$$\mathcal{R}_2 \quad \begin{cases} m\ddot{x} + T\dfrac{x}{L} = 0, \\[2mm] m\ddot{y} + T\dfrac{y}{L} + g = 0, \\[2mm] x^2 + y^2 - L^2 = 0. \end{cases}$$

Prove that the corresponding formally integrable system $\mathcal{R}_2^{(4)}$ has zero symbol. Show that each unknown x, y, T must satisfy one ODE of order 3 by itself. Modify the control system by considering an elastic pendulum with law $L - L_0 = kT$, and study this new situation in a similar way.

CHAPTER VIII

Continuum Physics

In 1909, the brothers Eugène and François Cosserat developed a specific variational calculus based on group theory in order to relate geometry (strain) and physics (stress) in the mechanics of continua with nonsymmetric stress tensor. At first they wanted to extend to continua the well-known variational techniques of rigid bodies and analytical dynamics. This evolution is particularly clear from the titles of their successive publications [11, 38, 39, 41, 42]. However, they also wanted to clearly define the concept of *torsor* as a generalization of the concept of *vector* distinct to that of *tensor*, while showing that *the form of the static or dynamic equations for stress only depends on the structure of the underlying group involved*. They succeeded. However, all this involved tedious computations, and when François Cosserat died in 1914, his brother did not publish anymore on this topic.

In order to extend the argument by enlarging the group, H. Weyl, in 1917 [196, 197], tried, like J.C. Maxwell before him, to incorporate electromagnetism in a kind of similar 'dynamics'. He failed, but did find the basis of modern gauge theory. A similar objective was previously hold by H. Helmholtz [28], who tried to establish a link between elastodynamics and thermodynamics, while E. Mach and G. Lippmann [3, 111, 116] tried to establish a link between thermodynamics and electrodynamics.

More recently, G. Birkhoff in 1954 [26] and V. Arnold in 1966 [12, 13] (with no reference to Birkhoff's work although using the same formulas!) had difficulties when trying to pass from the dynamics of rigid bodies (Lie group of transformations) to the hydrodynamics of incompressible fluids (Lie pseudogroup of volume-preserving transformations).

Here we shall merely sketch these methods, and apply them in the study of the nonlinear field theories encountered in the engineering sciences. We will obtain the following three, rather striking, results:

(1) The brothers Cosserat tried merely to compute the two first nonlinear Spencer operators for the group of rigid motions.
(2) This method immediately gives all the results conjectured by Arnold and, in particular, a method of passing from rigid to elastic materials.
(3) Thermoelasticity, thermoelectricity, photoelasticity, viscosity and streaming birefringence, which are well-known field–matter couplings at the level of the engineering sciences, all follow immediately from this procedure, which hence contradicts the distinctions usually made between *reversible phenomena* and

irreversible phenomena.

Our exposition will prove that a better understanding of *dynamics* in general, i.e. elasto-, thermo-, electro- in increasing order of complexity, *must* follow from the following conceptual chain:

DYNAMICS ON THE GROUP OF RIGID MOTIONS

∩

DYNAMICS ON A LIE GROUP

∩

DYNAMICS ON A LIE GROUPOID.

At the same time we will *prove* that most of the differential geometry of nowadays mathematical physics is based upon a confusion between the *Janet sequence* (1920) and the *Spencer sequence* (1970) on the one hand, and between the *gauge sequence* (1960) and the *Spencer sequence* (1970) on the other hand. Correcting this confusion by means of new homological methods will immediately bring about a need for revisiting the foundations of apparently well-established theories, including continuum mechanics, thermodynamics, electromagnetism, gauge theory and general relativity.

In particular, for the first time ever, we will give a modern exposition of the elasticity theory of E. and F. Cosserat along the lines of the work of D.C. Spencer, while also extending the latter to electromagnetism, along ideas of H. Weyl for using the conformal group of space–time. Finally, similar to the noncompatibility of Newton's theory of gravitation with group theory, we will *prove* that Yang–Mills theory of electromagnetism and Einstein's theory of gravitation are not compatible with pseudogroup theory. More precisely, *the Yang–Mills and Einstein equations are not compatible with the Spencer sequence.*

Before giving a mathematical exposition of the techniques involved, we will quote 10 famous problems, the solution to which is still a challenge in continuum physics. We will also comment on relations between these problems, and on the tentative solutions proposed up till now.

Photoelasticity, as an experimental coupling between deformation and electric field in any transparent elastic material, has been discovered by Brewster in 1815. The phenomenological laws have been proposed independently by F.E. Neumann and J.C. Maxwell in 1850. Until recently one used to rely on the mathematical formulation by Pöckels (1889), but nowadays much theoretical and experimental work can be found in the literature.

Later, around 1872, Maxwell and Mach claimed and proved the existence of *streaming birefringence*, a coupling between rate of deformation and electric field in certain transparent viscous fluids, such as the polymer–resin part of two-component glues. Roughly speaking, as we will see below, the idea is to use the same phenomenological formula for the dielectric tensor as for photoelasticity, and to substitute rate of deformation instead of deformation (both are symmetric tensors). This thus leads to the same formula, but with different coupling constants (experimental coefficients). Here we recall that the rate of deformation, i.e. the derivative with respect to time of the strain tensor, has components equal to half the symmetrized sum of the derivatives of velocity. In fact, we will see that the common underlying idea is to enlarge the *constitutive laws* of both elasticity and electromagnetism by introducing *crossed terms*, responsible for the coupling. Of course, this phenomenological description makes sense *if and only if* one can give a geometric origin of electromagnetism similar to that of elasticity. *Since elasticity rests on group theory, this is a first experimental reason for believing that electromagnetism also rests on group theory.* Hence we are left with two questions:

- How to choose the group?
- How to use the group?

However, looking at the way viscosity is explained in any contemporary textbook on continuum mechanics, we have to face our first problem:

Problem 1. Although photoelasticity is treated by means of *reversible thermodynamics*, using Taylor expansion of the thermodynamic free energy up to third-order terms for isotropic materials, streaming birefringence can only be treated by means of *irreversible thermodynamics*, as rate of deformation can only enter the dissipative function. How to avoid this conceptual contradiction and how to unify the two formulations?

Similarly, *elasticity*, *heat*, and *electromagnetism* are three phenomena, pairwise coupled by *thermoelasticity*, *photoelasticity*, and *thermoelectricity*. We have already discussed the second coupling, while the first coupling has been known also in very ancient times. As for the last coupling, the construction of thermocouples by soldering together two different metals, and their use to measure high temperatures that cannot be measured otherwise by ordinary thermometers, is a rather standard industrial device in any cooking apparatus. Hence we are led to our second problem, since it is well known that thermoelasticity arises from the Taylor expansion of the free energy up to order two, when regarded as a function of deformation and temperature.

Problem 2. How is it possible that the two first couplings are treated by means of reversible thermodynamics, while the third is treated by means of irreversible thermodynamics?

Of course, if this unification were possible, it would give a geometric origin to temperature and *absolute temperature*, in particular. Also, it would put a shadow on the framework (dissipative potential, so-called *fluxes* and *forces*, the purely statistical approach, ...) used to exhibit the *Onsager relations*, because thermoelectricity is one of the best examples of these laws, which relate to the creation of entropy, a process based on the existence of a gradient of temperature, or on viscosity, and thus not interfering with thermoelasticity and photoelasticity. Certain authors have even explicitly stated this unification as a future challenge for testing the validity of ideas of Truesdell or Prigogine [45, 46, 121]. In that domain we also notice that changing the sign of the magnetic field occurring in the Onsager coefficients, as a macroscopic consequence of changing the sign of time on the statistical microscopic level, is absolutely contradictory to the basic tensorial nature of elementary special relativity theory. Indeed $\vec{B} = \vec{\nabla} \wedge \vec{A}$ does not interfere with time [77].

However, this unification will make sense only if the mathematical formulation of continuum mechanics can be clearly stated, at least as far as elasticity or elastody-namics are concerned. Surprisingly, and despite many modern theoretical attempts (Truesdell, Noll, Coleman, Rivlin, Toujin, ...) the situation has not evolved much since the beginning of this century, as we will see [45, 184]. Moreover, even the com-putational aspect, nowadays widely known as the *finite-element method*, raises some questions.

For the nonspecialist reader, we give a very rough sketch of this method for the problem of heating a room with wall S_0 at constant (absolute) temperature $T = T_0$ and given heat flux $\vec{q} = \vec{q}_1$ on the window S_1. At any point the heat flux $\vec{q}$ is related to the gradient of temperature by Fourier's law $\vec{q} = \kappa \vec{\nabla} T$, where κ is the thermal conductivity, and the heat equation $\vec{\nabla} \cdot \vec{q} = \kappa \Delta T = 0$ holds when no heat is created in any volume element. The finite-element technique applies to any engineering problem with equations governed by a variational principle. Here we prove that the solution of the problem for temperature T in the volume V bounded by the surface $S = S_0 + S_1$ is equivalent to the variational problem

$$\delta \left(\int_V \frac{\kappa}{2} (\vec{\nabla} T)^2 \, dV - \int_{S_1} T \vec{q} \cdot \vec{n} \, dS \right) = 0.$$

Indeed, consider the expression

$$\int_V \frac{\kappa}{2} (\vec{\nabla}(T + \delta T))^2 \, dV - \int_{S_1} (T + \delta T) \vec{q} \cdot \vec{n} \, dS,$$

and subtract from it the expression

$$\int_V \frac{\kappa}{2} (\vec{\nabla} T)^2 \, dV - \int_{S_1} T \vec{q} \cdot \vec{n} \, dS.$$

A short computation and one integration by parts gives

$$\int_S \kappa (\vec{n} \cdot \vec{\nabla} T) \delta T \, dS + \int_V \frac{\kappa}{2} (\vec{\nabla} \delta T)^2 \, dV - \int_{S_1} \delta T \vec{q} \cdot \vec{n} \, dS.$$

However, on S_0 we have fixed $T = T_0$, and so $\delta T|_{S_0} = 0$. So, the only term left is

$$\int_V \frac{\kappa}{2} (\vec{\nabla} \delta T)^2 \, dV \geq 0,$$

and the effective distribution of temperature minimizes the expression

$$\int_V \frac{\kappa}{2} (\vec{\nabla} T)^2 \, dV - \int_{S_1} T \vec{q} \cdot \vec{n} \, dS.$$

Fixing n points inside V, we may compute the approximate values of the various derivatives of T by using the values $T_1, \ldots, T_n$ at these points. Accordingly, we arrive at the problem of minimizing an expression of the form

$$\frac{1}{2} \sum_{i,j} A_{ij} T_i T_j - \sum_i B_i T_i,$$

and we are led to solving the linear system

$$\sum_j A_{ij} T_j = B_i,$$

i.e. to inverting the $(n \times n)$-matrix $A = (A_{ij})$. The last step can be done by a computer, even for n quite large to obtain a more precise result.

Recapitulating the above method, we start with T, compute $\vec{\nabla}T$, then express the heat flux by Fourier's law, and finally take into account the fact that the heat equation comes from a variational problem.

Surprisingly, a completely similar scheme can be adopted for elasticity problems when the displacement is known on a part S_0 and the stress is known on a part S_1 of the surface, where $S = S_0 + S_1$ is the surface bounding the volume V of continuum under study.

In that case we start with displacement, compute the deformation tensor, then express the stress by Hooke's law for the stress/strain relation, and finally take into account the fact that the stress equation also comes from a variational problem.

We are immediately led to the following problem, which is, in our opinion, the most critical one in engineering science since it touches upon most fundamental concepts.

Problem 3. On a conceptual level it is clear that, in classical thermoelasticity, *deformation* and *temperature* are on the same level as variables in the free energy. On the other hand, we have just seen that, on the computational level, it is clear that *displacement* and *temperature* are on the same level as variables that have to be made discrete. Hence we may ask whether temperature has to do with displacement or with deformation, since it cannot interfere with both at the same time. Of course, if finite elements must be used as guide line, then the gradient of temperature must be used on the level of deformation, and we may have to revisit the distinction between reversible and irreversible phenomena!

Considering now the mathematical aspect, it is well known that the notion of *vector* is a particular case of that of *tensor*, i.e. of a *geometric object* described by components transforming in a given way depending on the derivatives of any coordinate change up to a certain order (order one for tensors, order two for connections, ...). However, the other generalization of vector, namely *torsor* and, in particular, *torsor field*, is still poorly understood, although it is believed that it may be linked to group theory by regarding the dual of the Lie algebra of the group of rigid motions [19]. Hence we are led to the following problem:

Problem 4. What is the difference between a torsor and a geometric object?

A basic feature of contemporary mathematical physics, especially gauge theory and particle physics, is to describe *field theory* by means of *group theory* in such a way that the larger the group $(U(1), SU(2), SU(3), \ldots)$, the larger the number of fields that can be described by this approach. Unfortunately, the larger a group of transformations, the smaller the number of (differential) invariants it admits, and this is one of the reasons why the Lie groups of particle physics are 'abstract' groups, i.e. groups not acting on the base manifold (space–time). The situation is particularly clear in continuum mechanics, where an increase in the group of rigid motions by means of (space) dilatations must be compensated for by using the deviator (traceless part) instead of the strain tensor. Also, as we will see in more detail in section B and although this crucial point is not dealt with in textbooks, if we use the standard invariant approach to continuum mechanics, founding its geometric part on the differential invariant of the group of rigid motions, the *only* geometric object that can be introduced is a symmetric tensor, the deformation tensor, and its dual has no reason to be a symmetric tensor density. Namely, symmetry of the stress tensor is a nontrivial property that derives from equilibrium of a ... torsor!

Hence, to avoid such a vicious circle, it seems that *starting with the same group, a new mathematical approach may exist*, to avoid invariant concepts. Of course, if such an approach does exist, it will surely not be found in any known contemporary work on the foundation of continuum mechanics or hydrodynamics, where only very classical tools of Lie group theory are used [12, 13, 25]. This criticism applies to the approach of both the Truesdell and the Arnold school, and we are led to the following problem:

Problem 5. How to deal with Lie pseudogroups without regarding them as infinite Lie groups.

The only possible deductive mathematical way for exhibiting such a group foundation of continuum mechanics has been proposed by E. and F. Cosserat in 1909, and is clearly linked to the solution of problems 4 and 5. However, despite many contemporary attempts [175, 177, 184], the use of vector calculus in combination with Lie group theory has been quite poor as regards understanding the mathematical formulation of their book [11, 38, 39, 41, 93]. Hence we are led to the following problem:

Problem 6. What is the mathematical key of this book; in more detail, what is the modern setting in which the large amount of local computations given in it, and which are clearly not of a tensorial nature, can be used?

In fact, we will see that the clever discovery of E. and F. Cosserat is to base elasticity theory not on the Janet sequence, but rather on the Spencer sequence. Therefore the *geometry* of elasticity is expressed by the Spencer operator, while the physics is expressed by its dual in the adjoint sense of Chapter VII. For this reason they have been able to obtain *in one block*, the full set of stress/couple-stress equations; namely

$$\begin{cases} \partial_i \sigma^{ij} = f^j, \\ \partial_r \mu^{rij} + \sigma^{ij} - \sigma^{ji} = m^{ij}, \end{cases}$$

where (σ^{ij}) is the stress, $(\mu^{rij} = -\mu^{rji})$ is the couple-stress, (f^j) the volume density of forces, and $(m^{ij} = -m^{ji})$ is the volume density of couples. According to these formulas, the symmetry of the stress tensor comes from the *strong* constitutive assumption

$$\mu^{rij} = 0, \qquad m^{ij} = 0,$$

which is, luckily, usually satisfied in most engineering materials (metal, concrete, ...). As a first byproduct, not only the *number* of equations depends only on the group, according to the principle of virtual work (as many forces f^j as translations, as many couples m^{ij} as rotations, expressed by a skewsymmetric matrix), but also the *form* of the equations depends only on the group, and even on its Lie algebra alone ... but no Lie group technique has to be used!

Although this is not very well known, between 1905 and 1910 E. and F. Cosserat have also tried to incorporate *thermodynamics* (the reference to a 'theory of temperature' in [41, pp. 6, 147, 211] speaks for itself!) and *electrodynamics* ([41, pp. 2, 209] or [39, Ch. V, p. 209]) into their *theory of deformable bodies*, as a kind of interaction between geometry and physics, by means of two analogies.

The first analogy is known as the *Helmholtz analogy* ([38, 39, III.2, p. 537; III.3, p. 994]). In analytical mechanics, if $L(t, q, \dot{q})$ is the Lagrangian of a mechanical system with generalized variables q that are functions of time t, one can obtain the Hamiltonian via

$$H = \dot{q}\frac{\partial L}{\partial \dot{q}} - L.$$

Similarly, in thermostatics, if F is the free energy of a system at absolute temperature T, then, in general, say for a perfect gas,

$$dU = -P\,dV + T\,dS,$$

from the first and second principles of thermodynamics, where U is the internal energy, P the pressure, V the volume, and S the entropy. The relation between U and F is $F = U - TS$, and so

$$dF = -P\,dV - S\,dT,$$

proving that F is a function of V and T that is much more amenable to engineering than U itself, which is a function of V and S, because S cannot be *measured*. Looking for the partial derivatives, one obtains

$$U = F - T\frac{\partial F}{\partial T}.$$

Therefore thermostatics admits a variational scheme, under the condition that $L = -F$ after having chosen q such that $\dot{q} = T$. Unfortunately, and despite many recent attempts [28, 90, 186], the meaning of such a variable is unclear, and we are led to the following problem:

Problem 7. How to set up the Helmholtz analogy properly?

Although this analogy seems clearly stated, it is related to a very delicate problem that is often not pointed out in textbooks and to which we will now turn.

In the axiomatic formulation of reversible thermodynamics, the evolution of a system described by internal state variables is characterized by a 1-form α of work exchange and a 1-form β of heat exchange between the system and its surrounding. These forms must satisfy

$$\begin{cases} \alpha \wedge \beta \neq 0, \\ d(\alpha + \beta) = 0 & \text{(first principle)}, \\ \beta \wedge d\beta = 0 & \text{(second principle)}, \end{cases}$$

with the standard notations of exterior calculus. One should point out that the constructive version of the second principle, although dating back to E. Cartan (1910) is not that often used in practice and cannot be found in textbooks, even though *it allows one to avoid the implicit existence of absolute temperature as an integrating factor*.

Using the local solvability of the exterior derivative, one may of course set

$$\alpha + \beta = dU, \qquad \beta = T\,dS,$$

as usual. It follows that the *Helmholtz analogy still depends on the Helmholtz postulate*, ensuring the existence of so-called '*normal*' state variables, namely, state variables including the absolute temperature T and such that α does not contain dT or, equivalently, such that $S = -\partial F/\partial T$, since otherwise this relation is wrong.

Contrary to the belief in the literature, we have proved [145, p. 713] that *the above postulate is in fact a theorem*, and even an example of the famous *equivalence problem*. We recall that the equivalence problem is concerned with the possibility of passing from a certain geometric object to a given model by a change of local coordinates. Therefore, in our opinion, the practical use of the first and second principles in the reversible (static) case is that there is a *reason* to *separate* the absolute temperature from the other variables in the free energy, which is therefore endowed with much more importance than any other thermodynamic function (similarly to the fact that the Lagrangian surpasses the hamiltonian in analytical dynamics).

The following comparison with analytical dynamics also explains why the choice of normal variables amounts to using the two principles, under the condition that it can be shown that *strain and absolute temperature have a geometric origin that stress and entropy do not have*. Both proofs can be given by induction on the number of variables, starting with two variables.

domain	dynamics	thermostatics
problem	Darboux	Helmholtz
object	2-form ω	1-forms α, β
algebraic condition	$\det \omega \neq 0$	$\alpha \wedge \beta \neq 0$
differential condition	$d\omega = 0$	$d(\alpha + \beta) = 0, \beta \wedge d\beta = 0$
model	$\omega = \sum dp \wedge dq$	$\beta = T\,dS, dT$ not in α
coordinates	canonical	normal

The second analogy is known as the *Mach–Lippmann analogy* [3, 111, 116]. In thermostatics, a thermodynamic engine, say a steam machine, following a Carnot cycle satisfies the relation

$$\oint dS = \oint \frac{dQ}{T} = 0 \qquad \text{(Clausius formula)},$$

where Q is the heat, S the entropy, and T the absolute temperature. In electrostatics, the energy of a condensor of capacity C and potential V is $W = \frac{1}{2}CV^2$, while its charge is $Q = CV$ with standard notations (be careful not to be confused between notations). Hence, in a cycle of charge/discharge we have

$$\oint dQ = \oint \frac{dW}{V} = 0 \qquad \text{(conservation of charge)}.$$

The analogy between T and V is confirmed by the thermoelectric Seebeck effect, which establishes a linear relation between their respective gradients. However, its extension to the 4-potential $(\vec{A}, V)$ or (A_i) is still a conjecture, and we are led to the following problem:

Problem 8. How to set up the Mach–Lippmann analogy properly?

One may nevertheless notice that, in order to strengthen this analogy, that the finite-element method for electromagnetism starts with the 4-potential, constructs the electromagnetic field $(\vec{B} = \vec{\nabla} \wedge \vec{A}, \vec{E} = -\vec{\nabla}V - \partial\vec{A}/\partial t)$ or $(F_{ij} = \partial_i A_j - \partial_j A_i)$, and then constructs the electromagnetic induction $(\vec{H}, \vec{D})$ via the constitutive law $\vec{H} = \vec{B}/\mu$, $\vec{D} = \epsilon\vec{E}$, which involves the experimental constants ϵ and μ. For vacuum, the values ϵ_0 and μ_0 of these constants satisfy $\epsilon_0\mu_0 c^2 = 1$, where c is the velocity of light in vacuum. In space–time formalism, the constitutive law takes the form proposed by H. Minkowski [135, 181, 196]: with $x^4 = ct$, $A_4 = -V/c$ we may set

$$\begin{pmatrix} F_{11} & F_{12} & F_{13} & F_{14} \\ F_{21} & F_{22} & F_{23} & F_{24} \\ F_{31} & F_{32} & F_{33} & F_{34} \\ F_{41} & F_{42} & F_{43} & F_{44} \end{pmatrix} = \begin{pmatrix} 0 & B^3 & -B^2 & \frac{1}{c}E^1 \\ -B^3 & 0 & B^1 & \frac{1}{c}E^2 \\ B^2 & -B^1 & 0 & \frac{1}{c}E^3 \\ -\frac{1}{c}E^1 & -\frac{1}{c}E^2 & -\frac{1}{c}E^3 & 0 \end{pmatrix},$$

$$\begin{pmatrix} \mathcal{F}^{11} & \mathcal{F}^{12} & \mathcal{F}^{13} & \mathcal{F}^{14} \\ \mathcal{F}^{21} & \mathcal{F}^{22} & \mathcal{F}^{23} & \mathcal{F}^{24} \\ \mathcal{F}^{31} & \mathcal{F}^{32} & \mathcal{F}^{33} & \mathcal{F}^{34} \\ \mathcal{F}^{41} & \mathcal{F}^{42} & \mathcal{F}^{43} & \mathcal{F}^{44} \end{pmatrix} = \begin{pmatrix} 0 & H^3 & -H^2 & -cD^1 \\ -H^3 & 0 & H^1 & -cD^2 \\ H^2 & -H^1 & 0 & -cD^3 \\ cD^1 & cD^2 & cD^3 & 0 \end{pmatrix},$$

and obtain the constitutive law

$$\mathcal{F}^{kl} = \frac{1}{\mu}(\omega^{ik} - (n^2 - 1)u^i u^k)(\omega^{jl} - (n^2 - 1)u^j u^l)F_{ij},$$

where u^i is the (normalized) relativistic velocity and n is the refraction index, such that the following holds: $\mathcal{F}_{ij} = F_{ij}/\mu_0$ in vacuum, if the indices are raised or lowered by means of the Minkowski metric of space–time.

Finally, it should be noticed that the second set of Maxwell equations, i.e.

$$\vec{\nabla} \wedge \vec{H} - \frac{\partial\vec{D}}{\partial t} = \vec{j}, \qquad \vec{\nabla} \cdot \vec{D} = \rho,$$

or simply

$$\partial_i \mathcal{F}^{ij} = J^j,$$

on space–time, derive from the variational principle

$$\delta \int_V \left(\frac{\epsilon}{2} \vec{E}^2 - \frac{1}{2\mu} \vec{B}^2 \right) dV = 0,$$

or simply

$$\delta \int \frac{1}{4\mu_0} F^{ij} F_{ij} \, dx^1 \wedge dx^2 \wedge dx^3 \wedge dx^4 = 0$$

in vacuum (see again the excellent reference [135] for details). We thus discover that, in the finite-element method, the *absolute temperature and the 4-potential are on the same conceptual level*, while the gradient of temperature and the electric field are on the same level with the deformation.

However, the finite-element method has only been considered as a way to numerically solve engineering problems, and the above analogy (*generalizing the Mach–Lippmann analogy*) has only been considered as purely accidental, with no theoretical support. It immediately follows that our results will give, for the first time ever, a group-theoretical unification and understanding of the finite-element method for elasticity, heat, electromagnetism, and their known couplings.

Two last, but not least, delicate and well-known problems originate from the heat equation. Again, much work has recently been done, but we do believe that none of these is quite convincing, since most of them rely on phenomenological correct terminology or 'ad hoc' artificial computations [83, 119]. The first problem is as follows:

Problem 9. The heat equation, as obtained from the principles of thermodynamics, is of parabolic type and should admit thermal waves of infinite velocity, a result contradicting the basic postulate of special relativity concerning the velocity of light.

The second problem, though not often quoted, is even more contradictory (in our opinion) with special relativity. Indeed, for time-dependent problems such as fast cooling in nuclear plants, the Laplacian of the absolute temperature must be corrected by a term containing the derivative of the absolute temperature with respect to time, and the heat equation becomes

$$\kappa \Delta T - \rho c \frac{\partial T}{\partial t} = 0,$$

where ρ is the mass per unit volume and c is the heat capacity per unit mass. Multiplying by δT we obtain

$$\Delta T \delta T = \vec{\nabla} \cdot (\delta T \vec{\nabla} T) - \delta \left(\tfrac{1}{2} (\vec{\nabla} T)^2 \right),$$

but no such decomposition can be found for

$$\frac{\partial T}{\partial t} \delta T = \frac{\partial}{\partial t} (T \delta T) - T \frac{\partial \delta T}{\partial t},$$

and we are led to the problem:

Problem 10. The term $\partial T / \partial t$ cannot come from a variational procedure, even if integration is done on space–time.

Of course, if the solution of Problem 9 would bring only slight deviations from experimental results in engineering problems, the solution to Problem 10 would make it possible to use finite-element techniques even for time-dependent problems, instead of using finite elements for the space behavior and finite-difference type methods for the time evolution.

The purpose of this Chapter is to present a tentative solution of these Problems by means of the new mathematics presented in Chapters I–V. The reader will notice that *all the material* contained in those Chapters is *absolutely necessary* for our purpose, although at first sight the contents of these Chapters may seem *very far* from the desired applications to continuum physics.

As a byproduct, because gauge theory derives from a mathematical model of electromagnetism (Maxwell equations) and general relativity derives from a mathematical model of continuum mechanics (the divergence condition for the Einstein tensor comes from the stress equations on space–time), it will follow that the Yang–Mills and Einstein equations will no longer be compatible with these group-theoretical unification.

However, since this unification is not based on any physical assumption but only depends on the mathematical description of the Spencer operator, *the contradiction encountered is inescapable from the purely mathematical point of view*, and brings about the need to revisit the foundations of both gauge theory and general relativity.

Of course, such an approach is far from being complete, since it has started too recently. However, we hope that it will give new impulses to mathematical physics.

A. Variational Calculus

We start this Section with a few comments about the classical variational calculus and the various generalizations that have been proposed up till now. Then we will describe, in a purely mathematical way, the new variational calculus on groupoids, discovered by E. and F. Cosserat. Motivation from mechanics and the justification will be given in the next Section.

Classical analytical dynamics deals with movements of rigid bodies (beams, tops, rings, etc.). The position of such bodies in space is described by a certain number of variables, usually denoted by q^i. Their movement is described by letting the variables depend on time t, so that we have $q^i(t)$, and the corresponding velocity is usually denoted by $\dot{q}^i$. By a *Lagrangian* we understand a function of all these variables that is usually denoted by $L(t, q, \dot{q})$. By integration with respect to time and taking the variation, we obtain

$$\delta \int L(t, q, \dot{q})\, dt = \int \left(\frac{\partial L}{\partial q} \delta q + \frac{\partial L}{\partial \dot{q}} \delta \dot{q} \right) dt$$

$$= \int \left(\frac{\partial L}{\partial q} \delta q + \frac{\partial L}{\partial \dot{q}} \frac{d}{dt} (\delta q) \right) dt$$

$$= \left[\frac{\partial L}{\partial \dot{q}} \delta q \right] - \int \left(\frac{d}{dt} \left(\frac{\partial L}{\partial \dot{q}} \right) - \frac{\partial L}{\partial q} \right) \delta q\, dt.$$

As a matter of fact, one can prove that Newton dynamics implies the following second-order *Euler–Lagrange equations* [105, 199]:

$$\frac{d}{dt}\left(\frac{\partial L}{\partial \dot{q}^i}\right) - \frac{\partial L}{\partial q^i} = 0,$$

or simply

$$\frac{d}{dt}\left(\frac{\partial L}{\partial \dot{q}}\right) - \frac{\partial L}{\partial q} = 0.$$

One of the simplest examples is the free fall of a particle of mass m in a central gravitational field, described by the Lagrangian:

$$L \equiv \frac{m}{2}(\dot{x}^2 + \dot{y}^2 + \dot{z}^2) + \frac{k}{r},$$

where the radius r is the distance to the attracting origin point and k is a constant describing the gravity. Accordingly, the movement is described by the ODE

$$\begin{cases} m\ddot{x} + \dfrac{k}{r^3}x = 0, \\[2mm] m\ddot{y} + \dfrac{k}{r^3}y = 0, \\[2mm] m\ddot{z} + \dfrac{k}{r^3}z = 0. \end{cases}$$

This elementary result has convinced people that dynamics is related to the Euler–Lagrange equations. Our purpose here is to prove that reality is not that simple.

In certain cases, generalized forces Q_i must be introduced by means of their virtual work $\sum_i Q_i \delta q^i$, and one may introduce them as a kind of right hand term, as follows:

$$\frac{d}{dt}\left(\frac{\partial L}{\partial \dot{q}}\right) - \frac{\partial L}{\partial q} = Q.$$

When these forces are given as functions $Q(t, q, \dot{q}, \ddot{q})$, an interesting mathematical problem is to recover the Lagrangian solution of the above system. The answer is given by the following set of necessary and sufficient conditions, also called the *Helmholtz conditions* [9, 99, 187, ...]:

$$\begin{cases} H_{ij}^0 \equiv \dfrac{\partial Q_j}{\partial \ddot{q}i} - \dfrac{\partial Q_i}{\partial \ddot{q}^j} = 0, \\[3mm] H_{ij}^1 \equiv \dfrac{\partial Q_j}{\partial \dot{q}i} + \dfrac{\partial Q_i}{\partial \dot{q}j} - 2\dfrac{d}{dt}\dfrac{\partial Q_i}{\partial \ddot{q}^j} = 0, \\[3mm] H_{ij}^2 \equiv \dfrac{\partial Q_j}{\partial q^i} - \dfrac{\partial Q_i}{\partial q^j} + \dfrac{d}{dt}\dfrac{\partial Q_i}{\partial \dot{q}j} - \dfrac{d^2}{dt^2}\dfrac{\partial Q_i}{\partial \ddot{q}^j} = 0, \end{cases}$$

which is easily seen to be compatible. This problem should not be mistaken for the problem already studied in Section III.D, although both problems are sometimes called *inverse problems of variational calculus*.

Finally, if $\Phi(t, q)$ is a *conserved quantity*, i.e. if

$$\frac{d}{dt}\Phi \equiv \frac{\partial\Phi}{\partial t} + \frac{\partial\Phi}{\partial q}\dot{q} = 0,$$

then we can introduce a right hand term $L(t, q, \dot{q})$ by setting

$$\frac{d}{dt}\Phi \equiv \frac{\partial\Phi}{\partial t} + \frac{\partial\Phi}{\partial q}\dot{q} = L,$$

and we obtain, in succession,

$$\frac{d}{dt}\left(\frac{\partial L}{\partial \dot{q}}\right) - \frac{\partial L}{\partial q} = \frac{d}{dt}\left(\frac{\partial\Phi}{\partial q}\right) - \frac{\partial^2\Phi}{\partial t\partial q} - \frac{\partial^2\Phi}{\partial q\partial q}\dot{q}$$

$$= \frac{\partial^2\Phi}{\partial t\partial q} + \frac{\partial^2\Phi}{\partial q\partial q}\dot{q} - \frac{\partial^2\Phi}{\partial t\partial q} - \frac{\partial^2\Phi}{\partial q\partial q}\dot{q}$$

$$= 0.$$

Collecting the results, we obtain a kind of differential sequence, called the *variational sequence*:

$$\Phi \xrightarrow{\text{CL}} L \xrightarrow{\text{EL}} Q \xrightarrow{\text{HC}} H,$$

where the successive operators are the total derivative expressing the conservation law (CL), the Euler–Lagrange equation (EL) and the Helmholtz condition (HC).

Roughly speaking, the whole evolution of variational calculus during the last fifty years can be said to have been motivated by the following two problems:

- How to generalize the sequence above to several independent variables and higher-order operators?
- How to give an intrinsic description of such a sequence?

We will now give a brief account of these two approaches, and make some comments.

First of all, retaining the single independent variable t and increasing the order of the derivatives appearing in the lagrangian, we can again use total derivatives and integration by parts to easily extend to this setting the two first operators in the sequence above. The search for the Helmholtz condition is more involved, but can be successfully completed. E.g., one can verify that, for any function $\Phi(t, q, \dot{q})$ the total derivative

$$\frac{d\Phi}{dt} \equiv \frac{\partial\Phi}{\partial t} + \frac{\partial\Phi}{\partial q}\dot{q} + \frac{\partial\Phi}{\partial \dot{q}}\ddot{q} = L$$

satisfies the extended Euler–Lagrange equation

$$\frac{d^2}{dt^2}\left(\frac{\partial L}{\partial \ddot{q}}\right) - \frac{d}{dt}\left(\frac{\partial L}{\partial \dot{q}}\right) + \frac{\partial L}{\partial q} = 0.$$

If we retain the first order but increase the number of variables to n, and consider a section of a fibered manifold $\mathcal{E}$ over X, we may introduce the various jets of sections $y^k = f^k(x)$, $y_i^k = \partial_i f^k(x)$ and consider the action W to be defined by the integral

$$W = \int w(x, f^k(x), \partial_i f^k(x))\, dx^1 \wedge \cdots \wedge dx^n,$$

for a function $w(x, y_1)$ of the first jets. There is no need for giving the integration limits, since we need only formal results. In fact, the question of integration limits is troublesome, since it gives the feeling that the equations appearing in the variational sequence must have a definition independent of the 'cooking' procedure appealing to an integration by parts. Nevertheless, at this stage we obtain as before, by taking variations and integrating by parts:

$$\delta W = \int \left(\frac{\partial w}{\partial y^k} \delta f^k + \frac{\partial w}{\partial y_i^k} \delta \partial_i f^k \right) dx^1 \wedge \cdots \wedge dx^n$$

$$= (-1)^{i-1} \int \frac{\partial w}{\partial y_i^k} \delta f^k \, dx^1 \wedge \cdots \wedge \widehat{dx^i} \wedge \cdots \wedge dx^n$$

$$- \int \left(d_i \left(\frac{\partial w}{\partial y_i^k} \right) - \frac{\partial w}{\partial y^k} \right) \delta f^k \, dx^1 \wedge \cdots \wedge dx^n,$$

where the symbol $\frown$ indicates omission. In the resulting Euler–Lagrange equations

$$d_i \left(\frac{\partial w}{\partial y_i^k} \right) - \frac{\partial w}{\partial y^k} = 0$$

we have to understand that we differentiate and take the formal derivative at the jet level, and jet substitute $j_2(f)$ for second-order jets. Equivalently, $j_2(f)$ is a section of $J_2(\mathcal{E})$, and we have to take the pullback of the Euler–Lagrange expression to obtain functions on X. In this and subsequent Sections, this procedure will be implicitly understood.

The intrinsic description of the preceding local arguments may be given by considering the Lagrangian as a map $w \colon J_1(\mathcal{E}) \to \bigwedge^n T^*$ that must be composed on the left with $j_1 \colon \mathcal{E} \to J_1(\mathcal{E})$. Any section f of $\mathcal{E}$ may make it possible to define a section of $\bigwedge^n T^*$ by composition:

$$X \xrightarrow{j_1(f)} J_1(\mathcal{E}) \xrightarrow{w} \bigwedge^n T^*$$

and an action W by integration.

It is crucial that the variation δf of f is such that $(f, \delta f)$ is a section of the vertical bundle $E = V(\mathcal{E})$. It then follows from the isomorphism $V(J_1(\mathcal{E})) \simeq J_1(V(\mathcal{E}))$ that we may set $\delta j_1(f) = j_1(\delta f)$, with a slight abuse of language. By composition and linearization we obtain the composition

$$0 \longrightarrow T^* \otimes E \longrightarrow J_1(E) \xrightarrow{V(w)} \bigwedge^n T^*$$

again with a slight abuse of language. We can consider the resulting *momentum map* as an element in

$$\bigwedge^n T^* \otimes T \otimes E^* \simeq \bigwedge^{n-1} T^* \otimes E^*,$$

in order to understand the integration by parts as a manner of dualizing j_1 along the results of Chapter VII whenever we are dealing with a vector bundle E over X instead of with a fibered manifold.

A natural way to deal with *higher-order variational calculus* seems to introduce an action map $w \colon J_q(\mathcal{E}) \to \bigwedge^n T^*$ and similarly obtain the composition

$$0 \longrightarrow S_q T^* \otimes E \longrightarrow J_q(E) \xrightarrow{V(w)} \bigwedge^n T^*.$$

However, *we have no means for exhibiting a momentum map* from the corresponding element of $\bigwedge^n T^* \otimes S_q T \otimes E^*$. Indeed, the Spencer map provides a monomorphism $S_q T^* \otimes E \to T^* \otimes S_{q-1} T^* \otimes E$ which can be dualized to an epimorphism $T \otimes S_{q-1} T \otimes E^* \to S_q T \otimes E^*$. The latter cannot be used in a composition, unless we can split it by adding an external object.

In the first-order case, we may consider a *constrained variational calculus* by introducing a fibered submanifold $\mathcal{R}_1 \subset J_q(\mathcal{E})$ and an action map $w \colon \mathcal{R}_1 \to \bigwedge^n T^*$. As before, introducing the symbol, we obtain a composition

$$0 \longrightarrow M_1 \longrightarrow R_1 \xrightarrow{V(w)} \bigwedge^n T^*,$$

but, again, *there is no way of exhibiting a momentum map* from the corresponding element of $\bigwedge^n T^* \otimes M_1^*$, because the monomorphism $M_1 \to T^* \otimes E$ can only be dualized to an epimorphism $T \otimes E \to M_1^*$, which cannot be used in a composition unless it can be splitted by adding an external object.

Despite the above comments, it is nevertheless always possible to, at least locally, perform many integrations by parts in order to obtain Euler–Lagrange equations of order $2q$ that can be written as follows:

$$\sum_{|\mu|=0}^{q} (-1)^{|\mu|} d_\mu \left(\frac{\partial w}{\partial y_\mu^k} \right) = 0.$$

These must be pulled back by a section $j_{2q}(f)$ of $J_{2q}(\mathcal{E})$. Instead of considering this system of PDE as a system of order $2q$, it is sometimes better to introduce function $\lambda_k^\mu = \partial w / \partial y_\mu^k$ as a local description of a section λ_q of $\bigwedge^n T^* \otimes J_q(E)^*$, and the *linear operator of order* q,

$$\lambda_q \to \sum_{|\mu|=0}^{q} (-1)^{|\mu|} \partial_\mu \lambda_k^\mu$$

is nothing but the dual, in the adjoint sense of Chapter VII, of the linear operator $j_q \colon E \to J_q(E)$. *This remark will play a crucial role below.*

A more sophisticated way of providing an intrinsic differential geometric setting for the variational sequence is to incorporate it in the so-called *variational bicomplex*. Efforts started in 1970, and from the numerous authors that have worked on this subject we mention Anderson [9], Tsujishita [185], Tulczyjew [187, 188, 189], and Vinogradov [193]. In fact, using condensed jet notations, a simple description of the variational bicomplex can be given in local coordinates, as follows.

Introduce the *fundamental forms* as in Chapter V.B:

$$\theta_\mu^k = dy_\mu^k - y_{\mu+1_i}^k \, dx^i,$$

and consider the exterior calculus on $J_\infty(\mathcal{E})$, *which is the only jet space that can be stabilized by the formal derivatives* d_i. Decompose the exterior derivative by setting $d = d_V + d_H$, where V stands for 'vertical' and H for 'horizontal'. The formulas

$$d_H \Phi = d_i \Phi \, dx^i, \qquad d_V \Phi = \frac{\partial \Phi}{\partial y_\mu^k} \theta_\mu^k$$

for functions can be extended to forms by setting

$$d_V dx^i = 0, \qquad d_H dx^i = 0,$$
$$d_H \theta^k_\mu = dx^i \wedge \theta^k_{\mu+1_i}, \qquad d_V \theta^k_\mu = 0,$$

because

$$d\theta^k_\mu = dx^i \wedge dy^k_{\mu+1_i} = dx^i \wedge \theta^k_{\mu+1_i} + y^k_{\mu+1_i+1_j} dx^i \wedge dx^j = dx^i \wedge \theta^k_{\mu+1_i}.$$

We can now decompose any form on $J_\infty(\mathcal{E})$, by setting

$$dy^k_\mu = \theta^k_\mu + y^k_{\mu+1_i} dx^i,$$

into forms r times containing dx and s times containing θ. We denote the corresponding set by $\Omega^{r,s}$. We leave it to the reader to verify the easy identities

$$d_H \circ d_H = 0, \qquad d_V \circ d_V = 0$$

and to obtain the following bicomplex with locally exact rows and columns, called the *variational bicomplex*:

$$
\begin{array}{ccccccccc}
0 & \longrightarrow & \Omega^{0,2} & \xrightarrow{d_H} & \cdots & \xrightarrow{d_H} & \Omega^{n-1,2} & \xrightarrow{d_H} & \Omega^{n,2} & \longrightarrow & \Omega^{n,2}/d_H\Omega^{n-1,2} & \longrightarrow & 0 \\
& & \Big\uparrow{\scriptstyle d_V} & & & & \Big\uparrow{\scriptstyle d_V} & & \Big\uparrow{\scriptstyle d_V} & & \Big\uparrow{\scriptstyle HC} & & \\
0 & \longrightarrow & \Omega^{0,1} & \xrightarrow{d_H} & \cdots & \xrightarrow{d_H} & \Omega^{n-1,1} & \xrightarrow{d_H} & \Omega^{n,1} & \longrightarrow & \Omega^{n,1}/d_H\Omega^{n-1,1} & \longrightarrow & 0 \\
& & \Big\uparrow{\scriptstyle d_V} & & & & \Big\uparrow{\scriptstyle d_V} & & \Big\uparrow{\scriptstyle d_V} & & \nearrow{\scriptstyle EL} & & \\
0 & \longrightarrow & \mathbb{R} & \longrightarrow & \Omega^{0,0} & \xrightarrow{d_H} & \cdots \xrightarrow{d_H} & \Omega^{n-1,0} & \xrightarrow{d_H} & \Omega^{n,0} & & & \\
& & & & & & & & \text{CL} & & & &
\end{array}
$$

where $\mathbb{R}$ stands for the locally constant functions and we have exhibited the variational sequence along an elementary chase left to the reader as an exercise.

To aid the reader, we will work out the computations for $n = 1$, $k = 1$, i.e. when there is one independent variable t and one unknown q. For simplicity, we regard $t, q, \dot{q}, \ddot{q}$ as jet variables, and set

$$\theta = dq - \dot{q}\, dt, \qquad \dot\theta = d\dot{q} - \ddot{q}\, dt.$$

We then have

$$\frac{d\Phi}{dt} = \frac{\partial\Phi}{\partial t} + \frac{\partial\Phi}{\partial q}\dot{q} \Rightarrow \frac{d\Phi}{dt}\, dt = d_H\Phi,$$

$$\frac{d}{dt}\left(\frac{\partial L}{\partial \dot{q}}\right) - \frac{\partial L}{\partial q} = \frac{\partial^2 L}{\partial t \partial \dot{q}} + \frac{\partial^2 L}{\partial q \partial \dot{q}}\dot{q} + \frac{\partial^2 L}{\partial \dot{q}\partial \dot{q}}\ddot{q} - \frac{\partial L}{\partial q},$$

and we see that the Helmholtz condition reduces to the single condition

$$\frac{d}{dt}\left(\frac{\partial Q}{\partial \ddot{q}}\right) - \frac{\partial Q}{\partial \dot{q}} = 0,$$

as can be readily verified by a direct calculation.

If $L(t, q, \dot{q}) \, dt \in \Omega^{1,0}$, we obtain (be careful with summation)

$$d_V(L \, dt) = \frac{\partial L}{\partial q} \theta \wedge dt + \frac{\partial L}{\partial \dot{q}} \dot{\theta} \wedge dt \in \Omega^{1,1}.$$

Now, any element in $\Omega^{0,1}$ can be written as

$$A(t, q)\theta + B(t, q)\dot{\theta},$$

and we obtain, by taking d_H:

$$\frac{dA}{dt} \, dt \wedge \theta + \left(A + \frac{dB}{dt}\right) dt \wedge \dot{\theta} + B \, dt \wedge \ddot{\theta} \in \Omega^{1,1}.$$

Comparing similar terms, we immediately obtain

$$A = -\frac{\partial L}{\partial \dot{q}}, \qquad B = 0,$$

and a representative element of $\Omega^{1,1}/d_H\Omega^{0,1}$ may be written as

$$\left(\frac{d}{dt}\left(\frac{\partial L}{\partial \dot{q}}\right) - \frac{\partial L}{\partial q}\right) dt \wedge \theta.$$

This gives the Euler–Lagrange operator.

Finally, let $Q(t, q, \dot{q}, \ddot{q}) \, dt \wedge \theta \in \Omega^{1,1}$ be a representative of an element in $\Omega^{1,1}/d_H\Omega^{0,1}$ as before, killed by HC. A chase shows that the image of this element under d_V, i.e.

$$\frac{\partial Q}{\partial \dot{q}} \dot{\theta} \wedge dt \wedge \theta + \frac{\partial Q}{\partial \ddot{q}} \ddot{\theta} \wedge dt \wedge \theta \in \Omega^{1,2},$$

must be the image under d_H of an element

$$A(t, q, \dot{q})\theta \wedge \dot{\theta} \in \Omega^{0,2},$$

i.e. an element of the form (be careful!)

$$\frac{dA}{dt} \, dt \wedge \theta \wedge \dot{\theta} - A\theta \wedge dt \wedge \ddot{\theta} \in \Omega^{1,2}.$$

Comparing similar terms, we obtain

$$A = \frac{\partial Q}{\partial \ddot{q}},$$

and the desired Helmholtz condition quoted above. Despite the importance of the full variational sequence

$$0 \longrightarrow \mathbb{R} \longrightarrow \Omega^{0,0} \xrightarrow{d_H} \cdots \xrightarrow{d_H} \Omega^{n,0} \xrightarrow{\text{EL}} \Omega^{n,1}/d_H\Omega^{n-1,1} \xrightarrow{\text{HC}} \Omega^{n,2}/d_H\Omega^{n-1,2}$$

already exhibited, two problems can be raised at this stage:

- The order of jets cannot be stabilized, and the necessity to use infinite jets does not allow one to apply symbolic manipulations for the various computations involved. Also, the quotients $\Omega^{n,1}/d_H\Omega^{n-1,1}$ and $\Omega^{n,2}/d_H\Omega^{n-1,2}$ are defined by means of sections and cannot be identified with fibered manifolds.

- *At first sight*, the variational sequence does not seem to have any relation with the differential sequences studied in Chapter IV.

The solution of the first problem will become clear while solving the second problem and taking into account recent work of D. Krupka [98, 99], which, for the first time ever, makes it possible to stabilize the order of jets in the variational sequence.

Once again, this is a good illustration of the fact that stabilization of jet orders has always been a source of inspiration in the study of differential sequences.

As already noticed, the Euler–Lagrange operator can be viewed as the dual of j_q. According to results of Chapter VII, we may consider the first Spencer sequence when the fibered manifold $\mathcal{E}$ becomes a vector bundle E over X, pass to the projective limit

$$0 \longrightarrow E \xrightarrow{j_\infty} J_\infty(E) \xrightarrow{D} T^* \otimes J_\infty(E) \xrightarrow{D} \cdots \xrightarrow{D} \wedge^n T^* \otimes J_\infty(E) \longrightarrow 0,$$

and dualize this sequence in such a way as to obtain the sequence

$$J_\infty(E)^* \xrightarrow{\tilde{D}} T^* \otimes J_\infty(E)^* \xrightarrow{\tilde{D}} \cdots \xrightarrow{\tilde{D}} \wedge^n T^* \otimes J_\infty(E)^* \xrightarrow{\tilde{j}_\infty} \wedge^n T^* \otimes E^*.$$

We have already identified $\tilde{j}_\infty$ with EL and we will identify the last $\tilde{D}$ with CL. Considering the image of the first Spencer operator D,

$$\begin{cases} \partial_r \xi^k - \xi_r^k = \eta_{,r}^k & A_k^r \\ \partial_r \xi_i^k - \xi_{ri}^k = \eta_{i,r}^k & A_k^{i,r} \\ \partial_r \xi_{ij}^k - \xi_{rij}^k = \eta_{ij,r}^k & A_k^{ij,r} \end{cases}$$

etc., we may multiply each line by the corresponding A and integrate by parts, to find (be careful with summation)

$$\begin{cases} \xi^k \to \partial_r A_k^r = B_k, \\ \xi_i^k \to \partial_r A_k^{i,r} + A_k^i = B_k^i, \\ \xi_{ij}^k \to \partial_r A_k^{ij,r} + A_k^{i,j} + A_k^{j,i} = B_k^{ij}, \end{cases}$$

and, more generally,

$$\xi_\mu^k \to \partial_r A_k^{\mu,r} + \sum_{\nu+1_i=\mu} A_k^{\nu,i} = B_k^\mu,$$

up to sign.

In the linear case, consider an $(n-1)$-form of the type

$$(-1)^{r-1}(A_k^r(x)y^k + A_k^{i,r}(x)y_i^k + A_k^{ij,r}(x)y_{ij}^k + \dots)\,dx^1 \wedge \cdots \wedge \widehat{dx^r} \wedge \cdots \wedge dx^n.$$

Taking the exterior derivative, we obtain

$$((\partial_r A_k^r)y^k + (\partial_r A_k^{i,r})y_i^k + (\partial_r A_k^{ij,r})y_{ij}^k + \dots)\,dx^1 \wedge \cdots \wedge dx^n$$

$$+ (-1)^{r-1}(A_k^r\,dy^k + A_k^{i,r}\,dy_i^k + A_k^{ij,r}\,dy_{ij}^k + \dots)\,dx^1 \wedge \cdots \wedge \widehat{dx^r} \wedge \cdots \wedge dx^n.$$

Passing to the residue modulo the contact forms, and paying attention only to the coefficients of $dx^1 \wedge \cdots \wedge dx^n$, we obtain

$$A_k^r y_r^k + A_k^{i,r} y_{ri}^k + \cdots + (\partial_r A_k^r)y^k + (\partial_r A_k^{i,r})y_i^k + (\partial_r A_k^{ij,r})y_{ij}^k + \cdots.$$

Factoring out the jets, we obtain (again, be careful with summation)

$$(\partial_r A_k^r)y^k + (\partial_r A_k^{i,r} + A_k^{,i})y_i^k + (\partial_r A_k^{ij,r} + A_k^{i,j} + A_k^{j,i})y_{ij}^k + \ldots.$$

We immediately see that, with the above sign conventions, CL *coincides with* $\tilde{D}$, and so the final part of the variational sequence just dualizes the initial part of the first Spencer sequence. Hence, if we want to stabilize the jet order in the variational sequence, we also have to stabilize the jet order in the dual sequence, i.e., according to results in Chapter IV, *we have to pass from the first to the second Spencer sequence.*

As part of the second Spencer sequence is also a Janet sequence made up by first-order involutive operators, the corresponding symbol sequence is exact and the same property holds for the dual sequence (by Chapter VII).

We will establish, for the first time ever, a link between the dual of the second Spencer sequence and the stable variational sequence found by Krupka, exactly as we have established a link between the dual of the first Spencer sequence and the variational sequence, following from the bicomplex approach. In particular, the use of Spencer δ-cohomology will greatly simplify the exposition as given in [98, 99], which seems too technical otherwise. Such a result is not evident at first sight, because most of the effective computations cannot be done by hand and are quite tricky.

We start our study with a brief sketch of the work done by Krupka and give a few comments.

When $\mathcal{E}$ is a fibered manifold over X, we denote by Ω^r the space of r-forms on $J_q(\mathcal{E})$, and by Θ^r the subspace generated, as an ideal, by the contact forms θ and their exterior derivatives $d\theta$, chosen from those that are defined on $J_q(\mathcal{E})$ for $1 \leq r \leq n$. In fact, we notice that the Θ^r for $2 \leq r \leq n$ are generated, as ideals, by contact forms θ only. So, to project Ω^r modulo Θ^r for $2 \leq r \leq n$, we have to replace each dy_μ^k by $y_{\mu+1_i}^k dx^i$ whenever this is possible, i.e. whenever $0 \leq |\mu| \leq q - 1$. As a byproduct, after reduction, no use/interpretation can be given to the forms still involving dy_μ^k with $|\mu| = q$. This difficulty will be avoided in our approach, since we will use 'horizontal' forms only, of the type $\Omega^{r,0}$ for $0 \leq r \leq n$.

The situation is much more complicated when $r \geq n + 1$, and one should consider only the so-called 'strongly' contact forms, involving (at least) $r - n + 1$ contact forms θ and $d\theta$ after reduction. For this reason we believe that the variational sequence for $r \geq n + 1$ is of little use, a point of view strenghtened by other remarks later on.

Introducing the quotient spaces $A^r = \Omega^r/\Theta^r$ and using the fact that the exterior derivative $d \colon \Omega^r \to \Omega^{r+1}$ restricts to $d \colon \Theta^r \to \Theta^{r+1}$, the main idea then is to introduce the following induced (stabilized) variational sequence of Krupka:

$$0 \longrightarrow \mathbb{R} \longrightarrow A^0 \longrightarrow \ldots \longrightarrow A^n \longrightarrow A^{n+1}.$$

One can prove that this sequence is locally exact, by adapting the Poincaré lemma.

Comparing this with the variational sequence obtained from the bicomplex approach, we see now that the main difficulty (which is hidden in the literature) is to use the correct injective/projective limits. Indeed, the above substitution argument immediately shows that the injective limit A^r as $q \to \infty$ is $\Omega^{r,0}$ when $0 \leq r \leq n$, if we set $A^0 = \Omega^0$. Also, this injective limit is the dual of the projective limit obtained from the first Spencer sequence.

For a vector bundle E over X, we recall the definition of the (trivial) Spencer bundles:

$$C_{q,r}(E) = \wedge^r T^* \otimes J_q(E) \big/ \delta(\wedge^{r-1}T^* \otimes S_{q+1}T^* \otimes E).$$

We denote these simply by $C_r(E)$ if no confusion is possible regarding the order of jets. There is a canonical projection

$$C_{q,r}(E) \to \wedge^r T^* \otimes J_{q-1}(E).$$

Hence $\wedge^r T^* \otimes J_\infty(E)$ is the projective limit of both $\wedge^r T^* \otimes J_q(E)$ and $C_{q,r}(E)$ as $q \to \infty$. By duality we can obtain injective limits for the bundles

$$\wedge^n T^* \otimes C_{q,r}(E)^* \subset \wedge^n T^* \otimes \wedge^r T \otimes J_q(E)^* \simeq \wedge^{n-r}T^* \otimes J_q(E)^*.$$

Finally, we also recall the splitting sequences

$$0 \longrightarrow \delta(\wedge^r T^* \otimes S_q T^* \otimes E) \longrightarrow C_{q,r}(E) \xrightarrow{\pi^q_{q-1}} \wedge^r T^* \otimes J_{q-1}(E) \longrightarrow 0,$$

which allow us to describe the Spencer bundles in actual practice.

We first prove that the operator $A^n \to A^{n+1}$ induces the Euler–Lagrange operator on horizontal forms.

For future purposes, we will give the proof in case $\mathcal{E}$ is a vector bundle E over X, by considering only linear horizontal forms on $J_q(E)$. The space of such forms is isomorphic to $\wedge^n T^* \otimes J_q(E)^*$, and thus to $\wedge^n T^* \otimes C_{q,0}(E)^*$. In the general case the proof is similar, and follows from the use of the vertical machinery. In the linear case under consideration, we may introduce the n-form (be careful with summation)

$$(a_k(x)y^k + a_k^i(x)y_i^k + a_k^{ij}(x)y_{ij}^k + \dots)\, dx^1 \wedge \dots \wedge dx^n.$$

Taking the exterior derivative, we obtain the $(n+1)$-form

$$a_k(x)\, dy^k \wedge dx^1 \wedge \dots \wedge dx^n + a_k^i(x)\, dy_i^k \wedge dx^1 \wedge \dots \wedge dx^n + \dots,$$

which cannot be reduced modulo contact forms ... since it already is a contact form:

$$a_k(x)\, \theta^k \wedge dx^1 \wedge \dots \wedge dx^n + a_k^i(x)\, \theta_i^k \wedge dx^1 \wedge \dots \wedge dx^n + \dots,$$

and we must deal with strongly contact forms.

We have the relation

$$d\left((-1)^r a_k(x)\,\theta^k \wedge dx^1 \wedge \dots \wedge \widehat{dx^r} \wedge \dots \wedge dx^n\right) = \partial_r a_k(x)\,\theta^k \wedge dx^1 \wedge \dots \wedge dx^n$$

$$+ (-1)^r a_k(x)\, d\theta^k \wedge dx^1 \wedge \dots \wedge \widehat{dx^r} \wedge \dots \wedge dx^n$$

$$= \partial_r a_k(x)\, dy^k \wedge dx^1 \wedge \dots \wedge dx^n$$

$$+ a_k(x)\, dy_r^k \wedge dx^1 \wedge \dots \wedge dx^n,$$

etc.

Reduction modulo these forms gives

$$(a_k(x) - \partial_i a_k^i(x) + \partial_{ij} a_k^{ij}(x) - \dots)\, dy^k \wedge dx^1 \wedge \dots \wedge dx^n,$$

i.e.

$$\left(\sum_{|\mu|=0}^{q} (-1)^{|\mu|} \partial_\mu a_k^\mu(x) \right) dy^k \wedge dx^1 \wedge \cdots \wedge dx^n,$$

and we recognize at once the Euler–Lagrange operator of order q that dualizes j_q.

More generally, for $0 \le r \le n$, a (linear) representative of $\bigwedge^r T^* \otimes J_q(E)^*$ may be

$$(a_{k,I}(x) y^k + a_{k,I}^i(x) y_i^k + a_{k,I}^{ij}(x) y_{ij}^k + \dots) \, dx^I,$$

i.e.

$$\sum_{|\mu|=0}^{q} a_{k,I}^\mu(x) y_\mu^k \, dx^I,$$

where $dx^I \in \bigwedge^r T^*$. The exterior derivative of the general terms is

$$\partial_i a_{k,I}^\mu(x) y_\mu^k \, dx^i \wedge dx^I + a_{k,I}^\mu(x) \, dy_\mu^k \wedge dx^I.$$

By reduction modulo the contact forms we obtain

$$(a_{k,I}^\mu(x) y_\mu^k + a_{k,I}^\mu(x) y_{\mu+1_i}^k) \, dx^i \wedge dx^I.$$

This expression is still linear in the jets, but the first term has order $|\mu|$ while the second term has order $|\mu| + 1$.

Accordingly, the image in $\bigwedge^{r+1} T^* \otimes J_q(E)^*$ is still well defined *if and only if* the term $a_{k,I}^\mu(x) y_{\mu+1_i}^k \, dx^i \wedge dx^I$ with $|\mu| = q$ disappears.

In view of the link with Krupka's approach, we will denote $\bigwedge^n T^* \otimes C_{q,r}(E)^*$ by $A_q^{n-r}(E)$, with a slight abuse of language. It follows that, for $0 \le r \le n$, we have

$$\bigwedge^r T^* \otimes J_{q-1}(E)^* \subset A_q^r(E) \subset \bigwedge^r T^* \otimes J_q(E)^*.$$

There are also exact sequences

$$\bigwedge^{r-1} T^* \otimes S_{q+1} T^* \otimes E \xrightarrow{\delta} \bigwedge^r T^* \otimes S_q T^* \otimes E \xrightarrow{\delta} \bigwedge^{r+1} T^* \otimes S_{q-1} T^* \otimes E$$

and dual exact sequences

$$\bigwedge^{r+1} T^* \otimes S_{q+1} T \otimes E^* \xleftarrow{\tilde{\delta}} \bigwedge^r T^* \otimes S_q T \otimes E^* \xleftarrow{\tilde{\delta}} \bigwedge^{r-1} T^* \otimes S_{q-1} T \otimes E^*,$$

where we have used r instead of $n - r$. The above inclusion is described by the conditions

$$\sum_{\mu+1_i=\nu} a_{k,I}^\mu(x) \, dx^i \wedge dx^I = 0, \qquad |\nu| = q+1,$$

which therefore define the kernel of the left map $\tilde{\delta}$ in the last sequence. Exactness of this sequence implies that they are satisfied if and only if the qth-order term in $A_q^r(E)$ is a linear combination of r-forms of the type $y_{\lambda+1_i}^k \, dx^i \wedge dx^J$, with $|\lambda| = q - 1$ and $dx^J \in \bigwedge^{r-1} T^*$, which go to zero under an exterior derivation followed by reduction modulo contact forms, because $y_{\lambda+1_i+1_j}^k \, dx^i \wedge dx^j = 0$. According to this remark, *the induced operator $A_q^r(E) \to A_q^{r+1}(E)$ is well defined*. Indeed, the terms of order $q + 1$ disappear, while those of order q have the desired type, because they contain terms of the form

$$\partial_j a_{k,J}^\lambda(x) y_{\lambda+1_i}^k \, dx^j \wedge dx^i \wedge dx^J = -\partial_j a_{k,J}^\lambda(x) y_{\lambda+1_i}^k \, dx^i \wedge dx^j \wedge dx^J,$$

coming from terms of order q in $A_q^r(E)$, and also terms of the form

$$a_{k,I}^\mu(x)y_{\mu+1_j}^k \, dx^j \wedge dx^I \qquad \text{with } |\mu| = q - 1,$$

coming from terms of order $q - 1$ in $A_q^r(E)$, which are all of the desired type. Since the above operator is induced from the exterior derivative, we have the following variational sequence:

$$0 \longrightarrow \mathbb{R} \longrightarrow A^0(E) \xrightarrow{\tilde{D}_n} A^1(E) \xrightarrow{\tilde{D}_{n-1}} \ldots \xrightarrow{\tilde{D}_1} A^n(E) \xrightarrow{\tilde{j}_q} \textstyle\bigwedge^n T^* \otimes E^* \longrightarrow$$

which is nothing else but the dual of the (trivial) second Spencer sequence

$$0 \longrightarrow E \xrightarrow{j_q} C_0(E) \xrightarrow{D_1} \ldots \xrightarrow{D_n} C_n(E) \longrightarrow 0,$$

where, in both sequences, we have suppressed the index q for simplicity.

To aid the reader, we will study in more detail the first operator $\tilde{D}_n \colon A^0(E) \to A^1(E)$. We can use local coordinates

$$a_k(x)y^k + a_k^i(x)y_i^k + \ldots,$$

or, more generally,

$$\sum_{|\mu|=0}^q a_k^\mu(x)y_\mu^k,$$

for a representative of an element of $A^0(E) \subset J_q(E)^*$. Taking the exterior derivative, we obtain

$$\sum_{|\mu|=0}^q (\partial_i a_k^\mu(x)y_\mu^k \, dx^i + a_k^\mu(x) \, dy_\mu^k).$$

After reduction modulo contact forms, we obtain

$$\sum_{|\mu|=0}^q (\partial_i a_k^\mu(x)y_\mu^k + a_k^\mu(x)y_{\mu+1_i}^k) \, dx^i,$$

and we must have $a_k^\mu = 0$ for $|\mu| = q$ in order not to have a term of order $q + 1$. It follows that $A^0(E) = J_{q-1}(E)^* \simeq \bigwedge^n T^* \otimes C_n(E)^*$ because $C_n(E) = \bigwedge^n T^* \otimes J_{q-1}(E)$.

Up till now we have tried to convince the reader of the existence of a link between variational calculus and the second Spencer sequence.

The most striking result of this Chapter concerns the fact that, even though field theory comes from a variational principle (see the example of the engineering sciences), the corresponding field equations are not obtained as Euler–Lagrange equations, but rather as conservation laws, i.e. they dualize the operator D_1 in the second Spencer sequence. This is why we have not paid too much attention to the Helmholtz conditions, since the operator $\tilde{j}_q$ is formally surjective. In the next Sections we will see that this shift by one step in the interpretation of the second Spencer sequence (*geometry*) and of its dual variational sequence (*physics*) is crucial for properly understanding field theory, and could not have been discovered, or even imagined, by means of classical methods from differential geometry.

We will end this Section with a new variational principle, first exhibited by E. and F. Cosserat in their book '*Théorie des corps déformables*' [41]. We will justify the principle *a posteriori*, by applications of it in the next Sections.

According to Poincaré [41, 42], field theory should consist of two parts:

- GEOMETRY;
- PHYSICS.

These two parts are related by an *interaction* which essentially depends on the symmetries of the field/matter coupling through *constitutive laws*. This is particularly clear for elasticity theory, which can be taught abstractly, presenting stress and strain separately, and where there is need for an experiment only to determine the two *Lamé constants* for ordinary isotropic materials. In particular, it is well known that the definition of strain, the compatibility conditions for strain, and the stress equations are the same for any material. This same comment can be made for the first and second sets of Maxwell equations in electromagnetism.

The geometry of field theory allows us to define the *potential*, the operator expressing the *field* in terms of the potential, and the corresponding compatibility conditions, called *field equations*. We can summarize this setting by a *differential sequence*, as follows:

$$\text{potential} \longrightarrow \text{field} \xrightarrow[\text{equation}]{\text{field}} ? \,.$$

According to Chapter IV, *there are only two possible such differential sequences*:

- JANET SEQUENCE;
- SPENCER SEQUENCE.

However, since field theory is often of a nonlinear nature, we have seen in Chapter V that a *nonlinear version of the above sequences can exist only in the context of* (*pseudo*)*group theory*. So, the Janet sequence is no longer convenient, since

- it depends on differential invariants, and as we have seen in the Introduction, the larger the (pseudo)group, the smaller the number of differential invariants and thus the smaller the number of geometric objects that can describe the field components.
- In general, the geometric objects are of high order, and this is not compatible with a proper intrinsic variational calculus, as we have seen at the beginning of this Section.

Hence we are left with the nonlinear Spencer sequence. In chapter V we have seen that, apart from the situation in which the corresponding Lie pseudogroup is of finite type, the second Spencer sequence always supersedes the first Spencer sequence, both in the linear and the nonlinear case. Also, since the Bianchi identities do not constitute an 'operator' in the true sense, the *only* mathematical possibility for obtaining a coherent physical interpretation is as follows:

$$0 \longrightarrow \underset{\substack{\text{'background'}\\\text{pseudogroup}}}{\Gamma} \xrightarrow{\ j_q\ } \underset{\text{potential}}{C_0} \xrightarrow{\ \overline{D}_1\ } \underset{\substack{\text{field}}}{C_1} \xrightarrow{\ \overline{D}_2\ } \underset{\substack{\text{field}\\\text{equations}}}{C_2}$$

We also notice that the question of whether a field satisfying the field equations comes from a potential ... is *exactly* the problem of the local solvability at C_1, ... which is *exactly* the equivalence problem!

The geometric aspect of a field theory being settled this way, it remains to turn towards the physical aspect. As usual in the finite-element method, we can introduce a Lagrangian formulation, which is meaningful only if the corresponding *inductions* (dual fields) satisfy *induction equations* not depending on the Lagrangian itself. We may even hope that the field equations *and* the induction equations depend on $\Gamma \subset$ aut(X) *only*.

The remainder of this Section is devoted to the construction of a variational principle on the nonlinear Spencer sequence. We will see that our above wishes will come true. Of course, this construction will be justified by its many applications.

Nevertheless, we stress once more that it is absolutely astonishing that E. and F. Cosserat had such a program in mind at the beginning of this century ... and achieved it! We may even say that, there is not a single mathematical/conceptual mistake in their work. Surprisingly, we will see that this was also the program of H. Weyl for studying electromagnetism by means of group theory in his beautiful book '*Space, time, matter*'.

Let X be a manifold and $\Gamma \subset$ aut(X) the transitive Lie pseudogroup of solutions of an involutive Lie groupoid $\mathcal{R}_q \subset \Pi_q(X, X)$ of order q. The state of a continuum is determined by a section f_q of $\mathcal{R}_q$, or a *potential*, up to a motion of Γ. Accordingly, a *Lagrangian* is a function defined on $\mathcal{R}_{q,1}$, with values in $\wedge^n T^*$, which is invariant under the action of Γ on the target.

First Fundamental Result (Equivalence principle). According to the formal exactness of the second nonlinear Spencer sequence, there is the following *factorization diagram*:

$$
\begin{array}{ccc}
\mathcal{R}_{q+1} & \xrightarrow{\overline{D}} & T^* \otimes \mathrm{R}_q \\
\downarrow{\scriptstyle \pi_q^{q+1}} & \downarrow & \searrow \\
\mathcal{R}_q & \xrightarrow{\overline{D}_1} \quad C_1 & \xrightarrow{w} \wedge^n T^*
\end{array}
$$

Here, a Lagrangian theory is essentially defined by an *action map* w, and we have to remember that the epimorphism $T^* \otimes \mathrm{R}_q \to C_1$ depends on a twist by the matrix $A \subset T^* \otimes T$ with inverse $B = A^{-1}$ and local coordinates $A_i^k(x) = g_r^k(x)\partial_i f^r(x)$. Indeed, if f_{q+1} is a section of $\mathcal{R}_{q+1}$ and we set $\chi_q = \overline{D} f_{q+1}$, then

$$
\chi_{\mu,i}^k = -g_l^k A_i^r f_{\mu+1_r}^l + \text{terms of order} \le |\mu|,
$$

and we may introduce $\tau_q = \chi_q \circ B$ in order to project onto $C_1 = T^* \otimes \mathrm{R}_q /\delta(\mathrm{M}_{q+1})$ by killing the dependence on the jets of strict order $q + 1$, since

$$
\tau_{\mu,i}^k = -g_l^k f_{\mu+1_i}^l + \text{terms of order} \le |\mu|.
$$

Also, since $\mathrm{R}_q \subset J_q(T)$ is the Lie algebroid of $\mathcal{R}_q$, we recall that R_q is associated with $\mathcal{R}_{q+1}$ and with $\mathcal{R}_{q,1}$ under the following composition:

$$
\xi_q \to \overline{\xi}_q = \xi_q + \chi_q(\xi) \to \eta_q = f_{q+1}(\overline{\xi}_q).
$$

In the diagram above, $T^* \otimes R_q$ is associated with $\mathcal{R}_{q+1,1}$ in the sense that R_q is associated with $\mathcal{R}_{q+1}$ while T^* is associated with $\mathcal{R}_{0,1}$.

If $f_{q+1} \in \mathcal{R}_{q+1}$ and $\chi_q \in T^* \otimes R_q$ is defined over the range of f, we recall the existence of *finite gauge transformations*

$$\chi_q \to \overline{\chi}_q = f_{q+1}^{-1} \circ \chi_q \circ j_1(f) + \overline{D} f_{q+1}$$

preserving both the image of $\overline{D}$ and the kernel of $\overline{D}'$. In such a gauge transformation, the association mentioned above is used in the first term on the right hand side.

Similarly, C_1 is associated with $\mathcal{R}_{q+1}$ and $\mathcal{R}_{q,1}$, but the induced finite gauge transformations depend only on the jets of order q. Indeed, for $\eta_q = f_{q+1}(\xi_q)$ we have

$$\eta^l_{r_1 \dots r_q} f^{r_1}_{i_1} \dots f^{r_q}_{i_q} + \dots = f^l_{k i_1 \dots i_q} \xi^k + \dots + f^l_k \xi^k_{i_1 \dots i_q}.$$

Writing out the gauge transformations in the form

$$f_{q+1}(\overline{\chi}_q - \overline{D} f_{q+1}) = \chi_q \circ j_1(f),$$

a straightforward application of the preceding transformation formula gives

$$f^l_{k i_1 \dots i_q}(\overline{\chi}^k_{,i} + \delta^k_i) + \dots + f^l_k \overline{\chi}^k_{i_1 \dots i_q, i} = \chi^l_{r_1 \dots r_q, j} f^{r_1}_{i_1} \dots f^{r_q}_{i_q} \partial_i f^j,$$

and so

$$f^l_{i i_1 \dots i_q} + \dots + f^l_k \overline{\tau}^k_{i_1 \dots i_q, i} = \tau^l_{r_1 \dots r_q, s} f^{r_1}_{i_1} \dots f^{r_q}_{i_q} f^s_i,$$

completing the proof.

Similarly, for $\xi_{q+1} \in R_{q+1}$ and $\chi_q \in T^* \otimes R_q$ we recall the existence of *infinitesimal gauge transformations*

$$\chi_q \to \chi_q + \delta \chi_q$$

with

$$\delta \chi_q = L(j_1(\xi_{q+1})) \xi_q + D \xi_{q+1}$$

and

$$\begin{aligned} \delta \chi_q(\zeta) &= \{\xi_{q+1}, \chi_{q+1}(\zeta)\} + i(\xi) D \chi_{q+1}(\zeta) - \chi_q([\xi, \zeta]) + i(\zeta) D \xi_{q+1} \\ &= [\xi_q, \chi_q(\zeta)] - \chi_q([\xi, \zeta]) + i(A(\zeta)) D \xi_{q+1}, \end{aligned}$$

where χ_{q+1} is a lift of χ_q (which can be found because R_q is assumed to be involutive). Since $\tau_q(\zeta) = \chi_q(B(\zeta))$, we have

$$\delta \tau_q(\zeta) = i(\zeta) D \xi_{q+1} + \text{terms in } j_1(\xi_q),$$

and this formula induces a formal Lie derivative on C_1 that depends only on the jets of order q. However, the corresponding formulas involve quotients and cannot be written out explicitly.

In chapter V we have seen how the variations of the χ_q depend on the variation of f_q. In fact, a variation δf_q of f_q is such that $(f_q, \delta f_q)$ is a section of $V(\mathcal{R}_q)$, and there is a well-known isomorphism $V(\mathcal{R}_q) \simeq \mathcal{R}_q \times_X \mathcal{R}_q$ fibered *over the target* and associating to $(f_q, \delta f_q)$ a couple (f_q, η_q) with $\eta_q \in R_q$ a section *over the target*. We thus give the following important

Definition 1. A field of *virtual displacement* is a section of R_q. A field of *real displacement* is the q-jet of a solution, i.e. and infinitesimal transformation. By duality, a *torsor* field is a section of R_q^*.

According to the isomorphism $J_1(V(\mathcal{R}_q)) \simeq V(J_1(\mathcal{R}_q))$ we may set $j_1(\delta f_q) = \delta j_1(f_q)$, and we are ready for starting a variational procedure as usual. We formally define the action

$$W = \int w(x, \chi_q(x))\, dx^1 \wedge \cdots \wedge dx^n$$

and introduce its variation *over the source*:

$$\delta W = \int \frac{\partial w}{\partial \chi_q} \delta \chi_q\, dx^1 \wedge \cdots \wedge dx^n.$$

Definition 2. We call

$$\mathcal{X}_q = \frac{\partial w}{\partial \chi_q} \in \wedge^n T^* \otimes T \otimes R_q^* \simeq \wedge^{n-1} T^* \otimes R_q^*$$

a *surface torsor* over the source.

Integration by parts can be performed by substituting for χ_q the formulas already given. In this way we obtain a result that should look like ordinary Euler–Lagrange equations. However, these equations look sufficiently complicated, and we would like to have a better insight into their structure.

A first way to simplify the result is to change the virtual displacement (always over the source), by setting

$$\overline{\xi}_{q+1} = \xi_{q+1} + \chi_{q+1}(\xi),$$

where χ_{q+1} is a lift of χ_q and ξ_{q+1} is a lift of ξ_q. We obtain the variation

$$\delta \chi_q = D\overline{\xi}_{q+1} + \{\chi_{q+1}(\cdot), \overline{\xi}_{q+1}\},$$

which depends effectively on χ_q only. Hence, the only integration by parts which is truly important is:

$$\int \langle \mathcal{X}_q, D\xi_{q+1} \rangle\, dx^1 \wedge \cdots \wedge dx^n,$$

and we discover that the corresponding integration by parts ... amounts to nothing else but a computation of the dual $\tilde{D}$, in the adjoint sense, of the Spencer operator D. This can be readily done by setting

$$\langle \tilde{D}\mathcal{X}_q, \xi_{q+1} \rangle = d\langle \mathcal{X}_q, \xi_q \rangle - \langle \mathcal{X}_q, D\xi_{q+1} \rangle.$$

We leave it to the reader to verify that each term is well defined.

So, we obtain an operator

$$\tilde{D}: \quad \wedge^{n-1} T^* \otimes R_q^* \quad \to \quad \wedge^n T^* \otimes R_{q+1}^*$$
$$\qquad\quad \text{surface torsor} \qquad\qquad \text{volume torsor.}$$

In local coordinates, the expression of $\tilde{D}$ is defined by the decomposition of the equations

$$\left(\partial_i \mathcal{X}_k^{\mu,i} + \sum_{\lambda+1_r=\mu} \mathcal{X}_k^{\lambda,r}\right)\xi_\mu^k = 0, \qquad 0 \le |\mu| \le q$$

for any section $\xi_q \in R_q$ *and* the additional equation

$$\sum_{\lambda+1_r=\mu} \mathcal{X}_k^{\lambda,r}\xi_\mu^k = 0, \qquad |\mu| = q+1,$$

for any section ξ_{q+1}^q of M_{q+1}.

To stabilize the order, we may pass to the second linear Spencer sequence and look for $\tilde{D}$ by duality. In fact, we have monomorphisms

$$\wedge^n T^* \otimes C_1^* \to \wedge^{n-1} T^* \otimes R_q^*, \qquad \wedge^n T^* \otimes R_q^* \to \wedge^n T^* \otimes R_{q+1}^*,$$

where the first inclusion is described by the last formula above. So the operator

$$\tilde{D}_1 : \wedge^n T^* \otimes C_1^* \to \wedge^{n-1} T^* \otimes R_q^* = \wedge^{n-1} T^* \otimes C_0^*$$

is induced by $\tilde{D}$ and defined by

$$\langle \tilde{D}_1 \mathcal{X}_q, \xi_q \rangle = d\langle \mathcal{X}_q, \xi_q \rangle - \langle \mathcal{X}_q, D_1 \xi_q \rangle.$$

A second way to simplify the result, which was found already by E. and F. Cosserat in [41], is to pass from the source to the target; namely, to use η_{q+1} instead of ξ_{q+1} or $\bar{\xi}_{q+1}$. Indeed, *by extraordinary luck*, we have obtained, in Chapter V, the (delicate) formula

$$\delta\chi_q = f_{q+1}^{-1} \circ D\eta_{q+1} \circ j_1(f),$$

where $j_1(f_{q+1})$ acts on $T^* \otimes R_q$ in such a way that f_{q+1} acts on R_q while $j_1(f)$ acts on T^*. Defining the source torsor $\mathcal{Y}_q$ *over the target* by the formula

$$\delta w = \langle \mathcal{X}_q, f_{q+1}^{-1} \circ D\eta_{q+1} \circ j_1(f) \rangle = \Delta\langle \mathcal{Y}_q, D\eta_{q+1}\rangle,$$

with $\Delta = \det(\partial_i f^k(x))$, we finally obtain

$$\delta W = \int \mathcal{X}_q \cdot \delta\chi_q \, dx^1 \wedge \cdots \wedge dx^n = \int \langle \mathcal{Y}_q, D\eta_{q+1}\rangle \, dy^1 \wedge \cdots \wedge dy^n,$$

i.e.

Second Fundamental Result (Variational principle). The induction equations are described, *over the target*, by the dual of the Spencer operator.

This explains 'a posteriori' our comments regarding the variational sequence.

B. Elasticity

The purpose of this Section is to compare the old and the new variational calculus which both lead to a mathematical model of elasticity. This will provide us the opportunity to give an explicit description of some of the abstract results in Section A and to justify them 'a posteriori'. In particular, this setting will give the best reasons for using the Spencer sequence instead of the Janet sequence.

At the end of this Section we will study the particular case of Lie groups of transformations and sketch a possible use of groupoid theory in robotics.

It will follow that *elasticity only depends on group theory via the Spencer operator*. As a byproduct, we will prove that relativistic dynamics has a purely geometrical origin depending only on the structure of the Poincaré group of space-time. Meanwhile the reader will discover that all the concepts needed to understand the foundation of elasticity are new and could not be imagined without the help of the formal calculus. The results obtained therefore explain why all attempts at understanding the work done by E. and F. Cosserat have thusfar led to nothing.

The classical approach, which can be found in any textbook on Continuum Mechanics, is based on the Janet sequence and can be summarized as follows. For simplicity we will deal with the static infinitesimal version. The geometry is clearly based on the Janet sequence for the group of rigid motions (studied in Chapter V.E):

$$0 \longrightarrow \Theta \longrightarrow T \xrightarrow{\ \mathcal{D}\ } S_2 T^* \xrightarrow{\ \mathcal{D}_1\ } F_1.$$

To describe this sequence we introduce the standard euclidean metric $\omega = (\omega_{ij})$ and define $\mathcal{D}$ to be half the Killing operator in mathematics, i.e. $\mathcal{D}\xi = \frac{1}{2}\mathcal{L}(\xi)\omega = \epsilon$, which defines also the so-called *small strain tensor*

$$\epsilon_{ij} = \tfrac{1}{2}(\omega_{rj}\partial_i\xi^r + \omega_{ir}\partial_j\xi^r + \xi^r\partial_r\omega_{ij}).$$

Since $\omega_{ij} = 1$ if $i = j$ and 0 otherwise, we may raise or lower the indices by means of the metric and its inverse, in order to obtain in practice

$$\epsilon_{ij} = \tfrac{1}{2}(\partial_i\xi_j + \partial_j\xi_i).$$

Sections of the corresponding system are defined by the equations

$$R_1 \subset J_1(T) \qquad \omega_{rj}\xi_i^r + \omega_{ir}\xi_j^r + \xi^r\partial_r\omega_{ij} = 0.$$

Accordingly, a basis for the sections of R_1 may be given by the equations

$$\xi_{i,j} + \xi_{j,i} = 0,$$

and is thus represented by a skewsymmetric matrix.

Recall that the above sequence is not 'exactly' a Janet sequence, because R_1 is formally integrable but not involutive (the symbol M_1 of R_1 is of finite type but not even 2-acyclic with $F_1 = H_1^2(M_1)$ and $\dim F_1 = n^2(n^2 - 1)/12)$. In particular, for plane elasticity $(n = 2)$ we have $\dim F_1 = 1$ and the single compatibility condition

$$\partial_{11}\epsilon_{22} + \partial_{22}\epsilon_{11} - 2\partial_{12}\epsilon_{12} = 0.$$

The compatibility conditions are of order two in general, because we need one prolongation to obtain an involutive system R_2 with zero symbol.

The nonlinear Janet sequence is

$$0 \longrightarrow \Gamma \longrightarrow \mathrm{aut}(X) \underset{\omega\pi}{\overset{\Phi o j_1}{\longrightarrow}} S_2 T^* \overset{I o j_2}{\underset{0}{\longrightarrow}} \mathcal{F}_1,$$

with the same restriction as before. We leave it to the reader to explore this sequence, since it is not our purpose here to give full study of nonlinear elasticity theory but rather to indicate the reasons for which one has to abandon the Janet sequence.

Correspondingly, the variation of the strain tensor is obtained as follows:

$$2\delta\epsilon_{ij} = \omega_{rj}\partial_i\delta\xi^r + \omega_{ir}\partial_j\delta\xi^r + \delta\xi^r\partial_r\omega_{ij},$$

since the underlying metric is untouched. Therefore we may define an action

$$W = \int w(\epsilon_{ij}) \, dx^1 \wedge \cdots \wedge dx^n$$

with variation

$$\delta W = \int \frac{\partial w}{\partial \epsilon_{ij}} \delta\epsilon_{ij} \, dx^1 \wedge \cdots \wedge dx^n.$$

It is at this point that the troubles start.

Indeed, according to their definition, there are only $n(n+1)/2$ independent components ϵ_{ij}, e.g. those with $i \leq j$. Hence w can only be a function of these independent components. As a byproduct, the derivatives

$$\sigma^{ij} = \frac{\partial w}{\partial \epsilon_{ij}}$$

are defined only for $i \leq j$ and *the decision to define the stress tensor (density) by setting $\sigma^{ij} = \sigma^{ji}$ is purely artificial.* This is why, *in all textbooks,* the variational aspect of elasticity is *always* introduced *after* the stress has been defined in a purely phenomenological way by arguments dating back to A.L. Cauchy (1828), while its symmetry is 'proved' by using the equilibrium of ... *torsors!* ([52, 69]). To escape from this vicious circle, E. and F. Cosserat decided to base elasticity on a new variational calculus. In any case, assuming that the stress tensor is symmetric, we obtain

$$\delta W = \int \sigma^{ij}\omega_{rj}\partial_i\xi^r \, dx^1 \wedge \cdots \wedge dx^n,$$

and we may lower the indices of σ or ξ. E.g., we can obtain

$$\delta W = \int (-1)^{i-1}\sigma_k^i\delta\xi^k \, dx^1 \wedge \cdots \wedge \widehat{dx^i} \wedge \cdots \wedge dx^n - \int (\partial_r\sigma_k^r)\delta\xi^k \, dx^1 \wedge \cdots \wedge dx^n,$$

exhibiting the well-known surface stress on one side and stress equation on the other side.

A few attempts have been made at avoiding the above difficulty, by introducing a kind of nonsymmetrical stress with n^2 components equal to the partial derivatives of the displacement vector ξ. In that case not only the group invariant aspect of stress is cut down, but also the symmetry of the resulting stress cannot be proved.

The idea of the Cosserat brothers has been to change the geometrical framework *before* changing the physical one, while retaining the same background group.

In our case, apart from the Janet sequence, we have only the Spencer sequence. Again to aid the reader, we will first deal with the infinitesimal point of view before providing the finite point of view.

First of all we have $M_2 = 0$, and so $R_2 \simeq R_1$. It follows that the first and second linear Spencer sequences are isomorphic to

$$0 \longrightarrow \Theta \xrightarrow{j_1} R_1 \xrightarrow{D_1} T^* \otimes R_1 \xrightarrow{D_2} \wedge^2 T^* \otimes R_1.$$

The potential is now a section (ξ^k, ξ_i^k) of R_1, and the field is described by the image of $D_1 = D$, as follows:

$$\chi_{,i}^k = \partial_i \xi^k - \xi_i^k, \qquad \chi_{j,i}^k = \partial_i \xi_j^k - \xi_{ij}^k = \partial_i \xi_j^k.$$

The compatibility conditions are described by $D_2 = D$:

$$\begin{cases} \partial_i \chi_{,j}^k - \partial_j \chi_{,i}^k + \chi_{j,i}^k - \chi_{i,j}^k = 0, \\ \partial_i \chi_{r,j}^k - \partial_j \chi_{r,i}^k = 0. \end{cases}$$

This is the new setting for the geometry of elasticity.

We will now present the new setting for the corresponding variational calculus. We introduce the action

$$W = \int w(\chi_{,i}^k, \chi_{j,i}^k) \, dx^1 \wedge \cdots \wedge dx^n$$

and set

$$\sigma_k^i = \frac{\partial w}{\partial \chi_{,i}^k}, \qquad \mu_k^{ji} = \frac{\partial w}{\partial \chi_{j,i}^k}.$$

The variation of the field is now

$$\delta \chi_{,i}^k = \partial_i \delta \xi^k - \delta \xi_i^k, \qquad \delta \chi_{j,i}^k = \partial_i \delta \xi_j^k,$$

and we successively obtain:

$$\delta W = \int (\sigma_k^i (\partial_i \delta \xi^k - \delta \xi_i^k) + \mu_k^{ji} \partial_i \delta \xi_j^k) \, dx^1 \wedge \cdots \wedge dx^n$$

$$= \int (-1)^{r-1} (\sigma_k^r \delta \xi^k + \mu_k^{jr} \delta \xi_j^k) \, dx^1 \wedge \cdots \wedge \widehat{dx^r} \wedge \cdots \wedge dx^n$$

$$- \int (\partial_r \sigma_k^r \delta \xi^k + \sigma_k^i \delta \xi_i^k + \partial_r \mu_k^{jr} \delta \xi_j^k) \, dx^1 \wedge \cdots \wedge dx^n.$$

In the last integral the sum $\sigma_k^i \delta \xi_i^k$ involves all i and k, while in the sum $\partial_r \mu_k^{jr} \delta \xi_j^k$ only the different ξ_j^k are involved. Raising or lowering the indices by using the metric and taking into account the conditions $\xi_{i,j} + \xi_{j,i} = 0$, the last integral becomes

$$\int (\partial_r \sigma_k^r) \delta \xi^k + (\partial_r \mu^{ijr} + \sigma^{ij} - \sigma^{ji}) \delta \xi_{i,j}) \, dx^1 \wedge \cdots \wedge dx^n,$$

where the second sum runs over $i < j$. In these integrals we immediately recognize, *without any experimental support*, the virtual work of the surface stress and couple stress:

$$\int (-1)^{r-1} (\sigma_k^r \delta \xi^k + \mu_k^{jr} \delta \xi_j^k) \, dx^1 \wedge \cdots \wedge \widehat{dx^r} \wedge \cdots \wedge dx^n$$

together with the stress and couple stress equations that can be written with a right hand term describing the volume stress and couple stress:

$$\begin{cases} \partial_r \sigma_k^r = f_k, \\ \partial_r \mu^{ijr} + \sigma^{ij} - \sigma^{ji} = m^{ij}. \end{cases}$$

These are *precisely* the equations found by E. and F. Cosserat in [41, p. 137], with the only difference that they also dealt with the case $n = 3$ and used to present any skewsymmetric matrix like ($\mu^{ijr} = -\mu^{jir}$, $m^{ij} = -m^{ji}$) by a 'vector'

$$A = \begin{pmatrix} 0 & -a_3 & a_2 \\ a_3 & 0 & -a_1 \\ -a_2 & a_1 & 0 \end{pmatrix}$$

so that $A\vec{v} = \vec{a} \wedge \vec{v}$. It is also clear that these equations *exactly* dualize D_1, and so represent $\tilde{D}_1$. Regarding $\tilde{D}_2$ we find, by multiplying the compatibility conditions by A_k^{ij}, respectively $A_k^{r,ij}$, and subsequently integrating by parts as usual, that (up to sign):

$$\begin{cases} \partial_i A_k^{ij} = \sigma_k^j, \\ \partial_i A^{rs,ij} + A^{rs,j} - A^{sr,j} = \mu^{rsj}, \end{cases}$$

and we can readily verify that $\tilde{D}_1 \circ \tilde{D}_2 \equiv 0$. The crucial point is to notice that not only the number but also the form of the induction equations depend on R_1 only, since the knowledge of R_1 only allows us to construct the Spencer sequence and its dual sequence. However, contrary to the situation in the variational sequence, here $\tilde{D}_1$ is formally integrable and locally surjective. Indeed, we can solve the first divergence equation for any f_k to find σ, and substitute this into the second equation to find μ. Hence there is no Euler–Lagrange-type operator, and this result strengthens our doubts regarding the classical variational calculus.

It is rather more delicate to study the nonlinear setting. In our case we have

$$\mathcal{R}_1 \qquad \omega_{kl} y_i^k y_j^l = \omega_{ij},$$

with the same underlying metric, and so we have $\mathcal{R}_2 \simeq \mathcal{R}_1$, because $M_2 = 0$ while the jets of order 2 are equal to zero. Accordingly, the first nonlinear Spencer sequence is isomorphic to the second, and becomes

$$0 \longrightarrow \Gamma \xrightarrow{j_1} \mathcal{R}_1 \xrightarrow{\overline{D}} T^* \otimes \mathcal{R}_1 \xrightarrow{\overline{D}'} \wedge^2 T^* \otimes \mathcal{R}_1,$$

with successively

$$\overline{D} \qquad \begin{cases} \chi_{,i}^k = g_r^k \partial_i f^r - \delta_i^k, \\ \chi_{j,i}^k = g_r^k \partial_i f_j^r, \end{cases}$$

$$\overline{D}' \qquad \begin{cases} \partial_i \chi_{,j}^k - \partial_j \chi_{,i}^k - \chi_{i,j}^k + \chi_{j,i}^k - (\chi_{,i}^r \chi_{r,j}^k - \chi_{,j}^r \chi_{r,i}^k) = 0, \\ \partial_i \chi_{l,j}^k - \partial_j \chi_{l,i}^k - (\chi_{l,i}^r \chi_{r,j}^k - \chi_{l,j}^r \chi_{r,i}^k) = 0. \end{cases}$$

As already noted in Chapter V, we may also use the notations

$$A_i^k = g_r^k \partial_i f^r \rightarrow \partial_i A_j^k - \partial_j A_i^k + A_j^r \chi_{r,i}^k - A_i^r \chi_{r,j}^k = 0.$$

We see that $\overline{D}$ exactly describes the so-called *differential parameters* of E. and F. Cosserat [41, p. 123], while $\overline{D}'$ describes the ordinary torsion and curvature. In our previous book [146] we expressed the belief that the Cosserat brothers did not take into account these compatibility conditions. It therefore came as a surprise when we recently discovered that they wrote a rather lengthy comment in [93], since they have never referred to it. The question to be solved, which has also bothered other specialists of Cosserat media [177], is as follows. In the nonlinear Janet sequence the compatibility conditions are of order two indeed. How then is it possible that, in the nonlinear Spencer sequence, *they are of order one?* Again, this is a typical example of the confusion between *classical curvature* (Janet sequence) and the *gauge curvature* (Spencer sequence), or between $\mathcal{D}_1$ and D_2 in the linear setting. The brothers Cosserat proved, for $n = 3$, that the 9 torsion conditions have rank 9 with respect to the 9 components $\chi_{j,i}^k$, which can therefore be expressed as functions of $j_1(A)$. Substituting this into the curvature condition, we finally find 6 independent conditions for $j_2(A)$.

We will prove this result for arbitrary $n \geq 3$. Since $\det A \neq 0$, the quadratic terms of the torsion can be written as

$$A_i^r A_j^s (\tau_{r,s}^k - \tau_{s,r}^k).$$

Locally we have an element in $\bigwedge^2 T^* \otimes T$ which is the image under δ of $\tau \in T^* \otimes \mathrm{M}_1$. Since $n \geq 3$, we know that $\mathrm{M}_2 = 0$, and so we have an exact δ-sequence

$$0 \longrightarrow T^* \otimes \mathrm{M}_1 \overset{\delta}{\longrightarrow} \bigwedge^2 T^* \otimes T \longrightarrow 0,$$

because both vector bundles have the same dimension $n^2(n-1)/2$. Hence the rank of δ is maximal and equal to $n^2(n-1)/2$. We can thus compute τ as a function of $j_1(A)$ and substitute this into the curvature. The reason why we obtain $n^2(n-1)/12$ conditions for $j_2(A)$ comes from the relation

$$\omega_{rs}(x)A_i^r A_j^s = \omega_{kl}(f(x))\partial_i f^k \partial_j f^l,$$

relating the Spencer sequence to the Janet sequence, and from the tensorial nature of classical curvature.

The variation of the field becomes [41, p. 124]

$$\begin{cases} \delta\chi_{,i}^k = (\partial_i \xi^k - \xi_{,i}^k) + (\xi^r \partial_r \chi_{,i}^k + \chi_{,r}^k \partial_i \xi^r - \chi_{,i}^r \xi_r^k) \\ \quad = (\partial_i \overline{\xi}^k - \overline{\xi}_{,i}^k) + (\chi_{r,i}^k \overline{\xi}^r - \chi_{,i}^r \overline{\xi}_r^k), \\ \delta\chi_{j,i}^k = (\partial_i \xi_j^k) + (\xi^r \partial_r \chi_{j,i}^k + \chi_{j,r}^k \partial_i \xi^r + \chi_{r,i}^k \xi_j^r - \chi_{j,i}^r \xi_r^k) \\ \quad = (\partial_i \overline{\xi}_j^k) + (\chi_{r,i}^k \overline{\xi}_j^r - \chi_{j,i}^r \overline{\xi}_r^k), \end{cases}$$

and the transformation law of the surface torsor is [41, p. 136]

$$\begin{cases} \Delta \mathcal{Y}_l^{\cdot k} = g_l^r \mathcal{X}_r^{\cdot i} \partial_i f^k, \\ \Delta \mathcal{Y}_l^{m,k} = g_l^r f_s^m \mathcal{X}_r^{s,i} \partial_i f^k. \end{cases}$$

Combining these results, we leave it to the reader to perform the tedious but straight-forward calculation leading to the relation:

$$\delta W = \int \delta w \, dx^1 \wedge \cdots \wedge dx^n = \int \left[\mathcal{Y}_l^{\prime k} \left(\frac{\partial \eta^l}{\partial y^k} - \eta_k^l \right) + \mathcal{Y}_l^{m,k} \frac{\partial \eta_m^l}{\partial y^k} \right] dy^1 \wedge \cdots \wedge dy^n.$$

This leads to the *same* induction equations as before. It is quite striking that, even for a nonlinear field theory (namely when $\overline{D}_1$ and $\overline{D}_2$ are *effectively* nonlinear operators), the induction equations on the target are nevertheless linear, since they are described by $\tilde{D}_1$ or $\tilde{D}$ like in the present situation.

In the special case of Lie groups of transformations, as considered in this Section, we will give an alternative approach, based on the isomorphic sequences

GAUGE SEQUENCE

$$X \times G \longrightarrow T^* \otimes \mathcal{G} \xrightarrow{\text{MC}} \wedge^2 T^* \otimes \mathcal{G}$$

$$\wr\mid \qquad\qquad \wr\mid \qquad\qquad \wr\mid$$

$$0 \longrightarrow \Gamma \xrightarrow{j_q} \mathcal{R}_q \xrightarrow{\overline{D}} T^* \otimes \mathrm{R}_q \xrightarrow{\overline{D}'} \wedge^2 T^* \otimes \mathrm{R}_q$$

SPENCER SEQUENCE

Recall that finding a map from X to G, i.e. 'gauging' G over X, amounts to intro-ducing a section f_q over $\mathcal{R}_q$ for sufficiently large q. Also, the above local isomorphism

$$X \times \mathcal{G} \to \mathrm{R}_q : \lambda^\tau(x) \to \lambda^\tau(x)\partial_\mu \xi_\tau^k(x)$$

depends on the choice of a basis of infinitesimal generators ξ_τ for $\tau = 1, \ldots, \dim \mathrm{R}_q$. Similarly, there is an isomorphism

$$T^* \otimes \mathcal{G} \to T^* \otimes \mathrm{R}_q : A_i^\tau(x) \to \chi_{\mu,i}^k(x) = A_i^\tau(x)\partial_\mu \xi_\tau^k(x).$$

It usually depends on $x \in X$.

A Lagrangian is a function of the first jets of maps $X \to G$ which is invariant under the action of G by composition. Accordingly, it factorizes through a function defined on $T^* \otimes \mathcal{G}$ as follows:

$$X \times G \longrightarrow \quad T^* \otimes \mathcal{G} \quad \xrightarrow{w} \wedge^n T^*$$

$$a \quad \longrightarrow a^{-1} \, da = A.$$

Whenever

$$A_i^\tau(x) = \omega_\rho^\tau(a(x))\partial_i a^\rho(x),$$

we may set

$$W = \int w(x, A(x)) \, dx^1 \wedge \cdots \wedge dx^n.$$

and start the usual variation procedure. It successively gives

$$\delta w = \frac{\partial w}{\partial A_i^\tau} \delta A_i^\tau = \mathcal{A}_\tau^i \delta(\omega_\rho^\tau \partial_i a^\rho)$$

$$= \mathcal{A}_\tau^i \partial_\sigma \omega_\rho^\tau \partial_i a^\rho \delta a^\sigma + \mathcal{A}_\tau^i \omega_\rho^\tau \partial_i \delta a^\rho$$

$$= \partial_i(\mathcal{A}_\tau^i \omega_\rho^\tau \delta a^\rho) - (\partial_i \mathcal{A}_\tau^i + c_{\tau\nu}^\sigma A_i^\nu \mathcal{A}_\sigma^i)\omega_\rho^\tau \delta a^\rho,$$

while taking into account the Maurer–Cartan (MC) compatibility conditions:

$$\partial_\rho \omega_\sigma^\tau - \partial_\sigma \omega_\rho^\tau + c_{\alpha\beta}^\tau \omega_\rho^\alpha \omega_\sigma^\beta = 0.$$

We thus obtain the so-called *relative equations*

$$\partial_i \mathcal{A}_\tau^i + c_{\tau\nu}^\sigma A_i^\nu \mathcal{A}_\sigma^i = 0,$$

possibly with a right hand term in $\wedge^n T^* \otimes \mathcal{G}^*$ representing a volume torsor. Since $\det \omega \neq 0$, we can use $\omega \delta a$ as a virtual displacement instead of δa, and we have a corresponding *surface torsor*: $\mathcal{A} \in \wedge^n T^* \otimes T \otimes \mathcal{G}^* \simeq \wedge^{n-1} T^* \otimes \mathcal{G}$. Notice that, *although they are equivalent*, a direct comparison between the relative equations and the induction equations leads to extremely tedious computations, and for this reason it has never been exhibited. Also notice that the relative equations depend on the choice of infinitesimal generators, i.e. on the explicit integration of a system of infinitesimal Lie equations, that, in practice, may sometimes be impossible.

To simplify the relative equations we may introduce the *adjoint matrix* $M_\rho^\tau(a)$ representing the action of $Ad(a)$ on $\mathcal{G}$. We know from Proposition V.A.10 that

$$\frac{\partial M_\nu^\tau}{\partial a^\mu} + c_{\rho\sigma}^\tau \omega_\mu^\rho M_\nu^\sigma = 0.$$

Defining

$$\mathcal{B}_\rho^i = M_\rho^\tau \mathcal{A}_\tau^i,$$

we obtain the *absolute equations*

$$\partial_i \mathcal{B}_\rho^i = 0.$$

We will prove that this result can be explained by duality, again. Indeed, setting $\xi_\mu^k(x) = \lambda^\tau(x)\partial_\mu \xi_\tau^k(x)$, we obtain

$$\partial_i \xi_\mu^k - \xi_{\mu+1_i}^k = (\partial_i \lambda^\tau)\partial_\mu \xi_\tau^k(x),$$

and therefore the isomorphism of sequences

$$\wedge^0 T^* \otimes \mathcal{G} \xrightarrow{\ d\ } \wedge^1 T^* \otimes \mathcal{G} \xrightarrow{\ d\ } \wedge^2 T^* \otimes \mathcal{G}$$

$$\wr | \qquad\qquad\qquad \wr | \qquad\qquad\qquad \wr |$$

$$0 \longrightarrow \Theta \xrightarrow{\ j_q\ } \wedge^0 T^* \otimes R_q \xrightarrow{\ D\ } \wedge^1 T^* \otimes R_q \xrightarrow{\ D\ } \wedge^2 T^* \otimes R_q$$

where the upper differential sequence is induced by the Poincaré sequence. As a byproduct, we see that the dual of the first d (gradient) is the last d (divergence): $\wedge^{n-1} T^* \otimes \mathcal{G}^* \to \wedge^n T^* \otimes \mathcal{G}^*$, which describes the absolute equations.

We make a few comments regarding the above approach. For $n = 1$ we immediately recognize the Birkhoff–Arnold dynamics of a rigid body, with time t as independent variable, or the Kirchhoff–Love theory of thin elastic beams, with curvilinear abcissa s along the beam as independent variable. As a byproduct, the transition to hydrodynamics must be done with the same group, introducing space–time variables. Note that in the Birkhoff–Arnold approach the ambient space does not appear, because G does not act on X. This is particularly clear when G is the group of rigid motions of *space* and X is *time*. On the other hand, the Cosserat approach essentially uses X as a space on which G acts as a transformation group. For this reason the nonlinear Spencer sequence starts to the left with Γ, while the gauge sequence does not start on the left with anything meaningful from the geometric point of view. In any case, it is clear that the field must be a section of $T^* \otimes \mathcal{G}$ or $T^* \otimes R_q$. Therefore it seems quite surprising that the same differential sequence appears in gauge theory ... with an interpretation *shifted to the right by one step*, as the field should now be a section of $\wedge^2 T^* \otimes \mathcal{G}$. In view of the previous interpretation, we arrive at a contradiction. This contradiction will become stronger in Section D when we work out the group-theoretical origin of electromagnetism.

We conclude this Section by making a few remarks regarding the application of groupoid theory in robotics.

Up till now Lie group theory has entered robotics to describe explicitly the various movements in the articulations relating the many arms of a robot (rotoid, screw, ...). However, engineers are sometimes interested in studying the space of mobility of the final arm of a robot, independently of the constituting mechanisms. In that case it is known from striking experiments that the final arm may perform tricky movements, not constituting a Lie group at all. Also, one could be interested in studying the degree of mobility of a closed chain of mechanisms, i.e. when the final arm of the robot is linked to the initial arm. To our knowledge, there does not exist a general way of studying such problems, and the method which we present here should give some new impulse to robotics. In any case, it is the first time ever that the results of Chapter V are applied to such a dynamics.

The basic idea is to notice that when joining two arms by an articulation of whatever kind, it is sometimes important to have a global description of the movement (rotation, translation, screw, ...) *independently* of the exact description of that movement by the parameters of the corresponding Lie subgroup of the group of displacements. E.g., a basic fact about human behavior is that all articulations can rotate. This is precisely the difference between an explicit description of a Lie group of transformations and and the description of the corresponding Lie groupoid. Recall that the simplest example is that of affine transformations $y = ax + b$, which are invertible solutions of the second-order ODE $y_{xx} = 0$.

As a byproduct, the comparison of a robot (two arms and an articulation) with a groupoid (manifold with two projections) in the following picture needs no comment!

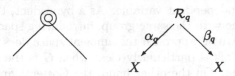

Let us study the case of rotation in a plane. We only have to fix a frame attached
to each arm in such a way that these frames coincide at the start of the movement.
Then the relations between the coordinates of a point attached to an arm with respect
to the two frames are described by the groupoid ($q = 1$, $n = 2$):

$$\mathcal{R}_1 \qquad \begin{cases} y^k = y_i^k x^i, \\ \omega_{kl} y_i^k y_j^l = \omega_{ij}. \end{cases}$$

Indeed, integrating the second set of PDE we get $y = Ax + B$ where A is an orthogonal
(2×2)-matrix and B is a vector. Substitution into the first set of PDE gives $B = 0$,
and so the 'pure' notation $y = Ax$.

When joining the '*target*' arm to the '*source*' arm of the second articulation, we
may repeat the above procedure while taking into account that the new articulation
occupies a specific position in the frame attached to the target arm of the first articulation. Finally, we only have to notice that the junction of successive arms in
robots ... corresponds to composition in the groupoid. In particular, the mechanical
description saying that the target arm of the first articulation is attached to the source
arm of the second articulation ... corresponds merely to the fibered product of the
composition in the groupoid.

We are now in a position to perform a general synthesis.

Definition 1. A *mechanism* with respect to a groupoid $\mathcal{R}_q$ over a base X is any
subsystem (take care!) of $\mathcal{R}_q$ obtained by composition of finitely many subgroupoids
of $\mathcal{R}_q$, in a specific order. A mechanism is said to be a *formally transitive mechanism*
if its projection on $X \times X$ is surjective.

A mechanism can be realized in several different ways. For two articulations we have
the following commutative diagram for $\mathcal{R}'_q, \mathcal{R}''_q \subset \mathcal{R}_q$ and their first prolongations:

$$\mathcal{R}'_{q+1} \times_X \mathcal{R}''_{q+1} \subset \mathcal{R}_{q+1}$$
$$\downarrow \qquad\qquad\qquad \downarrow$$
$$\mathcal{R}'_q \times_X \mathcal{R}''_q \ \subset \mathcal{R}_q$$

A chase projecting the composition at order $q + 1$ to that of order q proves that a
mechanism is formally integrable if it comes from formally integrable groupoids.

To study the mobility of a closed chain of mechanisms we only have to project
in each groupoid the inverse image of the q-jet of the identity (in $\mathcal{R}_q$) under the
composition. However, in general the composition need not be a subsystem of $\mathcal{R}_q$,

since it may fail to be a submanifold (Chinese wooden puzzles are nice examples of such a degenerate situation). Also, the above projections need not be formally integrable in general and, even if they are subsystems, only the formal theory of PDE makes it possible to find out the degrees of mobility (since one needs to know the *degrees of generality* of the systems).

Ordinary robotics corresponds to the above abstract setting when $n = 3$, $q = 1$ and $\mathcal{R}_1$ is the groupoid of isometries.

We will illustrate this new approach with a simple, but instructive, example.

Suppose a planar system ($n = 2$) is constituted by three rigid bars articulated in a way allowing only rotations with axis perpendicular to the plane, on the line through the points $(0,0)$, $(a,0)$, $(2a,0)$ in the plane. Accordingly, the corresponding groupoids are

$$\begin{pmatrix} y^1 \\ y^2 \end{pmatrix} = \begin{pmatrix} \partial y \\ \partial x \end{pmatrix} \begin{pmatrix} x^1 \\ x^2 \end{pmatrix}, \quad \begin{pmatrix} z^1 - a \\ z^2 \end{pmatrix} = \begin{pmatrix} \partial z \\ \partial y \end{pmatrix} \begin{pmatrix} y^1 - a \\ y^2 \end{pmatrix}, \quad \begin{pmatrix} x^1 - 2a \\ x^2 \end{pmatrix} = \begin{pmatrix} \partial x \\ \partial z \end{pmatrix} \begin{pmatrix} z^1 - 2a \\ z^2 \end{pmatrix},$$

where the (2×2) Jacobian matrices of 1-jets are orthogonal and 1-jets are used in order to indicate source and target. By composition we obtain

$$\begin{pmatrix} z^1 \\ z^2 \end{pmatrix} = \begin{pmatrix} \partial z \\ \partial y \end{pmatrix} \begin{pmatrix} y^1 \\ y^2 \end{pmatrix} - \begin{pmatrix} \partial z \\ \partial y \end{pmatrix} \begin{pmatrix} a \\ 0 \end{pmatrix} + \begin{pmatrix} a \\ 0 \end{pmatrix}$$

$$= \begin{pmatrix} \partial z \\ \partial y \end{pmatrix} \begin{pmatrix} \partial y \\ \partial x \end{pmatrix} \begin{pmatrix} x^1 \\ x^2 \end{pmatrix} - \begin{pmatrix} \partial z \\ \partial y \end{pmatrix} \begin{pmatrix} a \\ 0 \end{pmatrix} + \begin{pmatrix} a \\ 0 \end{pmatrix}$$

$$= \begin{pmatrix} \partial z \\ \partial y \end{pmatrix} \begin{pmatrix} \partial y \\ \partial x \end{pmatrix} \begin{pmatrix} \partial x \\ \partial z \end{pmatrix} \begin{pmatrix} z^1 \\ z^2 \end{pmatrix} - \begin{pmatrix} \partial z \\ \partial y \end{pmatrix} \begin{pmatrix} \partial y \\ \partial x \end{pmatrix} \begin{pmatrix} \partial x \\ \partial z \end{pmatrix} \begin{pmatrix} 2a \\ 0 \end{pmatrix}$$

$$+ \begin{pmatrix} \partial z \\ \partial y \end{pmatrix} \begin{pmatrix} \partial y \\ \partial x \end{pmatrix} \begin{pmatrix} 2a \\ 0 \end{pmatrix} - \begin{pmatrix} \partial z \\ \partial y \end{pmatrix} \begin{pmatrix} a \\ 0 \end{pmatrix} + \begin{pmatrix} a \\ 0 \end{pmatrix}.$$

All this must be an *identity* in (z^1, z^2), so we obtain

$$\begin{cases} \begin{pmatrix} \partial z \\ \partial y \end{pmatrix} \begin{pmatrix} \partial y \\ \partial x \end{pmatrix} \begin{pmatrix} \partial x \\ \partial z \end{pmatrix} = \begin{pmatrix} 1 & 0 \\ 0 & 1 \end{pmatrix}, \\ \begin{pmatrix} \partial z \\ \partial x \end{pmatrix} \begin{pmatrix} 2a \\ 0 \end{pmatrix} - \begin{pmatrix} \partial z \\ \partial y \end{pmatrix} \begin{pmatrix} a \\ 0 \end{pmatrix} = \begin{pmatrix} a \\ 0 \end{pmatrix}. \end{cases}$$

Since $a > 0$ is arbitrary, it follows that

$$2\frac{\partial z^1}{\partial x^1} - \frac{\partial z^1}{\partial y^1} = 1, \qquad 2\frac{\partial z^2}{\partial x^1} - \frac{\partial z^2}{\partial y^1} = 0.$$

Using the orthogonality conditions

$$\left(\frac{\partial z^1}{\partial x^1}\right)^2 + \left(\frac{\partial z^2}{\partial x^1}\right)^2 = 1, \qquad \left(\frac{\partial z^1}{\partial y^1}\right)^2 + \left(\frac{\partial z^2}{\partial y^1}\right)^2 = 1,$$

we obtain, after substitution:

$$\frac{\partial z^1}{\partial y^1} = 1, \quad \frac{\partial z^1}{\partial x^1} = 1, \quad \frac{\partial z^2}{\partial y^1} = 0, \quad \frac{\partial z^2}{\partial x^1} = 0, \Rightarrow \left(\frac{\partial z}{\partial x}\right) = \left(\frac{\partial z}{\partial y}\right) = \begin{pmatrix} 1 & 0 \\ 0 & 1 \end{pmatrix},$$

and so $x = y = z$.

This proves that the system is rigid *without integrating the articulations*, i.e. without using explicit rotation matrices.

C. Heat

The purpose of this Section is to prove that thermodynamics has a purely geometric origin that only depends on the structure of the Weyl group. In particular, we will explain how the Helmholtz analogy depends on the distinction between a jet of a bundle section and a section of a jet bundle.

Let ω be the standard Minkowski metric $(+++-)$ on space–time with $x^4 = ct$ and

$$ds^2 = (dx^1)^2 + (dx^2)^2 + (dx^3)^2 - (dx^4)^2.$$

We first recall, in terms of sections, the groupoid $\mathcal{R}_1$ and its first prolongation $\mathcal{R}_2$ that make it possible to define the Poincaré group:

$$\mathcal{R}_2 \begin{cases} \mathcal{R}_1 & \left\{ \omega_{kl}(f(x)) f_i^k(x) f_j^l(x) = \omega_{ij}(x), \right. \\ g_l^k(x)(f_{ij}^l(x) + \gamma_{rs}^l(f(x)) f_i^r(x) f_j^s(x)) = \gamma_{ij}^k(x). \end{cases}$$

We know that the symbol M_1 is neither 2-acyclic nor involutive, but it is of finite type with $M_2 = 0$.

Similarly, we recall, in terms of sections, the groupoid $\tilde{\mathcal{R}}_2$ (take care of the order!) allowing us to define the Weyl group:

$$\tilde{\mathcal{R}}_2 \begin{cases} \tilde{\mathcal{R}}_1 & \left\{ \omega_{kl}(f(x)) f_i^k(x) f_j^l(x) = a(x)\omega_{ij}(x), \right. \\ g_l^k(x)(f_{ij}^l(x) + \gamma_{rs}^l(f(x)) f_i^r(x) f_j^s(x)) = \gamma_{ij}^k(x). \end{cases}$$

Of course, with space–time coordinates, the Christoffel symbols γ are equal to zero and the second-order jets vanish because $\det(f_i^k(x)) \neq 0$. It follows that $M_2 = 0$ and that the Weyl group $\tilde{\Gamma}$ is obtained from the Poincaré group Γ by adding the dilatation $y = \lambda x$. We stress that this dilatation, acting on space *and* time *at the same time*, is not accessible to intuition.

We will prove that the *group inclusion*

$$\begin{array}{ccc} \text{Poincaré group} & \subset & \text{Weyl group} \\ \text{10 parameters} & & \text{11 parameters} \end{array}$$

implies the following *logical inclusion*:

$$\text{elastodynamics} \subset \text{thermodynamics}.$$

The basic idea in this Chapter is to explain how jet theory can lead to the concept of absolute temperature in this group-theoretical setting.

First of all, for a reader not familiar with jet theory, we will explain why we have used $a(x)$ instead of the $a = $ const appearing in the literature. Indeed, any dilatation $y = \lambda x$ implies $a = \lambda^2 = $ const. Conversely, any solution $y = f(x)$ of $\tilde{\mathcal{R}}_2$ satisfies

$$\omega_{kl}(f(x))\partial_i f^k(x)\partial_j f^l(x) = a(x)\omega_{ij}(x).$$

A straightforward calculation by direct differentiation shows that (for convenience, and where possible, n is kept arbitrary)

$$j_1(f)^{-1}(\omega) = a(x)\omega$$

$$\Rightarrow (j_2(f)^{-1}(\gamma))^k_{ij}(x) = \gamma^k_{ij}(x) + \delta^k_i a_j(x) + \delta^k_j a_i(x) - \omega_{ij}(x)\omega^{kr}(x)a_r(x),$$

with $a_i(x) = \frac{1}{2}\partial_i \log a(x)$, by computing the Christoffel symbol of $a(x)\omega$ over the source, since we are dealing with symmetric objects. Hence, $f \in \tilde{\Gamma} \Leftrightarrow \partial_i a(x) = 0 \Leftrightarrow a = $ const. However, things are completely different on the level of sections, where $a(x) \neq 0$ may, in effect, be an arbitrary nonzero *function* with

$$\det(f^k_i(x))^2 = a^n(x).$$

We define a function Θ *on the target* by

$$\Theta^2(f(x))\omega_{kl}(f(x))f^k_i(x)f^l_j(x) = \omega_{ij}(x),$$

so that

$$\Theta^n(f(x))\det(f^k_i(x)) = 1.$$

At the same time we introduce a 'vector' $u^k(y)$ *on the target* so that

$$f^k_4(x) = \frac{u^k(f(x))}{\Theta(f(x))},$$

and therefore

$$\omega_{kl}(y)u^k u^l = -1.$$

If we are dealing with the dilatation $y = \lambda x$ we have $\Theta = 1/\lambda$, but Θ may exist independently, on the jet level. We discover that introducing Θ just amounts to gauging the 1-parameter multiplicative group of dilatations. If we only care about the component of the identity, which is invariant under inversion, the topology is that of the positive real line:

For this reason Θ *will play the role of absolute temperature*. Similarly, if we impose the *constraint* $A^1_4 = A^2_4 = A^3_4 = 0$ on the moving frame, it follows that $f^k_4(x)$ is proportional to $\partial_4 f^k(x)$, and so u^k is the normalized velocity of special relativity while f^k_4 is the so-called *temperature vector* [16, 17, 83]. *No classical differential-geometry can allow this setting, which crucially depends on jet theory*. Also, *automatically* $\Theta = 1$ for the Poincaré group, and we notice the relation

$$\det A = \Theta^n \Delta,$$

where $\Delta = \det(\partial_i f^k(x))$ is the usual Jacobian determinant.

In special relativity the following quantities are essential for relativistic dynamics:

$$\begin{cases} \theta = \sqrt{1 - \dfrac{v^2}{c^2}}\,\partial_4 f_4 = \sqrt{-\omega_{kl}(y)\partial_4 f^k \partial_4 f^l}, \\[3mm] \rho = \sqrt{1 - \dfrac{v^2}{c^2}}\,\dfrac{\partial_4 f^4}{\Delta} = \dfrac{\theta}{\Delta}. \end{cases}$$

Taking into account the relation

$$\frac{\partial_4 f^4}{\Delta} = \frac{\partial(x^1, x^2, x^3)}{\partial(y^1, y^2, y^3)},$$

defined for $j_1(f)$, we understand that the above choice is the only way to separate *space* (ρ depends only on the choice of x^1, x^2, x^3 and not on that of x^4) from *time* (θ does not depend on the choice of x^1, x^2, x^3 and does depend on the choice of x^4; see [146] for more details). Since θ and ρ are clearly invariant under the action of the Poincaré group *on the target*, the *only possible* extension of this choice for the Weyl group, with the same invariance property, is to set (compare with [186])

$$\tilde{\theta} = \Theta\theta, \qquad \tilde{\rho} = \frac{\rho}{\Theta^3},$$

so that in both cases

$$\tilde{\theta} = \sqrt{-\omega_{rs}(x)A_4^r A_4^s}, \qquad \tilde{\rho} = \frac{\sqrt{-\omega_{rs}(x)A_4^r A_4^s}}{\det A},$$

since any function invariant under the action of the group on the target is either a differential invariant (and then the only possibility is to retain $\tilde{\rho}\tilde{\theta}^3 = \rho\theta^3$) or a function of the image of $\overline{D}$ in the Spencer sequence. We leave it to the reader to show that $\partial_i \xi^4$ does not appear in $\delta\rho$ or $\delta\tilde{\rho}$ [80, p. 506].

We should not forget that we are dealing with a section f_2 of $\tilde{\mathcal{R}}_2$, and so $y = f(x)$ is *arbitrary*. In particular, for a continuum at rest (thermostatics) we have $\tilde{\theta} = \Theta\partial_4 f^4$, while when all material points are being examined at the same moment (time slicing of space–time) we have $y^4 = x^4 + \text{const}$ and so $\partial_4 f^4 = 1$, *a possibility not given by special relativity*. We now understand that the Helmholtz analogy rests on a confusion between the effective derivative of final time y^4 with respect to initial time x^4 and the corresponding formal jet. As a byproduct, in a variational formulation we have to take care of $\partial_4 f^4$, perform the variation and integration by parts, and only *finally* set $\partial_4 f^4 = 1$, since *evaluation does not commute with derivation*. If the action density w only depends on $\tilde{\theta}$ and $\rho\theta^3$, then $\rho\theta^3$ will produce a traceless generalized stress tensor density with a diagonal rest form $(-p, -p, -p, 3p)$ allowing one to *define* the pressure p. Accordingly, the term similar to $\dot{q}(\partial L/\partial \dot{q}) - L$ that we are looking for will be $\tilde{\theta}(\partial w/\partial \tilde{\theta}) - w$, i.e. $\Theta(\partial w/\partial\Theta) - w$ after evaluation, although the interpretation is quite different.

We now prove that *the transition from Lagrangian to Hamiltonian just amounts to passing from target to source*.

The basic idea in classical analytical dynamics is to set $\delta q = \dot{q}\delta t$, so that

$$\left(\frac{d}{dt}\left(\frac{\partial L}{\partial \dot{q}}\right) - \frac{\partial L}{\partial q}\right)\dot{q}\,\delta t = \left(\frac{\partial L}{\partial t} + \frac{dH}{dt}\right)\delta t,$$

with $H = \dot{q}(\partial L/\partial \dot{q}) - L$. The natural generalization seems to be $\delta f_\mu^k = \xi^r \partial_r f_\mu^k \ldots$, but this is not convenient. Indeed, $f_q : X \to \mathcal{R}_q$ admits a tangent map $T(f_q) : T(X) \to T(\mathcal{R}_q)$ which *does not* provide an image in $V(\mathcal{R}_q)$ that can be used as a variation δf_q. Again the solution of this cannot be obtained without jet theory, because of the complexity involved.

First of all we recall the formula for computing $\chi_q = \overline{D}f_{q+1}$:

$$f_r^k \chi_{\mu,i}^r + \cdots + f_{\mu+1_r}^k \chi_{,i}^r = \partial_i f_\mu^k - f_{\mu+1_i}^k.$$

Setting $y_q = f_q(x)$, we want to obtain

$$(\sharp(\eta_q))_\mu^k = d_\mu \eta^k = f_r^k \overline{\xi}_\mu^r + \cdots + f_{\mu+1_r}^k \overline{\xi}^r = \delta f_\mu^k = \xi^i \partial_i f_\mu^k.$$

Multiplying the first formula by ξ^i and subtracting it from the second, we obtain

$$f_r^k(\overline{\xi}_\mu^r - \chi_{\mu,i}^r \xi^i) + \cdots + f_{\mu+1_r}^k(\overline{\xi}^r - \chi_{,i}^r \xi^i) = f_{\mu+1_i}^k \xi^i,$$

i.e.

$$f_r^k \xi_\mu^r + \cdots + f_{\mu+1_r}^k \xi^r = f_{\mu+1_i}^k \xi^i,$$

and so $(\flat(\xi_q))_\mu^k = 0$. Since $\det(f_i^k(x)) \neq 0$, by induction this leads to the equations

$$\overline{\xi}_\mu^k - \chi_{\mu,i}^k \xi^i = \xi_\mu^k = 0, \qquad \forall 1 \leq |\mu| \leq q.$$

We see that the variation we are looking for can be obtained if and only if there is a flat connection $(\xi) \to (\xi, 0, \ldots, 0)$, in particular, whenever the Lie pseudogroup Γ contains the translations (as in our case because $\partial_r \omega_{ij}(x) = 0$).

Finally, collecting the induction equations *over the target* and contracting with the variation $\delta f_q = \xi \cdot f_q$ amounts to looking *only* for the induction equations in ξ *over the source*. Since

$$\begin{cases} \delta A_i^k = \xi^r \partial_r A_i^k + A_r^k \partial_i \xi^r + \ldots, \\ \delta \chi_{\mu,i}^k = \xi^r \partial_r \chi_{\mu,i}^k + \chi_{\mu,r}^k \partial_i \xi^r + \ldots, & 1 \leq |\mu| \leq q, \end{cases}$$

by keeping only the terms in ξ and integrating by parts, we successively obtain

$$\delta w = \frac{\partial w}{\partial A_i^k}\delta A_i^k + \sum_{|\mu|=1}^q \frac{\partial w}{\partial \chi_{\mu,i}^k}\delta \chi_{\mu,i}^k$$

$$= \xi^r\left(\frac{\partial w}{\partial A_i^k}\partial_r A_i^k + \sum_{|\mu|=1}^q \frac{\partial w}{\partial \chi_{\mu,i}^k}\partial_r \chi_{\mu,i}^k\right)$$

$$+ d_i\left(\frac{\partial w}{\partial A_i^k}A_r^k \xi^r + \ldots\right) - \xi^r d_i\left(\frac{\partial w}{\partial A_i^k}A_r^k + \ldots\right).$$

Setting

$$T_r^i = \frac{\partial w}{\partial A_i^k} A_r^k + \sum_{|\mu|=1}^{q} \frac{\partial w}{\partial \chi_{\mu,i}^k} \chi_{\mu,r}^k - w \delta_r^i,$$

we finally obtain

$$\delta w = d_i \left(\frac{\partial w}{\partial A_i^k} A_r^k \xi^r + \ldots \right) - \xi^r (\partial_r w + d_i T_r^i)$$

and the general formula

$$\left(d_i \left(\frac{\partial w}{\partial \partial_i f_\mu^k} \right) - \frac{\partial w}{\partial f_\mu^k} \right) \partial_r f_\mu^k = \partial_r w + d_i T_r^i.$$

If $w = w(\tilde{\theta})$, the definition of $\tilde{\theta}$ gives $T_4^1 = T_4^2 = T_4^3 = 0$ and $T_4^4 = \tilde{\theta}(\partial w/\partial \tilde{\theta}) - w$, as required. We leave it to the reader to verify directly how tricky the computations can become, even for first-order jets.

Recapitulating the results above, we notice that the *absolute temperature* Θ *is not a field*, since the true corresponding field is $\tilde{\theta} = \theta \Theta$, but we can always impose the constraint $\theta = 1$ because the choice of x^4 is arbitrary. Also, over the source, we have

$$\alpha_i = \chi_{r,i}^r = -\frac{n}{\Theta} \partial_i \Theta,$$

i.e. the *gradient of temperature*. This result cannot be obtained by a standard procedure.

These two results solve our problem 3.

We will now study the infinitesimal point of view. We have the Lie algebroids

$$R_2 \quad \begin{cases} R_1 \quad \{ \omega_{rj} \xi_i^r + \omega_{ir} \xi_j^r + \xi^r \partial_r \omega_{ij} = 0, \\ \xi_{ij}^k + \gamma_{rj}^k \xi_i^r + \gamma_{ir}^k \xi_j^r - \gamma_{ij}^r \xi_r^k + \xi^r \partial_r \gamma_{ij}^k = 0, \end{cases}$$

$$\tilde{R}_2 \quad \begin{cases} \omega_{rj} \xi_i^r + \omega_{ir} \xi_j^r + \xi^r \partial_r \omega_{ij} = A(x) \omega_{ij}, \\ \xi_{ij}^k + \gamma_{rj}^k \xi_i^r + \gamma_{ir}^k \xi_j^r - \gamma_{ij}^r \xi_r^k + \xi^r \partial_r \gamma_{ij}^k = 0. \end{cases}$$

Multiplying the first-order equations of the second system by ω^{ij}, we obtain

$$2\xi_r^r + \xi^r \omega^{ij} \partial_r \omega_{ij} = nA,$$

and a way to eliminate A. Also, in our case $\xi_1^1 = \xi_2^2 = \cdots = \xi_r^r/n$ because of the special choice of the metric.

In the infinitesimal setting, the field is the image of the Spencer operator, i.e.

$$\chi_{\mu,i}^k = \partial_i \xi_\mu^k - \xi_{\mu+1_i}^k.$$

Hence, for the components of the field we have

$$\omega_{rj} \chi_{,i}^r + \omega_{ir} \chi_{,j}^r = \omega_{rj} (\partial_i \xi^r - \xi_i^r) + \omega_{ir} (\partial_j \xi^r - \xi_j^r)$$

$$= \omega_{rj} \partial_i \xi^r + \omega_{ir} \partial_j \xi^r - \frac{2}{n} \omega_{ij} \xi_r^r + \xi^r \partial_r \omega_{ij} - \frac{1}{n} \omega_{ij} \xi^r \partial_r \log |\det \omega|,$$

where the two last terms disappear. In practice we also have $\xi^4 = \mathrm{const}$, and we can combine this with $\chi^4_{,4} = \partial_4 \xi^4 - \xi^4_4$ to obtain the 'small' deformation and the relative change of temperature separately. The velocity v^i/c can be obtained from the above field when $j = 4$, $i = 1, 2, 3$, since we obtain

$$\omega_{ir}\partial_4\xi^r + \omega_{r4}\partial_i\xi^r - \frac{2}{n}\omega_{i4}\xi^r_r = \partial_4\xi^i \ll 1.$$

Hence, the factor of ξ^4 in the integration by parts of an isotropic action density linearly containing the four bilinear terms, with $i, j = 1, 2, 3$,

$$(\partial_4\xi^4 - \xi^4_4)(\partial_i\xi^i - \xi^i_i), \qquad (\partial_4\xi^4 - \xi^4_4)^2,$$
$$\omega^{ij}(\omega_{r4}\partial_i\xi^r + \omega_{ir}\partial_4\xi^r)\partial_j\xi^4_4, \qquad (\partial_4\xi^4 - \xi^4_4)\partial_4\xi^4_4$$

makes it possible to separately recover the four terms

$$\vec{\nabla}\cdot\vec{v}, \quad \frac{\partial T}{\partial t}, \quad \Delta T, \quad \frac{\partial^2 T}{\partial t^2}$$

in the heat equation. The nonlinear framework can be treated similarly by using α_i in order to bring the gradient of temperature into the generalized stress tensor [146]. Also, for $i = 1, 2, 3$ we have the formula

$$(\partial_i\xi^4_4 - \xi^4_{4i}) - (\partial_4\xi^4_i - \xi^4_{4i}) = \partial_i\xi^4_4 - \partial_4\xi^i_4 = \partial_i\xi^4_4 - \partial_{44}\xi^i,$$

mixing gradient of temperature and acceleration independently of second-order jets. These results agree with the phenomenological point of view adopted in the literature [54].

As already said, the gradient of temperature is a field because $\chi^r_{r,i} = \partial_i\xi^r_r$. Finally, for $i, j = 1, 2, 3$:

$$\omega_{rj}\chi^r_{4,i} + \omega_{ir}\chi^r_{4,j} = \omega_{rj}\partial_i\xi^r_4 + \omega_{ir}\partial_j\xi^r_4$$
$$= \partial_4(\omega_{rj}\partial_i\xi^r + \omega_{ir}\partial_j\xi^r),$$

taking into account the infinitesimal constraints $\partial_4\xi^i - \xi^i_4 = 0$ for $i = 1, 2, 3$ that identify velocity and rotation of the frame in space–time. So, the rate of deformation is also a field, and *we may incorporate in our theory concepts that are usually regarded as irreversible*.

To conclude this Section we notice that the combined use of the Weyl group and jet theory has made it possible to unify the mathematical models of elasticity and heat in a way that points out the necessity of revisiting almost entirely the foundations of irreversible thermodynamics.

D. Electromagnetism

The purpose of this Section is to prove that electromagnetism has a purely geometric origin that only depends on the structure of the conformal group of space–time and *not* on U(1). In particular, we will show that the Mach–Lippmann analogy comes from gauging the corresponding Lie group. This will automatically give a unification of elasticity, heat, and electromagnetism, in agreement with photoelasticity and the associated finite-elements approach ... but in contradiction with gauge theory.

We start with a mathematical study of the conformal group of transformations along the lines of Chapter V. We consider an arbitrary dimension n and we will only later on restrict to the particular case $n = 4$.

We associate to any $\omega \in S_2 T^*$ with $\det \omega \neq 0$ the Lie pseudogroup $\hat{\Gamma}$ of conformal isometries defined, in terms of sections, by the first-order groupoid

$$\widehat{\mathcal{R}}_1 \qquad \omega_{kl}(f(x)) f_i^k(x) f_j^l(x) = a(x) \omega_{ij}(x).$$

The first prolongation is defined by

$$\widehat{\mathcal{R}}_2 \begin{cases} \omega_{kl}(f(x)) f_i^k(x) f_j^l(x) = a(x) \omega_{ij}(x), \\ g_l^k(x)(f_{ij}^l(x) + \gamma_{rs}^l(f(x)) f_i^r(x) f_j^s(x)) \\ \qquad = \gamma_{ij}^k(x) + \delta_i^k a_j(x) + \delta_j^k a_i(x) - \omega_{ij}(x) \omega^{kr}(x) a_r(x). \end{cases}$$

As in the previous Section, again $a(x)$ and $a_i(x)$ are arbitrary functions satisfying $a(x) \neq 0$. Since $a = 1$ at the identity, below we assume that $a(x) > 0$. The finite Lie equations above are not in Lie form, since for this, one should eliminate $a(x)$ and $a_i(x)$. However, the corresponding first- and second-order geometric objects are easily seen to be

$$\begin{cases} \hat{\omega}_{ij} = \dfrac{\omega_{ij}}{|\det \omega|^{1/n}}, \\ \hat{\gamma}_{ij}^k = \gamma_{ij}^k - \dfrac{1}{n}(\delta_i^k \gamma_{rj}^r + \delta_j^k \gamma_{ri}^r - \omega_{ij} \omega^{ks} \gamma_{rs}^r), \end{cases}$$

so that $|\det \hat{\omega}| = 1$, $\hat{\gamma}_{ri}^r = 0$, and

$$\widehat{\mathcal{R}}_2 = \{ f_2 \in \Pi_2(X, X) | f_1^{-1}(\hat{\omega}) = \hat{\omega}, \ f_2^{-1}(\hat{\gamma}) = \hat{\gamma} \}.$$

Notice that $f_2^{-1}(\gamma) - \gamma \in S_2 T^* \otimes T$ is associated with $\Pi_1(X, X)$ and that $a_i(x)$ are thus the components of a 1-form that need not be closed. However, we leave it to the reader to prove that $f_2 = j_2(f)$ with $f \in \hat{\Gamma}$ implies $a_i = \frac{1}{2} \partial_i \log a$, which in turn implies $\partial_i a_j - \partial_j a_i = 0$ (this is an easy exercise on differential elimination). Below we will give another proof of this fact.

Passing to the infinitesimal point of view, the Lie algebra $\hat{\Theta}$ of infinitesimal conformal isometries is defined by the first-order system of conformal Killing equations

$$\hat{R}_1 \qquad \omega_{rj}(x) \xi_i^r + \omega_{ir}(x) \xi_j^r + \xi^r \partial_r \omega_{ij}(x) = A(x) \omega_{ij}(x),$$

obtained by eliminating the arbitrary function A. The first prolongation is

$$\hat{R}_2 \begin{cases} \omega_{rj}(x) \xi_i^r + \omega_{ir}(x) \xi_j^r + \xi^r \partial_r \omega_{ij}(x) = A(x) \omega_{ij}(x), \\ \xi_{ij}^k + \gamma_{rj}^k(x) \xi_i^r + \gamma_{ir}^k(x) \xi_j^r - \gamma_{ij}^k(x) \xi_r^k + \xi^r \partial_r \gamma_{ij}^k(x) \\ \qquad = \delta_i^k A_j(x) + \delta_j^k A_i(x) - \omega_{ij}(x) \omega^{kr}(x) A_r(x). \end{cases}$$

As before we notice that $L(\xi_2)\gamma \in S_2 T^* \otimes T$ and that $A_i(x)$ are the components of a 1-form that need not be closed. Similarly, we leave it to the reader to verify that $\xi_2 = j_2(\xi)$ with $\xi \in \hat{\Theta}$ implies $A_i = \frac{1}{2} \partial_i A$, which in turn implies $\partial_i A_j - \partial_j A_i = 0$, and we have

$$\hat{R}_2 = \{ \xi_2 \in J_2(T) | L(\xi_1)\hat{\omega} = 0, \ L(\xi_2)\hat{\gamma} = 0 \}.$$

We will now study the symbol $\widehat{M}_1$ of $\widehat{R}_1$ and its prolongations:

$$\widehat{M}_1 \qquad \omega_{rj}(x)\xi_i^r + \omega_{ir}(x)\xi_j^r = A(x)\omega_{ij}(x),$$

$$\widehat{M}_2 \qquad \xi_{ij}^k = \delta_i^k A_j(x) + \delta_j^k A_i(x) - \omega_{ij}(x)\omega^{kr}(x)A_r(x).$$

Eliminating $A(x)$ and $A_i(x)$, we obtain

$$\widehat{M}_1 \qquad \omega_{rj}(x)\xi_i^r + \omega_{ir}(x)\xi_j^r - \frac{2}{n}\omega_{ij}(x)\xi_r^r = 0,$$

$$\widehat{M}_2 \qquad \xi_{ij}^k = \frac{1}{n}(\delta_i^k\xi_{rj}^r + \delta_j^k\xi_{ri}^r - \omega_{ij}(x)\omega^{ks}(x)\xi_{rs}^r),$$

and so

$$\widehat{M}_3 \qquad \xi_{ijt}^k = \frac{1}{n}(\delta_i^k\xi_{rjt}^r + \delta_j^k\xi_{rit}^r - \omega_{ij}(x)\omega^{ks}(x)\xi_{rst}^r).$$

Contracting with respect to k and t gives

$$(n-2)\xi_{rij}^r + \omega_{ij}(x)\omega^{kl}(x)\xi_{rkl}^r = 0.$$

Contracting now by $\omega^{ij}(x)$ gives

$$2(n-1)\omega^{ij}(x)\xi_{rij}^r = 0,$$

and so $\widehat{M}_3 = 0$, $\forall n \geq 3$. In that case the 2-acyclicity of $\widehat{M}_2$ depends only on the injectivity of δ in the sequence

$$0 \longrightarrow \Lambda^2 T^* \otimes \widehat{M}_2 \xrightarrow{\delta} \Lambda^3 T^* \otimes T^* \otimes T.$$

We prove this in local coordinates. Indeed, suppose that

$$\xi_{ir,st}^k + \xi_{is,tr}^k + \xi_{it,rs}^k = 0.$$

Contracting with respect to k and i gives

$$\xi_{ur,st}^u + \xi_{us,tr}^u + \xi_{ut,rs}^u = 0,$$

and so the first relation implies

$$\delta_r^k\xi_{ui,st}^u + \delta_s^k\xi_{ui,tr}^u + \delta_t^k\xi_{ui,rs}^u - \omega^{kj}(\omega_{ir}\xi_{uj,st}^u + \omega_{is}\xi_{uj,tr}^u + \omega_{it}\xi_{uj,rs}^u) = 0.$$

Contracting with respect to k and r gives

$$(n-3)\xi_{ui,st}^u - \omega^{rj}(\omega_{is}\xi_{uj,tr}^u + \omega_{it}\xi_{uj,rs}^u) = 0.$$

Contracting now by ω^{is} finally gives

$$2(n-2)\omega^{is}\xi_{ui,st}^u = 0.$$

Hence $\widehat{M}_2$ is 2-acyclic, $\forall n \geq 4$.

By the Criterion for formal integrability (Chapter III), for $n \geq 4$ we find that $\widehat{R}_2$ is formally integrable if and only if $\pi_2^3 : \widehat{R}_3 \to \widehat{R}_2$ or $\pi_2^3 : \widehat{R}_3 \to \widehat{R}_2$ is surjective. This condition involves only $j_2(\omega)$ or $j_2(\widehat{\omega})$. The details are left to the reader, since the procedure is rather similar to the one adopted in Chapter V for the Killing equations. If ρ_{lij}^k is the Riemann tensor, we can introduce the Weyl tensor as follows:

$$\tau_{lij}^k = \rho_{lij}^k - \frac{1}{n-2}(\delta_j^k \rho_{li} - \delta_i^k \rho_{lj} + \omega_{li}\rho_j^k - \omega_{lj}\rho_i^k) + \frac{1}{(n-1)(n-2)}(\delta_j^k \omega_{li} - \delta_i^k \omega_{lj})\rho,$$

and the integrability condition is $\tau_{lij}^k = 0$. In particular, we notice that the constant Riemannian curvature condition $\rho_{lij}^k = c(\delta_j^k \omega_{li} - \delta_i^k \omega_{lj})$ *automatically* implies zero Weyl tensor. We can also deduce that the Weyl tensor is a section of $\mathrm{H}_1^2(\widehat{\mathrm{M}}_1)$, similarly as done in Chapter V.

Under the above condition we are now in a position to construct the linear and nonlinear first/second Spencer sequences. Let us start with he nonlinear framework.

Since we have nonzero second-order jets, we have a more complicated expression for the torsion and curvature describing $\overline{D}'$; namely:

$$\begin{cases} \partial_i A_j^k - \partial_j A_i^k - A_i^r \chi_{r,j}^k + A_j^r \chi_{r,i}^k = 0, \\ \partial_i \chi_{l,j}^k - \partial_j \chi_{l,i}^k - \chi_{l,i}^r \chi_{r,j}^k + \chi_{l,j}^r \chi_{r,i}^k - A_i^r \chi_{lr,j}^k + A_j^r \chi_{lr,i}^k = 0. \end{cases}$$

We immediately understand why it is usual to have zero 'torsion' and nonzero 'curvature', since we may indeed write the curvature condition as

$$\begin{aligned} \partial_i \chi_{l,j}^k - \partial_j \chi_{l,i}^k - \chi_{l,i}^r \chi_{r,j}^k + \chi_{l,j}^r \chi_{r,i}^k &= A_i^r \chi_{lr,j}^k - A_j^r \chi_{lr,i}^k \\ &= A_i^r A_j^s (\tau_{lr,s}^k - \tau_{ls,r}^k) \\ &= \varphi_{l,ij}^k. \end{aligned}$$

Setting $\chi_{r,i}^r = \alpha_i$ while contracting with respect to k and l, we obtain

$$\varphi_{r,ij}^r = \varphi_{ij} = \partial_i \alpha_j - \partial_j \alpha_i,$$

which is *the key relation establishing a link with electromagnetism*. Also, since $\overline{D}'$ is formally integrable and provides all compatibility conditions for $\overline{D}$, i.e. all the field equations, we must also have

$$\partial_i \varphi_{jk} + \partial_j \varphi_{ki} + \partial_k \varphi_{ij} = 0.$$

The intrinsic version of these results is stated in the two following propositions [146].

Proposition 1. *There is a short exact sequence*

$$0 \longrightarrow T^* \otimes \mathrm{R}_1 \longrightarrow T^* \otimes \widehat{\mathrm{R}}_1 \xrightarrow{\alpha} T^* \longrightarrow 0.$$

PROOF. There is a well-defined map

$$J_1(T) \to S_2 T^* : \xi_1 \to \tfrac{1}{2}L(\xi_1)\omega$$

that can be contracted by ω^{-1} and tensored by T^* to give a map $T^* \otimes J_1(T) \to T^*$ with kernel $T^* \otimes \mathrm{R}_1$, by definition. The restriction to $T^* \otimes \widehat{\mathrm{R}}_1$ defines the above short exact sequence by counting dimensions, since $\dim \widehat{R}_1 - \dim R_1 = \dim \widehat{M}_1 - \dim M_1 = 1$. In local coordinates we have

$$
\begin{aligned}
\alpha_s &= \tfrac{1}{2}\omega^{ij}(\omega_{rj}\chi^r_{i,s} + \omega_{ir}\chi^r_{j,s} + \chi^r_{,s}\partial_r\omega_{ij}) \\
&= \chi^r_{r,s} + \tfrac{1}{2}\chi^r_{,s}\omega^{ij}\partial_r\omega_{ij} \\
&= \chi^r_{r,s} + \gamma^t_{rt}\chi^t_{,s}.
\end{aligned}
$$

Notice that the connection γ has been used to separate 1-jets, even though $\gamma = 0$ in practice. $\square$

Proposition 2. *There is a (nonexact) sequence*

$$
0 \longrightarrow T^* \otimes R_2 \longrightarrow T^* \otimes \widehat{R}_2 \longrightarrow \wedge^2 T^*
$$

PROOF. There is a well-defined map

$$
J_2(T) \to S_2 T^* \otimes T : \xi_2 \to L(\xi_2)\gamma
$$

that can be tensored by T^* and composed with δ to give a map $T^* \otimes J_2(T) \to \wedge^2 T^* \otimes T^* \otimes T$ that can be contracted to $\wedge^2 T^*$ and twisted by A. The above sequence then follows from the definition of R_2. In local coordinates we successively have

$$
\begin{aligned}
\beta^k_{lr,s} &= \tau^k_{lr,s} + \gamma^k_{ur}\tau^u_{l,s} + \gamma^k_{lu}\tau^u_{r,s} - \gamma^u_{lr}\tau^k_{u,s} + \tau^u_{,s}\partial_u\gamma^k_{lr}, \\
\varphi^k_{l,ij} &= A^r_i A^s_j(\beta^k_{lr,s} - \beta^k_{ls,r}), \\
\varphi_{ij} &= \varphi^r_{r,ij}.
\end{aligned}
$$

Again, we notice that the connection γ has been used to separate 2-jets, even though $\gamma = 0$ in practice. $\square$

Combining these Propositions, we successively obtain

$$
\begin{aligned}
\varphi_{ij} &= A^r_i A^s_j(\tau^u_{ur,s} - \tau^u_{us,r}) + A^r_i A^s_j\gamma^u_{ut}(\tau^t_{r,s} - \tau^t_{s,r}) \\
&\quad + A^r_i A^s_j(\tau^u_{,s}\partial_u\gamma^k_{kr} - \tau^u_{,r}\partial_u\gamma^k_{ks}) \\
&= \partial_i\chi^r_{r,j} - \partial_j\chi^r_{r,i} + \gamma^k_{kt}(\partial_i\chi^t_{,j} - \partial_j\chi^t_{,i}) \\
&\quad + A^r_i\chi^u_{,j}\partial_u\gamma^k_{kr} - A^s_j\chi^u_{,i}\partial_u\gamma^k_{ks} \\
&= \partial_i\chi^r_{r,j} - \partial_j\chi^r_{r,i} + \gamma^k_{kt}(\partial_i\chi^t_{,j} - \partial_j\chi^t_{,i}) \\
&\quad + \chi^r_{,i}\chi^u_{,j}\partial_u\gamma^k_{kr} - \chi^s_j\chi^u_{,i}\partial_u\gamma^k_{ks} + \chi^u_{,j}\partial_u\gamma^k_{ki} - \chi^u_{,i}\partial_u\gamma^k_{kj} \\
&= \partial_i\chi^r_{r,j} - \partial_j\chi^r_{r,i} + \gamma^k_{kt}(\partial_i\chi^t_{,j} - \partial_j\chi^t_{,i}) + \chi^u_{,j}\partial_i\gamma^k_{ku} - \chi^u_{,i}\partial_j\gamma^k_{ku} \\
&= \partial_i\alpha_j - \partial_j\alpha_i
\end{aligned}
$$

in an intrinsic way. It follows that the identity

$$
\partial_i\varphi_{jk} + \partial_j\varphi_{ki} + \partial_k\varphi_{ij} = 0
$$

also has an intrinsic meaning, and we obtain the following important Theorem:

Theorem 3. *The nonlinear Spencer sequence for the conformal group of transformations projects onto a part of the Poincaré sequence, according to the following commutative and exact diagram:*

$$0 \longrightarrow \hat{\Gamma} \xrightarrow{j_2} \widehat{\mathcal{R}}_2 \xrightarrow{\overline{D}} T^* \otimes \widehat{R}_2 \xrightarrow{\overline{D}'} \wedge^2 T^* \otimes \widehat{R}_2$$

$$\wedge^2 T^* \xrightarrow{d} \wedge^3 T^*$$

This result (mathematically) contradicts classical gauge theory. Indeed, since the nonlinear Spencer sequence is isomorphic to its corresponding gauge sequence, we see that we have a way of obtaining a 2-form from a 1-form with values in a Lie algebroid, and *not* from the curvature 2-form. Again, no classical differential geometry can bring about such a result, which was conjectured by Mie [120] and Weyl [196, 197, 198]. Moreover, we know from Section A that the associated variational calculus will depend on an action density (Lagrangian) defined on $T^* \otimes \widehat{R}_2$ and *not* on $\wedge^2 T^* \otimes \widehat{R}_2$. *This shift by one step is the main result in the present Section.*

Introducing Θ as in the previous Section and setting $\gamma = 0$, we obtain

$$\alpha_i = \chi^r_{r,i} = -n \left(\frac{1}{\Theta} \partial_i \Theta + A^r_i a_r \right).$$

It follows that φ_{ij} does not depend on Θ anymore, but on the matrix A and the *second-order jets* a_r which play the role of 4-potential, as conjectured by Weyl. Note that the gradient of temperature (derivative of 1-jet) cannot be measured independently of the 4-potential (2-jet), although we do not know the value of the physical constant reducing the 4-potential to M^{-1} unit in the MKSA system. If a solenoid with magnetic field B and surface S passes through the circuit made up by two metal wires of a thermo-electric circuit, the classical thermocouple difference of potential must be modified by a term proportional to the circulation of the spatial part of the 4-potential; namely, to BS. However, it seems quite delicate to realize such an experiment. Also, we see that the situation is rather similar to that in the previous Section, where the temperature (1-jet) could not be measured independently of dynamic quantities (derivative of 0-jets).

Looking for the variations, we notice that

$$\delta \alpha_i = (\partial_i \xi^r_r - \xi^r_{ri}) + \xi^r \partial_r \alpha_i + \alpha_i \partial_i \xi^r - \chi^s_{,i} \xi^r_{rs}$$

$$= \left[\left(\frac{\partial \eta^r_r}{\partial y^k} - \eta^k_{rk} \right) - n g^s_l a_s \left(\frac{\partial \eta^l}{\partial y^k} - \eta^l_k \right) \right] \frac{\partial f^k}{\partial x^i},$$

$$\delta \varphi_{ij} = \partial_i \delta \alpha_j - \partial_j \delta \alpha_i.$$

For a medium at rest we can set

$$y^k = f^k(x) = x^k, \qquad y^k_i = f^k_i(x) = \delta^k_i \Rightarrow \chi^k_{,i} = 0.$$

The simplified variation then becomes

$$\delta \alpha_i = (\partial_i \xi^r_r - \xi^r_{ri}) + \xi^r \partial_r \alpha_i + \alpha_r \partial_i \xi^r.$$

We recognize:

- The variation $\partial_i \xi_r^r$ introduced by Weyl in [196].
- The variation ξ_{ri}^r that amounts to varying the 4-potential itself and which corresponds to the δA_i of engineers.
- The variation $\xi^r \partial_r \alpha_i + \alpha_r \partial_i \xi^r$ that takes into account the fact that $\alpha_i(x)\, dx^i$ is a 1-form.

The combination of the first two variations into the Spencer operator is new and could not have been discovered by Weyl.

Before looking for the induction equations, we will prove that the above result necessitates a revisit of a basic point of classical gauge theory ([1]).

Recall how in classical analytical dynamics ($n = 3$) the Lorentz force can follow from a modification of the Lagrangian of a free particle (mass m, charge e). Indeed, from the Lagrangian

$$L = \tfrac{1}{2} m \omega_{ij} \dot{x}^i \dot{x}^j + e \dot{x}^i A_i$$

we obtain the Euler–Lagrange equations

$$\omega_{ij} \ddot{x}^j + e \dot{x}^j (\partial_j A_i - \partial_i A_j) = 0,$$

i.e. in standard notation

$$\frac{d\vec{v}}{dt} = e\vec{v} \wedge \vec{B}, \qquad \vec{B} = \vec{\nabla} \wedge \vec{A}.$$

The Hamiltonian becomes

$$H = \frac{1}{2m} \omega^{ij} (p_i - eA_i)(p_j - eA_j),$$

and the transition from the uncharged to the charged particle amounts to the 'transformation'

$$p_i \rightarrow p_i - eA_i.$$

The basic idea of gauge theory is to transform the momentum p_i to the operator $-i\hbar\partial_i$, according to the correspondence principle of quantum mechanics, and to transform A_i to a connection. Quite contrary to this point of view, we have:

Corollary 4. *The transformation* $p_i \rightarrow p_i - eA_i$ *has a purely group-theoretical origin.*

PROOF. The basic idea is to study the action of the second-order jets of $\Pi_2(X, X)$ on $J_1(T)$ and to dualize it. For an action density $w(\chi_{,i}^k, \alpha_i)$ we have:

$$\delta w = \frac{\partial w}{\partial \chi_{,i}^k} \delta\chi_{,i}^k + \frac{\partial w}{\partial \alpha_i} \delta\alpha_i = \mathcal{X}_k^{i} \delta\chi_{,i}^k + \mathcal{X}^i \delta\alpha_i$$

$$= \mathcal{X}_s^{i} g_i^s \left(\frac{\partial \eta^l}{\partial y^k} - \eta_k^l \right) \frac{\partial f^k}{\partial x^i} + \mathcal{X}^i \left[\left(\frac{\partial \eta_r^r}{\partial y^k} - \eta_{rk}^r \right) - n g_i^s a_s \left(\frac{\partial \eta^l}{\partial y^k} - \eta_k^l \right) \right] \frac{\partial f^k}{\partial x^i}$$

$$= \Delta \mathcal{Y}_l^{ik} \left(\frac{\partial \eta^l}{\partial y^k} - \eta_k^l \right) + \Delta \mathcal{Y}^k \left(\frac{\partial \eta_r^r}{\partial y^k} - \eta_{rk}^r \right),$$

and therefore

$$\begin{cases} \Delta \mathcal{Y}_l'^k = \mathcal{X}_s'^i g_l^s \partial_i f^k - n \mathcal{X}^i g_l^s a_s \partial_i f^k, \\ \Delta \mathcal{Y}^k = \mathcal{X}^i \partial_i f^k. \end{cases}$$

Setting

$$\Delta \mathcal{Y}_{l\text{pure}}'^k = \mathcal{X}_s'^i g_l^s \partial_i f^k,$$

we finally obtain

$$\mathcal{Y}_l'^k = \mathcal{Y}_{l\text{pure}}'^k - n g_l^s a_s \mathcal{Y}^k.$$

Hence the action of second-order jets *only* is

$$\mathcal{Y}_l'^k \to \mathcal{Y}_l'^k - n a_l \mathcal{Y}^k.$$

The Corollary now follows from the special case $k = 4$, since $\mathcal{Y}^4$ is the charge density. $\square$

For the Lorentz force we have

Corollary 5. *The Lorentz force has a purely group-theoretical origin.*

PROOF. Caring only about the terms in ξ, we have

$$\mathcal{X}^i \delta \alpha_i = \xi^r \mathcal{X}^i \partial_r \alpha_i + \mathcal{X}^i \alpha_r \partial_i \xi^r + \dots$$
$$= \partial_i (\xi^r \alpha_r \mathcal{X}^i) + \xi^r \left[\mathcal{X}^i (\partial_r \alpha_i - \partial_i \alpha_r) - \alpha_r \partial_i \mathcal{X}^i \right] + \dots .$$

However, from the proof of the previous Corollary we deduce that

$$\mathcal{X}^i A_i^s \xi_{rs}^r = 0, \quad \det A \neq 0 \Rightarrow \mathcal{X}^i = 0 \Rightarrow \mathcal{Y}^k = 0.$$

Hence, in order to have a density of electric current, the action density w must also depend on the electromagnetic field φ_{ij}. We can set $\partial w / \partial \varphi_{ij} = \mathcal{X}^{ij}$ (be careful, since this definition is not coherent with the use of χ_2) and use the variation $\delta \varphi_{ij}$ to extend the dual transformation formula in Corollary 4 as follows, with $b_k = g_k^s a_s$:

$$\begin{cases} \Delta \mathcal{Y}_l'^k = \mathcal{X}_s'^i g_l^s \partial_i f^k - n \mathcal{X}^i b_l \partial_i f^k + n \mathcal{X}^{ij} \dfrac{\partial b_l}{\partial y^s} \partial_i f^k \partial_j f^s, \\ \Delta \mathcal{Y}^{rs} = \mathcal{X}^{ij} \partial_i f^r \partial_j f^s. \end{cases}$$

We finally obtain

$$\mathcal{Y}_l'^k = \mathcal{Y}_{l\text{pure}}'^k - n \mathcal{Y}^k b_l + n \mathcal{Y}^{ks} \dfrac{\partial b_l}{\partial y^s}.$$

Taking into account the second set of Maxwell equations,

$$\dfrac{\partial \mathcal{Y}^{rs}}{\partial y^r} - \mathcal{Y}^s = 0$$

(we will derive these in the sequel), we finally obtain (exercise)

$$\mathcal{Y}_{\text{pure}}^{k,l} - \mathcal{Y}_{\text{pure}}^{l,k} = 0, \qquad \mathcal{Y}_{k\text{pure}}'^k = 0, \qquad \dfrac{\partial \mathcal{Y}_l'^k}{\partial y^k} = \dfrac{\partial \mathcal{Y}_{l\text{pure}}'^k}{\partial y^k}.$$

This important result proves that the Lorentz force indeed comes from the dynamical terms A_i^k contained in the action density. (See the end of this Section.) $\square$

We will now produce the induction equations, by looking directly at the infinitesimal point of view. The reader will immediately see how the Spencer operator and the structure of the conformal group *automatically* produce the basic results of electromagnetism.

Local coordinates for the Spencer operator can only be:

$$\chi_{,i}^k = \partial_i \xi^k - \xi_i^k, \qquad \chi_{j,i}^k = \partial_i \xi_j^k - \xi_{ij}^k, \qquad \chi_{rj,i}^r = \partial_i \xi_{rj}^r,$$

since $\widehat{M}_2 \simeq T^*$ and $\widehat{M}_3 = 0$ for $n \geq 3$. We now also assume that $n \geq 4$ in order to have involutive operators. More specifically, we introduce

$$\alpha_i = \chi_{r,i}^r = \partial_i \xi_r^r - \xi_{ri}^r, \qquad \varphi_{ij} = \chi_{ri,j}^r - \chi_{rj,i}^r = \partial_i \alpha_j - \partial_j \alpha_i = \partial_j \xi_{ri}^r - \partial_i \xi_{rj}^r.$$

For an action density $w(\chi_{,i}^k, \alpha_i, \varphi_{ij})$ we will set

$$\mathcal{X}^i = \frac{\partial w}{\partial \alpha_i}, \qquad \mathcal{X}^{ij} = \frac{\partial w}{\partial \varphi_{ij}}$$

and obtain (be careful with the summations):

$$\delta W = \int (\mathcal{X}_k^{\prime i}(\partial_i \delta \xi^k - \delta \xi_i^k) + \mathcal{X}^i(\partial_i \delta \xi_r^r - \delta \xi_{ri}^r) + \mathcal{X}^{ij}\partial_j \delta \xi_{ri}^r) \, dx^1 \wedge \cdots \wedge dx^n$$

$$= \int \partial_i (\mathcal{X}_k^{\prime i}\delta \xi^k + \mathcal{X}^i \delta \xi_r^r + \mathcal{X}^{si} \delta \xi_{rs}^r) \, dx^1 \wedge \cdots \wedge dx^n$$

$$- \int \left((\partial_i \mathcal{X}_k^{\prime i})\delta \xi^k + \sum_{i<k} \mathcal{X}_k^{\prime i}\delta \xi_i^k + \left(\partial_i \mathcal{X}^i + \frac{1}{n}\mathcal{X}_i^{\prime i} \right)\delta \xi_r^r + (\partial_j \mathcal{X}^{ij} + \mathcal{X}^i)\delta \xi_{ri}^r \right)$$

$$dx^1 \wedge \cdots \wedge dx^n.$$

The induction equations are seen to be well-known equations from electrodynamics:

$$\begin{cases} \partial_i \mathcal{X}_k^{\prime i} = 0, & \mathcal{X}^{i,j} - \mathcal{X}^{j,i} = 0, \\[2mm] \partial_i \mathcal{X}^i + \frac{1}{n}\mathcal{X}_i^{\prime i} = 0, \\[2mm] \partial_i \mathcal{X}^{ij} - \mathcal{X}^j = 0. \end{cases}$$

Note that the last two equations imply the trace condition $\mathcal{X}_i^{\prime i} = 0$, which is well-known in the case of pure electromagnetism. For the standard space–time generalized stresss $\rho u^i u^j$, the trace becomes $-\rho$, and electromagnetism cannot be separated from gravitation as we will see in the next Section.

Summarizing the above results, the linear Spencer sequence projects onto the Poincaré sequence (*with a shift*) as follows:

$$0 \longrightarrow \widehat{\Theta} \xrightarrow{\ j2\ } \widehat{R}_2 \xrightarrow{\ D\ } T^* \otimes \widehat{R}_2 \xrightarrow{\ D\ } \wedge^2 T^* \otimes \widehat{R}_2$$

$$\qquad\qquad\qquad \downarrow \qquad\qquad\quad \downarrow \qquad\qquad\qquad \downarrow$$

$$\qquad\qquad\quad T^* \xrightarrow{\ d\ } \wedge^2 T^* \xrightarrow{\ d\ } \wedge^3 T^*$$

This provides the first set of Maxwell equations, and we can obtain the second set by duality. This is exactly the scheme conjectured by Weyl, but it could not be discovered before 1970 since the uses of second-order jets and of the Spencer operator are essential. The contradiction with classical gauge theory needs no comment!

Now that we have obtained geometry and physics, we make some comments regarding the constitutive relations of electromagnetism, especially in vacuum.

First of all we prove that the use of the 4-potential in the study of electromagnetic waves is not necessary. Indeed, the classical approach works via the substitution

$$A_i \to F_{ij} \to \mathcal{F}^{ij}$$

in the second set of Maxwell equations, and provides the second-order wave equation for the potential *if and only if* we introduce the so-called Lorentz divergence-free condition for the potential. This is a hidden constitutive law, since the potential is a 1-form and not an $(n-1)$-form! (This remark is never made in textbooks.) One can avoid this approach as follows.

Recall that the constitutive law in vacuum is

$$\mathcal{F}_{ij} = \frac{1}{\mu_0} F_{ij},$$

while the first set of Maxwell equations is

$$\partial_i F_{jk} + \partial_j F_{ki} + \partial_k F_{ij} = 0.$$

Applying the operator $\omega^{kl}\partial_l$, we get

$$\partial_{il} F_j^l - \partial_{jl} F_i^l + \omega^{kl}\partial_{kl} F_{ij} = 0.$$

Taking into account the second set of Maxwell equations in vacuum, we finally obtain the required wave equations:

$$\omega^{kl}\partial_{kl} F_{ij} = 0.$$

The second comment concerns the invariance of the Maxwell equations. Indeed, most people believe that the conformal group is the largest group of invariance of the Maxwell equations. *This is wrong*: we will prove that the set of Maxwell equations is invariant under *any* transformation while, *on the other hand*, the conformal group is the largest group of invariance of the constitutive relations in vacuum.

The first set of Maxwell equations admits the intrinsic form $F = dA \Rightarrow dF = 0$, and is hence invariant under any transformation. For the second set we have

$$\begin{cases} \mathcal{F}'^{kl}(\varphi(x)) = \dfrac{1}{\Delta}\partial_i\varphi^k\partial_j\varphi^l\mathcal{F}^{ij}(x), \\ x' = \varphi(x), \end{cases}$$

and so

$$\frac{\partial \mathcal{F}'^{kl}}{\partial x'^k} = \partial_j \varphi^l \mathcal{F}^{ij} \frac{\partial(\frac{1}{\Delta}\partial_i \varphi^k)}{\partial x'^k} + \frac{1}{\Delta}\partial_i(\mathcal{F}^{ij}\partial_j \varphi^l)$$

$$= \frac{1}{\Delta}\mathcal{F}^{ij}\partial_{ij}\varphi^l + \frac{1}{\Delta}\partial_i \mathcal{F}^{ij}\partial_j \varphi^l$$

$$= \frac{1}{\Delta}\partial_i \mathcal{F}^{ij}\partial_j \varphi^l.$$

This result, which is compatible with the existence of a density of current, highly depends on the fact that F, and so $\mathcal{F}$, is a skewsymmetric object.

For the constitutive relations in vacuum we have more precisely:

$$\mathcal{F}^{kl} = \frac{1}{\mu_0}\hat{\omega}^{ik}\hat{\omega}^{jl}F_{ij},$$

where the use of $\hat{\omega}$ instead of ω is essential since F is a tensor but $\mathcal{F}$ is a density. Accordingly, the constitutive relations are invariant under a transformation $x' = \varphi(x)$ *if and only if*

$$\frac{1}{\Delta}\hat{\omega}_{ru}(\varphi(x))\hat{\omega}_{sv}\varphi(x)\partial_i\varphi^r\partial_j\varphi^s\partial_k\varphi^u\partial_l\varphi^v = \hat{\omega}_{ik}(x)\hat{\omega}_{jl}(x).$$

Contraction by $\hat{\omega}^{jl}$ gives an equation of the form

$$a(x)\Delta^{-1/2}\hat{\omega}_{rs}(\varphi(x))\partial_i\varphi^r(x)\partial_j\varphi^s(x) = \hat{\omega}_{ij}(x).$$

Taking the respective determinants, we obtain the condition $a(x) = 1$ and therefore *exactly* the PDE defining the conformal group of space–time.

In the modern framework we can introduce

$$A_{ij}(x) = \omega_{rs}(x)A_i^r(x)A_j^s(x) = \Theta^2 \omega_{kl}(f(x))\partial_i f^k(x)\partial_j f^l(x),$$

which is a conformally flat metric having zero Weyl tensor, and consider the following action density *over the source*:

$$w = \frac{1}{4\mu_0}B^{ir}B^{js}\varphi_{ij}\varphi_{rs}\det A,$$

where (B^{ij}) is the inverse matrix of (A_{ij}). We leave it to the reader to prove that w describes the classical EM Lagrangian *over the target* and does no longer depend on Θ, which thus leads to the traceless Maxwell tensor and the above constitutive relations, both with the Lorentz force of Corollary 5.

We finally notice that the use of a diagonal tensor $(1, 1, 1, -n^2)$ instead of $\omega^{rs}(x)$ in the definition of A_{ij} *exactly* leads to the Minkowski constitutive relations for a moving medium of refraction index n as given in [135]. Alternatively, we notice that a dilatation of space *or* time *separately* is *not* a conformal transformation and thus offers a way to modify the index of refraction.

E. Gravitation

The purpose of this Section is to prove that gravitation has a purely geometric origin that only depends on the structure of the conformal group of space–time. In particular, we will exhibit the basic concepts of gravitation (Newton law, Poisson equation) similarly as we have done for electromagnetism (Lorentz force, Maxwell equations) in the previous Section, i.e. *independently of any experimental framework*.

We start by recalling the striking analogy between electromagnetism and gravitation, which exists in classical or relativistic continuum mechanics.

Continuum mechanics deals with gravitation by introducing two symmetric 2-tensors: the *Eötvös tensor* $\tau_{ij} = \partial_{ij}\phi$, made up from the second-order derivatives of the gravitational potential ϕ and the *Abraham tensor* [2], defined in local coordinates by

$$\nu_{ij} = \partial_i\phi\partial_j\phi - \tfrac{1}{2}\omega_{ij}\omega^{rs}\partial_r\phi\partial_s\phi.$$

Notice, however, that *both tensors are in fact defined in terms of the gravitational field* $\partial_i\phi$.

Because of the advance of technology, a modern electrostatic gyroscope is sensitive to the derivatives of the gravitational field (look at the gravitational torque acting on the gyroscope). Using the fact that the gravitational field itself is the gradient of the gravitational potential, naturally leads to the Eötvös tensor.

> *In gravitation, the Abraham tensor will play the role that the Maxwell tensor plays in electromagnetism.*

Indeed, looking at the generalized stress equation $\partial_i\sigma^{ij} = f^j$, it is well known that the Maxwell tensor

$$\mu_j^i = \mathcal{F}^{ir}F_{rj} + \frac{1}{4}\delta_j^i\mathcal{F}^{rs}F_{rs}$$

makes it possible to write the case when f^j is a Lorentz force in a single divergence equation:

$$\partial_i(\sigma^{ij} - \mu^{ij}) = 0.$$

The reader immediately sees that *the computation crucially uses the two sets of Maxwell equations*. Similarly, the case when f^j is the gravitational force can be written as a single divergence equation:

$$\partial_i(\sigma^{ij} - \nu^{ij}) = 0.$$

The reader immediately sees that *the computation crucially uses the fact that the gravitational field comes from a potential, as well as the Poisson equation*

$$\operatorname{tr}\tau = \omega^{ij}\partial_{ij}\phi = \kappa\rho.$$

Notice that both the Maxwell and the Abraham tensor can be written in terms of the respective *fields*.

> *It remains to explain and justify this analogy.*

In particular, we have to understand the relation between the second set of Maxwell equations and the Poisson equation.

In special relativity the Lorentz or Poincaré group is only used as a working assumption, and the conformal group could be used instead. This is particularly clear in Einstein's famous 1905 mémoire on the electrodynamics of moving bodies [55]. Indeed, the Michelson–Morley experiment implies that

$$ds^2 = 0 \Leftrightarrow d\bar{s}^2 = 0$$

when selecting two inertial frames, using the constancy of the velocity of light c. We may assume that

$$d\bar{s}^2 = a(x)\,ds^2, \qquad a(x) > 0,$$

and most textbooks try to prove in vain that $a(x) = 1$ when space–time is considered as homogeneous and isotropic (no electromagnetism, no gravitation). So, otherwise, people should be fair to admit that the conformal factor $a(x)$ has 'something' to do with electromagnetism and/or gravitation ([67, 135, 181]).

Finally, in general relativity the way to 'absorb' the gravitational force is to use a covariant rather than an ordinary derivative, and to write

$$\nabla_i \sigma^{ij} = 0.$$

However, Einstein himself has proved that such an equation is equivalent to a pure divergence-type equation, obtained by adding a gravitational term to σ, but this result is purely 'artificial' since it is not intrinsic [138, 181].

Meanwhile, the single Poisson equation is replaced by the 10 Einstein equations

$$\rho_{ij} - \tfrac{1}{2}\omega_{ij}\rho \sim \sigma_{ij},$$

where ρ_{ij} is the Ricci tensor, in such a way that there is zero (covariant) divergence for both terms. However, the equations above depend on the classical curvature and *not* on the gauge curvature. In any case, the gauge curvature is a 2-form with values in the Lie algebra of rotations, while the stress only involves translation [209]. Also, along the Hilbert variational scheme of general relativity, the Lagrangian should be defined on the first Janet bundle or on the second Spencer bundle, contradicting the results of the previous Sections. *Last but not least*, the gauge curvature cannot be expressed in terms of a metric in the Spencer sequence, since no symmetric metric-like tensor exists in this sequence.

As a first step towards a solution of the above problems, we notice that the results of the previous Section imply that the electromagnetic field $\wedge^2 T^*$ should *not* be considered as a curvature form, but rather as an object *induced* from the first Spencer bundle. By the analogy between electromagnetism and gravitation, we may hope that the Ricci tensor is in fact also an object *induced* from the first Spencer bundle, in a way specifically related to the existence of second-order jets. More precisely, we will prove that gravitation comes from the second-order jets of the conformal group, exactly as we have proved the similar result for electromagnetism in the previous Section ([146, 149, 152]).

We start by revisiting the relations between the Riemann and the Weyl tensor.

First of all, the inclusion $R_1 \subset \widehat{R}_1$ induces an inclusion $M_1 \subset \widehat{M}_1$, and we obtain the following commutative diagram with exact columns induced by the maps

$$T^* \otimes S_2 T^* \otimes T \to T^* \otimes T^* : \xi^k_{lj,i} \to \xi^r_{rj,i},$$

$$\wedge^2 T^* \otimes T^* \otimes T \to \wedge^2 T^* : \xi^k_{l,ij} \to \xi^r_{r,ij},$$

respectively, for $n \geq 3$:

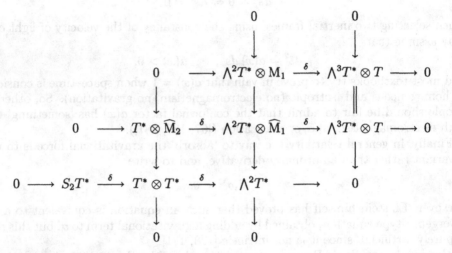

We can define the *Weyl bundle* $H^2_1(\widehat{M}_1)$ in a way exactly similar to the definition of the *Riemann bundle* $H^2_1(M_1)$ in Chapter V.E. The previous diagram hence induces the following commutative and exact diagram

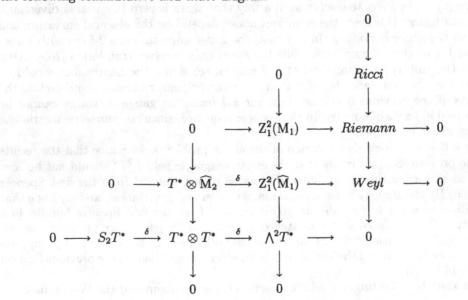

The *Ricci bundle*, the kernel of the canonical projection of the *Riemann bundle* onto the *Weyl bundle*, is isomorphic to S_2T^*, as is well known. However, this diagram automatically points out the confusion that has arisen between the splitting of the Riemann tensor into the Weyl tensor and the Ricci tensor in the Janet sequence, and the splitting of $T^* \otimes T^*$ into $\wedge^2T^*$ (*electromagnetism*) and S_2T^* (*gravitation*) in the Spencer sequence.

The following Proposition, describing the additional fields brought about by the transition from the Poincaré to the conformal group, proves that the *bottom row of the last diagram is related to $\widehat{C}_1$ and not to $\widehat{C}_2$*, contradicting the basic assumptions leading to Einstein or Yang–Mills equations ([182]).

Proposition 1. *There is a short exact sequence*

$$ 0 \longrightarrow C_1 \longrightarrow \hat{C}_1 \longrightarrow T^* \times_X S_2T^* \times_X \wedge^2T^* \longrightarrow 0. $$

PROOF. First of all $C_1 = T^* \otimes R_2 \simeq T^* \otimes R_1$ and $\widehat{C}_1 = T^* \otimes \hat{R}_2$ project, respectively, onto $T^* \otimes R_1$ and $T^* \otimes \hat{R}_1$. Hence the projection onto T^* follows from Proposition D.1. It remains to prove that there is an isomorphism

$$ 0 \longrightarrow T^* \otimes \widehat{M}_2 \longrightarrow S_2T^* \times_X \wedge^2T^* \longrightarrow 0. $$

This is easily seen by defining the two projections

$$ \begin{cases} T^* \otimes \widehat{M}_2 \rightarrow S_2T^* : \xi^r_{ri,j} \rightarrow \frac{1}{2}(\xi^r_{ri,j} + \xi^r_{rj,i}), \\ T^* \otimes \widehat{M}_2 \rightarrow \wedge^2T^* : \xi^r_{ri,j} \rightarrow \xi^r_{ri,j} - \xi^r_{rj,i}), \end{cases} $$

and by counting the dimensions. $\square$

Using the infinitesimal point of view, we immediately obtain from the image of the Spencer operator D defined in the previous Section:

$$ \begin{cases} \varphi_{ij} = \chi^r_{ri,j} - \chi^r_{rj,i} = \partial_i\xi^r_{rj} - \partial_j\xi^r_{ri}, \\ \tau_{ij} = \frac{1}{2}(\chi^r_{ri,j} + \chi^r_{rj,i}) = \frac{1}{2}(\partial_i\xi^r_{rj} + \partial_j\xi^r_{ri}). \end{cases} $$

When there is no electromagnetic field, we can find a function ϕ such that $\xi^r_{ri} = \partial_i\phi$, and we find $\tau_{ij} = \partial_{ij}\phi$, i.e. *exactly the Eötvös tensor*.

The study of the finite point of view is more subtle.

The starting point here is the full expression of the zero curvature condition, written in the unusual form

$$ \partial_i\chi^k_{l,j} - \partial_j\chi^k_{l,i} + \chi^r_{l,j}\chi^k_{r,i} - \chi^r_{l,i}\chi^k_{r,j} = A^r_i\chi^k_{lr,j} - A^r_j\chi^k_{lr,i} $$
$$ = A^r_i A^s_j(\tau^k_{lr,s} - \tau^k_{ls,r}). $$

The main problem is that the direct computation of $\tau^k_{lr,s}$ is absolutely very tedious, since it involves the third-order jets f^k_{lrs}, which are nonzero in the finite setting. But now, *unlike in general relativity*, the total number of components of the curvature is n^2. Since we have already used $n(n-1)/2$ of them in the last Section, namely the combinations $\tau^r_{ri,j} - \tau^r_{rj,i}$, we are left with only $n(n+1)/2$ other components, which we have chosen to be $\tau_{ij} = (\tau^r_{ri,j} + \tau^r_{rj,i})/2$. The main trick is to explicitly compute the left hand term of the curvature by using the expressions for $\chi_{j,i}^k$ given in Chapter V, and twist by A in order to compute the difference $\tau_{lr,s}^k - \tau_{ls,r}^k$. Since contraction with respect to k and l has already been used in the previous Section, we can only use contraction in k and r or s, *the same as the one used for obtaining the Ricci tensor from the Riemann tensor, but with a completely different meaning* since we are now dealing with the Spencer sequence and *not* with the Janet sequence.

After a tedious but straightforward computation, setting $a_i = f_i^k b_k$ and taking into account the conditions, *over the target*,

$$\frac{\partial b_l}{\partial y^k} - \frac{\partial b_k}{\partial y^l} = 0 \Leftrightarrow \varphi_{ij} = 0,$$

we finally obtain *over the source*

$$\rho_{ij} = \tau_{ri,j}^r - \tau_{ij,r}^r$$

$$= f_i^k f_j^l \left[(n-2)\frac{\partial b_k}{\partial y^l} + \omega_{kl}(y)\omega^{rs}(y)\frac{\partial b_r}{\partial y^s} + (n-2)b_k b_l - (n-2)\omega_{kl}(y)\omega^{rs}(y)b_r b_s \right],$$

and so

$$\rho = \omega^{ij}(x)\rho_{ij} = \frac{1}{\Theta^2}(n-1)\left[2\omega^{kl}(y)\frac{\partial b_k}{\partial y^l} - (n-2)\omega^{kl}(y)b_k b_l \right].$$

In this case, setting $\tau = \omega^{ij}(x)\tau_{ij}$, we obtain

$$n\rho_{ij} = n\tau_{ri,j}^r - \delta_i^r \tau_{sj,r}^s - \delta_j^r \tau_{si,r}^s + \omega_{ij}(x)\omega^{rt}(x)\tau_{st,r}^s$$

$$= (n-2)\tau_{ij} + \omega_{ij}(x)\tau,$$

i.e.

$$n\rho = 2(n-1)\tau \Rightarrow \tau_{ij} = \frac{n}{n-2}\rho_{ij} - \frac{n}{2(n-1)(n-2)}\omega_{ij}(x)\rho,$$

and therefore

$$\tau_{ij} = n f_i^k f_j^l \left[\frac{\partial b_k}{\partial y^l} + b_k b_l - \tfrac{1}{2}\omega_{kl}(y)\omega^{rs}(y)b_r b_s \right],$$

which is a particularly simple formula leading to

$$\tau = \frac{n}{\Theta^2}\left[\omega^{kl}(y)\frac{\partial b_k}{\partial y^l} - \frac{n-2}{2}\omega^{kl}(y)b_k b_l \right].$$

We see that the principal part of τ_{ij} is the Eötvös tensor, up to a constant (physical) factor, and its trace τ must provide the (unique) Poisson equation. Hence, to have a Lagrangian that is well defined *over the target*, we must choose a gravitational action density proportional to $\tau \det A$.

We have the general variational formulas *over the source*:

$$\delta \tau_{lj,i}^k = B_i^r \partial_r \xi_{lj}^k + \xi^r \partial_r \tau_{lj,i}^k$$

$$+ \tau_{lj,r}^k \xi_i^r + \tau_{lr,i}^k \xi_j^r + \tau_{rj,i}^k \xi_l^r - \tau_{lj,i}^r \xi_r^k$$

$$+ \tau_{r,i}^k \xi_{lj}^r - \tau_{l,i}^r \xi_{rj}^k - \tau_{j,i}^r \xi_{lr}^k,$$

$$\delta \det A = \xi^r \partial_r \det A + \det A(\partial_r \xi^r - \xi_r^r),$$

and we obtain by contraction

$$\delta(\tau \det A) = \partial_r(\xi^r \tau \det A + \omega^{ij}(x)B_i^r \det A \, \xi_{sj}^s)$$
$$- \frac{n-2}{n}\tau \det A \, \xi_r^r + \frac{n-2}{n}\omega^{ij}(x)\tau_{r,i}^r \det A \, \xi_{sj}^s.$$

We will prove that the corresponding induction equations provide the following results:

$$\begin{cases} \xi_{ri}^r \to & \text{gravitational potential,} \\ \xi_r^r \to & \text{Poisson equation,} \\ \xi^r \to & \text{Newton law.} \end{cases}$$

As we have seen in Section C, the simplest dynamical action density is a function of A_i^k. Recalling the variation,

$$\delta A_i^k = \xi^r \partial_r A_i^k + A_r^k \partial_i \xi^r - A_i^r \xi_r^k,$$

we obtain the induction equations

$$\xi^r \to \partial_i(\mathcal{X}_k^{ri} A_r^k) - \mathcal{X}_k^{ri} \partial_r A_i^k = 0,$$
$$\xi_i^k \overset{k \neq i}{\to} \mathcal{X}_k^{ri} A_i^r \xi_r^k \equiv \overline{\mathcal{X}}_k^{rr} \xi_r^k = 0.$$

The last equation proves that the *dynamical action density must be a function of A_{ij}*. Setting, as in Section D,

$$\mathcal{Y}_{k\text{pure}}^{rl} = \frac{1}{\Delta}g_k^r \mathcal{X}_r^i \partial_i f^l,$$

we have $\mathcal{Y}^{k,l} - \mathcal{Y}^{l,k} = 0$ and we successively obtain (where we drop the word 'pure' for simplicity):

$$\xi^r \left[\partial_i \left(\frac{\partial y^s}{\partial x^r} \frac{\partial x^i}{\partial y^l} \Delta \mathcal{Y}_s^l \right) - \Delta f_s^k \mathcal{Y}_k^l \frac{\partial x^i}{\partial y^l} \partial_r A_i^s \right] = 0,$$

i.e.

$$\Delta \eta^k \frac{\partial \mathcal{Y}_k^{rl}}{\partial y^l} + \xi^r \overline{\mathcal{X}}_s^{rk} g_t^s \partial_r f_k^t = 0.$$

The following tricky computation essentially uses the symmetry of the stress tensor $\overline{\mathcal{X}}$ and the fact that ω is constant:

$$\overline{\mathcal{X}}_s^{rk} g_t^s \partial_r f_k^t = \overline{\mathcal{X}}^{u,v}\omega_{us}(x)g_t^s \partial_r f_v^t$$
$$= \overline{\mathcal{X}}^{u,v}\Theta^2 \omega_{st}(y)f_u^s \partial_r f_v^t$$
$$= \frac{1}{2}\overline{\mathcal{X}}^{u,v}\Theta^2 \omega_{st}(y)(f_u^s \partial_r f_v^t + f_v^s \partial_r f_u^t)$$
$$= \frac{1}{2}\overline{\mathcal{X}}^{u,v}\Theta^2 \partial_r(\omega_{st}(y)f_u^s f_v^t)$$
$$= \frac{1}{2}\overline{\mathcal{X}}^{u,v}\Theta^2 \partial_r \left(\frac{\omega_{uv}(x)}{\Theta^2} \right)$$
$$= -\text{tr}\,\overline{\mathcal{X}}\frac{1}{\Theta}\partial_r\Theta.$$

Collecting the results, we obtain the following dynamical equations *over the target*:

$$\eta^k \left[\frac{\partial \mathcal{Y}_k^{ll}}{\partial y^l} - \mathrm{tr}\, \mathcal{Y} \frac{1}{\Theta} \frac{\partial \Theta}{\partial y^k} \right] = 0.$$

This is Newton's law if we can prove that Θ, or rather $\log \Theta$, has something to do with the usual gravitational potential. However, we have

$$\xi_{ri}^r \to \tau_{r,i}^r = 0 \Leftrightarrow \alpha_i = 0 \Leftrightarrow b_k = -\frac{1}{\Theta} \frac{\partial \Theta}{\partial y^k},$$

and for $n = 4$ *only*:

$$\xi_r^r \to 2n\Theta\omega^{kl}(y) \frac{\partial^2 \Theta}{\partial y^k \partial y^l} = \mathrm{tr}\, \mathcal{Y}$$

(up to a constant factor) ([56, 138]). If we use $w = \tilde{\theta}$, as in Section C, then

$$\mathrm{tr}\, \mathcal{Y} = -\rho\Theta,$$

and we obtain the Poisson equation. As a byproduct, we may choose $\Theta = 1 - \phi/c^2$, where ϕ is the usual gravitational potential.

We will now exhibit the Abraham tensor.

From the general theory of Section A we know that, *over the target*, the final dynamic equation *must* be

$$\mathcal{Y}_k^{ll} = \mathcal{Y}_{k\mathrm{pure}}^{ll} + \mathcal{Y}_{k\mathrm{add}}^{ll} \Rightarrow \frac{\partial \mathcal{Y}_k^{ll}}{\partial y^l} = 0,$$

where $\mathcal{Y}_{k\mathrm{add}}^{ll}$ is only produced by the variation, *over the target*, of $\tau \det A$. We obtain, *over the target*:

$$\delta(\tau \det A) = -n\Theta^2 \Delta \omega^{kl}(y) \frac{\partial b_k}{\partial y^r} \frac{\partial \eta^r}{\partial y^l} + \tau \det A \frac{\partial \eta^l}{\partial y^l} + \dots,$$

and so

$$\mathcal{Y}_{k\mathrm{add}}^{ll} = \tau\Theta^4 \delta_k^l - n\Theta^2 \omega^{rl}(y) \frac{\partial b_r}{\partial y^k}.$$

It follows that we have, in succession,

$$\frac{\partial \mathcal{Y}_{k\mathrm{add}}^{ll}}{\partial y^l} = \frac{\partial}{\partial y^l} \left(-n\Theta\omega^{rs}(y) \frac{\partial^2 \Theta}{\partial y^r \partial y^s} \delta_k^l + n\Theta\omega^{rl}(y) \frac{\partial^2 \Theta}{\partial y^r \partial y^k} - n\omega^{rl}(y) \frac{\partial \Theta}{\partial y^r} \frac{\partial \Theta}{\partial y^k} \right)$$

$$= -2n \frac{\partial \Theta}{\partial y^k} \omega^{rs}(y) \frac{\partial^2 \Theta}{\partial y^r \partial y^s}$$

$$= -2n \frac{\partial}{\partial y^l} \left(\omega^{rl}(y) \frac{\partial \Theta}{\partial y^r} \frac{\partial \Theta}{\partial y^k} - \frac{1}{2} \omega^{rs}(y) \frac{\partial \Theta}{\partial y^r} \frac{\partial \Theta}{\partial y^s} \delta_k^l \right)$$

$$= -\mathrm{tr}\, \mathcal{Y} \frac{1}{\Theta} \frac{\partial \Theta}{\partial y^k} \qquad \text{(from Poisson)}.$$

We will give another short proof of this result by exhibiting the so-called 'energy of gravitation'. Indeed, over the target we have, for $n = 4$ only:

$$-\int \tau \det A \, dx^1 \wedge \cdots \wedge dx^4 = \int 4\Theta \omega^{kl}(y) \frac{\partial^2 \Theta}{\partial y^k \partial y^l} \, dy^1 \wedge \cdots \wedge dy^4$$

$$= \int \frac{\partial}{\partial y^l} \left(4\Theta \omega^{kl}(y) \frac{\partial \Theta}{\partial y^k} \right) dy^1 \wedge \cdots \wedge dy^4$$

$$- \int 4\omega^{kl}(y) \frac{\partial \Theta}{\partial y^k} \frac{\partial \Theta}{\partial y^l} \, dy^1 \wedge \cdots \wedge dy^4,$$

and the variation of the last integral immediately produces the required Abraham tensor.

It may seem very strange to the reader that the absolute temperature of Section C becomes the gravitational potential in Section E. The explanation of this surprising fact, known in the literature as the *Duhamel analogy* [52], is also an important result in this Chapter.

At the end of Section D we have seen that the (standard) EM action density *in vacuum* did not contain the factor Θ anymore so that its variation did not depend explicitly on jets of strict order 1. Accordingly, the constitutive laws were seen to be invariant under the largest possible group of invariance: the conformal group. If we inquire into this property in the case of gravitation, we cannot avoid the factor Θ^2. Hence, if we want to preserve *at least* the conformal invariance of the 2-tensor density that must be used to contract τ_{ij} to a scalar, we can only arrive at an action density of the form $\Omega^2 \tau \det A$. It follows that Θ should be replaced by $\Theta\Omega$ in the variation over the target. Accordingly,

$$b_k + \frac{\partial \log(\Theta\Omega)}{\partial y^l} = 0,$$

$$\frac{\partial y_k^{'l}}{\partial y^l} - \text{tr}\, \mathcal{Y} \frac{1}{\Theta} \frac{\partial \Theta}{\partial y^k} - \text{tr}\, \mathcal{Y} \frac{1}{\Omega} \frac{\partial \Omega}{\partial y^k} = 0,$$

and we may even replace $\omega_{kl}(y)$ by $\Omega^2 \omega_{kl}(y)$ in the dynamical part of the action in order to recover exactly the same result as before, with Ω instead of Θ if we set $\Theta = 1$. Nevertheless, Similarly as in Section C, we have to take care of the fact that *variation does not commute with evaluation*. Finally, the only important thing to keep in mind is that *gravitation only depends on the second-order jets*, in whatever way these are introduced. Indeed, taking into account $\alpha_r = 0$ we have

$$\frac{\partial y_k^{'l}}{\partial y^l} + \text{tr}\, \mathcal{Y}\, b_k = 0,$$

using $\bar{\xi}_2$ instead of ξ_2.

As already seen in Section C, the simplest density is A_4^4 under the constraints $A_4^1 = A_4^2 = A_4^3 = 0$. Accordingly, in the constrained variational calculus initiated by E. and F. Cosserat we may consider the action density $\lambda_k(x) A_4^k$. We find

$$\delta(\lambda_k(x) A_4^k) = \xi^r \lambda_k(x) \partial_r A_4^k + \lambda_k(x) A_r^k \partial_4 \xi^r - \lambda_k(x) A_4^r \xi_r^k.$$

Retaining the gravitational action density $\tau \det A$, from the first-order jets we deduce the conditions

$$\lambda_1 = \lambda_2 = \lambda_3 = 0.$$

The remaining zero-order induction equations are

$$\partial_4(\lambda_k(x)A_i^k) - \lambda_k(x)\partial_i A_4^k = 0,$$

i.e.

$$A_i^k \partial_4 \lambda_k(x) + \lambda_k(x)(A_4^r \chi_{r,i}^k - A_i^r \chi_{r,4}^k) = 0.$$

For $i = 4$ we obtain $\partial_4 \lambda_4 = 0$, and we are left with

$$A_4^4 \chi_{4,i}^4 - A_i^r \chi_{r,4}^4 = 0.$$

Since $\alpha_i = 0$ (from the variation of the second-order jets), we finally obtain $\chi_{k,4}^4 = 0$, or $\chi_{4,4}^k = 0$, for $k = 1, 2, 3$.

We recapitulate the conditions obtained thus far on the infinitesimal level:

$$A_4^1 = A_4^2 = A_4^3 = 0 \rightarrow \partial_4 \xi^k - \xi_4^k = 0, \qquad k = 1, 2, 3,$$

$$\alpha_i = 0 \rightarrow \partial_i \xi_r^r - \xi_{ri}^r = 0,$$

$$\chi_{4,4}^k = 0 \rightarrow \partial_4 \xi_4^k - \xi_{44}^k = 0, \qquad k = 1, 2, 3,$$

and so finally,

$$\partial_{44} \xi^k - \omega^{kr} A_r = 0 \qquad \text{with } A_i = \frac{1}{2}\partial_i A.$$

However, the conformal factor A does not satisfy the $n(n+1)/2$ equations $\partial_{ij} A = 0$, and $\omega^{ij} \partial_{ij} A$ is not equal to zero everywhere, because of the Poisson equation. *The above conditions for a freely falling particle justify the equivalence principle*, and we can even say, with a slight abuse of language, that an *accelerometer merely helps measuring the image of the Spencer operator*.

We will now examine a few postulates of general relativity theory.

The first postulate concerns the definition of geodesics *independently of the Einstein equations*. With standard notations, the variation of $(g_{rs}(f(x))\partial_4 f^r \partial_4 f^s)^{1/2}$ introduces the Newton law in first approximation, via the gradient $\partial g_{44}/\partial y^k$ which comes from the variation of g *over the target*. Such a term *cannot* exist in the new dynamics, because now we have a quadratic form in the A_4^k with metric defined *over the source* and hence not varied. Similarly, in the linear choice $\lambda_k(x)A_4^k$ the terms $\lambda_k(x)$ are not varied.

The second postulate concerns the way to produce the Einstein equations by the variation δg of the metric in the Hilbert variational procedure. (For a fascinating study of the many tentatives of Einstein, see the excellent biography [137]). Again, this clearly contradicts the new dynamics, because the jet-potentials that have to be varied are in one-to-one correspondence with the parameters of a Lie group. The only possibility to agree is to consider a conformally flat metric A_{ij} depending only on ω and $j_1(f)$, but this assumption (first brought about by Nördström [129, 138]) is not valid in general relativity. Finally, it should be noticed that even if one accepts the Einstein equations, then the choice of the dynamical action is not provided by the theory and no rule is given for selecting the metric solutions [113, 114, 207].

If we try to follow the scheme adopted by general relativity, we have to consider *over the target* an action density of the form (compare with [60])

$$g^{kl}(y) \left(\frac{\partial b_k}{\partial y^l} + b_k b_l - \frac{1}{2} \omega_{kl}(y) \omega^{rs}(y) b_r b_s \right).$$

To obtain the full variation we have to use the formula

$$\delta \tau_{ij} = B_i^r \partial_r \xi_{sj}^s + \xi^r \partial_r \tau_{ij} + \tau_{ir} \xi_j^r + \tau_{rj} \xi_i^r - \tau_{i,j}^r \xi_{sr}^s$$

and obtain

$$\delta(\tau_{ij} \det A) = \partial_r(\xi^r \tau_{ij} \det A) + (\tau_{ir} \xi_j^r + \tau_{rj} \xi_i^r - \tau_{ij} \xi_r^r) \det A + \ldots.$$

To avoid second-order jets in the variation we could directly vary b alone in the action density. We readily obtain the conditions:

$$(2g^{kl}(y) - \omega_{rs}(y) g^{rs}(y) \omega^{kl}(y)) b_l = \frac{\partial g^{kl}(y)}{\partial y^l}.$$

Hence, the 'divergence' term on the right hand side *cannot* vanish, as in the Logounov theory [113], because this is just the way of producing nonzero second-order jets. In this interpretation the symmetric tensor density g is a kind of (generalized) *'gravitational potential'*. The simplest choice $g^{kl} = \Theta^2 \omega^{kl}(y)$ only leads to one scalar potential Θ and to the gravitational field $b_k = -\partial \log \Theta / \partial y^k$, as in the source approach considered above.

We end this Section with a direct study of the infinitesimal point of view. The conformal origin of gravitation will strikingly emerge, similarly to the emergence of the conformal origin of electromagnetism from a similar calculation in Section D.

Consider the linear action density:

$$\int [\rho(x)(\partial_4 \xi^4 - \xi_4^4) + g^i(x)(\partial_i \xi^r - \xi_{r_i}^r) + g^{ij}(x)(\partial_i \xi_{r_j}^r - 0)] \, dx^1 \wedge \cdots \wedge dx^n.$$

After variation and integration by parts, we are left with the following induction equations for a static gravitational field:

$$\xi^4 \rightarrow \partial_4 \rho = 0 \Rightarrow \rho = \rho_0,$$

$$\xi_4^4 \rightarrow n \partial_i g^i + \rho = 0,$$

$$\xi_{4i}^4 \rightarrow \partial_i g^{ij} + g^j = 0.$$

If we choose a diagonal g^{ij} with $g^{11} = g^{22} = g^{33}$, we see that g^i is the gravitational field (up to a constant factor), and finally we obtain the Poisson equation, written in the form $n \partial_{ij} g^{ij} = \rho_0$. However, the reader may wonder why we add two terms to the gravitational action and why g^i can be identified with the gravitational field.

The explanation of these facts is that the variation of the action is particularly simple *over the target*. To aid the reader, we provide all details since the calculation is far from any classical calculation in this domain.

Recall the relations

$$na_i = n f_i^k b_k = g_i^t f_{ti}^l, \qquad \delta g_k^i = -g_r^i \eta_k^r.$$

Accordingly, we obtain successively

$$\delta f_{ij}^k = \eta_r^k f_{ij}^r + \eta_{rs}^k f_i^r f_j^s,$$

$$n\delta a_i = f_{ti}^l \delta g_l^t + g_l^t \delta f_{ti}^l = f_i^r \eta_{sr}^s,$$

$$n\delta b_k = n g_k^i \delta a_i + n a_i \delta g_k^i = \eta_{rk}^r - n b_r \eta_k^r.$$

Setting now $\tau_{ij} = f_i^k f_j^l T_{kl}$, we obtain

$$\delta \tau_{ij} = \eta_r^k f_i^r f_j^l T_{kl} + \eta_s^l f_i^k f_j^s T_{kl} + f_i^k f_j^l \delta T_{kl}$$

$$= f_i^k f_j^l (T_{rl} \eta_k^r + T_{ks} \eta_l^s + \delta T_{kl}).$$

Although the variation of T_{kl} seems very complicated, many terms cancel out and we finally obtain

$$\delta \tau_{ij} = f_i^k f_j^l \left[-n \frac{\partial b_r}{\partial y^k} \left(\frac{\partial \eta^r}{\partial y^l} - \eta_l^r \right) - n b_r \left(\frac{\partial \eta_k^r}{\partial y^l} - \eta_{kl}^r \right) + \left(\frac{\partial \eta_{rk}^r}{\partial y^l} - 0 \right) \right].$$

We can verify that we have indeed a linear combination of the Spencer operator, in agreement with the general theory. By a direct identification, using the formula

$$\delta \det A = \left(\frac{\partial \eta^r}{\partial y^r} - \eta_r^r \right) \det A,$$

we obtain the correspondence

$$\text{source} \quad g^{ij}(x)(\partial_i \xi_{rj}^r - 0) \leftrightarrow g^{kl}(y) \left(\frac{\partial \eta_{rk}^r}{\partial y^l} - 0 \right) \quad \text{target},$$

$$\text{source} \quad g^i(x)(\partial_i \xi_r^r - \xi_{ri}^r) \leftrightarrow -n g^{kl}(y) b_r \left(\frac{\partial \eta_k^r}{\partial y^l} - \eta_{kl}^r \right) \quad \text{target}.$$

Hence, in a first approximation we have the identification

$$g^i(x) \leftrightarrow -n \omega^{kl}(y) b_k$$

which we have been looking for.

Let us now recover the induction equations

$$\eta_{kl}^r \rightarrow -\frac{\partial g^{kl}}{\partial y^l} \eta_{rk}^r + n g^{kl} b_r \eta_{kl}^r = 0.$$

Since we successively have

$$n g^{kl} b_r \eta_{kl}^r = g^{kl} b_r (\delta_k^s \eta_{sl}^s + \delta_l^s \eta_{sk}^s - \omega_{kl} \omega^{rt} \eta_{st}^s)$$

$$= g^{kl} b_k \eta_{sl}^s + g^{kl} b_l \eta_{sk}^s - \omega_{rs} g^{rs} \omega^{kl} b_l \eta_{sk}^s$$

$$= (2 g^{kl} - \omega_{rs} g^{rs} \omega^{kl}) b_l \eta_{uk}^u,$$

we recover the expression for b from $j_1(g)$, given already over the target;

$$\eta_k^r \rightarrow \left[\frac{\partial}{\partial y^l} (g^{kl} b_r) + g^{kl} \frac{\partial b_l}{\partial y^r} \right] \eta_k^r \sim \eta_s^s.$$

Indeed, developing and substituting the preceding relation, we obtain

$$g^{kl}(T_{rl}\eta_k^r + T_{kr}\eta_l^r) \sim \eta_s^s.$$

This amounts to

$$g^{rl}\eta_r^k + g^{kr}\eta_r^l \sim \eta_s^s,$$

which is indeed only possible if $g \sim \omega$. Of course, this must be true if and only if the variation of the purely dynamical part is also proportional to η_s^s.

Under this assumption we have already proved that the zero-order factor of η^k is just the Newton force. More precisely, we obtain from the above formulas:

$$\eta^r \to -\frac{\partial g^{kl}}{\partial y^r}T_{kl},$$

and we only have to choose $g^{kl}(y) = \Theta^2 \omega^{kl}(y)$ in order to recover the result.

We will now examine the possibility of conformal symmetry breaking.

The main reason for this comes from the post-Newtonian approximation. It is known from [183] that the following is a classical Lagrangian giving the same prevision as general relativity:

$$L = m\left(\frac{v^2}{2} + \frac{1}{8}\frac{v^4}{c^2} + \phi + \frac{3}{2}\frac{\phi}{c^2}v^2 - \frac{1}{2}\frac{\phi^2}{c^2}\right).$$

From the corresponding Euler–Lagrange equations we should obtain

$$\left(1 + 3\frac{\phi}{c^2}\right)\frac{d\vec{v}}{dt} = \left(1 - \frac{\phi}{c^2}\right)\vec{\nabla}\phi + \dots,$$

although, with $\Theta = 1 - (\phi/c^2)$ we have obtained

$$\frac{d\vec{v}}{dt} = -\frac{c^2}{\Theta}\vec{\nabla}\Theta + \dots = \left(1 + \frac{\phi}{c^2}\right)\vec{\nabla}\phi + \dots,$$

where we recall that $\phi = GM/r$ for a particle of mass m at a distance r from a central attractive mass $M \gg m$ with gravitational constant G.

If we would like to use an arbitrary generalized potential g, then T_{kl} should be replaced by

$$T_{kl} - \frac{1}{n-2}\omega_{kl}T = n\left(\frac{\partial b_k}{\partial y^l} - \frac{1}{n-2}\omega_{kl}\omega^{rs}\frac{\partial b_r}{\partial y^s} + b_k b_l\right)$$

with $T = \omega^{kl}T_{kl}$, in order to obtain the simpler condition

$$2g^{kl}b_l = \frac{\partial g^{kl}}{\partial y^l} - \frac{1}{n-2}\frac{\partial(\omega_{rs}g^{rs})}{\partial y^l}\omega^{kl}.$$

E.g., with $g = (1,1,1,-(1+4\frac{\phi}{c^2}))$ endowed with refraction index $n \simeq 1 + 2\frac{\phi}{c^2}$, we should have $\vec{b} = -\vec{\nabla}\frac{\phi}{c^2}$ as a correct approximation indeed, but this choice is rather artificial.

In any case, whatever approach is taken, we should obtain for the resulting curvature and its various contractions:

$$n\rho_{lij}^k = n(\tau_{li,j}^k - \tau_{lj,i}^k)$$
$$= \delta_l^k \tau_{ri,j}^r + \delta_i^k \tau_{rl,j}^r - \omega_{li}\omega^{ks}\tau_{rs,j}^r$$
$$- \delta_l^k \tau_{rj,i}^r - \delta_j^k \tau_{rl,i}^r + \omega_{lj}\omega^{ks}\tau_{rs,i}^r,$$
$$n\rho_{ij} = n\rho_{irj}^r$$
$$= (n-1)\tau_{ri,j}^r - \tau_{rj,i}^r + \omega_{ij}\omega^{st}\tau_{rs,t}^r,$$
$$n\rho = 2(n-1)\omega^{ij}\tau_{ri,j}^r.$$

We leave it to the reader to verify that *the corresponding Weyl tensor vanishes* (this is a tedious exercise involving contraction and identification). This result contradicts general relativity, although it is *necessarily* implied by the use of the Spencer sequence.

To conclude this Section, we may say that, although we have not been able to recover the classical relativistic tests, we nevertheless hope to have convinced the reader that *the common geometric origin of electromagnetism and gravitation is surely not the one usually accepted nowadays*. In particular, the use of second-order jets allows us to prove that *electromagnetism and gravitation in vacuum have the same conformal origin*.

Bibliography

[1] E.S. Abers, B.W. Lee, Gauge theories, *Physics Reports Sect. C of Physics Letters* **9**, 1 (1973), 1–59.

[2] M. Abraham, *Jb. Radioakt.* **11** (1914), 470.

[3] F.W. Adler, Über die Mach–Lippmannsche Analogie zum zweiten Hauptsatz, *Ann. der Physik* **22** (1907), 587.

[4] Y. Aharonov, D. Bohm, *Phys. Rev.* **115** (1959), 485. *Phys. Rev.* **125** (1962), 2192.

[5] I.J.R. Aitchison, A.J.G. Hey, *Gauge theories in particle physics*, A. Hilger, Bristol, 1982. 341 p.

[6] V. Aldaya, J.A. de Azcarraga, Variational principle on r^{th} order jets of fibre bundles in field theory, *J. Math. Phys.* **19** (1978), 1869–1975.

[7] V. Aldaya, J.A. de Azcarraga, Geometric formulation of classical mechanics and field theory, *Rev. del Nuovo Cimento* **3**, 10 (1980), 66 p.

[8] U. Amaldi, *Congrès de la Société Italiénne pour le progrès des sciences*, Parma, Italy, 1907. (In Italian.)

[9] I. Anderson, G. Thompson, The inverse problem of the calculus of variations for ordinary differential equations, *Memoirs Amer. Math. Soc.* **473**, 98 (1992), 11 p.

[10] H. Andrillat, *Introduction à l'étude des cosmologies*, Collection Intersciences, A. Colin, Paris, 1970. 224 p.

[11] P. Appell, *Traité mécanique rationelle*, **III** Note sur la théorie de l'action euclidienne, Gauthier-Villars, Paris, 1909. 557–629.

[12] V. Arnold, Sur la géométrie des groupes de Lie de dimension infinie et ses applications à l'hydrodynamique des fluides parfaits, *Ann. Inst. Fourier (Grenoble)* **16**, 1 (1966), 319–361.

[13] V. Arnold, *Méthodes mathématiques de la mécanique classique*, **Appendice 2** (géodésiques des métriques invariantes à gauche sur des groupes de Lie et hydrodynamique du fluide parfait) Mir, Moscow, 1974, 1976. 318 p.

[14] E. Artin, *Galois theory*, Notre Dame Univ. Publications, 1942.

[15] H. Arzelies, *Relativité généralisée, gravitation*, **I** Gauthier-Villars, Paris, 1961. 377 p.

[16] H. Arzelies, *Thermodynamique relativiste et quantique*, Gauthier-Villars, Paris, 1968. 704 p.

[17] H. Arzelies, *Fluides relativistes*, Masson, Paris, 1971. 209 p.

[18] M.F. Atiyah, I.G.MacDonald, *Introduction to commutative algebra*, Addison-Wesley, 1969.

[19] Y. Bamberger, J.P. Bourgignon, Torseurs sur un espaces affine, *Rapport M25.0470, Centre de Mathématiques de l'École Polytechnique* (1970).

[20] L. Baulieu, M. Bellon, R. Grimm, Some remarks on the gauging of the Virasoro algebra, *Phys. Lett. B* **260** (1991), 63.

[21] J. Beckers, J. Harnad, M. Perroud, P. Winternitz, Tensor fields invariant under subgroups of the conformal group of space-time, *J. Math. Phys.* **19**, 10 (1978), 2126–2153.

[22] M.D. di Benedetto, J.W. Grizzle, C.H. Moog, Computing the differential output rank of a nonlinear system, *Proc. 26th Conf. on Decision and Control, Los Angeles* (1987), 142–145.

[23] M.D. di Benedetto, J.W. Grizzle, C.H. Moog, Rank invariants of nonlinear systems, *SIAM J. Control Optim.* **27** (1989), 658–672.

[24] A. Bialynicki-Birula, On Galois theory of fields with operators, *Amer. J. Math.* **84** (1962), 89–109.

[25] B.A. Bilby, Geometry and continuum mechanics, In: *Proc. IUTAM Symp. 'Mechanics of Generalized Continua', Stuttgart, Germany 1967*, Springer, Berlin, (1968), 180–199.

[26] G. Birkhoff, *Hydrodynamics*, Princeton Univ. Press, Princeton, New Jersey, 1954. [French translation: *Hydrodynamique*, Dunod, Paris, 1955. 228 p.]

[27] D. Bleecker, Gauge theory and variational principle, In: *Global Analysis, Pure and Applied*, 1 (Advanced Book Program, World Science Division XVIII), Addison-Wesley, Reading, Mass., 1981. 179 p.

[28] L. de Broglie, *La thermodynamique de la particule isolée*, Gauthier-Villars, Paris, 1964. 61, 95.

[29] R.L. Bryant, S.S. Chern, R.B. Gardner, H. Goldschmidt, *Exterior differential systems*, Math. Sciences Research Inst. Publ. **18**, Springer, Berlin, 1991.

[30] B. Buchberger, Gröbner bases: an algorithmic method in polynomial ideal theory, In: *Multidimensional Systems*, N.K. Bose (ed.), Reidel, Dordrecht, 1985. 184–232.

[31] D. Canarutto, M. Modugno, Ehresman's connections and the geometry of energy-tensors in Lagrangian field theories, *Tensor, N.S.* **42** (1985), 112–120.

[32] E. Cartan, Sur la structure des groupes infinis de transformations, *Ann. Ec. Norm. Sup.* **21** (1904), 153–206.

[33] E. Cartan, Sur une généralisation de la notion de courbure de Riemann et les espaces à torsion, *C.R. Acad. Sci. Paris* **174** (1922), 522.

[34] E. Cartan, Sur les variétés à connexions affines et la théorie de la relativité généralisées, *Ann. Ec. Norm. Sup.* **40** (1923), 325–412. *Ann. Ec. Norm. Sup.* **41** (1924), 1–25. *Ann. Ec. Norm. Sup.* **42** (1925), 17–88.

[35] R.G. Chambers, *Phys. Letters* **5** (1960), 3.

[36] S. Chandrasekhar, *Hydrodynamic and hydromagnetic stability*, Dover, 1970.

[37] T.P. Cheng, Ling-Fong Li, Gauge invariance, Ressource Letter, *Amer. J. Physics* **56**, 7 (1988), 586–599.

[38] O.D. Chwolson, Traité de physique, **I** Note sur la dynamique du point etdu corps invariable, Hermann, Paris, 1914. 236–273. [French translation by E. Davaux]

[39] O.D. Chwolson, Traité de physique, **V. Fasc. 1.** Hermann, Paris, 1914. 209–214. [French translation by E. Davaux]

[40] P.M. Cohn, *Lie groups*, Cambridge Tracts **46**, Cambridge Univ. Press, 1965.

[41] E. Cosserat, F. Cosserat, *Théorie des corps déformables*, Hermann, Paris, 1909. 226 p.

[42] E. Cosserat, F. Cosserat, Principes de la mécanique rationnelle, In: Encycl. des Sciences Mathématiques **IV**, 1 Gauthier-Vilars, Paris, 1910. [From the German survey of A. Voss.]

[43] J. Davenport, Y. Siret, E. Tournier, *Calcul formel*, Masson, Paris, 1986.

[44] J. Descusse, J.F. Lafay, M. Malabre, On the structure at infinity of block-decouplable systems, the general case, *IEEE Trans. Autom. Control* **28** (1983), 1115–1118.

[45] J.J.D. Domingos, M.N.R. Nina, J.H. Whitelaw, Foundations of continuum thermodynamics, In: *Proc. Bussaco Symp. Portugal, 1973* MacMillan Press, London, (1974), 337 p.

[46] R.J. Donelly, R. Herman, I. Prigogine, Non-equilibrium thermodynamics, variational techniques and stability, In: *Proc. Symp. Univ. Chicago, 1965* Univ. Chicago Press, (1967), 313 p.

[47] B. Doubrovine, S. Novikov, A. Fomenko, *Géométrie contemporaine, méthodes et applications*, **I** Mir, Moscow, 1979. 420 p.

[48] J. Douglas, Solution of the inverse problem of the calculus of variations, *Proc. Nat. Acad. Sci. USA* **25** (1939), 631–637.

[49] J. Douglas, Solution of the inverse problem of the calculus of variations, *Trans. Amer. Math. Soc.* **50** (1941), 71–128.

[50] J. Drach, Thèse de Doctorat: Essai sure une théorie générale de l'intégration et sur la classification des transcendantes, *Ann. Ecole Norm. Sup.* **15** (1898), 243–384.

[51] W. Drechsler, M.E. Mayer, *Fiber bundle techniques in gauge theories*, Lecture notes in Physics **67**, Springer, Berlin, 1977. 248 p.

[52] J. Duc, D. Bellet, Mécanique des solides réels, élasticité, *Cours de l'École Nationale Sup. de l'Aéronautique et de l'Espace*, CEPADUES ed., Toulouse, (1977), 211 p.

[53] S.V. Duzhin, T. Tsujishita, Conservation laws of the BBM equations, *J. Phys. A. Math. gen.* **17** (1984), 3267–3276.

[54] C. Eckart, The thermodynamics of irreversible processes, I. Single fluid, *Physical Review* **58** (1940), 267–269. II. Fluid mixtures, *Physical Review* **58** (1940), 269–275. III. Relativistic theory of the single fluid, *Physical Review* **58** (1940), 919–924.

[55] A. Einstein, Electrodynamik bewegter Körper, *Ann. der Physik* **17** (1905), 891–921.

[56] A. Einstein, A.D. Fokker, *Ann. der Physik* **44** (1914), 321.

[57] A. Einstein, Die grundlagen der allgemeine Relativitätstheorie, *Ann. der Physik* **49** (1916), 769–822.

[58] L.P. Eisenhart, *Riemannian geometry*, Princeton Univ. Press, 1926.

[59] M. Ferraris, M. Francaviglia, On the global structure of Lagrangian and Hamiltonian formalisms in higher order calculus of variations, In: *Proc. Internat. Meeting 'Geometry and Physics', Florence, Italy, Oct. 1982* Pitagora Editrice, Bologna, 1983. 278 p.

[60] M. Ferraris, J. Kijowski, On the equivalence of the relativistic theories of gravitation, *General Relativity and Gravitation* **14**, 2 (1982), 165–179.

[61] M. Francaviglia, D. Krupka, The Hamilton formalism in higher order variational problems, *Ann. Inst. H. Poincaré* **XXXVII** (1982), 295–315.

[62] M. Fliess, Décomposition en cascade des systèmes automatiques et feuilletages invariants, *Bull. Soc. Math. France* **113** (1985), 285–293.

[63] M. Fliess, A new approach to the structure at infinity of nonlinear systems, *Systems Control Letters* **7** (1986), 419–421.

[64] M. Fliess, Automatique et corps différentiéls, *Forum Math.* **1** (1989),

[65] M. Fliess, Controllability revisited, In: *The influence of R.E. Kalman*, A.C. Antoulas (ed.), Springer, 1991. 463–474.

[66] K. Forsman, Constructive commutative algebra in nonlinear control theory, *Linköping Studies in Science and Technology, Dissertation Linköping Univ.* **261** (1991), [Linköping Univ. S-58183].

[67] J. Foster, J.D. Nightingale, *A short course in general relativity*, Longman, London, 1979. 192 p.

[68] P.L. Garcia, J. Munoz, On the geometric structure of higher order variational calculus, In: *Proc. IUTAM-ISIMM Symp. on Modern Developments in Anal. Mechanics, Torino, Italy, June 7–11, 1982*, S. Benenti, M. Francaviglia, A. Lichnerowicz (eds.) Technoprint, Bologna, 1983.

[69] P. Germain, *Mécanique des milieux continus*, **I** Masson, Paris, 1973. 417 p.

[70] A. Glumineau, C.H. Moog, Essential orders and the nonlinear decoupling problem, *Int. J. Control* **58** (1989), 1825–1834.

[71] I.I. Goldenblat, Some problems of the mechanics of deformable continua, Noordhoff, Groningen, 1960. 300 p. [Translated from the Russian by Z. Mroz.]

[72] H. Goldschmidt, Prolongations of linear partial differential equations, I. A conjecture of E. Cartan, *Ann. Sc. École Norm. Sup.* **1** (1968), 417–444. II. Inhomogeneous equations, *Ann. Sc. École Norm. Sup.* **1** (1968), 617–625.

[73] H. Goldschmidt, Integrability criterion for systems of nonlinear partial differential equations, *J. Differential Geom.* **1** (1969), 269–307.

[74] H. Goldschmidt, Sur la structure des équations de Lie, I. Le troisième théorème fondamental, *J. Differential Geom.* **6** (1972), 357–373. II. Équations formellement transitives, *J. Differential Geom.* **7** (1972), 67–95.

[75] H. Goldschmidt, J. Gasqui, *Déformations infinitésimales des structures conformes plates*, Birkhäuser, Boston, 1984. 226 p.

[76] H. Goldschmidt, D.C. Spencer, On the nonlinear cohomology of Lie equations I, II, *Acta Math.* **136** (1976), 103–239.

[77] S.R. de Groot, The Onsager relations: theoretical basis, In: *Proc. Bussaco Symp. 'Foundations of Continuum Thermodynamics, 1973*, MacMillan Press, London, (1974), 159–183.

[78] A. Haddak, Differential algebra and controllability, In: *Proc. IFAC Symp. on Nonlinear Control and Systems Designs, June 14-16, 1989, Capri, Italy*, 434–437.

[79] D.F. Hays, Variational formulation of the heat equation, In: *Proc. Chicago Symp. 'Non-Equilibrium Thermodynamics', 1965*, Univ. Chicage Press, Chicago, (1966), 17–43.

[80] G. Herglotz, Über die mechanik des deformierbaren Körpers vom Standpunkt der Relativitätstheorie, *Ann. der Physik* **36** (1911), 493–517.

[81] J.M. Hill, *Differential equations and group methods for scientists and engineers*, CRC Press, Boca Raton, Florida, 1992. 201 p.

[82] J.E. Humphreys, *Linear algebraic groups*, Springer, 1975.

[83] P. Iglesias, Essai de thermodynamique rationnelle des milieux continus, *Ann. Inst. H. Poincaré* **A 34** (1981), 1–24.

[84] A. Isidori, *Nonlinear control systems: an introduction*, Lecture notes in Control and Inform. Sciences **92**, Springer, Berlin, 1985. 297 p.

[85] M. Janet, Les systèmes d'équations aux dérivées partielles, *J. Math. Pure Appl.* **3** (1920), 65.

[86] M. Janet, *Leçons sur les systèmes d'équations aux dérivées partielles*, Cahiers Scientifiques **IV**, Gauthier-Villars, Paris, 1929.

[87] J. Jezierski, J. Kijowski, Une description Hamiltonienne du frottement et de la viscosité, *C.R. Acad. Sci. Paris Sér. 4* **301** (1985), 221.

[88] H.H. Johnson, Classical differential invariants and applications to partial differential equations, *Math. Ann.* **148** (1962), 308–329.

[89] I. Kaplansky, *An introduction to differential algebra*, Hermann, Paris, 1976. 64 p.

[90] J. Kijowski, W.M. Tulczyjew, Relativistic hydrodynamics of isoentropic flows, *Atti Acad. Sci. Torino* (1987).

[91] A. Kitapci, L.M. Silverman, System structure at infinity, *System Control Letters* **3** (1983), 123–131.

[92] S. Kobayashi, K. Nomizu, *Foundations of differential geometry*, **I-II** J Wiley, New York, 1963–1969.

[93] G. Koenigs, Leçons de cinématique [avec notes par G. Darboux, E. et F. Cosserat], Hermann, Paris, 1897. 391–417.

[94] E.R. Kolchin, S. Lang, Algebraic groups and the Galois theory of differential fields, *Amer. J. Math.* **80** (1958), 103–110.

[95] E.R. Kolchin, *Differential algebra and algebraic groups*, Acad. Press, New York, 1973. 446 p.

[96] E.R. Kolchin, *Differential algebraic groups*, Acad. Press, New York, 1984.

[97] I.S. Krasilshchik, A.M. Vinogradov, Nonlocal symmetries and the theory of coverings, *Acta Applic. Math.* **2** (1984), 79–96.

[98] D. Krupka, Variational sequences on finite order jet spaces, In: *Proc. Conf. 'Differential Geometry and Its Applications'*, Aug. 27-Sept. 2, 1989, Brno, Czechoslovakia, World Scientific, Singapore, 1990. 236–254.

[99] D. Krupka, Topics in the calculus of variations: variational sequences, In: *Proc. Conf. 'Differential Geometry and Its Applications'*, Aug. 1992, Opava, Czechoslovakia, to apear.

[100] A. Kumpera, Invariants différentiels d'un pseudogroupe de Lie, *J. Differential Geom.* **1** (1975), 347–416.

[101] A. Kumpera, D.C. Spencer, *Lie equations*, Ann. Math. Studies **73**, Princeton Univ. Press, Princeton, New Jersey, 1972. 293 p.

[102] E. Kunz, *Introduction to commutative algebra and algebraic geometry*, Birkhäuser, 1985. 238 p.

[103] M. Kuranishi, *Lectures on involutive systems of partial differential equations*, Publ. Soc. Math. São Paulo, 1967.

[104] L. Landau, E. Lifschitz, *Hydrodynamics*, Mir, Moscow, 1971. 670 p.

[105] L. Landau, E. Lifschitz, *Mechanics*, Mir, Moscow, 1981. 242 p.

[106] L. Landau, E. Lifschitz, *Elasticity*, Mir, Moscow, 1967. 206 p.

[107] L. Lang, *Algebra*, Addison-Wesley, 1971.

[108] M. von Laue, *Jb. RadioAkt.* **14** (1917), 263.

[109] S. Lie, Die Grundlagen für die Theorie der unendlichen Gruppen, I, *Leipziger Berichte* (1891), 391. II, *Leipziger Berichte* (1895), 282. [Also: *Gesamm. Abh.*, **VI**.]

[110] J. Lifermann, *Systèmes linéaires et variables d'état*, Masson, Paris, 1972. 227 p.

[111] G. Lippmann, *C.R. Acad. Sci. paris* **82** (1876), 1425. Über die Analogie zwischen absoluter Temperatur und elektrischen Potential (Erwiedering an F.W. Adler), *Ann. der Physik* **23** (1907), 994.

[112] J.L. Lions, Are there connections between turbulence and controllability?, In: *9th Internat. Conf. on Analysis and Optimization of Systems*, INRIA, June 12–15, Antibes, France.

[113] A.A. Logunov, Yu.M. Loskutov, M.A. Mestvirishvili, Relativistic theory of gravity, *Int. J. of Modern Physics* **3**, 9 (1988), 2067–2099.

[114] A.A. Logunov, Yu.M. Loskutov, Once more on the nonuniqueness of the predictions of the general theory of relativity, *Theoret. Math. Phys.* **76** (1988), 779–783.

[115] F.S. Macaulay, *The algebraic theory of modular systems*, Cambridge Tracts in Math. and Math. Physics **19**, Cambridge Univ. Press, 1916; 1964. 112 p.

[116] E. Mach, Die Geschichte unde die Würzel des Satzes von der Erhaltung der Arbeit, *Prag. Calve* (1872), 54. *Prinzipien der Wärmelehre*, J.A. Barth, Leizig, 1900. 330 p.

[117] K. Mackenzie, *Lie groupoids and Lie algebroids in differential geometry*, London Math. Soc. Lecture Notes **124**, Cambridge Univ.Press, (1987), 327 p.

[118] P. Medolaghi, Sulla teoria dei gruppi infiniti continui, *Ann. Mat. Pura Appl.* **25** (1897), 179–218.

[119] C. Marle, Sur l' établissement des équations de l'hydrodynamique des fluides relativistes dissipatifs, *Ann. Inst. H. Poincaré* **10**, 2 (1969), 127–194.

[120] G. Mie, Grundlagen einer Theorie der Materie, *Ann. der Physik* **37** (1912), 511. **39** (1912), 1. **40** (1913), 1.

[121] D.G. Miller, The Onsager relations: experimental evidence, In: *'Foundations of Continuum Thermodynamics'*, *Proc. Bussaco Symp., 1973*, J.J. Delgado Domingo, M.N.R. Nina, J.H. Whitelaw (eds.), Macmillan Press, London, 1974. 337 p.

[122] M.C. Mintchev, V.B. Petkova, I.T. Todorov, Conformal invariance in quantum field theory, *Publ. Scuola Normale Sup. Pisa, Cl. di Scienze* (1978), 275 p.

[123] C.H. Moog, Linear algebra and nonlinear control, In: *Proc. Conf. 'New Trends in Nonlinear Control'*, *Nantes, June 1988*, Lecture notes in control and inform. sciences **122**, Springer, Berlin, 1989.

[124] A.S. Morse, Structural invariants of linear multivariable systems, *SIAM J. Control Optim.* **11** (1973), 446–465.

[125] J.I. Neimark, N.A. Fufaev, Dynamics of nonholonomic systems, *Amer. math. Soc. Translations of Math. Monographs* **33** (1972), 518 p.

[126] A. Nijenhuis, Natural bundles and their general properties (geometric objects revisited), In: *Differential Geometry in Honour of K. Yano*, Kinokuniya, Tokyo, (1972), 317–334.

[127] H. Nijmeijer, J.M. Schumacher, Zeros at infinity for affine nonlinear control systems, *IEEE Trans. Autom. Control* **30** (1985), 566–573.

[128] H. Nijmeijer, A.J. van der Schaft, *Nonlinear dynamical control systems*, Springer, Berlin, (1990), 417 p.

[129] G. Nordström, *Phys.Z* **13** (1912), 1126. *Ann. der Physik* **40** (1913), 856. *Ann. der Physik* **42** (1913), 533. *Ann. der Physik* **43** (1914), 1101.

[130] D.G. Nortcott, *An introduction to homological algebra*, Cambridge Univ. Press, 1966.

[131] D.G. Northcott, *Lectures on rings, modules and multiplicities*, Cambridge Univ.Press, 1968.

[132] R.W. Ogden, On Eulerian and Lagrangian objectivity in continuum mechanics, *Warszawa Arch. Rech.* **36**, 2 (1984), 207–218.

[133] P. Olver, *Applications of Lie groups to differential equations*, Springer, Berlin, 1986.

[134] G.A. Oravas, L. McLean, Historical development of energetical principles in elastomechanics, *Applied Mechanics Reviews* **19**, 8 (1966), 647–658. *Applied Mechanics Reviews* **19**, 11 (1966), 919–933.

[135] V. Ougarov, *Théorie de la relativité restreinte*, Mir, Moscow, 1969, 1979. 304 p.

[136] L.V. Ovsiannikov, *Group analysis of differential equations*, Mir, Moscow, 1978. 400 p. [English translation by W.F. Ames, Acad. Press, New York, 1982, 416p.]

[137] A. Pais, The science and the life of Albert Einstein, Clarendon Press, Oxford, 1982. 552 p.

[138] W. Pauli, *Theory of relativity*, Pergamon Press, London, New York, 1958. 241 p.

[139] E.V. Pankratev, Computations in differential and difference modules, *Acta Applic. Math.* **16** (1989), 167–189.

[140] E. Picard, Sur les équations différentielles et les groupes algébriques de transformations, *Ann. Fac. Sc. Univ. Toulouse* **1** (1887), 1–15.

[141] E. Picard, *Traité d'analyse*, **3** Chapt. 17, Gauthier-Villars, Paris, 1898, 1928. [Published under separate cover as *Analogies entre la théorie des équations différentielles linéaires et la théorie des équations algébriques*, Gauthier-Villars, 1936, 1976.]

[142] F.A.E. Pirani, D.C. Robinson, W.F. Shadwick, *Local jet bundle formulation of Bäcklund transformations*, Math. Physics Studies **1**, D. Reidel, Dordrecht, 1979. 132 p.

[143] Y. Pironneau, On nonholonomic nonlinear constraints, In: *Proc. IUTAM Symp. on Modern Developments in Analytic Mechanics, Torino, Italy* 1982.

[144] J.F. Pommaret, *Systems of partial differential equations and Lie pseudogroups*, Gordon and Breach, London, New York, 1978. 402 p. [Translated into Russian by A.M. Vinogradov, Mir, Moscow, 1983; see Math. Reviews 81f:58046.]

[145] J.F. Pommaret, *Differential Galois theory*, Gordon and Breach, London, New York, 1983. 760 p.

[146] J.F. Pommaret, *Lie pseudogroups and mechanics*, Gordon and Breach, London, New York, 1988. 590 p. [See Math. Reviews 90e:58166.]

[147] J.F. Pommaret, La structure de la mécanique des milieux continus, *C.R. Acad. Sci. Paris* **296** (1983), 517–520.

[148] J.F. Pommaret, La structure des théories de jauge, *C.R. Acad. Sci. Paris* **296** (1983), 989–992.

[149] J.F. Pommaret, La structure de l'electromagnétisme et de la gravitation, *C.R. Acad. Sci. Paris* **297** (1983), 493–496.

[150] J.F. Pommaret, La structure de la dynamique analytique, *C.R. Acad. Sci. Paris* **299** (1984), 815–818.

[151] J.F. Pommaret, Géométrie différentielle algébrique et théorie du contrôle, *C.R. Acad. Sci. Paris* **302** (1986), 547–550.

[152] J.F. Pommaret, Einstein equations and Spencer sequences, In: *Atti del VI Convegno Naz. di Relativita Generale et Fisica della Gravitazione, Firenze, Italy, Oct. 10–13, 1984*, Pitagora Ed., Bologna, 1986. 191–198.

[153] J.F. Pommaret, Structure at infinity revisited, *Systems and Control Letters* **20** (1993), 327–333.

[154] J.F. Pommaret, New perspectives in control theory for partial differential equations, *IMA J. Math. Control and Information* **9** (1992), 305–330.

[155] J.F. Pommaret, A. Haddak, Effective methods for systems of algebraic PDE,

In: *Effective methods in Algebraic Geometry*, T. Mora, C. Traverso (eds.), Birkhäuser, 1991. 411–426.

[156] J.F. Pommaret, S. Lazzarini, Lie pseudogroups and differential sequences: new perspectives in two-dimensional conformal geometry, *J. Geometry and Physics* **10** (1993), 47–91.

[157] Ch. Riquier, *Les systèmes d'équations aux dérivées partielles*, Gauthier-Villars, Paris, 1910.

[158] Ch. Riquier, *La méthode des fonctions majorantes et les systèmes d'équations aux dérivées partielles*, Mémorial Sci. Math. **XXXII**, Gauthier-Villars, Paris, 1910.

[159] J.F. Ritt, *Differential equations from the algebraic standpoint*, Amer. Math. Soc. Colloquium Publ. **14**, Amer. Math. Soc., 1932.

[160] J.F. Ritt, *Differential algebra*, Dover, 1966. 184 p.

[161] V.V. Rumiantsev, On Hamilton principle for nonholonomic systems, *PPM* **42** (1978), 387–399.

[162] D.J. Saunders, *The geometry of jet bundles*, London Math. Soc. Lecture notes **142**, Cambridge Univ. Press, 1989.

[163] A. Seidenberg, An elinimation theory for differential algebra, *Univ. Cal. Publ. Math. (N.S.)* **3** (1956), 31–66.

[164] I.R. Shafarevich, *Basic algebraic geometry*, Spinger, 1977. (Translated from the Russian.)

[165] A.J. van der Schaft, Representing a nonlinear state spec system as a set of higher-order differential equations in the inputs and outputs, *Systems and Control Letters* **12** (1989), 151–160.

[166] A.J. van der Schaft, Duality for linear systems, In :   *Analysis of Controlled Dynamical Systems*, B. Bernard, B. Bride, J.P. Gauthier, I. Krupka (eds.), Progress in Systems and Control Theory, Birkhäuser, 1991. 393–403.

[167] H.G. Schopf,   Allgemeinrelativistische Prinzipien der Kontinuummechanik, *Ann. der Physik* **12** (1964), 377–395.

[168] F. Schwarz, The Riquier–Janet theory and its application to nonlinear evolution equations, *Physica* **110** (1984), 243–251.

[169] F. Schwarz, Monomial orderings and Gröbner bases, *SIGSAM Bull.* **95**, 25 (1991), 10–23.

[170] S.N. Singh, A modified algorithm for invertibility in nonlinear systems, *IEEE Trans. Autom. Control* **26** (1981), 595–598.

[171] W. Slebodzinski, *Formes extérieures et leurs applications*, PWN, Warszawa, 1963.

[172] D.C. Spencer, Overdetermined systems of partial differential equations, *Bull. Amer. math. Soc.* **75** (1965), 1–114.

[173] S. Sternberg, *Lectures on differential geometry*, 2nd ed., Chelsea, New York, 1983. 442 p.

[174] I. Stewart, *Galois theory*, Chapman and Hall, 1973.

[175] J. Sudria, *L'action euclidienne de déformation et de mouvement*, Mémorial Sci. Math. **29**, Gauthier-Villars, Paris, 1935. 56 p.

[176] A.H. Taub, General relativistic variational principle for perfect fluids, *Physical Review* **94** (1954), 1468–1470.

[177] P.P. Teodorescu, *Dynamics of linear elastic bodies*, Ed. Acad. Bucaresti, 1972. [Also: Turnbridge Wells, Kent, 1975. 411 p.]

[178] J.M. Thomas, Riquier's theory, *Ann. of Math.* **30** (1929); *Ann. of Math.* **35** (1935).

[179] J.M. Thomas, *Systems and roots*, W. Byrd Press, 1962.

[180] I.T. Todorov, *Conformal description of spinning particles*, Trieste notes in Physics, Springer, Berlin, 1986. 74 p.

[181] R.C. Tolman, *Relativity, thermodynamics and cosmology*, Clarendon Press, Oxford, 1934, 1966. 500 p.

[182] A. Trautman, Yang–Mills theory and gravitation, a comparison, In: *Geometric Techniques in Gauge Theory, Proc. 5th Conf. Differential Equations, Scheveningen, Netherlands, 1981*, Lecture notes in math. **926**, Springer, Berlin, 1982. 179–189.

[183] H.-J. Treder, H.-H. von Borzeszkowski, A. van der Merwe, W. Yourgrau, *Fundamental principles of general relativity theories*, Plenum Press, New York, 1980. 216 p.

[184] C. Truesdell, W. Noll, The nonlinear field theories of mechanics, *Handbuch der Physik* **III/3** Springer, Berlin, 1965.

[185] T. Tsujishita, On variational bicomplexes associated to differential equations, *Preprint Osaka Univ.* (1980), ; *Osaka J. math.* **19**, 2 (1982), 311–363.

[186] W.M. Tulczyjew, Relativistic hydrodynamics is a symplectic field theory, In: *Geometrodynamics Symp. 1983*, Technoprint, Bologna, 1984. 91–99.

[187] W.M. Tulczyjew, The Lagrange complex, *Bull. Soc. Math. France* **105** (1977), 419–431.

[188] W.M. Tulczyjew, The Euler–Lagrange resolution, *Lecture notes in math.* **836** Springer, Berlin, 1980. 22–48.

[189] W.M. Tulczyjew, Cohomology of the Lagrange complex, *Ann. Scuola Norm. Sup. Pisa* (1988), 217–227.

[190] E. Vessiot, Thèse Doctorat: Sur l'intégration des équations différentielles ordinaires, *Ann. Ec. Norm. Sup.* **9** (1892), 197–282.

[191] E. Vessiot, Sur la théorie des groupes infinis, *Ann. Ec. Norm. Sup.* **20** (1903), 411–451.

[192] E. Vessiot, Sur la théorie de Galois et ses diverses généralisations, *Ann. Ec. Norm. Sup.* **21** (1904), 9–85.

[193] A.M. Vinogradov, The C-spectral sequence: Lagrangian formalism and conservation laws, I-II, *J. Math. Anal. Applications* **100**, 1 (1984), 1–129.

[194] B.L. van der Waerden, *Algebra*, F. Ungar, 1938, 1970.

[195] F.W. Warner, *Foundations of differentiable manifolds and Lie groups*, Scott, Foresmann and Co., 1971.

[196] H. Weyl, *Space, time, matter*, Springer, Berlin, 1918, 1958. 234 p. [Also: Blanchard, Paris, 1958. 290 p.]

[197] H. Weyl, *Sitzungsber. Preuss. Akad. Wissenschaft.* (1918), 65. *Math. Z.* **2** (1918), 384. *Ann. der Physik* **59** (1919), 101.

[198] H. Weyl, *Z. Physik* **56** (1929), 330.

[199] E.T. Whittaker, *Analytical dynamics*, Dover, New York, 1944. 436 p.

[200] J.C. Willems, From time series to linear systems, I, *Automatica* **22** (1986), 561–580. II, *Automatica* **22** (1986), 675–694. III, *Automatica* **23** (1987), 87–115.

[201] J.C. Willems, Models for dynamics, *Dynamics Reported* **2** (1989), 171–269.

[202] Wu Wen-Tsun, On the foundations of algebraic differential geometry, *Mathematics Mechanization, Research Preprints, Inst. Systems Science*, Acad. Sinica, Beijing, **3** (1989), 1–26.

[203] Wu Wen-Tsun, Mechanical theorem proving of differential geometries and some of its applications in mechanics, *J. Automated Reasoning* **7** (1991), 171–191.

[204] C.N. Yang, Magnetic monopoles, fiber bundles and gauge fields, *Ann. New York Acad. Sci.* **294** (1977), 86.

[205] C.N. Yang, Geometry and physics, In: *Jerusalem Einstein Centennial Symp. on Gauge Theories and Unification of Physical Forces*, Y. Ne'iman (ed.), Addison-Wesley, 1981. 3.

[206] C.N. Yang, R.L. Mills, Conservation of isotropic spin and isotropic gauge invariance, *Phys.Review* **96** (1954), 191–195.

[207] H. Yilmaz, Present status of gravitation theories, *Hadronic J.* **9**, 6 (1986), 233–238.

[208] O. Zariski, P. Samuel, *Commutative algebra*, Chapman and Hall, 1973.

[209] Zou ZhenLong, Some researches on gauge theories of gravitation, *Scientia Sinica* **22**, 6 (1979), 628–636.

Index

Other *Mathematics and Its Applications* titles of interest:

V.I. Istratescu: *Fixed Point Theory. An Introduction.* 1981, 488 pp.
out of print, ISBN 90-277-1224-7

A. Wawrynczyk: *Group Representations and Special Functions.* 1984, 704 pp.
ISBN 90-277-2294-3 (pb), ISBN 90-277-1269-7 (hb)

R.A. Askey, T.H. Koornwinder and W. Schempp (eds.): *Special Functions: Group Theoretical Aspects and Applications.* 1984, 352 pp. ISBN 90-277-1822-9

A.V. Arkhangelskii and V.I. Ponomarev: *Fundamentals of General Topology. Problems and Exercises.* 1984, 432 pp. ISBN 90-277-1355-3

J.D. Louck and N. Metropolis: *Symbolic Dynamics of Trapezoidal Maps.* 1986, 320 pp. ISBN 90-277-2197-1

A. Bejancu: *Geometry of CR-Submanifolds.* 1986, 184 pp.
ISBN 90-277-2194-7

R.P. Holzapfel: *Geometry and Arithmetic Around Euler Partial Differential Equations.* 1986, 184 pp. ISBN 90-277-1827-X

P. Libermann and Ch.M. Marle: *Sympletic Geometry and Analytical Mechanics.* 1987, 544 pp. ISBN 90-277-2438-5 (hb), ISBN 90-277-2439-3 (pb)

D. Krupka and A. Svec (eds.): *Differential Geometry and its Applications.* 1987, 400 pp. ISBN 90-277-2487-3

Shang-Ching Chou: *Mechanical Geometry Theorem Proving.* 1987, 376 pp.
ISBN 90-277-2650-7

G. Preuss: *Theory of Topological Structures. An Approach to Categorical Topology.* 1987, 318 pp. ISBN 90-277-2627-2

V.V. Goldberg: *Theory of Multicodimensional (n+1)-Webs.* 1988, 488 pp.
ISBN 90-277-2756-2

C.T.J. Dodson: *Categories, Bundles and Spacetime Topology.* 1988, 264 pp.
ISBN 90-277-2771-6

A.T. Fomenko: *Integrability and Nonintegrability in Geometry and Mechanics.* 1988, 360 pp. ISBN 90-277-2818-6

L.A. Cordero, C.T.J. Dodson and M. de Leon: *Differential Geometry of Frame Bundles.* 1988, 244 pp. out of print, ISBN 0-7923-0012-2

E. Kratzel: *Lattice Points.* 1989, 322 pp. ISBN 90-277-2733-3

E.M. Chirka: *Complex Analytic Sets.* 1989, 396 pp. ISBN 0-7923-0234-6

Kichoon Yang: *Complete and Compact Minimal Surfaces.* 1989, 192 pp.
ISBN 0-7923-0399-7

A.D. Alexandrov and Yu.G. Reshetnyak: *General Theory of Irregular Curves.* 1989, 300 pp. ISBN 90-277-2811-9

Other *Mathematics and Its Applications* titles of interest:

B.A. Plamenevskii: *Algebras of Pseudodifferential Operators*. 1989, 304 pp.
ISBN 0-7923-0231-1

Ya.I. Belopolskaya and Yu.L. Dalecky: *Stochastic Equations and Differential Geometry*. 1990, 288 pp. ISBN 90-277-2807-0

V. Goldshtein and Yu. Reshetnyak: *Quasiconformal Mappings and Sobolev Spaces*. 1990, 392 pp. ISBN 0-7923-0543-4

A.T. Fomenko: *Variational Principles in Topology. Multidimensional Minimal Surface Theory*. 1990, 388 pp. ISBN 0-7923-0230-3

S.P. Novikov and A.T. Fomenko: *Basic Elements of Differential Geometry and Topology*. 1990, 500 pp. ISBN 0-7923-1009-8

B.N. Apanasov: *The Geometry of Discrete Groups in Space and Uniformization Problems*. 1991, 500 pp. ISBN 0-7923-0216-8

C. Bartocci, U. Bruzzo and D. Hernandez-Ruiperez: *The Geometry of Supermanifolds*. 1991, 242 pp. ISBN 0-7923-1440-9

N.J. Vilenkin and A.U. Klimyk: *Representation of Lie Groups and Special Functions. Volume 1: Simplest Lie Groups, Special Functions, and Integral Transforms*. 1991, 608 pp. ISBN 0-7923-1466-2

A.V. Arkhangelskii: *Topological Function Spaces*. 1992, 206 pp.
ISBN 0-7923-1531-6

Kichoon Yang: *Exterior Differential Systems and Equivalence Problems*. 1992, 196 pp. ISBN 0-7923-1593-6

M.A. Akivis and A.M. Shelekhov: *Geometry and Algebra of Multidimensional Three-Webs*. 1992, 358 pp. ISBN 0-7923-1684-3

A. Tempelman: *Ergodic Theorems for Group Actions*. 1992, 400 pp.
ISBN 0-7923-1717-3

N.Ja. Vilenkin and A.U. Klimyk: *Representation of Lie Groups and Special Functions, Volume 3. Classical and Quantum Groups and Special Functions*. 1992, 630 pp. ISBN 0-7923-1493-X

N.Ja. Vilenkin and A.U. Klimyk: *Representation of Lie Groups and Special Functions, Volume 2. Class I Representations, Special Functions, and Integral Transforms*. 1993, 612 pp. ISBN 0-7923-1492-1

I.A. Faradzev, A.A. Ivanov, M.M. Klin and A.J. Woldar: *Investigations in Algebraic Theory of Combinatorial Objects*. 1993, 516 pp. ISBN 0-7923-1927-3

M. Puta: *Hamiltonian Mechanical Systems and Geometric Quantization*. 1993, 286 pp. ISBN 0-7923-2306-8

V.V. Trofimov: *Introduction to Geometry of Manifolds with Symmetry*. 1994, 326 pp. ISBN 0-7923-2561-3